Positive Psychology and Disability Series

The **Positive Psychology and Disability** book series provides comprehensive coverage of research and practice issues pertaining to the application of constructs and principles from the discipline of positive psychology across the disability context. In addition, books in the series address the application of strengths-based approaches to understanding disability and designing and implementing supports to enable people with disabilities to live, learn, work, and play as meaningful participants in their communities. Drawing from traditional areas of focus in positive psychology, series books:

- Provide cutting-edge research and practice that can improve the quality of life and well-being of people with disabilities.
- Address theory and research across such areas as hope, optimism, self-determination, and character strengths.
- Examine research- and evidence-based practices to promote involvement in goal setting, problem solving, and decision making.
- Synthesize research and practice from multiple disciplines and apply strengths-based approaches to disability.

Volumes published in this series are must-have resources for researchers, professors, clinicians and related professionals, and graduate students in developmental and positive psychology, special education, social work, child and school psychology, and other allied disciplines.

More information about this series at http://www.springer.com/series/15589

Ishita Khemka • Linda Hickson
Editors

Decision Making by Individuals with Intellectual and Developmental Disabilities

Integrating Research into Practice

Springer

Editors
Ishita Khemka
School of Education
St. John's University
Queens, NY, USA

Linda Hickson
Teachers College
Columbia University
New York, NY, USA

Positive Psychology and Disability Series
ISBN 978-3-030-74674-2 ISBN 978-3-030-74675-9 (eBook)
https://doi.org/10.1007/978-3-030-74675-9

This Springer imprint is published by the registered company Springer Nature Switzerland AG
The registered company address is: Gewerbestrasse 11, 6330 Cham, Switzerland

We dedicate this book to our cherished friends, Aikaterini Chatzistyli and Gay Culverhouse, long-time collaborators and supporters of our research, who lost their lives during the global pandemic and who had relentlessly advocated throughout their lives for the right and dignity of all individuals with disabilities to make their own decisions and to become autonomous, independent, and fully participating members of their societies.

Foreword

Enormous changes in the macro- and micro-systems of psychology and education have led to a more sophisticated and positive regard in the understanding and treatment of persons with intellectual and developmental disabilities (IDD). Most of these changes have resulted in longer and more productive lives for adults with IDD, for their families, and for the men and women who care for them and teach them in their schools, their places of employment, and their homes. This book (hereafter referred to as *Decision Making*) is a reflection of the progress that the changes have propelled as well as a testimony to the scientific and policy advances that we are likely to see in the future.

The editors of *Decision Making*, Drs. Ishita Khemka and Linda Hickson, are uniquely qualified to have overseen its creation. Dr. Hickson spent most of her career at Teachers College, Columbia University (TC), as a professor, a researcher, and an administrator. Dr. Khemka, currently Associate Professor and Program Coordinator of Special Education at St. John's University, was a graduate student at TC and collaborated with Dr. Hickson, first as a mentee and, subsequently and currently, as a partner. Together they developed the *Pathways Model of Decision Processing* (updated in chapter "Theoretical Perspectives on Decision Making" by Khemka) to guide decision-making research and practice for individuals with IDD and created evidence-based curricula—PEER-DM and ESCAPE-DD—to teach adolescents and adults decision making–based strategies for abuse prevention. *Decision Making* documents that work and its impact but expands substantially into related and relevant topics. Authors are a *Who's Who* of scientists and educators who have made and continue to make groundbreaking contributions to decision making by persons with IDD. They are respected leaders in their fields, with reputations that signify the influence that they have had. As one example of their impact, six authors of seven chapters in this book overlap with ones who have written chapters in the recently published two-volume *American Psychological Association Handbook of Intellectual and Developmental Disabilities* (Glidden, 2021).

The contents of *Decision Making* are wide-ranging and thorough, with a broad scope in theory, research, and practice. Its focus includes not only programs from the USA, but authors who bring perspectives from six other countries—Australia,

Canada, Japan, the Netherlands, Switzerland, and the UK—thus providing an international vision. *Decision Making* is appropriately organized into five parts, with chapters divided approximately equally among them. Foundations and theory are emphasized in the first two parts, whereas chapters on applications and practices dominate in the subsequent three parts. Authors end chapters with summary sections that provide conclusions and suggestions for future directions, a format that is useful for the reader who can quickly capture the main themes and return for a more thorough reading as time and interest permits.

The readership audience for this book will be wide-ranging and will likely include psychologists, special education teachers and administrators, social workers, lawyers and judges, police personnel, health professionals, and, of course, parents and other family members who care for and about their relatives with IDD. Parts and chapters will be differentially relevant to readers, depending on their backgrounds and interests. For example, *Decision Making and Culturally Diverse Individuals with IDD and their Families: A Call for Research* is likely to be of interest to most readers. In this chapter, Vanegas, Magaña, Zeng, and Pavon propose a socio-ecological model of decision making in culturally diverse individuals with IDD, building on the Bronfenbrenner (1977) social-ecological model, but including a variable of culture that exists in the intersection of the micro-, meso-, and macrosystems. The increasing diversity of the American population as well as those of European countries ensures that this chapter is not only highly relevant now, but likely will continue to be for at least the next decade.

In addition to its attention to cultural diversity, *Decision Making* also addresses the diversity of etiology and functioning. In Part II, individual chapters focus on persons with IDD with different causation and characteristics. For example, Goscicki, Josol, Fisher, Dykens, and Hodapp describe a general decision-making model and illustrate etiological differences among individuals with Williams, Prader-Willi, and Down syndrome, based on the clusters of features associated with each of these etiologies. Focusing on syndrome-specific characteristics, the authors note that individuals with Williams syndrome tend to be hypersocial and may be vulnerable to maltreatment by others with intention to harm (also addressed in chapter "Decision Making and Vulnerability to Maltreatment" in Part IV by Hickson and Khemka). Parts III, *Profiles of Decision Making*, and IV, *Applications of Decision Making*, are likely to be of particular interest to families. They contain chapters with a focus on specific ages and stages, such as adolescence and transition to adulthood, and also on particular types of decision making such as financial. For example, in the Murphy and Clare chapter on financial decision making, the authors carefully describe the research by Suto and colleagues (e.g., Suto et al., 2007) demonstrating poorer financial decision making among participants with ID in comparison to those without ID. Despite these differences, both groups showed similar patterns of performance, and it was recommended that more practice in financial decision making was essential for improvement regardless of intellectual ability or disability. Parts III and IV also focus on group and individual differences including different disability diagnoses such as autism and ADHD, and different types of decision making such as social functioning (chapter "Social Functioning and Decision

Making: From Group to Individual Differences Across the Autism Spectrum"—Levin, Gaeth, Levin, & Chen) and cognitive, emotional, and moral decision making (chapter "Cognitive, Emotional, and Moral Decision Making in Adolescents and Adults with ASD"—Komeda)

The necessity of practice and both behavioral and strategy-based interventions are a focus in Parts IV, *Applications of Decision Making* and V, *Positive Psychology Interventions for Decision Making*. In these parts, authors describe multiple approaches and emphasize the necessity of both strategic intervention and practice. For example, in chapter "From Social Vulnerability Assessment to Active Prevention Measures: A decision-making Perspective", Petipierre and Tabin explain their modifications of the Hickson/Khemka ESCAPE-DD program for French-speaking Swiss participants. This active prevention strategy embodies the importance of individual knowledge and responsibility that are essential for self-protection and, ultimately, successful independent living.

Decision-making strategies must be learned, and although some may be acquired in the natural environment, others must be explicitly taught. Techniques used in teaching decision making strategies are varied, but behavioral methods play a major role, as described by Uher, Josol, and Fisher in chapter "Behavioral Approaches to Teaching decision-making to Individuals with IDD". They describe the power of these methods in a focus on high-risk decision-making situations involving social victimization—including bullying, sexual assault, and abduction—all of which persons with IDD are more likely to experience than those without disabilities. Instruction, modeling, rehearsal, and feedback are useful techniques, and an *in situ* assessment to maximize generalization skills can solidify the training, leading to a reduction in the likelihood that persons with IDD will be victimized.

In the final chapter "Decision Making and IDD: Future Directions for Research and Practice", Khemka summarizes the current status of decision making and IDD, and is optimistic about the greater autonomy that it has promoted, an autonomy that is likely to remain a goal for persons with IDD. I heartily agree and, furthermore, believe that this book will be part of the impetus that propels it and therefore an important contribution to the IDD field. The publisher—Springer—and the Positive Psychology Series editors—Michael Wehmeyer and Karrie Shogren (also chapter authors)—are to be congratulated for yet another volume that is likely to be widely read and to motivate programmatic change important for persons with IDD and those who live and work with them.

American Psychological Association
St. Mary's City, MD, USA

Laraine Masters Glidden

Laraine Masters Glidden, Ph.D., is Distinguished Professor Emerita at St. Mary's College of Maryland, the liberal arts public honors college of the State of Maryland. She is well known for her federally funded longitudinal research of families rearing adopted and birth children with IDD, demonstrating that families who initially experience negative reactions to the realization that their babies have IDD adapt over time and demonstrate well-being that is similar to adoptive families who knowingly chose to rear children with IDD. Dr. Glidden has extensive editing experience, having served as Editor of the *International Review of Research in Developmental Disabilities* for 11 years, overseeing the publication of 16 volumes. Most recently, she was Editor-in-Chief of the two-volume *American Psychological Association Handbook of Intellectual and Developmental Disabilities*, published in 2021.

References

Bronfenbrenner, U. (1977). Toward an experimental ecology of human development. *American Psychologist, 32*, 513–531.

Glidden, L. M. (2021). *APA Handbook of Intellectual and Developmental Disabilities*. American Psychological Association.

Suto, W. M. I., Clare, I. C. H., & Holland, A. J. (2007). *Financial Decision-making: Guidance for supporting financial decision making by people with learning disabilities*. British Institute of Learning Disabilities.

Preface

As part of Springer's series on positive psychology, this book on research and practice in decision making by individuals with intellectual and developmental disabilities (IDD) focuses on strengths-based approaches. For many years, IDD was viewed through the lens of a medical model, with an emphasis on finding cures and remediating deficits to address problems that were assumed to reside in the individual (see Hickson et al., 1995 for this history). However, the current convergence of movements such as person-centered planning, self-determination, quality of life, and positive psychology has inspired a widespread shift to a social-ecological model of IDD. This model was adopted by the American Association on Intellectual and Developmental Disabilities in 1992 (then the American Association on Mental Retardation) (Luckasson et al., 1992). The social-ecological model views IDD as a mismatch between the capabilities of the individual and the demands of their environment which can be ameliorated by systems of supports targeted to build upon a person's strengths to improve the person-environment fit. Because decision making plays a pivotal role in all aspects of life, increasing the decision-making capacity and/or opportunities of a person with IDD can have a profound impact on their quality of life. Seligman (1998), an early advocate of positive psychology, defined its mission as "to measure, understand and then build the human strengths and the civic virtues" (p. 2). Seligman and Csikszentmihalyi (2000) pointed to the centrality of decision making and viewed people "as decision-makers, with choices, preferences, and the possibility of becoming masterful, efficacious…" (p. 8).

Decision making is closely related to the construct of *problem solving*. Wehmeyer and Shogren (2017, p. 217) have defined problem solving as a "task, activity, or situation for which a solution is not immediately identified, known, or obtainable." According to Bransford and Stein (1993, p. 7), "a problem exists when there is a discrepancy between an initial state and a goal state, and there is no ready-made solution for the problem solver." Although there are many disparate definitions of these terms, we see problem solving as differing from decision making in that "decision-making tasks typically lack a clearly defined goal state and rules on how to proceed" (Hickson & Khemka, 2013, p. 201), making decision making more open ended than problem solving. According to dual process theories (e.g., Kahneman,

2011; Stanovich, 1999, 2010), decision making may involve either a rapid, automatic intuitive process or a slower, effortful deliberative process, encompassing multiple steps, such as: (1) framing the problem, (2) identifying alternatives, (3) evaluating possible consequences, and (4) selecting a course of action (Hickson & Khemka, 2013).

The terms *choice* and *decision making* are often used interchangeably, but we believe that it is important to highlight a key difference between the terms as they are typically used. Decision making almost always involves the generation of possible alternatives, while choice (more akin to preference) typically implies selecting among given alternatives. Decision making is sometimes considered a *social skill*, albeit a complex one consisting of multiple components. Finally, decision making is often listed as a necessary component of *self-determination*, which has been defined as having causal agency in one's life (Wehmeyer & Little, 2013).

The term *intellectual and developmental disabilities* in the title of this book is used in its most inclusive sense. It includes individuals with intellectual disability (ID), defined by the American Association on Intellectual and Developmental Disabilities as "significant limitations both in intellectual functioning (IQ of approximately 70 or below) and adaptive behavior as expressed in conceptual, social, and practical adaptive skills, and origination during the developmental period" (Schalock et al., 2021, p.23). The overlapping condition of *developmental disabilities* (DD) is defined as a "severe, chronic disability manifested before age 22 resulting from mental and/or physical impairments which are likely to continue indefinitely" (Anderson et al., 2019, p. 423). For an updated discussion of these terms, see Schalock and Luckasson (2021). Also included under the IDD definition is *autism spectrum disorder* (ASD), which is defined as consisting of persistent deficits in social communication and social interaction and restricted, repetitive patterns of behavior that emerge during the early developmental period (American Psychiatric Association, 2013).

Although most of the chapters in this book address decision making by individuals whose disabilities clearly fall under the IDD umbrella, a few chapters draw upon relevant work with individuals with closely related conditions that often co-occur with IDD, such as *anxiety* and *attention deficit hyperactivity disorder* (ADHD). Elevated rates of anxiety and other mental disorders are frequently reported in individuals with IDD (e.g., Einfeld et al., 2011). ADHD is a neurodevelopmental disorder, involving difficulties with planning and attention that have been associated with distinct patterns of decision making (e.g., Evans, 2003, 2008). It has been estimated that from 30% to 50% of individuals with ASD show ADHD symptoms and two thirds of individuals with ADHD have features of ASD and that both frequently co-occur with ID (e.g., Leitner, 2014; Miyasaka et al., 2018). Both groups typically share deficits in executive function, which affects the ability to organize thoughts and activities (Craig et al., 2016). We also consider the role of specific etiologies associated with IDD (e.g., Down syndrome, Williams syndrome, Prader-Willi syndrome) wherever there is relevant research pertaining to decision making.

The primary focus of this book is on adolescents and adults, who to date have been the participants in most of the existing research on decision making involving individuals with IDD and related disabilities. However, available information on decision making by children has been included in some chapters when it has implications for the later life stages.

In recruiting chapter authors for this book, we cast a broad net, not only geographically, with authors from all over the world, but conceptually including diverse populations as well as definitions and approaches to the study of decision making. Our own work has been focused on independent decision making in interpersonal situations in a variety of contexts, including interactions with strangers as well as interactions with family members, peers, and intimate partners—often in situations with a risk of abuse or victimization (e.g., Khemka et al., 2005, 2016). The scope of this book, however, is much more expansive, drawing upon international research that includes decision making in time-sensitive emergency situations and personal decisions in multiple domains. The book also addresses decision making in the context of a variety of laboratory decision-making tasks designed to reveal various aspects of the decision processes in relation to individual differences and IDD.

In addition to independent decision making, several chapter authors discuss aspects of supported and/or shared decision making where supports are provided by trusted friends, family members, or others to augment the decision-making capabilities of individuals with IDD, often in situations with serious consequences. A commitment to the implementation of the United Nations *Convention on the Rights of Persons with Disabilities* (CRPD) has provided a strong basis for replacing old models of guardianship/conservatorship, in the case of individuals deemed to have limited decision-making capacity, and replacing them with a self-determined decision making model where the self is the primary decision-maker—especially for health-care and other major life decisions. Providing individuals with disabilities with opportunities for supported decision making, as opposed to substituted decision making, or guardianship, is a key mandate of the CRPD, which was adopted in 2006. The CRPD came into force in 2008, with 119 nations ratifying and 153 additional nations signing it and committing to implementing all of its provisions (Werner, 2012). As of 2019, 177 countries had ratified (Szmukler, 2019). Even for countries that did not ratify or sign it (including the United States), the CRPD has set an international standard for the human rights of people with IDD, as well as increasing international interest in the study of decision making.

Article 12 of the CRPD bestows all persons with disabilities (and mental disorders) with the right to *legal capacity*, which provides them with a legal right to experience equal human dignity and the freedom to make their own decisions in all domains of life. By adopting a social model of disability, the CRPD obligates participating societies to provide the highest level of accommodations and appropriate supports, as needed, to facilitate a person's legal right to autonomously express her/his will and preferences.

In keeping with the legal and mental capacity directives of the CRPD for individuals with IDD, our own focus, and that of this book, has been to promote research and practice aimed at expanding opportunities for decision making and developing evidence-based strategies for improving the independent decision-making competencies needed for the effective and independent decision-making participation of all individuals with IDD. The chapters included in this book address this priority from multiple perspectives and allow for a wide-ranging discussion for how best to meet the directives set forward in the CRPD.

The purpose of *Decision Making by Individuals with Intellectual and Developmental Disabilities: Integrating Research into Practice* is to stimulate an exchange of ideas and thoughtful consideration of various aspects of decision making among researchers and practitioners from across the globe in the interest of ultimately expanding meaningful resources for individuals with IDD. The chapters of the book are organized into five parts.

The focus of Part I is on *Opportunities and Supports for Decision Making.* The emphasis of these five chapters is upon expanding the rights and opportunities for individuals with IDD to make their own decisions. Wehmeyer (chapter "Decision Making and Self-Determination") situates decision making within the over-arching framework of self-determination. Shogren et al. (chapter "Supported Decision-Making") explicate and advocate for supported decision-making as a replacement for substituted decision making in the form of guardianship. Sullivan et al. (chapter "Promoting Decision-Making Capabilities in the Healthcare of Adults with Intellectual and Developmental Disabilities: Ethics and Practice") discuss the applications of supported decision-making for healthcare decisions by individuals with IDD. The fourth chapter, by Bigby and Douglas (chapter "Examining the Complexities of Support for Decision-Making Practice"), examines the complexities of providing decision support for individuals with IDD. The final chapter in Part I, by Burke et al. (chapter "Parent and Sibling Roles in Decision Making with Individuals with Intellectual and Developmental Disabilities"), examines the roles of family members as decision-making supporters of individuals with IDD.

The five chapters in Part II examine a variety of *Influences and Considerations in Decision Making.* Khemka (chapter "Theoretical Perspectives on Decision Making") begins this part with a broad review of theoretical perspectives on decision making, based on research with individuals without IDD, as a context for presenting an updated version of the *Pathways Model of Decision Processing.* A chapter by Goscicki et al. (chapter "Etiology and Decision Making: Considerations and Implications") examines considerations and implications of etiology for decision making. A chapter by Adams and Malone (chapter "The Impact of Anxiety on Decision Making in Individuals with Intellectual and Developmental Disabilities or a Diagnosis on the Autism Spectrum") focuses on the role of anxiety in decision making, and one by Bexkens and Mueller (chapter "Affective Decision Making and Peer Influence in Youth with Intellectual Disability") focuses on affective decisions and peer influence in youth with ID. The role of cultural diversity in decision making by individuals with IDD and their families is considered by Vanegas et al. in the final chapter of Part II (chapter "Decision Making and Culturally Diverse Individuals with IDD and Their Families: A Call for Research").

The five chapters in Part III, *Profiles of Decision Making*, present profiles of the decision-making patterns of various subgroups of individuals with IDD. This part begins with an overview by Khemka (chapter "Understanding Decision Making and Intellectual and Developmental Disabilities") of some of the patterns and processes that characterize the decision making of individuals across the range of IDD disorders. This overview is followed by a chapter on a common sense model of decision making by individuals with ID by Greenspan (chapter "Common Sense in Persons with Intellectual Disabilities: The Challenge of Deciding to Say *No* to Social Manipulation") and one on judgment and decision making in individuals with ADHD by Lyon et al. (chapter "Judgment and Decision-Making Paradigms in Adolescents and Adults with ADHD and Associations with Cognitive Abilities"). Chapters "Social Functioning and Decision Making: From Group to Individual Differences Across the Autism Spectrum" (Levin et al.) and "Cognitive, Emotional, and Moral Decision Making in Adolescents and Adults with ASD" (Komeda) focus on patterns of decision making as they relate to group and individual differences in ASD.

Part IV, *Applications of Decision Making*, includes four chapters that focus on applications of decision making in real-life contexts. The chapters include decision making during transition planning by Mello and Sanderson (chapter "Decision Making During Transition to Adulthood") and a chapter on financial decision making by Murphy and Clare (chapter "Financial Decision Making and People with Intellectual Disabilities"). Also explored in this part is the role of decision making in vulnerability to maltreatment by Hickson and Khemka (chapter "Decision Making and Vulnerability to Maltreatment") and one on decision making by individuals with IDD in criminal justice situations by Clare and Murphy (chapter "Criminal Justice Decisions and People with Intellectual Disabilities").

The three chapters in Part V explore the potential of *Positive Psychology Interventions for Decision Making.* These chapters examine the effectiveness of decision making–based interventions designed to reduce the social vulnerability of adolescents and/or adults with IDD through strengths-based approaches. The efficacy of these interventions is assessed using rigorous approaches that include Behavioral Skills Training (BST) and randomized controlled designs. A chapter by Petitpierre and Tabin (chapter "From Social Vulnerability Assessment to Active Prevention Measures: A Decision-Making Perspective") explores issues in assessing vulnerability and active prevention of victimization. Uher et al. (chapter "Behavioral Approaches to Teaching Decision Making to Individuals with IDD") explore the potential of BST to increase the safety skills of individuals with IDD. Khemka and Hickson examine the effectiveness of strategy-based decision-making interventions in the final chapter (chapter "Strategy-Based Interventions for Effective Interpersonal Decision Making").

Part VI, *New Directions in Decision-Making Research*, contains a final chapter by Khemka (chapter "Decision Making and IDD: Future Directions for Research and Practice") in which she pulls together common themes of the chapters and their implications for research and practice. She expresses optimism and enthusiasm for the future of research in the positive psychology of decision making and IDD.

New York, NY, USA   Linda Hickson

References

American Psychiatric Association. (2013).
Anderson, L. L., Larson, S. A., MapelLentz, S., & Hall-Lande, J. (2019). A systematic review of U. S. studies on the prevalence of intellectual or developmental disabilities since 2000. *Intellectual and Developmental Disabilities, 57*(5), 421–438.
Bransford, J. D., & Stein, B. S. (1993). *The IDEAL problem solver* (2nd ed.). W. H. Freeman.
Craig, F., Legrottaglie, A. R., Palumbi, R., de Giambattista, C., & Margari, L. (2016). A review of executive function deficits in autism spectrum disorder and attention-deficit/hyperactivity disorder, *Neuropsychiatric Disease and Treatment, 12*, 1191–1202.
Einfeld, S. L., Ellis, L. A., & Emerson, E. (2011). Comorbidity in intellectual disability and mental disorder in children and adolescents: A systematic review. *Journal of Intellectual & Developmental Disability, 36*(2), 137–143. https://doi.org/10.1080/13668250.2011.572548
Evans, J. St. B. T. (2008). Dual-processing accounts of reasoning, judgment and social cognition. *Annual Review of Psychology, 59*, 255–278.
Evans, J. St. B. T. (2003). In two minds: Dual process accounts of reasoning. *Trends Cognitive Science, 7*, 454–450.
Hickson, L., Blackman, L. S., & Reis, E. M. (1995). *Mental retardation: Foundations of educational programming*. Allyn & Bacon.
Hickson, L., & Khemka, I. (2013). Problem solving and decision making. In M. L. Wehmeyer (Ed.), *The Oxford handbook of positive psychology and disability*. Oxford University Press.
Kahneman, D. (2017). *Thinking fast and slow*. Farrar, Straus, & Giroux.
Khemka, I., Hickson, L., & Mallory, S. B. (2016). Evaluation of a decision making curriculum for teaching adolescents with disabilities to resist negative peer pressure. *Journal of Autism and Developmental Disorders, 46*(7), 2372–2384.
Khemka, I., Hickson, L., & Reynolds, G. (2005). Evaluation of a decision making curriculum designed to empower women with mental retardation to resist abuse. *American Journal on Mental Retardation, 110*930, 193–204.
Leitner, Y. (2014). The co-occurrence of autism and attention deficit hyperactivity disorder in children – what do we know? *Frontiers in Human Neuroscience, 8*, 1–8.
Miyasaka, M., Kajimura, S., & Nomura, M. (2018). Biases in understanding attention deficit Hyperactivity Disorder and Autism spectrum disorder in Japan. *Frontiers in Psychology, 9*, 1–25.
Schalock, R. L., & Luckasson, R. (2021). Intellectual disability, developmental disabilities, and the field of intellectual and developmental disabilities. In L. M. Glidden (Ed.) *APA Handbook of intellectual and developmental disabilities* (pp. 31–45). American Psychological Association.
Schalock, R. L., Luckasson, R., & Tasse, M. J. (2021). *Intellectual disability: Definition, diagnosis, classification, and systems of supports* (12th ed.). American Association on Intellectual and Developmental Disabilities.
Seligman, M. E. P. (1998). Building human strength: Psychology's forgotten mission. *APA Monitor, 29*(1), 2.
Seligman, M. E. P., & Csikszentmihalyi, M. (2000). Positive psychology: An introduction. *American Psychologist, 55*(1), 5–14.
Stanovich, K. E. (2010). *Decision making and rationality in the modern world*. Oxford University Press.
Stanovich, K. E. (1999). *Who is rational? Studies of individual differences in reasoning*. Erlbaum.
Szmukler, G. (2019). "Capacity", "best interests", "will and preferences" and the UN Convention on the Rights of Persons with Disabilities. *World psychiatry, (8)*, 34–41.
Wehmeyer, M. L. & Little, T. D. (2013). Self-determination. In M. L. Wehmeyer (Ed.), *The Oxford handbook of positive psychology and disability* (pp. 116–136). Oxford University Press.

Wehmeyer, M. L., & Shogren, K. A. (2017). Problem solving and self-advocacy. In K. A. Shogren, M. L. Wehmeyer, & N. N. Singh (Eds.), *Hankbook of positive psychology in intellectual and developmental disabilities: Translating research into practice* (pp. 217–245. Springer.
Werner, S. (2012). Individuals with intellectual disabilities: A review of the literature on decision making since the Convention on the Rights of People with Disabilities (CRPD). *Public Health Reviews, 34*, 1–27.

Acknowledgments

I am grateful to my mother Chitralekha Khemka, my husband Mohit, and my daughters Iyra and Riya for their love, support, and encouragement. Thank you!

—Ishita Khemka

I am deeply grateful to Penelope Morgan Colman for her limitless enthusiasm and support for this work and to our children and grandchildren for all they do to enrich our lives.

—Linda Hickson

Contents

Editor Bios

Ishita Khemka, Ph.D., is Associate Professor and Coordinator of the Programs in Special Education at St. John's University, New York. She has extensive experience in teacher preparation, program evaluation, and implementing grant-funded training projects. With a strong commitment to inclusion and equity for individuals with intellectual and developmental disabilities, she has initiated several innovative programs and community partnerships. Thc most notable has been the creation of a college immersion program in New York City for young adults with intellectual and developmental disabilities with an emphasis on the arts, social justice, and self-determination. Her research has significantly advanced the study of interpersonal decision making in individuals with intellectual and developmental disabilities, including autism spectrum disorders. She has been the lead author of two decision making–based prevention programs (*ESCAPE-NOW* and *PEER-DM*) for reducing social vulnerability and maltreatment in adults and adolescents with intellectual and developmental disabilities. Developed as part of a series of grant projects with her collaborator, Dr. Linda Hickson, these prevention programs are widely used in many countries and exemplify high-quality, evidence-based practices for abuse prevention.

Linda Hickson, Ph.D., Professor Emerita served as Professor of Education at Teachers College, Columbia University from 1979 to 2013. As long-time coordinator of the programs in intellectual disability/autism, she taught and mentored many masters and doctoral students preparing for careers in the education of people with disabilities. In addition, she directed several international initiatives to enhance the educational opportunities of people with disabilities in countries such as Ireland, Qatar, and Cyprus. She has published numerous chapters and research articles on a wide array of topics, including theoretical and applied aspects of decision making in people with disabilities. More recently, her research has focused on finding ways to reduce the social vulnerability of adolescents and adults with intellectual disabilities and autism by increasing their ability to make effective self-protective decisions. Dr. Hickson, with her collaborator, Dr. Ishita Khemka, conducted a series of projects, funded by the Joseph P. Kennedy, Jr. Foundation, the Williams Syndrome

Association, and the New York State Developmental Disabilities Planning Council, that resulted in the development and evaluation of *ESCAPE-NOW*—a decision-making curriculum to decrease the vulnerability of women and men with intellectual and developmental disabilities— and *PEER-DM*—a decision-making program to decrease the peer victimization of adolescents with intellectual disabilities and autism.

Part I
Opportunities and Supports for Decision Making

Chapter 1
Decision Making and Self-Determination

Michael L. Wehmeyer

Introduction

Over the course of this book, readers will have the opportunity to learn about definitions and theories of decision making, applications of decision making in the disability context, and strategies to promote the decision-making skills of people with intellectual and developmental disabilities (IDD). The intent of this chapter is to provide a frame for thinking about decision making within a broader psychological construct, self-determination, in which decision making plays a critical role. This, in turn, can provide a theoretical umbrella within which to "design interventions to enable people with IDD to acquire a repertoire of effective decision-making skills" (Khemka & Hickson, 2017, p. 146)

Of course, there are many models of decision making or, as Khemka and Hickson position it in their chapters in this book, multiple pathways. Different skills and abilities may come into play depending upon the context or the type of decision to be made. Some decision-making models emphasize stepwise approaches that incorporate problem-solving and means-end thinking. Researchers in motivational psychology have defined-decisions as "goal-directed behavior made by the individual in response to a certain need, with the intention of satisfying the motive that the need occasions" (Jabes, 1978, p. 86). It is evident, then, that any overarching framework within which to situate decision making must be broad and able to encompass a wide array of types of decisions and decision-making circumstances. Research on the self-determination construct fits these requirements and has been applied extensively in the disability context, thus providing a useful overarching theoretical perspective within which to situate decision making for the purposes of this chapter.

M. L. Wehmeyer (✉)
University of Kansas, Lawrence, KS, USA
e-mail: wehmeyer@ku.edu

I. Khemka, L. Hickson (eds.), *Decision Making by Individuals with Intellectual and Developmental Disabilities*, Positive Psychology and Disability Series,
https://doi.org/10.1007/978-3-030-74675-9_1

Theories of Self-Determination

The first use of the term "self-determination" was by the seventeenth-century English philosopher John Locke (1690) in his *Essay Concerning Human Understanding*. Essentially, Locke proposed that human thought and action were caused by processes internal to the person (e.g., thought, reflection) rather than being imposed, externally. Thus, self-determination (or self-determinism) has its roots in the philosophical doctrine of determinism, which suggests that all action, including human behavior and actions, are effects of preceding causes (Wehmeyer et al., 2017). Based upon an early work in psychology by personality (Angyal, 1941) and motivational (Deci, 1980; Deci & Ryan, 1985) psychologists, Wehmeyer (1992) proposed that to develop interventions to promote self-determination, it was useful to think of being self-determined as acting as a *causal agent*, that is, making or causing things to happen in one's life.

Self-Determination Theory

The earliest applications of the self-determination construct to psychology emerged in personality psychology and then in theoretical formulations of intrinsic motivation. In an early personality psychology text, Angyal (1941) proposed that a central feature of the science of personality was the study of two essential determinants of or causes of human behavior: "autonomous-determinism" (or self-determined or self-caused action) and "heteronomous-determinism" (other-determined or other-caused action) (p. 21). Early motivational and cognitive psychologists incorporated elements of determinism and self-determination into theories of intrinsic motivation (White, 1959; effectance motivation), personal causation (DeCharms, 1959), and locus of causality (Heider, 1958). Particularly, DeCharms (1959) noted that people strive "to be a causal agent, to be the primary locus of causation for, or origin of…behavior" (p. 269).

The most visible role for self-determination in motivation psychology has been set forth in the work of Edward Deci and Richard Ryan (Ryan & Deci, 2000, 2017) on self-determination theory (SDT), which is "an empirically based, organismic theory of human behavior and personality development" (Ryan & Deci, 2017, p. 3). The theory focuses on identifying "factors, both intrinsic to individual development and within social contexts, that facilitate vitality, motivation, social integration, and well-being…" (p. 3).

Consistent with the theoretical formulations of Angyal, DeCharms, and White on which it was built, SDT presumes that "humans are active organisms, motivated to assimilate and integrate knowledge and capacities in both their physical and social environments" (Adams et al., 2017, p. 47). SDT posits the existence of three basic psychological needs—autonomy, competence, and relatedness—fulfillment of which leads to intrinsic and autonomous motivation, optimal growth and development, and positive well-being. The basic psychological need for *autonomy* is satisfied when a person experiences choice and acts volitionally, thus perceiving themselves as the origin of their actions. Autonomous actions are those that are self-endorsed and congruent with one's values and interests (Vansteenkiste et al., 2010). The basic psychological need for *competence* refers not to skills or skill levels, but to a person's need to perceive themselves as effective within environments (Deci & Vansteenkiste, 2004)—to perceptions of competence and mastery and to the experience of perceived mastery and effectiveness (Deci et al., 2013). The basic psychological need for *relatedness* refers to the need to feel connected with other people and to feel a sense of social belonging, to care for others, and to be cared for by others (Ryan & Deci, 2017).

SDT was, in turn, instrumental in formulating a second theoretical framework intended to facilitate the development of interventions to promote self-determination in the education of learners with disabilities: causal agency theory.

Causal Agency Theory

Beginning with work in SDT on autonomy and with DeCharms' suggestion that people strive to be causal agents in their lives, Wehmeyer and colleagues (Wehmeyer, 1992, 1999; Wehmeyer et al., 1996) proposed a functional model of self-determination in which self-determined action was understood as acting in ways that enable a person to make or cause things to happen in their life. The functional model attempted to answer the question "What is self-determination?" to provide a framework to develop interventions to promote self-determination. The functional model defined self-determination as "acting as the primary causal agent in one's life and making choices and decisions regarding one's quality of life free from undue external influence or interference" (Wehmeyer, 1999, p. 56). Further, the functional model proposed a series of component elements of self-determined behavior that consisted of knowledge, skills, and abilities that would enable people to act as a causal agent in their lives. Decision-making skills were among these component elements.

In 2004, Wehmeyer suggested that there was a need to shift the question asked from *What is self-determination?* to *How does someone become self-determined?* Wehmeyer and colleagues (Wehmeyer, 2004; Wehmeyer & Mithaug, 2006) proposed the *causal agency theory* to answer the "how does" question. In 2015, Shogren et al. expanded and refined the causal agency theory, incorporating research in SDT and positive psychology to better describe the development of self-determination across the life span (Shogren et al., 2015; Wehmeyer et al., 2017).

Causal agency theory defines self-determination as a:

> *...dispositional characteristic manifested as acting as the causal agent in one's life. Self-determined people (i.e., causal agents) act in service to freely chosen goals. Self-determined actions function to enable a person to be the causal agent is his or her life. (Shogren et al., 2015, p. 258)*

Within this theoretical framework, self-determined action refers to "the degree to which action is self-caused, volitional and agentic, driven by beliefs about the relationships between actions (or means) and ends" (p. 258). Causal agency is, logically, at the heart of causal agency theory. Wehmeyer (2004) discussed causal agency as such:

> *The adjective 'causal' is defined as expressing or indicating cause; showing the interaction of cause and effect. The term 'agent' is a noun that means one who acts or has the authority to act or, alternatively, a force or substance that causes change. The noun 'agency' refers to one through whom something, typically change, is accomplished. Self-determined people are causal agents in their lives. They act 'with authority' to make or cause something to happen in their lives. Causal agency implies more, however, than just causing action; it implies that the individual who makes or causes things to happen in his or her life does so with an eye toward causing an effect to accomplish a specific end or to cause or create change. Causal agency, as opposed to implying strictly that an individual simply caused some event to happen, implies that action was purposeful or performed to achieve an end. (pp. 351–352)*

Self-determined actions, thus, are purposeful, planned, and intentional and serve to enable the person to support self- (versus other-) determination. To understand how people become self-determined, Wehmeyer and Mithaug (2006) proposed a series of operators for causal events, causal behavior, or causal actions that function as a means for the person (the causal agent) to achieve valued goals, to exert control in their life, and, ultimately, to become more self-determined.

Figure 1.1 provides an overview of these "operators," describing how people become causal agents in their lives. The primary operators in this causal action sequence are (1) the *capability* to perform causal actions or behaviors, subdivided into *causal capability* and *agentic capability* and (2) challenges to the organism's self-determination, through *opportunities* or *threats*, that serve as a catalyst to causal action.

Quite simply, *capability* refers to the condition of being capable—of having certain capacities. People who are causal agents have certain capacities that enable them to respond to challenges to their self-determination. Wehmeyer (2004) proposed two types of capabilities, *causal capability* and *agentic capability*, which are central to the process of acting as a causal agent. These capabilities represent two components of causal action: (1) causing something to happen (causal capability) and (2) directing that action toward a preferred end (agentic capability). These two types of capabilities involve both mental and physical capacities. Wehmeyer (2004) noted, however, that "limitations to the number or complexity of such capacities that might otherwise hinder causal or agentic action can, in fact, be mitigated by a wide array of supports, including technological devices, social networks and supports, and so forth" (p. 353).

Acting in such a way as to cause something to happen (causal capability) involves skills knowledge and beliefs that one is capable and that if one acts, one can reasonably expect positive outcomes. Examples of causal capabilities include goal-setting, problem-solving, and decision-making skills. Agentic capability involves the knowledge and skills needed to direct causal action and perceptions or beliefs about

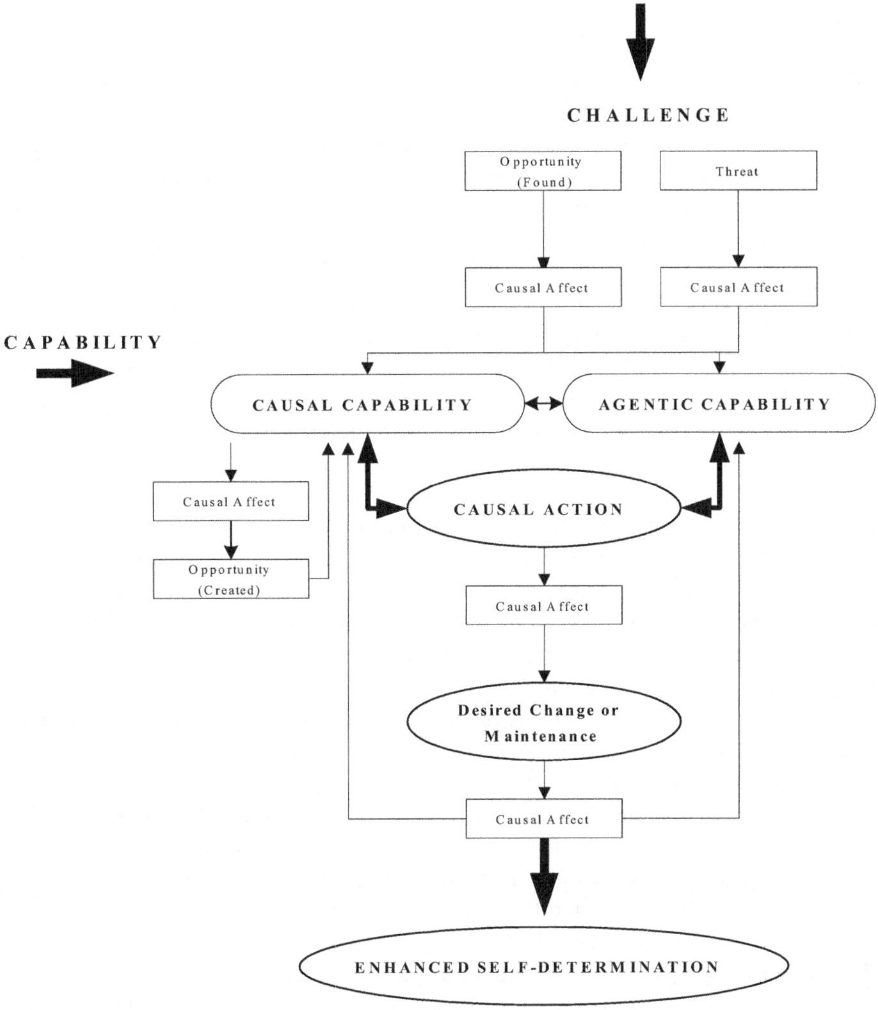

Fig. 1.1 Overview of causal agency theory. (Reprinted with permission from Wehmeyer & Mithaug, 2002)

oneself and one's environment that enable one to act. These skills primarily involve self-regulatory and self-management skills.

Causal agency theory hypothesizes that people who are causal agents engage in causal action (implementing causal and agentic capabilities) in response to *challenges* that serve as a catalyst for such action. A challenge is "any circumstance under which one has to engage in the full use of one's abilities or resources to resolve a problem or threat or to achieve a goal or objectives" (Wehmeyer, 2004, p. 354). Such challenges emerge in a person's life as either *opportunities* or *threats*. Opportunity refers to situations or circumstances that provoke the person to create change or make something happen, based upon one's preferences and interests. Opportunities can be *found*, which refers to unanticipated circumstances that did not involve effort on the part of the person, or *created*, in which the person consciously acts to create a favorable circumstance.

The second challenging condition, *threat*, involves situations or circumstances that "threaten the organism's self-determination and provoke the organism to exercise causal action to maintain a preferred outcome or to create change that is consistent with one's own values, preferences, or interests, and not the values, preferences or interests of others" (Wehmeyer, 2004, pp. 354–355).

This early version of causal agency theory also proposed that affective or emotional aspects regulate behavior in support or hinderance of causal actions. So, the basic operators in causal action involve causal and agentic action in response to challenges (opportunities, found or created, or threats) that enable persons to direct their behavior to achieve either the desired change or maintain a preferred circumstance or situation.

As noted, Fig. 1.1 provides an overview of *operators* in causal agency theory describing how people become causal agents in their lives. Figure 1.2 provides a schema for causal action within causal agency theory. Accordingly, in response to challenges (opportunities or threats), people who are causal agents begin with a goal generation process. This process includes a self-analysis of one's strengths, needs, limitations, preferences, and interests as they pertain to the challenge circumstance. That process leads to the identification of actions needed to act on the challenge that, in turn, are prioritized based on relative importance to addressing the challenge. When actions have been prioritized, the person frames the most important action(s) in terms of a goal.

Once the goal is stated, the person engages in a *goal discrepancy analysis* process, the gist of which is an analysis of the person's current state pertaining to the goal and the (desired) goal state. That analysis enables the person to state the *goal discrepancy problem*. Importantly for the current chapter, this often parallels a decision-making process in which the first step is to identify the problem and generate alternatives or options from which the person must decide.

Without going into too much detail (see Wehmeyer & Mithaug, 2006, for more such detail), the causal action schema moves from setting the goal discrepancy problem to analyzing one's capacities to solve the problem presented by the goal discrepancy analysis and examines the degree to which the challenge will support goal attainment. In this process, the person must adjust their capacity (i.e., acquire new or refine knowledge and skills pertaining to the problem) or adjust the challenge itself (modify some aspect of the challenge) to create a *just-right match* between capacity and challenge so as to optimize the person's probability of solving the goal discrepancy problem. Again, in most models of decision making, after one identifies the problem and what one's options or alternatives from which to make a decision are, the person evaluates each option to see how it aligns with preferences and values and what consequences might derive from each such option upon taking action. This is, in function, similar to the capacity-challenge discrepancy analysis phase of the causal action schema.

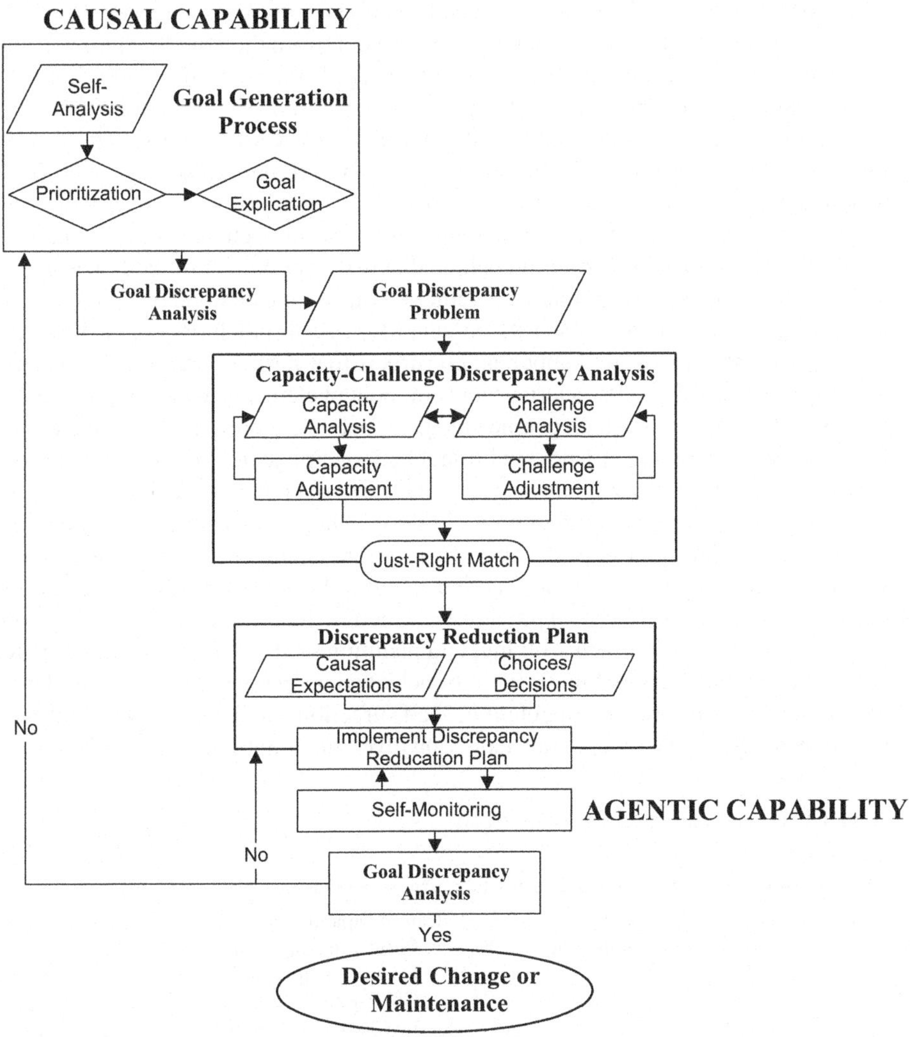

Fig. 1.2 Causal action schema. (Reprinted with permission from Wehmeyer & Mithaug, 2002)

Once the person has created a just-right match, that person creates a *discrepancy reduction plan* by setting causal expectations (What can I expect to achieve?), making choices and decisions about strategies and methods to reduce the discrepancy between current status state and the goal state and then finalizing and implementing a discrepancy reduction plan. This plan, equivalent to the action planning process in the goal setting and attainment sequence, is iterative, with the person comparing their current state with the desired goal state. When that *goal discrepancy analysis process* results in the closure of the gap between the person's current state and the goal state, then the desired causal action has been completed.

In 2015, Shogren and colleagues refined and updated the causal agency theory. A primary purpose for this was to better align research from the functional model and causal agency theory with research that had emerged since the early 2000s in SDT, so as to be able to better describe the development of self-determination. Table 1.1 depicts the updated framework for self-determined (or causal) action. The theory proposes three *essential characteristics* of self-determined action—volitional action, agentic action, and action-control beliefs—that contribute to causal agency and the development of self-determination. These essential characteristics refer not to specific actions performed or the beliefs that drive action, but to the function the action serves for the person; that is, whether the action enabled the person to act as a causal agent and enhances the development of self-determination. For each essential characteristic, there are component constructs that further operationalize (theoretically) each essential characteristic and component elements, which are the skills, beliefs, and activities in which people engage to become more self-determined. The first two essential characteristics, volitional and agentic action, directly mirror the causal and agentic capability constructs in the earlier version of the theory.

Volitional action is action based on conscious choices that reflect one's preferences. Component constructs of volitional action are autonomy and self-initiation. Component elements of volitional action involve the initiation and activation of causal capabilities, as described previously, which include choice-making, decision making, goal-setting, problem-solving, and planning skills. *Agentic action* refers to the means by which something is done or achieved. Component constructs related to agentic action are self-regulation, self-direction, and pathways thinking. Component elements of agentic action involve the initiation and activation of

Table 1.1 Components of causal agency theory

Essential characteristics	Component constructs	Component elements
Volitional action	Autonomy Self-initiation	Causal capabilities Choice-making skills Decision making skills Goal-setting skills Problem-solving skills Planning skills
Agentic action	Self-regulation Self-direction Pathways thinking	Agentic capabilities Self-management skills (self-monitoring, self-evaluation, etc.) Goal attainment skills Problem-solving skills Self-advocacy skills
Action-control beliefs	Psychological empowerment Self-realization Control expectancy Agency beliefs Causality beliefs	Self-awareness Self-knowledge

Reprinted with permission from Shogren, Raley, Burke, and Wehmeyer (2019a)

agentic capabilities, as described previously, and which include self-management, goal attainment, and self-advocacy skills.

A third essential characteristic in the updated version of the causal agency theory expands on the area of the affective or emotional aspects that regulate behavior in support or hinderance of causal actions, referred to as *causal affect* in Fig. 1.1. To account for these affective elements, causal agency theory incorporates basic tenets of action-control theory (Little et al., 2002). Action-control theory hypothesizes three types of the so-called action-control beliefs: (1) *control expectancy beliefs*, which refer to beliefs about the link between the self and the goal; (2) *capacity beliefs*, which refer to beliefs about the link between the self and the means for achieving the goal; and (3) *causality beliefs*, which are beliefs about the utility or usefulness of a given means for attaining a goal. Action-control beliefs form the third essential element in causal agency theory. Psychological empowerment, self-realization, control expectancy, agency beliefs, and causality beliefs form the component constructs of this essential characteristic, with self-awareness and self-knowledge forming the component elements.

The Development of Self-Determination

As noted previously, the functional model of self-determination and, subsequently, both iterations of causal agency theory were formulated and tested to provide frameworks for understanding self-determination or more specifically the development and emergence of self-determination, so as to provide a theoretically sound framework upon which to build assessments to measure self-determination and interventions to promote self-determination, particularly within the disability context. The most recent iteration of causal agency theory provided, for the first time, a framework that could address all aspects of development and allowed us to better model this process. Figure 1.3 provides a flow chart illustrating this process.

As reflected in Fig. 1.3 and as consistent with proposals from causal agency theory (Wehmeyer, 2004), self-determination develops as a function of the interaction between the threats or demands and opportunities in one's environment and attempts by the person to meet basic psychological needs of autonomy, competence, and relatedness (Shogren, Little et al., 2017a). These interactions between the person's need to be autonomous, competent, and maintain meaningful relationships, as framed by SDT, energizes a causal action sequence that involves volitional and agentic action, which in turn are mediated by action-control beliefs about one's ability to act as a causal agent (Mumbardó-Adam et al., 2018). Put more simply, the developmental model derived from causal agency theory proposes that from early in childhood and onward, humans are motivated to act volitionally (e.g., based upon autonomous or intrinsic motivation) and employ a causal action sequence involving volitional and agentic action mediated by action-control beliefs that enable them to act as a causal agent in their lives (e.g., make things happen). Repeated experiences of causal agency, in turn, result in enhanced self-determination.

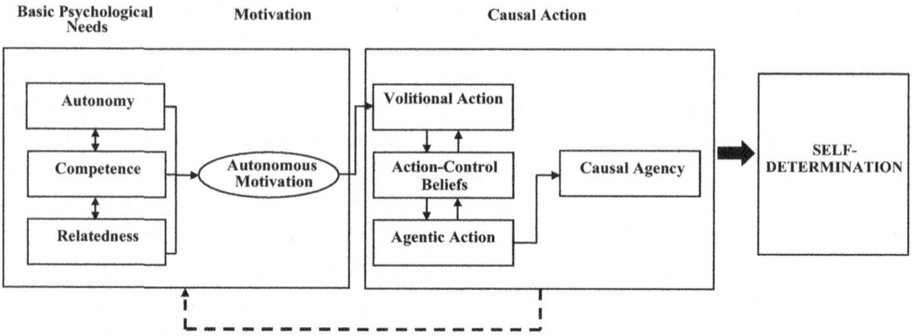

Fig. 1.3 Model of the development of self-determination. (Reprinted with permission from Shogren & Wehmeyer, 2016)

Decision Making and Self-Determination

The utility to situating decision making within the broader construct of self-determination lies in the power it provides to understand the interrelationships between various skills that are involved in making a decision, independent of the decision-making process employed and by providing a means to understand the development of critical skills involved in decision making. Further, making decisions is inextricably linked to ideas of autonomy and self-regulation, which in turn can be better understood within the contexts of theories of self-determination. Also, as mentioned previously, taxonomies of the decision-making process do not exist within a vacuum. One cannot really understand decision-making processes, independent of what decision-making model one uses, without also understanding the role of skills such as problem-solving, goal-setting and attainment, planning, choice-making, self-awareness, self-knowledge, and others in the type of decision and the context. If we, as a field, are to provide supports to people with IDD to promote effective participation in decision making, those activities cannot be disconnected, discrete skills training on steps in any decision-making process, but must include a more holistic approach that takes into account the array of skills, abilities, and beliefs that comprise the process. Situating decision making within the broader construct of self-determination provides that framework.

First, it is important to understand the reciprocal relationship between self-determination and decision making. Causal agency theory situates decision making among the component elements of volitional action, so it is clearly a critical skill and process in promoting self-determination. But one can easily hypothesize that the relationship between self-determination and decision making runs the other way as well. Making and engaging in decisions promotes the development of self-determination. People who are more self-determined are also quite likely to take greater ownership over and involvement in the decision-making process. There is evidence of a similar reciprocal relationship between self-determination and student involvement in educational planning. Williams-Diehm et al. (2008) found that the involvement of students with disabilities in transition planning predicted higher levels of self-determination, but that the relationship went both ways, with higher self-determination predicting higher levels of involvement. This is likely the case for self-determination and decision making as well.

Second, the hallmark of an autonomous individual in our societies is that a person makes their own decisions. Of course, what that really means is not that every person makes every decision completely independently, but instead that people are, essentially, the causal agents in decisions that impact their lives. There are a myriad of issues that will play into how much the person makes the decision individually, from the complexity of the decision to be made to the degree to which that decision impacts others. But, if we understand an autonomous person to be someone who acts based upon their own preferences and interests, and not necessarily as someone who does things on their own, then we see the critical relationship between decision making, causal agency, and self-determination.

Third, specific to the topic of this book, there has been almost three decades worth of research, assessment development, and intervention development and evaluation in self-determination in the disability context. Thus, there is a wealth of information available specific to the disability context from which researchers and practitioners interested in decision making and people with disabilities can draw. A necessarily brief summary of that research and its implications for decision making specifically as it pertains to people with IDD concludes this chapter.

Self-Determination and Decision Making by People with Intellectual and Developmental Disabilities

Self-Determination and People with Intellectual and Developmental Disabilities

It is beyond the scope of this chapter to provide a comprehensive review of the research and knowledge base pertaining to people with IDD and self-determination. There are several recent such reviews available (Shogren et al., 2017b; Wehmeyer & Shogren, 2017). But it is worth summarizing this research (from Shogren et al., 2017b) to provide a snapshot of what is known about self-determination and people with IDD.

1. *People with IDD are less self-determined than their nondisabled peers and, often, less self-determined than peers with other types of disabilities.* This is a well-documented research finding. Although there are a myriad of factors serving as barriers to the enhanced self-determination of people with IDD, research points to the low expectations others hold about IDD and the limited opportunities people with IDD have to learn and practice skills related to self-determination. People with IDD remain among the most frequently segregated populations in school and society, and the research supports the contention that the environments in which people with IDD live, learn, work, and play remain highly segregated and limit opportunities for choice, problem-solving, decision making, and so forth.

2. *When provided the opportunity to learn and practice skills and have experiences leading to self-determination, people with IDD can and do acquire these skills and become more self-determined.* There is now a compelling body of evidence that shows that if youth and young people are provided instruction to promote self-determination, provided opportunities to practice those skills, and provided opportunities to have experiences that teach them about their preferences and abilities, they will, indeed, acquire these skills and become more self-determined. One of the frequently identified barriers to learning and acquisition is IQ, and perceived intelligence is one of the factors that limit the opportunities for people with IDD to learn and practice skills and have experiences leading to enhanced self-determination. However, the research is clear that IQ has very little impact on relative self-determination status and that much stronger predictors of self-determination are choice opportunities, participation in problem-solving and decision making, and environments that honor preferences and interests.
3. *There are a number of evidence- and research-based practices that have been shown to contribute to the self-determination of youth and young adults with IDD.* There are now numerous evidence-based practices to promote student self-regulated problem-solving and goal setting, involvement in educational and life planning, and to promote the acquisition of decision-making skills related to the component elements of self-determined action that have been shown to benefit youth and adults with IDD. Most of the focus in this area has been with the youth and young adults of transition age, but as work by Khemka and Hickson described elsewhere in this book shows, this is true across the life span.
4. *Self-determination has been linked to more positive school and adult outcomes and more positive quality of life and life satisfaction for people with IDD.* Finally, at least with regard to this brief summarization, there is strong evidence that young people with IDD who are more self-determined achieve more positive school and adult outcomes, including improved educational and transition goal attainment, more positive work and employment outcomes, and improved community inclusion. Further, multiple studies provide links between higher self-determination and more positive quality of life, independent decision making, and self-reported life satisfaction for youth and adults with IDD.

The knowledge base in this area is clear. Youth and adults with IDD can become more self-determined if provided with learning opportunities and experiences that emphasize their strengths and interests, and when they become more self-determined, they achieve more positive school and adult outcomes and experience a more positive quality of life and life satisfaction.

Promoting the Decision-Making Skills of People with Intellectual and Developmental Disabilities

Throughout this book readers will learn about a variety of approaches and interventions to facilitate decision-making skills and opportunities for people with IDD, many of which involve instruction and support pertaining to component elements of self-determined action, such as goal setting, problem-solving, planning, and so forth. To conclude this chapter, a model of support (the Self-Determined Decision-making Model or SDDMM), to promote meaningful participation in the decision-making process by people with IDD that we developed based on our prior efforts to promote self-determination, is described as one example of how efforts to promote self-determination can support efforts to promote decision making.

SDDMM is one of a family of interventions to promote self-determination and self-regulated problem-solving to set and attain goals. The initial such model was the *Self-Determined Learning Model of Instruction* (SDLMI) (Mithaug et al., 1998; Shogren et al., 2019b; Wehmeyer et al., 2000), which was designed to enable teachers to teach students to self-regulate problem-solving to set a goal, create an action plan to achieve that goal, and self-monitor and self-evaluate progress toward the goal, revising the action plan or goal as needed so as to attain a relevant goal. Implementation of the SDLMI consists of a three-phase instructional process. Each phase presents a problem to be solved by the student (What is my goal? What is my plan? What have I learned?), and the student solves the respective problem by answering a series of four questions per phase that students learn, modify to make their own, and apply to reach self-set goals. Each question is linked to a set of teacher objectives, and each instructional phase includes a list of educational supports that teachers can use to teach or support students to self-direct learning.

The student questions are constructed to guide a student through a self-regulated problem-solving sequence to solve the problem in the phase. The questions differ by phase, but they represent parallel steps in the problem-solving sequence: (a) identify the problem, (b) identify potential solutions to the problem, (c) identify barriers to solving the problem, and (d) identify consequences of each solution. These are the basic steps in any problem-solving process, and they form the means-end problem-solving sequence represented by the student questions in each phase of the model and enable the student to solve the problem posed in each phase.

The SDLMI has strong evidence of its efficacy in promoting self-determination as well as more positive school and adult outcomes (see Wehmeyer et al., 2017). The SDLMI was developed for use by teachers within schools. A second member of the self-determined goal setting intervention family, the *Self-Determined Career Design Model* (SDCDM; Wehmeyer et al., 2003), was designed for use to promote self-determined career decision making for adults with IDD. Specifically, the SDCDM was designed to enable vocational counsellors and employment support professionals to support youth, young adults, and adults with disabilities to self-regulate problem-solving to set and attain career, life design, and employment goals. The process for the SDCDM is identical to that of the SDLMI, with the exception that the questions are focused on career, life design, and employment-related outcomes. Research using the SDCDM has shown that participants who received supports from rehabilitation professionals using the SDCDM showed significantly enhanced autonomous functioning as a result (Shogren et al., 2016).

SDDMM (Shogren et al., 2019b) is the third member of the self-regulated goal-setting family. The SDDMM is intended to enable support people, family members, and trusted friends to support people with disabilities to self-regulate problem-solving to set and attain goals related to the person's involvement in the decision-making process. As has been discussed, the first step in the decision-making process is a problem-solving process to identify the options from which a person can select. The SDDMM is a process through which trusted friends, allies, family members, support professionals, and others identified by a person with a disability can support that person to set goals that enable them to solve problems associated with the decision-making process.

The SDDMM process is identical to that of the SDLMI and the SDCDM. There are three phases, each with a problem for the person to solve. The person is guided to solve the problem in each phase by answering a series of four questions (per phase). There are objectives for support persons to use to assist the person to answer each question and a set of decision support strategies that may assist in supporting the person to answer questions. Tables 1.2, 1.3, and 1.4 provide the questions, support objectives, and support strategies for each of the three phases in the SDDMM.

Table 1.2 Questions, objectives, and decision supports for phase 1 of the Self-Determined Decision-making Model (SDDMM)

Problem to solve: What is my decision-related goal?	
Question 1: *What decision do I need to make?* Objectives: Enable person to identify decisions that need to be made Enable person to identify preferences/interests/beliefs/values related to decisions that need to be made Enable person to identify strengths and needs related to decisions that need to be made Enable and support person to prioritize decisions that need to be made and select most important decision to make Question 2: *What do I know about it now?* Objectives: Enable person to identify his or her current status in relation to prioritized decision to make Enable person to identify knowledge/skills/needs of decision to make Assist person to gather information about opportunities and barriers in their environments pertaining to prioritized decision to make Question 3: *What must change for me to make a decision?* Objectives: Support person to prioritize needs related to decision to make Enable person to choose primary need and decide if action needs to be focused toward capacity building, modifying the environment or both Question 4: *What can I do to make this happen?* Objectives: Teach/support person to state goals/objectives related to prioritized decision to make	*Decision supports* Awareness training Self-assess preferences, interests, and abilities pertaining to decision area Problem-solving instruction Choice-making instruction Goal-setting instruction

Reprinted with permission from Shogren and Wehmeyer (2017)

Table 1.3 Questions, objectives, and decision supports for phase 2 of the Self-Determined Decision making Model (SDDMM)

Problem to solve: What is my plan?	
Question 1: *What actions can I take to reach my decision-related goal?* Objectives: Enable person identify alternatives to achieve decision-related goal Assist person to gather information on consequences of alternatives Enable person to select best action alternatives Question 2: *What could keep me from taking action?* Objectives: Support person to identify barriers to implementing action alternatives Support person to identify actions to remove barriers Question 3: *What can I do to remove these barriers?* Objectives: Assist person in identifying appropriate decision supports to implement selected action alternative Teach person knowledge/skills needed to implement selected supports Question 4: *When will I take action?* Objectives: Enable person to determine schedule for action plan to remove barriers and implement supports Support and enable person to implement the action plan Enable person to self-monitor their progress	*Decision supports* Exploration of community resource/support Problem-solving instruction Self-scheduling training Self-instruction training Picture-cue training Self-advocacy instruction Assertiveness training Self-monitoring instruction

Reprinted with permission from Shogren and Wehmeyer (2017)

Table 1.4 Questions, objectives, and decision supports for phase 3 of the Self-Determined Decision-making Model (SDDMM)

Problem to solve: What have I achieved?	
Question 1: *What actions have I taken?* Objectives: Enable person to self-evaluate and articulate progress toward goal Question 2: *What barriers have been removed?* Objectives: Assist person to compare progress with their desired outcomes Question 3: *What has changed to enable me to make the decision that needs to be made?* Objectives: Support person to reevaluate goal if progress is insufficient Assist person to decide if goal remains the same or changes Collaborate with person to identify if the action plan is adequate or inadequate given revised or retained goal Assist person to change action plan if necessary Question 4: *Have I achieved what I want to achieve?* Objectives: Enable person to decide if progress is adequate, inadequate, or if goal has been achieved If this goal has been achieved, enable person to decide if a different goal is needed to make the decision that needs to be made	*Decision supports* Self-evaluation instruction

Reprinted with permission from Shogren and Wehmeyer (2017)

Conclusions

The intent of this book is to examine theoretical considerations in the study of decision making as well as practical applications in social interpersonal domains for adolescents and adults with IDD. Situating decision making, theoretically, within the broader construct of self-determination has multiple benefits. Decision making, no matter what model or process is used, can be complex and multifaceted, and conceptualizing that process within theoretical models of self-determination allows for a richer understanding of the different processes involved in making a decision and provides information on the development of these processes.

Further, decision making is viewed as an essential component of living an autonomous life, and as such, thinking about decision making as solely a set of disconnected, discrete skills has the potential to limit the empowering nature of making one's decisions. Finally, far too many people view people with IDD as lacking the capacity to solve complex problems and make complex decisions and thus exclude them from doing so, even in the context of their own lives. Situating decision making in the context of self-determination illuminates the fact that what is important is not that any person makes a decision completely independently, but that one is the causal agent in one's life.

There is a vast body of literature that substantiates the fact that promoting self-determination has positive benefits across multiple spheres of life for people with IDD. Causal agency theory has provided a useful framework for describing the development of self-determination and understanding how people become self-determined, which in turn provides a useful frame within which to design interventions to promote self-determination, such as the Self-Determined Decision-Making Model.

With the exception of some of the research highlighted in this book, there has been far too little emphasis on promoting decision-making skills and opportunities for people with IDD. The uncertainty and complexity of decision making makes the process more difficult, often for people with IDD, who need more intensive and different types of supports to successfully participate in such decisions. But as strengths-based approaches to understanding disability replace deficits-based approaches of the past, and as people with IDD become more and more a part of the mainstream of life, a focus on promoting and supporting decision making becomes essential, not just to ensure successful functioning but to ensure that people are empowered to live full, rich lives in their communities. Many of the models and processes described in this book provide direction for the field in achieving such outcomes, and approaching these tasks by understanding decision making as a part of the broader context of promoting and enhancing self-determination will ensure that we, as a field, contribute to this empowerment focus.

References

Adams, N., Little, T. D., & Ryan, R. M. (2017). Self-determination theory. In M. L. Wehmeyer, K. A. Shogren, T. D. Little, & S. J. Lopez (Eds.), *Development of self-determination through the life-course* (pp. 47–54). Springer.

Angyal, A. (1941). *Foundations for a science of personality*. Harvard University Press.

Deci, E. L. (1980). *The psychology of self-determination*. Lexington Books.

Deci, E. L., & Ryan, R. (1985). *Intrinsic motivation and self-determination in human behavior*. Plenum Press.

Deci, E. L., & Vansteenkiste, M. (2004). Self-determination theory and basic need satisfaction: Understanding human development in positive psychology. *Ricerche di Psichologia, 27*, 17–34.

Deci, E. L., Ryan, R. M., & Guay, F. (2013). Self-determination theory and actualization of human potentials. In D. M. McInerney, H. W. Marsh, R. G. Craven, F. Guay, & D. M. McInerney (Eds.), *Theory driving research: New wave perspectives on self-processes and human development* (pp. 109-133). Information Age Publishing.

Jabes, J. (1978). *Individual processes in organizational behavior*. Harlam Davidson, Inc.

Khemka, I., & Hickson, L. (2017). The role of positive psychology in interpersonal decision making. In K. A. Shogren, M. Wehmeyer, & N. N. Singh (Eds.), *Handbook of positive psychology in intellectual and developmental disabilities: Translating research into practice* (pp. 145–171). Springer.

Little, T. D., Hawley, P. H., Henrich, C. C., & Marsland, K. (2002). Three views of the agentic self: A developmental synthesis. In E. L. Deci & R. M. Ryan (Eds.), *Handbook of self-determination research* (pp. 389–404). University of Rochester Press.

Locke, J. (1690). *An essay on human understanding*. https://www.gutenberg.org/files/10615/10615-h/10615-h.htm

Mithaug, D., Wehmeyer, M. L., Agran, M., Martin, J., & Palmer, S. (1998). The self-determined learning model of instruction: Engaging students to solve their learning problems. In M. L. Wehmeyer & D. J. Sands (Eds.), *Making it happen: Student involvement in educational planning, decision making and instruction* (pp. 299–328). Paul H. Brookes.

Mumbardó-Adam, C., Guàrdia-Olmos, J., & Giné, C. (2018). Assessing self-determination in youth with and without disabilities: The Spanish version of the AIR self-determination scale. *Psicothema, 30*(2), 238–243.

Ryan, R. M., & Deci, E. L. (2000). Self-determination theory and the facilitation of intrinsic motivation, social development, and well-being. *American Psychologist, 55*, 68–78.

Ryan, R. M., & Deci, E. L. (2017). *Self-determination theory: Basic psychological needs in motivation, development, and wellness*. Guilford Press.

Shogren, K. A., & Wehmeyer, M. L. (2016). *Relationship between basic psychological needs, motivation, causal action, and self-determination*. Kansas University Center on Developmental Disabilities.

Shogren, K. A., & Wehmeyer, M. L. (2017). *Self-determination interventions*. Kansas University Center on Developmental Disabilities.

Shogren, K. A., Wehmeyer, M. L., Palmer, S. B., Forber-Pratt, A., Little, T., & Lopez, S. (2015). Causal agency theory: Reconceptualizing a functional model of self-determination. *Education and Training in Autism and Developmental Disabilities, 50*(3), 251–263.

Shogren, K. A., Gotto, G. S., Wehmeyer, M. L., Shaw, L., Seo, H., Palmer, S., … Barton, K. N. (2016). The impact of the self-determined career development model on self-determination outcomes. *Journal of Vocational Rehabilitation, 45*, 337–350.

Shogren, K. A., Little, T. D., & Wehmeyer, M. L. (2017a). Human agentic theories and the development of self-determination. In M. L. Wehmeyer, K. A. Shogren, T. D. Little, & S. Lopez (Eds.), *Development of self-determination across the life course* (pp. 17–26). Springer.

Shogren, K. A., Wehmeyer, M. L., & Burke, K. (2017b). Self-determination. In K. A. Shogren, M. L. Wehmeyer, & N. N. Singh (Eds.), *Handbook of positive psychology in intellectual and developmental disabilities: Translating research into practice* (pp. 49–64). Springer.

Shogren, K. A., Raley, S. K., Burke, K. M., & Wehmeyer, M. L. (2019a). *The self-determined learning model of instruction teacher's guide*. Kansas University Center on Developmental Disabilities.

Shogren, K. A., Wehmeyer, M. L., Martinis, J., & Blanck, P. (2019b). *Supported decision making: Theory, research, and practice to enhance self-determination and quality of life*. Cambridge University Press.

Vansteenkiste, M., Niemiec, C., & Soenens, B. (2010). The development of the five mini-theories of self-determination theory: An historical overview, emerging trends, and future directions. In T. Urdan & S. Karabenick (Eds.), *Advances in motivation and achievement, vol. 16: The decade ahead*. Emerald Publishing.

Wehmeyer, M. L. (1992). Self-determination and the education of students with mental retardation. *Education and Training in Mental Retardation, 27*, 302–314.

Wehmeyer, M. L. (1999). A functional model of self-determination: Describing development and implementing instruction. *Focus on Autism and Other Developmental Disabilities, 14*, 53–61.

Wehmeyer, M. L. (2004). Beyond self-determination: Causal agency theory. *Journal of Developmental and Physical Disabilities, 16*, 337–359.

Wehmeyer, M. L., & Mithaug, D. (2002). *Causal agency theory: How causal capabilities affect self-determination*. Proposal to the U.S. Department of Education, Office of Special Education Programs, Field Initiated Research Projects grant competition (CFDA 84.324C).

Wehmeyer, M. L., & Mithaug, D. (2006). Self-determination, causal agency, and mental retardation. In L. M. Glidden (Series Ed.) & H. Switzky (Vol. Ed), *International review of research in mental retardation: Vol. 31 Current perspectives on individual differences in personality and motivation in persons with mental retardation and other developmental disabilities* (pp. 31–71). Academic.

Wehmeyer, M. L., & Shogren, K. A. (2017). Applications of the self-determination construct to disability. In M. L. Wehmeyer, K. A. Shogren, T. D. Little, & S. Lopez (Eds.), *Development of self-determination through the life-course* (pp. 111–123). Springer.

Wehmeyer, M. L., Kelchner, K., & Richards, S. (1996). Essential characteristics of self-determined behavior of individuals with mental retardation. *American Journal on Mental Retardation, 100*, 632–642.

Wehmeyer, M. L., Palmer, S., Agran, M., Mithaug, D., & Martin, J. (2000). Promoting causal agency: The self-determined learning model of instruction. *Exceptional Children, 66*, 439–453.

Wehmeyer, M., Lattimore, J., Jorgensen, J., Palmer, S., Thompson, E., & Schumaker, K. M. (2003). The self-determined career development model: A pilot study. *Journal of Vocational Rehabilitation, 19*, 79–87.

Wehmeyer, M. L., Shogren, K. A., Little, T. D., & Lopez, S. J. (2017). Introduction to the self-determination construct. In M. L. Wehmeyer, K. A. Shogren, T. D. Little, & S. J. Lopez (Eds.), *Development of self-determination through the life-course* (pp. 3–16). Springer.

Williams-Diehm, K., Wehmeyer, M. L., Palmer, S., Soukup, J. H., & Garner, N. (2008). Self-determination and student involvement in transition planning: *A multivariate analysis*. *Journal on Developmental Disabilities, 14*, 25–36.

Chapter 2
Supported Decision-Making

Karrie A. Shogren, Evan E. Dean, Brad Linnenkamp, Sheida K. Raley, Jonathan Martinis, and Peter Blanck

Conceptualizations of disability changed in significant and positive ways beginning in the last quarter of the twentieth century, creating opportunities and needs for changes in disability services and supports. Prior to this time, disability was viewed from a deficit lens, characterized as a form of difference or a deviation from normality (Wehmeyer, 2013). Such conceptualizations led to a range of responses from charitable and benevolent at best to discriminatory and segregationist at worst. Recent disability rights and self-advocacy movements, however, have challenged prevailing notions that disability was something to eliminate or remediate when "worthy" of remediation (Blanck, 2014, 2020).

Modern disability movements have led to the emergence of new ways to conceptualize disability that emphasize the role of self-determination, strengths-based approaches, individualized supports, environmental modifications, contextual analysis, and community participation (Shogren et al., 2017b). As such, efforts to recognize, celebrate, and enforce the civil and human rights of people with disabilities are increasingly reflected in law and public policy, although the actualization of such changes in practice remains more vexing (Shogren et al., 2009; Shogren et al., 2015). However, ongoing efforts to define and actualize innovative supports that are rooted in strengths-based conceptualizations of disability are occurring. Defining

K. A. Shogren (✉) · E. E. Dean · B. Linnenkamp · S. K. Raley
Kansas University Center on Developmental Disabilities, University of Kansas, Lawrence, KS, USA
e-mail: shogren@ku.edu; Edean2@ku.edu; Blinn@ku.edu; Raley@ku.edu

J. Martinis · P. Blanck
Burton Blatt Institute, Syracuse University, Syracuse, NY, USA
e-mail: JGMartin@Law.Syr.Edu; pblanck@law.syr.edu

I. Khemka, L. Hickson (eds.), *Decision Making by Individuals with Intellectual and Developmental Disabilities*, Positive Psychology and Disability Series,
https://doi.org/10.1007/978-3-030-74675-9_2

disability as a natural part of the human experience pushes forward conceptualizations of disability that recognize the interaction between a person and their context and the need for personal and environmental supports to enhance functioning consistent with the values and vision of the person with a disability (Thompson et al., 2009; Thompson et al., 2014).

One arena where changes are occurring is in the ways in which we, as a society, support people with intellectual and developmental disabilities (IDD) in making major life decisions. Prevailing practices, such as plenary guardianship, are rooted in outdated, deficit-based models of disability (Martinis & Gustin, 2017; Martinis et al., in press). Emerging models, such as supported decision-making, are rooted in an understanding of the human rights and legal capacity of people with IDD and the ability to concurrently promote self-determination and human agency while also providing needed supports (Blanck & Martinis, 2015; Shogren & Wehmeyer, 2015).

This chapter will provide an overview of supported decision-making and its emergence in the disability field. We describe how it aligns with the broader rights-based movement in the disability field that adopts strengths-based, social-ecological models of disability and how it challenges other-directed decision-making models, such as guardianship. Next, we describe emerging directions and innovations in building systems of supports for decision making and highlight strategies for making this happen across the life course, using examples of ways that people with disabilities articulate the value of supported decision-making.

Supported Decision-Making: Definition and Emergence in the Field of Intellectual and Developmental Disabilities

Supported decision-making is receiving attention in law, policy, research, and practice throughout the world to actualize changing conceptualizations of disability that enable people with IDD to remain at the center of decision-making processes in their lives. Supported decision-making provides a framework for thinking more broadly about how we, in policy and practice, support decision making across the life course (Shogren & Wehmeyer, 2015; Shogren et al., 2019). Supported decision-making has been identified as an alternative to overbroad or undue guardianship, typically court-ordered in the form of plenary or total other-directed decision making (Kohn et al., 2013). As such, the emergence of supported decision-making advances theory, research, policy, and practice that contribute to enhanced self-determination and the quality of life for people with disabilities (Martinis & Blanck, 2019; Shogren et al., in press).

Guardianship and People with Disabilities

Although supported decision-making is receiving significant attention in the disability field, the predominant "support" for the decision-making needs of adults with disabilities in the United States and throughout the world remains plenary guardianship. Guardianship is a "legal process where a court removes some or many of the legal and decision-making rights from an individual and transfers all or some of them to another person" (National Council on Disability, 2019, p. 3). While guardianship laws vary from jurisdiction to jurisdiction, in general, Americans are ordered under guardianship when a state court decides that they cannot direct their lives or make some or all decisions due to having a disability or condition that impacts their ability to do so.

An individual under guardianship is determined to be incapacitated in the eyes of the law. A third party, usually called a *guardian*, but in some states a *conservator*, is then appointed to direct the person's life and make decisions for the person in areas where the court feels he or she cannot (Quality Trust for Individuals with Disabilities, 2013). In cases in which the court directs that all life decisions are to be determined by the guardian, an order of *plenary* guardianship is issued, and where only some decisions are covered, it is called a *limited* guardianship. Although data on guardianship is limited, data from a national survey followed by in-depth interviews and site visits with public guardianship programs across the United States suggested that plenary guardianship is ordered in the vast majority of cases. Further, public guardianship programs rarely collect outcome data and have uneven oversight and accountability (Teaster et al., 2007).

Guardianship has been characterized as "a humanitarian response to the vulnerability of the incompetent" (Frolik, 1998, p. 350). However, this view is rooted in a deficit-based model of disability that assumes that people with disabilities need to be cared for because they are incapable of making decisions. This runs counter to modern conceptualizations of disability that hold that with appropriate, individualized supports, *all people* can engage in decision-making processes, and the rights of citizenship need not be removed.

Guardianship, however, has a long history in law and policy, and there have been few alternatives. In fact, although guardianship should generally be considered as a "last resort" after exhausting all other legally recognized options (National Council on Disability, 2019), the use of guardianship has been increasing. For example, the estimated number of Americans under guardianship has tripled since 1995, with most under plenary guardianship (Reynolds, 2002; Schmidt, 1995; Uekert & Van Duizend, 2011). However, there is very limited information and evaluation of whether and when these individuals require guardianship or, once guardianship is ordered, "whether the protected person continue[s] to need or benefit" from it (Wright, 2010). These factors have led to the identification of a critical need for change in the disability and international human rights community, including alternative approaches such as supported decision-making.

United Nations Convention on the Rights of Persons with Disabilities

A strengths-based, social-ecological perspective of disability was adopted in the United Nation's Convention on the Rights of People with Disabilities (CRPD; United Nations, 2006).

Strengths-based social-ecological models of disability, which have been adopted in multiple disability classification frameworks, including the International Classification of Functioning (World Health Organization, 2007) and the American Association on Intellectual and Developmental Disabilities' terminology and classification manual on intellectual disability (Schalock et al., 2010), define disability as a state of functioning resulting from the interaction between personal capacities and environmental or contextual demands. People with disabilities experience a mismatch between their personal capacities and environmental demands, which creates a need for supports (Thompson et al., 2009). Through understanding support needs, specific supports can be identified that reduce the gap between personal capacity and the demands of the environment and enhance human functioning and quality of life. This shift toward understanding people by their support needs as opposed to their deficits has significant implications for supports planning, including planning for supports for decision making as will be described later in this chapter.

The CRPD represented a major step forward in recognizing a social-ecological approach to understanding and advancing the international human rights of people with disabilities. One major focus of the CRPD was to change attitudes and approaches to disability across the world. Article 12 of the CRPD addresses the rights of people with disabilities to engage in and have supports for legal decision making about their lives, stating that "persons with disabilities have the right to recognition everywhere as persons before the law," and "enjoy legal capacity on an equal basis with others in all aspects of life" (CRPD, Art. 12 (1), (2)). These declarations that people with disabilities, like everyone, have an inherent right to direct their own lives and receive the support they need to do so, challenged the use of overbroad or undue guardianship that occurs in many jurisdictions, including the United States. To date, 181 countries around the world have ratified the CRPD. Although this does not include the United States, CRPD and its vision for advancing human rights for people with disabilities have had worldwide impacts on policy and practice, including in the United States.

Significantly, in asserting the right to legal capacity, the CRPD acknowledged that people with disabilities sometimes may require support and assistance to exercise their legal capacity and that signatories to the treaty ("State Parties") should provide such support in their domestic laws (CRPD, Art. 12). This assertion is based on the premise that needing supports does not take away one's inherent right to engage in decision making and that new models for providing such supports are needed, such as supported decision-making. In recommendations for how to effectively implement Article 12, supported decision-making was introduced and defined as a:

> *Broad term that encompasses both informal and formal support arrangements, of varying types and intensity. For example, persons with disabilities may choose one or more trusted support persons to assist them in exercising their legal capacity for certain types of decisions, or may call on other forms of support, such as peer support, advocacy (including self-advocacy support), or assistance with communication...Support can also constitute the development and recognition of diverse, non-conventional methods of communication, especially for those who use non-verbal forms of communication to express their will and preferences. (Committee on the Rights of Persons with Disabilities, 2014)*

The CRPD emphasized that supported decision-making must be "available to all" including those that need a high degree of support. Further, the CRPD emphasized that the need for support in decision making should not be used to deny other human and civil rights, including voting, reproductive rights, and parental rights. When supported decision-making is utilized as an alternative to guardianship arrangements, there must be protections in place to enable persons with disabilities to "terminate or change the support relationship at any time" ensuring they retain agency over their supported decision-making arrangements and the persons providing support, indicating a significant difference from guardianship arrangements which are controlled by the court.

Jenny Hatch and Supported Decision-Making in the United States

Firstly, supported decision-making provides multiple benefits both for the person and for the society. At the person level, supported decision-making has the potential to increase self-determination and engagement in life decisions, which research has shown can have collateral benefits in multiple areas of life (Blanck & Martinis, 2015; Shogren & Wehmeyer, 2015). Secondly, supported decision-making helps people avoid the social isolation from overbroad or undue guardianship, allowing them to be equally integrated into their communities, consistent with the CRPD, the Americans with Disabilities Act, 1990, and other laws and legal cases (e.g., Developmental Disabilities and Bill of Rights Act, 2006; Olmstead v. L.C., 1999; Rehabilitation Act of 1973, 2006).

The benefits of supported decision-making are illustrated by the experiences of Margaret "Jenny" Hatch, one of the first people in the United States who advocated and ultimately used supported decision-making as an alternative to guardianship. When Jenny Hatch was 29, she lived in her community, had a job, and had the supports she needed to be successful through friends and her church (Hatch, 2015).

Unfortunately, Jenny, who happened to have Down syndrome, was hit by a car while riding her bicycle and required an operation on her back to address her injuries. Shortly thereafter, Jenny's mother and stepfather filed a petition seeking full *guardianship*, asking the local state court to appoint them to take control over

Jenny's life and to make all decisions for her. After a short hearing on the petition, the court placed Jenny under a temporary guardianship, pending a full trial, thereby removing her right to make all decisions about her life.

Jenny's guardians were given the legal power to decide, at least for the time being, where she lived, who she saw, how she spent her time and money, what (if any) healthcare she received, and what type of education, living, and employment services she was provided (Ross et al. v. Hatch, 2013). As Jenny wrote about this period in her life:

> *I was forced to live in a group home. I did not like it there. I didn't feel safe there. I would tell everyone, but no one would listen. They took away everything that was important to me. I was not allowed to work at my job or see my friends. They would not even let me go to my church. My computer and my cell phone were taken away and I was not allowed to even call my friends…They told me I had to work at [a segregated 'sheltered' workshop]. I snapped snaps together all day and it was boring. I did not like it. I wanted my life back, but they told me to forget about my old life and move on with my new life. (Hatch et al., 2015, p. 66)*

At her guardianship trial, Jenny presented evidence that she could make her own decisions, with the support of her friends and professionals she knew and worked with. She stated "I don't need a [guardian]…I need help" (Vargas, 2013). To prove this, Jenny demonstrated to the court that she used supported decision-making relationships the same way most people, including people with and without disabilities, do to make decisions: They "seek advice, input, and information from knowledgeable friends, family, or professionals so they may make their own informed choices" without the need for a guardian (Quality Trust for Individuals with Disabilities, 2013). In addition, professionals called to testify in this case acknowledged that Jenny needs "assistance" to make legal, financial, and life decisions but that she could benefit from supported decision-making.

After hearing the evidence, the judge denied the petition for permanent, plenary guardianship. Instead, the judge restored all of Jenny's right to make decisions in all areas of her life with the exception of health and safety decisions. For those decisions, the judge appointed Jenny's friends (not her mother and step-father) to be her limited guardians, for 1 year, to "assist [her] in making and implementing decisions we have termed 'supported decision-making'" (Ross et al. v. Hatch, 2013). Jenny's limited guardianship expired in August of 2014. Since that time, she has lived and worked where she wanted and become the inspiration and face of the Jenny Hatch Justice Project, which works to increase knowledge and access to supported decision-making across the country.

Jenny Hatch became known as "the rock that started the avalanche" of supported decision-making (Vargas, 2013). Her case was followed by many others (e.g., Commonwealth Council on Developmental Disabilities, 2017; King, 2020). In addition, there has been a marked increase in state legislation and policy recognizing and expanding access to supported decision-making.

In 2014, the Virginia General Assembly directed the state Secretary of Health and Human Services to "recommend strategies to improve the use of supported decision-making in the Commonwealth and ensure that individuals...are consistently informed about and receive the opportunity to participate in their important life decisions" (H.J.Res. 190, Reg. Sess. (Va. 2014)). In 2015, Texas amended its guardianship law to give legal recognition to supported decision-making as an alternative to guardianship (Tex. Est. Code § 1002.031). The law requires courts to find by *clear and convincing evidence* that a person cannot make decisions using supports and services before appointing a guardian (Tex. Est. Code § 1101.101(a)(D) and (E)). The National Guardianship Association (NGA), which represents over 1000 guardians, conservators, and fiduciaries across the United States, has said "[s]upported decision making should be considered for the person before guardianship, and the supported decision making process should be incorporated as a part of the guardianship if guardianship is necessary" (National Guardianship Association, 2015).

At the time this chapter was written, 11 states and the District of Columbia had passed laws formally recognizing supported decision-making as an alternative to guardianship and authorizing people with disabilities to enter into supported decision-making arrangements (Martinis et al., in press). In addition, the federally funded National Resource Center for Supported Decision-making has engaged in and supported research and education in supported decision-making in every state (Blanck & Martinis, 2015).

Social-Ecological Approaches to Supported Decision-Making

As highlighted, there has been significant movement in law and policy to articulate the fundamental right of people with disabilities to be involved in decisions about their lives. This is preventing situations like the one experienced by Jenny Hatch. It also has promoted valued outcomes of social-ecological models of disability, including self-determination and interdependence.

However, supported decision making is– as noted in the CRPD – a broad term that can include a range of formal and informal supports. As such, there is a critical need for frameworks to conceptualize how to design and deliver supports for decision making that are aligned with social-ecological models of disability. Such frameworks should recognize that support needs are defined by the interaction between the person, their current capabilities, and contextual demands.

Shogren and Wehmeyer (2015) proposed a framework for research and intervention that addresses the diverse ways that supported decision-making can be implemented. In this context, Shogren and Wehmeyer (2015) described supported decision-making "as a personalized process that changes over the life course and involves an array of supports that are based on, at any given point in time, the person's abilities and environmental demands for decision making" (p. 19). They proposed that assessment and interventions to promote maximal participation in decision making must address three domains of emphasis: (a) decision-making abilities, (b) environmental demands for decision making, and (c) support needs for decision making.

Consistent with a social-ecological model, this framework for supported decision-making defines disability as a mismatch between personal competencies and current environmental demands, creating a need for individualized supports to achieve valued outcomes. Mismatches can occur, then, when the demands of medical, financial, and personal/social situations are not well-aligned with current resources and supports for a person with a disability. Such a mismatch creates a need for planned, thoughtful supports, which may involve technology, education, the involvement of others, or universal design in information and the environment (Thompson et al., 2009).

Therefore, it is the responsibility of the systems that support people with disabilities to address such a mismatch. The systems must ensure that the person with IDD is consistently at the center of decision making about their lives, building legal capacity, promoting human agency, and providing needed supports. These supports should be individualized and strengths-based, not overly broad like guardianship, which is the focus of supported decision-making arrangements (Martinis & Blanck, 2019).

Emerging Directions in Supported Decision-Making: Building Systems of Supports for Decision making Across the Life Course

So far, this chapter has highlighted emerging theory and legal directions as they relate to enabling supported decision-making. The next sections highlight emerging directions for actualizing supported decision-making in practice. This will include how assessment can be used to plan for the supports needed for decision making as well as the ways in which systems of supports can be created that enable people with IDD to self-determine the supports they need for decision making. Further, the subsequent sections highlight the role of people with disabilities as well as their supporters in the decision-making process, including family members, adult service providers, educators, legal professionals, and healthcare professionals.

Finally, we provide a case study written by a person with IDD who practices supported decision-making. In this case, he describes how he made a health-related decision and how he chooses his supports for decision making. This highlights the application of supported decision-making in practice and ways that systems of support can be created to enable supported decision-making.

Assessment to Guide Supports Planning

Building on the social-ecological framework for understanding supported decision-making, we have engaged in efforts to devise new strategies to enable support teams, in which the person with a disability is at the center, to identify effective supports needed for decision making. Shogren et al. (2017c) described the development and refinement of the *Supported Decision-Making Inventory* (SDMI). Aligned with the three domains of a social-ecological model (decision-making abilities, environmental demands for decision making, and support needs for decision making) (Shogren & Wehmeyer, 2015), the SDMI includes three inventories that, together, form a system that can be used to identify and plan for supports needed in decision making. The result is to help enable people with disabilities to act meaningfully in supported decision-making arrangements.

Table 2.1 provides excerpts from each of the SDMI inventories. Section 1 is the *Supported Decision-Making Personal Factors Inventory*. Items in this section ask about feelings and experiences that a person has had that influence decision making and consists of six items. Section 2 is the *Supported Decision-Making Environmental Demands Inventory*. It includes ten items that enable support teams to determine the degree to which supports and opportunities are available for the person to engage in decisions about health, legal matters, financial matters, social/friendship relationships, and independent/community living. Section 3, the *Decision-Making Supports Inventory* measures the extent to which the person needs supports in aspects of the decision-making process and consists of 12 items.

In developing the SDMI, systematic reviews of the literature were undertaken (Shogren et al., 2017a, c; Uyanik et al., 2017) and used to guide item development within each of the three inventories. It is intended that the SDMI be used on a routine basis to collect and regularly evaluate information about needed supports for decision making, particularly as there are changes in environmental demands. Examples of such changes could include the emergence of a new employment opportunity that impacts a person's finances or the onset of a new health condition that impacts medical decisions.

In reviewing the literature, existing definitional frameworks for decision-making abilities were synthesized, as were factors identified in the literature as influencing decision making. Further, issues related to environmental demands, particularly the risks and complexities associated with decisions in specific domains including health, legal, financial, social relationships, and community living, were summarized. Research and existing assessment practices from the fields of special education, psychology, social welfare, legal studies, and gerontology informed the development of the assessment (Shogren et al., 2017a).

The SDMI is completed as an interview with a person with a disability, with supporters chosen by the person who is engaged in the process of providing support for responses and an understanding of contextual factors. The goal of the SDMI is to provide a meaningful assessment tool that may be utilized by planning teams in any context to attempt to understand and support agency and self-determination in

Table 2.1 Excerpted items from the Supported Decision-Making Inventory

Section 1: ***Supported Decision-Making Personal Factors Inventory***

Part A: Feelings

1. People in my life care about me.
   Not at All True 1 ------------ 2 ------------ 3 ------------ 4 ------------ 5 *Very True*

Part B: Experiences

1. To what degree have you had chances to be involved in decisions?
   Low 1 ------------- 2 ------------- 3 ------------- 4 ------------- 5 *High*

Section 2: ***Supported Decision-Making Environmental Demands Inventory***

1. To what degree are there opportunities available for you to make high risk and/or complicated decisions about your health?
   Low 1 ------------- 2 ------------- 3 ------------- 4 ------------- 5 *High*
2. To what degree are there supports available to enable you to make high risk and or/ complicated decisions about your health?
   Low 1 ------------- 2 ------------- 3 ------------- 4 ------------- 5 *High*

Section 3: ***Decision Making Supports Inventory***

1. I understand why a decision needs to be made.
   Level of Support Needed (the support you need to do this)
   Low 1 ------------- 2 ------------- 3 ------------- 4 ------------- 5 *High*
2. I identify options for my decision that I know about.
   Level of Support Needed (the support you need to do this)
   Low 1 ------------- 2 ------------- 3 ------------- 4 ------------- 5 *High*

decision making. Based on responses, supporters and planning teams can better understand needs, environmental demands, and ways to enhance autonomy and agency over decision making, selecting supports aligned with the person's preferences, interests, values, and needs. After obtaining an understanding of the needs and environmental demands in the decision-making process, the next step is to identify the types, intensity, duration, and frequency of supports that may be put in place to address those areas of needed supports and the roles of the person with disabilities and their supporters.

Role of the Person with Intellectual and Developmental Disabilities in Supported Decision-Making

Decision making in a supported decision-making framework starts and ends with the person who is making the decision. It is important to remember that while all people need support to make decisions, people with IDD may need additional supports for decision making. However, such support needs do not change the right and ability of people with IDD to be engaged in this process. Additionally, all too often, people with IDD have not had the opportunity to make decisions and may need supports throughout the life course to access and engage with opportunities to learn and apply the steps involved in making decisions.

When designing interventions and systems of supports, it is critical to support people with IDD to evaluate what they need to make and enact their decisions. Assessment tools such as the SDMI can assist people with disabilities and their supporters in identifying areas of need (Shogren et al., 2019). For people with IDD to identify needed supports, a systematic exploration of their preferences, interests, values, and beliefs related to the decision may also be necessary, particularly if their decision-making opportunities have been limited. Table 2.2 presents a process that can be used to support people with IDD in the decision-making process consistent with Shogren and Wehmeyer's (2015) framework (described previously).

Table 2.2 Decision making and support planning process

Social-ecological framework	Decision making and supports planning process
Person making the decision	1. Recognize circumstances that indicate a decision needs to be made 2. Identify the goal which the decision is in service of 3. Identify interests, strengths, and values pertaining to the decision. (added for support planning)
Context of the decision	4. Assess the complexity of the decision. (added for support planning) 5. Generate options that would address the circumstances which necessitate making the decision 6. Determine the consequences of each option and the probability that the consequence will occur 7. Understand the importance of each option to the person making the decision and the degree to which each consequence is acceptable 8. Select the most attractive option based on the analysis 9. Assess environmental factors that could impact making or enacting the decision (living arrangement, opportunity to make decisions, family attitudes about decision making; added for support planning)
Supports needed for the decision	10. Evaluate the demands of the decision and the degree to which abilities of the person match (added for support planning) 11. Identify supports necessary to bridge the gap between the demands of the decision and the abilities of the person (added for support planning)

Role of Families in Supported Decision-Making

Family members play a critical role in promoting self-determination and supported decision-making in two ways. First, at younger ages, family members can provide opportunities for the person with IDD to make decisions in a safe environment (Grolnick, 2009; Palmer et al., 2013). In this role, family members can support a person's decision-making capacity development by enabling children and adolescents with IDD to make decisions and experience the result of the decisions, with supports and feedback.

The family member can then reflect with the person with IDD to support them in learning from the decision and considering how they can apply what they have learned to future decisional opportunities. For example, if a person has a goal to meet other kids their age and makes a decision to play on a baseball team to make new friends, the family member could support reflection by saying, "I'm proud of you for giving baseball a try. Did it help you meet new kids?" Conversely, if the person ends up not enjoying baseball and does not interact with other kids, the family member could still acknowledge the risk by saying, "I'm proud of you for trying baseball," but could then support developing other options to achieve the goal by saying, "It didn't seem like you enjoyed playing baseball, though. Maybe we can find a different club based on your interest in animals where you can make friends."

The second way family members can support decision making is by ensuring that the person with IDD has a broad range of experiences throughout the life course on which to base decisions (Grolnick, 2009; Palmer et al., 2013). If the person needs to gauge possible outcomes from different decisions, families can promote the experiences necessary to analyze the possible outcomes. In the above example, family members were ensuring the person had these opportunities by encouraging trying baseball, and then encouraging the person to try something different based on their other interests when that was not the best fit and reminding the person that there were other ways to achieve the goal of making friends during a social activity. Such opportunities and supports for understanding preferences, exploring options, and making increasingly complex decisions can be supported by families through the life course, promoting engagement in decision making and greater community participation outcomes (Millar, 2003; Shogren & Turnbull, 2006; Shogren & Wittenburg, in press).

Role of Support and Service Providers in Supported Decision-Making

For many people with IDD, interaction with service and support providers occurs on a daily basis. Support providers, including direct support professionals, case managers, or other people involved in planning for and delivering individualized supports to meet the needs of people with IDD are well-positioned to play a role in providing supports needed to engage people with disabilities in decision making and to promote self-determination (Hewitt & Nye-Lengerman, 2019). However, support providers often lack training to do so (Hewitt et al., 2013). Additionally, provider organizations and staff often view their role as that of a caregiver rather than as a promoter of self-determination, which complicates efforts for supporting decision making (McConkey & Collins, 2010) and necessitates new and different training and support frameworks for service and support providers.

When support providers are trained to promote self-determination, evidence suggests that people with IDD have better outcomes related to enacting their decisions and are engaged in more meaningful community activities (Cudré-Mauroux et al., 2020; Mansell et al., 2008). Ongoing work is needed to align emerging directions in supported decision-making with training for direct support professionals, case managers, and others who are responsible for ensuring a person with a disability receives the supports they need and want. Current efforts are underway to conceptualize such training, which focus on (a) how to support the person with IDD to follow the decision-making and support planning process outlined in Table 2.2, (b) how to support the person to enact the decision they made even if the supporter does not agree with the decision, and (c) how to support the person to reflect on the decision made and the outcomes.

Role of Educators in Supported Decision-Making

As students who receive special education services reach the age of majority (which is usually 18, but can vary by state), the Individuals with Disabilities Education Act (IDEA, 2004) requires schools to notify parents/guardians that the educational rights previously afforded to them will transfer to the student once they reach the age of majority (National Council on Disability, 2019). If the student continues to receive special education services after they reach the age of majority, they will automatically become the legal decision maker for all educational decisions, including consenting to re-evaluations, agreeing to changes in educational placement, and signing the Individualized Education Program (IEP) put forth by their educational team.

If students who have reached the age of majority need support to make educational decisions, options that do not require guardianship exist, such as having an educational representative. However, research suggests that educational professionals (e.g., special educators, transition coordinators) often recommend that parents/guardians seek guardianship during the transfer-of-rights period, particularly for students with intellectual and developmental disabilities (Jameson et al., 2015; Kanter, 2015; National Council on Disability, 2019). As mentioned previously, there is limited data on guardianship, limiting the identification of definitive reasons for the threefold increase in guardianship since 1995 (Reynolds, 2002; Schmidt, 1995; Uekert & Van Duizend, 2011). Some hypothesize that one of the factors relates to IDEA's transfer-of-rights requirements that necessitate that educational professionals merely notify parents/guardians instead of also informing them of the full range of decision making options, including supported decision-making (National Council on Disability, 2018).

Under this hypothesis, parents may tend to accept the suggestion from educational professionals and seek guardianship arrangements to avoid losing the right to be involved in making educational decisions for their child. Patterns in recent data suggest this is plausible, as young adults aged 18–22 with developmental disabilities receiving state supports and services were the most likely (more so than even adults over 75) to be under guardianship arrangements (National Association of State Directors of Developmental Disabilities Services and Human Services Research Institute, 2017). This evidence suggests a "school-to-guardianship" pipeline (National Council on Disability, 2019, p. 29). As with support and service providers, these findings are likely due to a lack of information and training. This explanation is supported by themes that emerged from a focus group study that included special educators who worked with transition-age youth with disabilities, who suggested they did not learn about guardianship or other options during their teacher education programs and did not know how to communicate about alternatives (Millar, 2007).

Researchers and others have advocated for an amendment to IDEA that would require schools to provide students and their families with information about the full range of decision-making options during the transfer-of-rights period, including supported decision-making (Raley et al., in press). This policy change would prompt teacher education programs as well as schools to provide training to pre- and in-service educators on alternatives to guardianship, including supported decision making, to support students with IDD and their families.

Additionally, some school districts have taken action to interrupt the school-to-guardianship pipeline and promote the use of supported decision-making. For example, the District of Columbia Public Schools offers students and families supported decision making as an option as a part of their educational planning and supports for students as young as those in pre-K. Educational professionals are encouraged to collaborate with the student and their family to identify supports needed for success across educational environments, develop a plan to promote decision-making skills, and enable the student to make decisions with the support of people they trust.

However, in the absence of a district-wide protocol for enhancing supported decision-making, educational professionals can encourage their students with IDD to engage in supported decision-making. This can be accomplished by discussing the student's strengths and abilities regarding decision making and through supports for families to learn about alternatives to guardianship that can provide the support the student needs without removing their legal capacity and right to decision making. Strategies for facilitating these conversations are being developed, but need ongoing exploration and evaluation (Millar, 2014).

Role of Legal Professionals in Supported Decision-Making

When family members or others petition to become a guardian of a person with IDD, legal professionals are in an ideal position to educate and provide information on least restrictive alternatives, such as the use of supported decision-making. Across most states, guardianship laws state that guardianship should be used as a last resort (Martinis & Blanck, 2019), implying that other options to support the decision-making process have been tried and shown not to work. Yet, the rising number of people under guardianship in the United States calls this into question.

To address this area of need, the American Bar Association, with assistance from the National Resource Center for Supported Decision-Making, created the PRACTICAL guide to support attorneys in analyzing guardianship cases and consider least-restrictive alternatives (American Bar Association, 2016). This tool was designed to be used in client interviews to determine supports that could be used to enable the person with IDD to make their own decisions and maintain their legal right to decision making.

PRACTICAL is an acronym for the steps attorneys should take to support their client in identifying the least-restrictive alternative to guardianship: (a) presume guardianship is not needed; (b) reason (clearly identify the reasons for concern); (c) ask if a concern may be caused by a temporary condition; (d) community (determine if concerns can be addressed with community resources, including family supports); (e) team (ask the person if they have a team to support making-decisions), (f) identify the person's strengths in decision making, (g) challenges (screen for concerns related to identified supporters); (h) appoint a legal supporter who has the person's trust; and (g) limit an unnecessary guardianship arrangement (American Bar Association, 2016). Legal professionals can use and apply this framework in their work and in conversations with people with IDD, their families, and their supporters to ensure that the range of options is understood and considered and that alternatives such as supported decision-making are explored prior to pursuing guardianship arrangements.

Role of Healthcare Professionals in Supported Decision-Making

The previous two subsections on educators and legal professionals mainly focused on policies and practices to limit the use of guardianship and promote the use of supported decision-making, given the role of educators and legal professionals in this process. While these issues can also be relevant for healthcare professionals who are involved in discussions about guardianship, healthcare professionals also have a key role to play in engaging people with IDD in decision making about medical- and health-related issues. Therefore, we focus in this section on how healthcare professionals can use supported decision-making in practice. Some of these practices may also be relevant for educators and legal professionals.

Medical and health decisions can be complex, and certain decisions can be associated with a high degree of risk and complexity, particularly with regard to weighing different consequences or options. Additionally, medical information, if not presented in plain language, can be difficult to understand, and many people both with and without disabilities often need support from a trusted loved one or friend to process the complexity of medical decisions and the information provided. People with IDD are similar in this respect. However, there are practices that healthcare professionals can adopt to support people with IDD to make their own medical and health decisions (Martinis & Blanck, 2019; Shogren et al., 2017a).

One critical first step is to ask the person with an IDD how they would like to be supported as they interact with health professionals and make health and medical decisions. People with IDD may prefer to have a trusted supporter attend and participate in health-related visits. Depending on the age of the person, healthcare professionals may be hesitant to allow supporters in the exam room based on privacy laws such as the Health Insurance Portability and Accountability Act (HIPPA). However, while HIPPA protects a person's right to privacy, a person may waive that right and identify people with whom medical information can be shared, and supports should be readily available to enable these options.

If the person agrees to share information with their trusted supporter, healthcare professionals can have less concern about violating HIPPA. If professionals are still hesitant, a person can draft, sign, and notarize their own HIPPA release or authorization stating that they allow their trusted supporter to have access to medical and health information. One consideration for large medical facilities is where to store such a HIPPA release. Many electronic health records do not store this information, or it is difficult for healthcare professionals to retrieve it, so healthcare professionals may need to put procedures in place to ensure a person's support preferences are honored. Finally, people with IDD can be encouraged to be a full partner in medical and health decisions if healthcare professionals ask questions and talk directly to the person with IDD rather than the supporter, even if the supporter may provide the answer to questions.

Cross-System Supported Decision-Making: Culture of Coordinated Support Model

An emerging best practice in supporting people with disabilities to make their own decisions and identify and receive the services and supports they need to do so is the *Culture of Coordinated Support Model* (Martinis & Gustin, 2017; Shogren et al., 2019). In the *Culture of Coordinated Support Model*, people with disabilities, their families, and professionals work together (such as those listed above) and collaborate and coordinate their efforts by holding joint planning meetings and jointly developing individualized support plans with cross-system goals, objectives, and services based on the person's strengths, needs, and interests (Martinis & Gustin,

2017). This brings together the multiple stakeholders that play a role in building systems of supports for decision making and the use of supported decision-making in practice.

Supported decision-making is a critical component of the model because most, if not all, supports and services provided to and used by people with disabilities require forms of supported decision-making. For example, the student-led Individualized Education Program (IEP) in special education, informed choice in the vocational rehabilitation Individualized Plan for Employment process, and person-centered planning in Medicaid waiver individualized support planning, each require people with disabilities, families, and professionals to work together to ensure that the person is aware of his or her situation and options so that the person may choose his or her goals, objectives, and supports, which are the central tenets of supported decision making (Martinis & Blanck, 2019). By enabling the cross-system use of supported decision-making and sharing of information and resources, the *Culture of Coordinated Support Model* empowers people with IDD and professionals to coordinate and collaborate on their efforts, increasing the breadth of supports available while increasing system-wide efficiency and effectiveness (Martinis & Blanck, 2019).

One Vermont high school, vocational rehabilitation agency, and Medicaid waiver provider joined together to create a unified Culture of Coordinated Support system empowering people with IDD to use supported decision-making to identify, plan for, and receive their educational, employment, and community living supports and services (Martinis et al., 2019). After creating joint support plans, the person, his or her family, and agencies coordinated plan implementation and regularly communicated to address problems as they arose and updated goals and supports as needed. After 1 year, the agencies surveyed program participants – including students, parents, teachers, counselors, and other professionals – for their impression of the model and its impact. The initial results were positive, although more research is needed that allows for more rigorous evaluation of the impact of the *Culture of Coordinated Support Model* on student outcomes:

- 86% of respondents strongly agreed or agreed that the *Culture of Coordinated Support Model* helped prepare students for life after high school.
- 100% strongly agreed or agreed that the *Culture of Coordinated Support Model* improved the supports provided to students.
- 73% strongly agreed or agreed that the model was an easy process.
- 93% strongly agreed or agreed that the model has increased the amount of supports and services that students received.
- 100% strongly agreed or agreed that the model helped identify student needs.
- 100% strongly agreed or agreed that the model provided better supports to meet student needs.
- 93% strongly agreed or agreed that the *Culture of Coordinated Support Model* is a good use of planning meetings.

Case Example of Supported Decision-Making in Practice

Putting each of the pieces of the social-ecological model into place to enable meaningful applications of supported decision-making can be complex in the day-to-day lives of people with disabilities and those that support them. However, there are clear examples of this occurring (Hatch, 2015; Hatch et al., 2015; Martinis & Blanck, 2019; Shogren et al., 2019). In the sections that follow, our co-author who has lived experience with IDD shares how he makes decisions and chooses supports.

Brad specifically discusses how he uses supported decision-making in making decisions about his eating habits to meet his goal of better health. He also describes ways his supporters help him enact his decisions. While Brad chose and had been closely connected to his supporters for years before this specific decision, the case also highlights the traits Brad finds important in a supporter, such as being nonjudgmental, recognizing that occasional deviations from the plan are acceptable, and offering advice but recognizing that Brad is the final decision-maker. It also shows how Brad has evaluated the supports that enable him to implement his decision so that he can use what he learns to inform future decisions and supports.

Brad's story is organized with headers that reflect the social-ecological model and the decision making and support planning process proposed in Table 2.2. The text in this section consists of Brad's words in italics. Occasionally, Brad's co-authors have inserted comments, with Brad's permission, in brackets to highlight concepts described in the chapter.

Person Making the Decision

Decision making process Steps 1 and 2: Brad recognizes that a decision needs to be made and identifies the goal that his decision is in the service of.

> *I had a friend of mine pass away suddenly. He still ate the bad stuff up to the day that he passed. He would just live off of frozen food. It took me a long time to get to the point to say I needed to do something [about my eating habits], but that event helped me. Then, it finally clicked one day and I said "if I don't do anything now things [my health] will get worse." I had to stop drinking soda and start eating healthier. I had one particular day at home when I was looking in my recycle bin and I realized that I had like 10 or 15 pop bottles and I said "I've got to stop. I've got to do something different." I've tried different diet things as I was growing up, but nothing ever seemed to stick because I can never see the weight actually coming off. So, my two main supporters helped me look around for something that I thought would work. The food choices I make now have helped me take off weight and I feel a lot better. So, once you see something that really starts working that really motivates you to continue.*

Step 3: Brad identifies his interests, strengths, and values pertaining to the decision.

> *After my friend died, living longer and healthier really became important to me, so I got motivated. I didn't think about it much before. I like eating a lot of different kinds of food, so I knew that if I found something that I thought worked, I could stick with it. One other*

thing, I'm a really social person and I have a lot of people I like to hang out with. Getting my morning coffee is important to me. And, I wanted to still go out to eat with my friends. I knew I couldn't do it as much as I was, but if this was going to work, I still needed to eat with my friends.

Context of the Decision

Step 4: Brad describes the complexity of finding and maintaining a healthy lifestyle that would work for him.

As I said, I tried a lot of different diets in my life, but nothing ever stuck. I couldn't ever see the weight coming off, and it was too much work to keep up with everything. I needed something that was easy. That I didn't have to spend all of my time thinking about. There's so much out there about diets and a lot of them wouldn't work for me, so I needed people who knew me really well and who knew what was important to me to help narrow down my choices. It had to be something where I could go to the store and find things and that the people who help me cook could make. Also, I still wanted to go out to eat sometimes.

Steps 5–8: Brad generates options, identifies the consequences and importance of each option, and selects the most attractive option.

I knew that if I didn't do anything or if I didn't find something I could stick to, I would end up like my friend, so I really wanted something I could stick to. Finally, I found an option where I liked the food I could eat, and I talked with the person who helps me cook and she could make the recipes, so it seemed like a good one to try.

Step 9: Brad assesses environmental factors that can impact decision making. Brad recognizes that the supporters that help him enact decisions are a key part of his environment. And, he describes his approach for identifying people he will ask for support with decisions. Finding supports that will honor decisions and support the decision-making process is critical to him.

In general, it's important to me to pick people who I can get along with and who I think will support my decisions – not try to talk me into their decisions. I think the way that I make that decision is thinking about, "What did they [the supporter] do for me?" Sometimes there are certain supports that you can ask certain people for and then if you ask another person they might not be receptive to it. I think that's how I make my decisions about supporters – just by looking at the reaction [from someone] and [observing] how they respond when you ask specific questions. I think part of it is a feeling I get. I use my sense of humor a lot to get me through a lot of things and so I like it when people understand my sense of humor and don't get offended. So, it's kind of like you get a feel for the person and if they are resistant and don't want to do anything for you or they come up with excuses about not doing something. And that also gives me a sign as well.

I have a small group of people that I really trust who support me with my decisions. I can bounce ideas off of them and get their feedback, but I end up making the final decision for myself. I like keeping my group small because I don't like a bunch of people knowing my business on a day-to-day basis. For day-to-day health decisions, like maintaining my diet, I usually rely on one person. She's the person who has been with me the longest and still supports me with direct care, so I see her a lot. What's cool is that she and one other person who I've known a long time are going to be the two people that, if I pass away or if I end up in a situation where I need a medical decision made and I cannot do it, they've agreed to do it for me.

Supports Needed for the Decision

Step 10: Brad evaluates the demands of the decision and the degree to which his abilities match.

> *My diet is part of a lifestyle change that involves staying away from rice, bread, pasta, and potatoes. With this lifestyle, the thing I like about it is that the decisions are easy – no potatoes, no rice, no bread, no pasta. Once you get those four groups out of your head there's plenty of alternatives. I've learned to replace my rice with cauliflower rice or cauliflower and then there's some others as well. But, a lot of times my supporter helps me with my meals. Even if the main decisions are easy, I need help finding the right stuff in the store. Also, I can do my own shopping, but it's easier to buy the bad stuff, and I like to have some help. I'm also not good around stoves, so need some help cooking.*

Step 11: Brad identifies supports necessary to bridge the gap between the demands of the decision and the abilities of the person. Brad's supports include everyday supports like shopping for the right foods and cooking and also psychosocial supports.

> *My main supporter, who has been with me the longest, is the one that usually helps me with most of my grocery shopping and making sure that I get the right foods. Not that I can't go buy my own things, but we set it up to where we'll buy enough so I can get by a couple weeks without having to go every week and get groceries. She is really helping me recognize the good stuff and it's funny because when I first moved out on my own, I was eating all the bad stuff because it's easy to throw it in the microwave. One of the things that I struggle with is that I'm not very good around hot stoves and stuff like that. We'll cook once and store food for a couple of weeks, so I know I'll have food. It's also good to have just a certain number of meals a week that I could just take out and put in the oven.*
>
> *Also, sometimes I really crave the bad stuff and eat it anyway. Then I start to feel pretty down about myself and need someone to talk about things with. Another thing she helps me with is to not get down on myself. One day when she came over I finally had to say, "Hey, I've started to pick up caffeine again (pop)" and I was feeling like, man, I did all this work and I'm certain I'll get the craving again for pop and she told me don't worry about it. Having one every once in a while is not the end of the world. So that helped me to understand moderation. It's just one of those things and so right now I just don't feel so bad that if I have picked up something that I shouldn't. All of these supports together help me make the decisions I want to, on a regular basis and move me toward my goal of being healthy.*

Conclusions

This chapter provided an overview of supported decision-making and its emergence in the disability field, including the role of strengths-based, social-ecological models of disability in enabling increased attention to supported decision-making. We highlighted theory and legal directions as they relate to actualizing supported decision-making in practice and how supported decision-making challenges current other-directed decision-making models, such as guardianship. In highlighting ways that multiple stakeholders including people with disabilities, family members, educators, legal professionals, health professionals, and support and service providers can enable supported decision-making, we emphasized ways that systems of supports can be coordinated to promote not only supported decision-making but also enhanced self-determination, quality of life, and community participation. Through the voice of a self-advocate, we emphasize the real and meaningful impacts this has in people's lives, and what it means to keep the person with a disability at the center of the process in the short- and long-term.

In moving forward, it is critical to continue to ensure that support systems, and the stakeholders that act as supporters, are prepared to support and enable the decision making of people with IDD. Resources have been put in place (e.g., supported decision-making forms that can be used by school systems, the PRACTICAL guide, the *Culture of Coordinated Support Model*, the *Supported Decision-Making Inventory*), but more work is needed to ensure that the least restrictive alternative to guardianship is available and implemented to protect the person's basic human and civil right to make decisions in their lives. Research that identifies the most effective strategies to enable supported decision-making, including the most effective combination of supports that target people with disabilities as well as their supporters, is also critically needed. Researchers and practitioners can focus on developing and evaluating resources that are informed by the experiences of self-advocates to enable them to promote supported decision-making on a day-to-day basis. The integration of these options into legislation is critically needed to make supported decision-making an option in legal practice (i.e., as an alternative to guardianship) and in systems of supports (i.e., building systems of support for supported decision-making). For example, policy changes are needed that enhance the legal authority of supported decision-making (e.g., amendments to state guardianship laws, IDEA), and educational materials for legal professionals, educators, health professionals, and service and support providers on cases like Jenny Hatch's that demonstrate the power of self-advocacy with support from trusted professionals and the feasibility of alternatives to guardianship are needed. Research is then needed on the impacts of such policy changes and educational materials on access to supported decision-making arrangements.

Relatedly, ongoing work is needed to tell the story of people with IDD who use supported decision-making in their lives to position supported decision-making as the default option for the decision-making process instead of plenary guardianship, such as the perspective shared in this chapter. Such efforts will enhance what is known from the research literature and enable all members of our society to better understand the day-to-day practices used by people with IDD, by family members, and by supporters to actualize supported decision-making across life domains. Researchers must build on this knowledge and experience to develop and test supports derived from a strengths-based, social-ecological model, identifying the most effective ways to jointly address decision-making abilities, environmental demands, and support needs for decision making. Such work should be applied to continue to build available resources and tools for enhancing participation and self-determination in supported decision-making arrangements and to systematically evaluate the impact of supported decision-making on self-determination, participation, and quality of life outcomes.

Acknowledgments The work on this chapter was supported by the US Department of Health and Human Services, Administration on Community Living, Administration on Intellectual and Developmental Disabilities through Cooperative Agreement #90DM0001-01-00, and National Resource Center for Supported Decision-Making, awarded to Quality Trust, with a subcontract to the University of Kansas and Syracuse University. The research was also supported by the US Department of Health and Human Services, Administration on Community Living, National Institute on Disability, Independent Living, and Rehabilitation Research through Grant 90DP0076-01-00 awarded to Syracuse University with a subcontract to the University of Kansas and grants from the Kansas Council on Developmental Disabilities (KCDD-2020-KUCFR-SDM-01) and the WITH Foundation to the University of Kansas.

The views herein reflect only the authors' and not those of any funding agency, university, or other entity.

References

American Bar Association. (2016). *PRACTICAL tool for lawyers: Steps in supporting decision making*. https://www.americanbar.org/groups/law_aging/resources/guardianship_law_practice/practical_tool/

Americans with Disabilities Act of 1990, 42 U.S.C. § 12101 et seq.

Blanck, P. (2014). *The struggle for web accessibility by persons with cognitive disabilities*. Cambridge University Press.

Blanck, P. (2020). *Disability law and policy*. Foundation Press.

Blanck, P., & Martinis, J. G. (2015). The right to make choices: The National Resource Center for supported decision making. *Inclusion, 3*, 24–33. https://doi.org/10.1352/2326-6988-3.1.24

Committee on the Rights of Persons with Disabilities. (2014). General Comment No. 1. Article 12: Equal recognition before the law, U.N. Doc. CRPD/C/GC/1. Adopted April 11, 2014.

Commonwealth Council on Developmental Disabilities. (2017). Freed form guardianship. A Kentucky first: Woman wins her rights in court using supported decision making. Exceptional Family Kentucky.

Cudré-Mauroux, A., Piérart, G., & Vaucher, C. (2020). Partnership with social care professionals as a context for promoting self-determination among people with intellectual disabilities. *Research in Developmental Disabilities, 99*, 103602. https://doi.org/10.1016/j.ridd.2020.103602

Developmental Disabilities Assistance and Bill of Rights, 42 U.S.C 144.

Frolik, L. (1998). Guardianship reform: When the best is the enemy of the good. *Stanford Law and Policy Review, 9*, 347–355.

Grolnick, W. S. (2009). The role of parents in facilitating autonomous self-regulation for education. *Theory and Research in Education, 7*, 164–173.

Hatch, J. (2015). My story. *Inclusion, 3*(1), 34–34. https://doi.org/10.1352/2326-6988-3.1.34

Hatch, J., Crane, S., & Martinis, J. (2015). Unjustified isolation is discrimination: The Olmstead case against overbroad and undue organizational and public guardianship. *Inclusion, 3*, 65–74.

Hewitt, A. S., Nord, D., Bogenschutz, M., & Reinke, J. (2013). Community Living. Inclusion, 1(1), 17–27. https://doi.org/10.1352/2326-6988-1.1.017

Hewitt, A., & Nye-Lengerman, K. (Eds.). (2019). *Community living and participation for people with intellectual and developmental disabilities*. American Association on Intellectual and Developmental Disabilities.
Individuals with Disabilities Education Act, as amended in 2004, 20 U.S.C. § 1400 et seq.
Jameson, J. M., Riesen, T., Polychronis, S., Trader, B., Mizner, S., Martinis, J., & Hoyle, D. (2015). Guardianship and the potential of supported decision making with individuals with disabilities. *Research and Practice for Persons with Severe Disabilities, 40*, 36–51. https://doi.org/10.1177/1540796915586189
Kanter, A. S. (2015). Guardianship for young adults with disabilities as a violation of the purpose of the individuals with disabilities education improvement act. *Journal of International Aging Law and Policy, 8*, 1–68.
King, S. J. (2020). Our Journal of Supported Decision making for Ryan. *Impact, 32*(1). https://publications.ici.umn.edu/impact/32-1/our-journey-of-supporteddecision making-for-ryan
Kohn, N. A., Blumenthal, J. A., & Campbell, A. T. (2013, October; 2012). Supported decision making: A viable alternative to guardianship? *Penn State Law Review, 117*, 1111–1157.
Mansell, J., Beadle-Brown, J., Whelton, B., Beckett, C., & Hutchinson, A. (2008). Effect of service structure and organization on staff care practices in small community homes for people with intellectual disabilities. *Journal of Applied Research in Intellectual Disabilities, 21*(5), 398–413. https://doi.org/10.1111/j.1468-3148.2007.00410.x
Martinis, J., & Blanck, P. (2019). *Supported decision making: From justice for Jenny to justice for all!* Something Else Solutions Press.
Martinis, J., & Gustin, J. (2017). Supported decision making as an alternative to overbroad and undue guardianship. *The Advocate, 60*(6), 41–46.
Martinis, J., Cassidy, K., Gustin, J., Nadeau, J., & Robinson, D. (2019). Creating a culture of coordinated support in Vermont: The successful transitions program. *Impact, 32*(1), 19–21.
Martinis, J., Harris, J., Fox, D., & Blanck, P. (in press). State guardianship laws and supported decision making in the United States: Analysis and implications for future research, education, and advocacy. *Journal of Disability Policy Studies*.
McConkey, R., & Collins, S. (2010). The role of support staff in promoting the social inclusion of persons with an intellectual disability. *Journal of Intellectual Disability Research, 54*, 691–700.
Millar, D. S. (2003). Age of majority, transfer of rights and guardianship: Considerations for families and educators. *Education and Training in Developmental Disabilities, 38*(4), 378–397.
Millar, D. S. (2007). "I never put it together": The disconnect between self-determination and guardianship – Implications for practice. *Education and Training in Developmental Disabilities, 42*, 119–129.
Millar, D. S. (2014). Addition to transition assessment resources: A template for determining the use of guardianship alternatives for students who have intellectual disability. *Education and Training in Autism and Developmental Disabilities, 49*, 171–188.
National Association of State Directors of Developmental Disabilities Services and Human Services Research Institute. (2017). *National core indicators*. http://www.nationalcoreindicators.org/charts/
National Council on Disability. (2018). *Beyond guardianship: Toward alternatives that promote greater self-determination*. Author. https://ncd.gov/publications/2018/beyond-guardianship-toward-alternatives
National Council on Disability. (2019). *Turning rights into reality: How guardianship and alternatives impact the autonomy of people with intellectual and developmental disabilities*. Author. https://ncd.gov/publications/2019/turning-rights-into-reality
National Guardianship Association. (2015). *Position statement on guardianship, surrogate decision making, and supported decision making*. Retrieved from: http://www.guardianship.org/documents/NGA_Policy_Statement_052016.pdf
Olmstead v. L. C. (98-536) 527 U.S. 581 (1999).

Palmer, S. B., Summers, J. A., Brotherson, M. J., Erwin, E. J., Maude, S. P., Haines, S. J., Stroup-Rentier, V. L., & Chu, S. Y. (2013). Foundations for self-determination in early childhood intervention: Fidelity, feasibility, and outcomes. *Journal of Early Intervention, 33*, 38–47.

Quality Trust for Individuals with Disabilities. (2013). *Supported decision making: An agenda for action*. http://jennyhatchjusticeproject.org/node/264

Raley, S. K., Shogren, K. A., Martinis, J., & Wehmeyer, M. L. (in press). Age of majority and alternatives to guardianship: A necessary amendment to the individuals with disabilities education act of 2004. *Journal of Disability Policy Studies*.

Rehabilitation Act, 29 U.S.C. 701. (2006).

Reynolds, S. L. (2002). Guardianship primavera: A first look at factors associated with having a legal guardian using a nationally representative sample of community-dwelling adults. *Aging and Mental Health, 6*, 109–120.

Ross et al. v. Hatch. (2013). Circuit Court for the City of Newport News.

Schalock, R. L., Borthwick-Duffy, S., Bradley, V., Buntix, W. H. E., Coulter, D. L., Craig, E. P. M., Gomez, S. C., Lachapelle, Y., Luckasson, R. A., Reeve, A., Shogren, K. A., Snell, M. E., Spreat, S., Tasse, M. J., Thompson, J. R., Verdugo, M. A., Wehmeyer, M. L., & Yeager, M. H. (2010). *Intellectual disability: Definition, classification, and systems of support* (11th ed.). American Association on Intellectual and Developmental Disabilities.

Schmidt, W. C. (1995). *Guardianship: Court of last resort for the elderly and disabled*. Carolina Academic Press.

Shogren, K. A., & Turnbull, A. P. (2006). Promoting self-determination in young children with disabilities: The critical role of families. *Infants and Young Children, 19*, 338–352.

Shogren, K. A., & Wehmeyer, M. L. (2015). A framework for research and intervention design in supported decision making. *Inclusion, 3*, 17–23. https://doi.org/10.1352/2326-6988-3.1.17

Shogren, K. A., & Wittenburg, D. (in press). Improving outcomes of transition age youth with disabilities: A life course perspective. *Career Development and Transition for Exceptional Individuals, 43*(1), 18–28.

Shogren, K. A., Bradley, V., Gomez, S. C., Yeager, M. H., Schalock, R. L., with, Borthwick-Duffy, S., Buntinx, W. H. E., Coulter, D. L., Craig, E. P. M., Lachapelle, Y., Luckasson, R. A., Reeve, A., Snell, M. E., Spreat, S., Tasse, M. J., Thompson, J. R., Verdugo, M. A., & Wehmeyer, M. L. (2009). Public policy and the enhancement of desired outcomes for persons with intellectual disability. *Intellectual and Developmental Disabilities, 47*, 307–319.

Shogren, K. A., Luckasson, R., & Schalock, R. L. (2015). Using context as an integrative framework to align policy goals, supports, and outcomes in intellectual disability. *Intellectual and Developmental Disabilities, 53*, 367–376. https://doi.org/10.1352/1934-9556-53.5.367

Shogren, K. A., Wehmeyer, M. L., Lassman, H., & Forber-Pratt, A. J. (2017a). Supported decision making: A synthesis of the literature to guide assessment and intervention development. *Education and Training in Autism and Developmental Disabilities, 52*, 144–157.

Shogren, K. A., Wehmeyer, M. L., & Singh, N. N. (Eds.). (2017b). *Handbook of positive psychology in intellectual and developmental disabilities: Translating research into practice*. Springer.

Shogren, K. A., Wehmeyer, M. L., Uyanik, H., & Heidrich, M. (2017c). Development of the supported decision making inventory system. *Intellectual and Developmental Disabilities, 55*, 432–439. https://doi.org/10.1352/1934-9556-55.6.432

Shogren, K. A., Wehmeyer, M. L., Martinis, J., & Blanck, P. (2019). *Supported decision making: Theory, research and practice to enhance self-determination and quality of life*. Cambridge University Press.

Shogren, K. A., Rifenbark, G. G., Wehmeyer, M. L., Dean, E. E., Killeen, M. B., & Karsevar, J. (in press). Refining the supported decision making inventory. *Journal of Policy and Practice in Intellectual Disabilities*.

Teaster, P. B., Wood, E. F., Lawrence, S. A., & Schmidt, W. C. (2007). Wards of the state: A national study of public guardianship. *Stetson Law Review, 37*, 193–242.

Thompson, J. R., Bradley, V., Buntinx, W. H. E., Schalock, R. L., Shogren, K. A., Snell, M. E., Wehmeyer, M. L., with, Borthwick-Duffy, S., Coulter, D. L., Craig, E. P. M., Gomez, S. C.,

Lachapelle, Y., Luckasson, R. A., Reeve, A., Spreat, S., Tasse, M. J., Verdugo, M. A., & Yeager, M. H. (2009). Conceptualizing supports and the support needs of people with intellectual disability. *Intellectual and Developmental Disabilities, 47*, 135–146.

Thompson, J. R., Schalock, R. L., Agosta, J., Teninty, L., & Fortune, J. (2014). How the supports paradigm is transforming the developmental disabilities service system. *Inc, 2*, 86–99. https://doi.org/10.1352/2326-6988-2.2.86

Uekert, B. K., & Van Duizend, R. (2011). *Adult guardianships: A "best guess" national estimate and the momentum for reform*. National Center for State Courts.

United Nations. (2006). *Convention on the rights of persons with disability*. http://www.un.org/disabilities/default.asp?navid=14&pid=150

Uyanik, H., Shogren, K. A., & Blanck, P. (2017). Supported decision making: Implications from positive psychology for assessment and intervention in rehabilitation and employment. *Journal of Occupational Rehabilitation, 27*, 498–506.

Vargas, T. (2013). Woman with down syndrome seeks power to live the way she wants. *The Washington Post*. Retrieved from: https://www.washingtonpost.com/local/virginia-woman-with-down-syndrome-seeks-power-to-live-the-way-she-wants/2013/07/20/76102a82-d789-11e2-a9f2-42ee3912ae0e_story.html?utm_term=.5d04c0dd656b.

Wehmeyer, M. L. (Ed.). (2013). *The story of intellectual disability*. Brookes.

World Health Organization. (2007). *International classification of functioning, disability and health: Children and youth version*. Author.

Wright, J. (2010). Guardianship for your own good: Improving the well-being of respondents and wards in the USA. *International Journal of Law and Psychiatry, 33*, 350–368.

Chapter 3
Promoting Decision Making Capabilities in the Healthcare of Adults with Intellectual and Developmental Disabilities: Ethics and Practice

William F. Sullivan, John Heng, and Michael Bach

Introduction

Adults with intellectual and developmental disabilities (IDD) are likely to be found on the margins of healthcare systems. They experience high rates of poverty, exploitation, and trauma (Brucker & Nord, 2016; Byrne, 2018; Hughes et al., 2012). Many are socially isolated (Hurd et al., 2018). Their health needs, which often are multiple and complex, can go unrecognized and unaddressed (Hughes-McCormack et al., 2018; Sullivan et al., 2018; Williamson et al., 2017).

These challenges are compounded by experiences that such adults have of being on the periphery of decision making regarding their own health care (Hoole & Morgan, 2011). Potvin et al. (2019) found that support to navigate health systems and person-centered care (e.g., respect for decision-making capabilities, communication, rapport, and advocacy) were what adults with IDD said they needed to better access primary care assessments. When deemed to lack capacity to make healthcare

W. F. Sullivan (✉)
St Michael's Hospital Department of Family and Community Medicine, University of Toronto, Toronto, ON, Canada

Surrey Place, Toronto, ON, Canada
e-mail: bill.sullivan@utoronto.ca

J. Heng
Departments of Philosophy, Thanatology, and Disability Studies, King's University College at Western University, London, ON, Canada
e-mail: jheng@uwo.ca

M. Bach
Managing Director, Institute for Research and Development on Inclusion and Society, Toronto, ON, Canada

Adjunct Professor of Disability Studies, Ryerson University, Toronto, ON, Canada
e-mail: mbach@irisinstitute.ca

I. Khemka, L. Hickson (eds.), *Decision Making by Individuals with Intellectual and Developmental Disabilities*, Positive Psychology and Disability Series,
https://doi.org/10.1007/978-3-030-74675-9_3

decisions on their own, adults with IDD usually have these decisions made for them by healthcare providers and caregivers, often informally, and by legally recognized substitute decision-makers or surrogates.

Recently, supported decision making has been proposed as an approach to healthcare decision making that challenges fundamental premises of contemporary Western bioethics and health law, as well as prevalent healthcare practices. Supported decision making is sometimes confused with shared decision making, which is a well-established aspect of person-centered healthcare. The main difference between these two approaches is that, in shared decision making, the person needing healthcare is assessed to meet the criteria of legal capacity for healthcare decision making, whereas in supported decision making, the person requires the help of one or more supporters to meet those criteria.

This chapter (a) elaborates on supported decision making in healthcare, (b) discusses some possible philosophical and ethical underpinnings of this approach, (c) concludes that a supported decision making approach affirms the inherent dignity of persons with IDD and can enable them to make and authorize healthcare decisions in a way that is legally recognized, and (d) will give examples of and helpful tools for implementing supported decision making in healthcare practices. Further inquiry is needed, however, regarding specific barriers to and facilitators of these practices.

Supported Decision making in Healthcare

There is no universally accepted definition of supported decision making in healthcare publications (Davidson et al., 2015). The term is sometimes used, broadly, to refer to an aspect of person-centered approaches to healthcare, which aim to be *supportive* of persons when they make healthcare decisions. Understood in this way, supported decision making overlaps with what good clinicians are accustomed to doing when they determine with persons their unique health needs and collaborate with persons and their caregivers in shared decision making concerning the person's goals of care and treatment options (Simmons & Gooding, 2017).

In disability law and policy, however, advocates mean by supported decision making something specific. It is an approach that recognizes the authority of persons to make decisions with the help of close persons whom they trust and who know them well (let us call them *decision-making supporters*) (Bach & Kerczner, 2010; Dinerstein, 2012). This understanding of supported decision making has only recently been applied in health care (Sullivan & Heng, 2018). Both shared decision making and supported decision making aim to enhance communication and promote collaboration between healthcare providers and persons needing health care to reach agreed-upon goals and plans of care; however, in shared decision making, consideration should also be given to how adults with IDD might need to be accommodated and supported in order to participate in this process. The *Primary Care of Adults with Intellectual and Developmental Disabilities: 2018 Consensus Guidelines* state that "Capacity for decision making is relational. Many patients with IDD can participate to some extent in decision making if provided with accommodations and supports by health professionals and family and other committed caregivers" (Sullivan et al., 2018, p. 259).

The significance of supported decision making is both ethical and legal. Recognizing someone's capabilities and authority to make decisions is associated with respecting both their moral agency and status as a person equal to other persons before the law. Both of these are linked to respecting the social standing of persons with disabilities and their full inclusion and participation in society.

A major impetus to adopting supported decision making in law has been the United Nations' *Convention on the Rights of Persons with Disabilities* (CRPD), hereafter abbreviated to *Convention*. A total of 182 states are party to the CRPD as of July 2021. Article 12(1) obliges states party to the *Convention* to recognize the right of persons with disabilities "to enjoy legal capacity on an equal basis with others in all aspects of life" (United Nations, 2006, p. 10). Article 12(3) of the *Convention* stipulates that persons with disabilities should have "access to the support they may require in exercising their legal capacity" (United Nations, 2006, p. 10). According to the *Convention*, therefore, legal capacity is not only a right that persons with disabilities hold and should enjoy in common with other human beings. It is also a right that can be exercised and acted upon by all persons with disabilities, *regardless of the type or severity of their impairments*. The decisions that they arrive at, with supports as needed, should be legally recognized.

In its *General Comment No. 1*, the United Nations Committee on the Rights of Persons with Disabilities states that, under the CRPD, "States parties' obligation to replace substitute decision-making regimes by supported decision making requires both the abolition of substitute decision-making regimes and the development of supported decision-making alternatives. The development of supported decision-making systems in parallel with the maintenance of substitute decision making regimes is not sufficient to comply with Article 12 of the Convention" (United Nations Committee on the Rights of Persons with Disabilities, 2014, article 28, p. 6).

Several jurisdictions have changed or are beginning to change their legal frameworks to recognize supported decision-making approaches, e.g., Peru, Colombia, some states in the USA, and some provinces in Canada. In these jurisdictions, supported decision making is being recognized as a complement or an alternative to guardianship and other substitute, surrogate, or proxy decision-making arrangements in healthcare. Substitute decision-making arrangements, while well-entrenched in health law and health care in most countries of the world, take the authority for healthcare decisions away from persons who are deemed *incapable* or *incompetent* to make such decisions independently. In contrast, supported decision making would enable such persons to make healthcare decisions *interdependently*, i.e., with a range of accommodations, including those to address environmental sensitivities, stress, or barriers to communicating, training, and help from one or more decision-making supporters. Supporters can be trusted family members or others who know the person well enough to interpret the person's authentic values and to help affirm, select, and execute decisions regarding treatment options that best align with these values.

Decision making in healthcare is a process, the outcome of which is a decision to consent to or refuse treatments proposed by healthcare professionals. Some healthcare practices already involve persons as much as possible in the decision-making process regarding their healthcare. Nevertheless, as understood by the United Nations' *Convention*, supported decision making entails a radical rethinking of who should authorize the treatment decisions of persons with IDD. It questions prevalent practices such as assessing persons for decision-making capacity or competence without considering their need to be accommodated or supported, substitute decision-makers or surrogates making decisions for persons who are deemed to lack capacity or competence when appropriate accommodations and supports are not offered to those persons, and involuntary medical treatments for such persons in their *best interests* when they are capable of refusing such treatments if provided accommodations and supports.

It is true that, in terms of outcomes, a person's supported decision could resemble a decision made on behalf of the person through a person-centered substitute decision-making process. For example, Daniel Sulmasy and Lois Snyder (Sulmasy & Snyder, 2010, p. 1947) have proposed a *substituted interests* model of substitute decision making that "promotes the patient as a unique person, in the context of his or her relationships, applying the patient's authentic values, wishes, and real interests, as best they can be known." Section 4 of the *Mental Capacity Act* of the United Kingdom (2005) stipulates that decision-makers on behalf of persons who are deemed to lack capacity under the law should act on those persons' best interests. The *Mental Capacity Act* stipulates that, in so doing, substitute decision-makers should "permit and encourage the person [i.e., the patient] to participate, or to improve his ability to participate, as fully as possible in any act done for him and any decision affecting him." Moreover, the substitute decision-maker should ascertain, as far as is reasonable, "the person's past and present wishes and feelings (and, in particular, any relevant written statement made by him when he had capacity), the beliefs and values that would be likely to influence his decision if he had capacity, and the other factors that he would be likely to consider if he were able to do so." Legislation governing healthcare in several countries has similar provisions; however, most countries do not.

For advocates of supported decision making, there is a fundamental ethical and legal difference between supported and substitute decision making approaches in healthcare. In supported decision making, the agency and legal authority of the person who is supported to make a healthcare decision is retained and recognized. In substitute decision-making approaches, that agency and legal authority for the decision ultimately rests with the substitute decision-maker *acting on behalf of* the person. However much this substitute decision-maker involves the person in the decision-making process, and however closely the decision matches what the person would have wanted based on the person's authentic values, it is the authority of *another person than the patient* that is being recognized as having ethical and legal weight and significance. Supported decision-making practice cultivates the ethical, legal, and social recognition of a person's agency in healthcare decision making, even if extensive decision-making supports are needed, including the interpretive support of family and friends. To the extent that substitute decision making, however well-intentioned, diminishes a person's legal recognition, it also diminishes these forms of recognition. Anderson and Honneth (2005) and Honneth (2007) contend that substitute decision making constitutes systemic patterns of recognition that devalue a person's inherent dignity and equality in the eyes of those who must legally address the person as an object of another's care and decisions. They do not recognize such persons as subjects in their own right, however challenging and complex this may be.

It is important for healthcare providers, policy makers, and ethicists to attend to the perspectives of many persons with disabilities and their caregivers regarding this point. The claim is often made that designating persons with IDD to be *incapable* or *incompetent* to make certain healthcare decisions in no way undermines respect for their personhood or inherent dignity. What many persons with disabilities and their advocates are saying, however, is that they experience substitute decision making differently. They can see how substitute decision making organizes the social *mis*-recognition of a person's agency.

Audrey Cole, who has a son with a severe intellectual disability (ID), was an early advocate for legally recognizing supported decision making. Her persistent and pioneering work, together with that of her associates, lay the groundwork for Article 12 of the United Nations' *Convention*. She writes: "Supported decision making is a formalization of the natural process that most of us use throughout our lives when making decisions: if we need help, we seek it. Few of us could say that we have never sought help or advice in making a decision. As long as our decision-making capacity is not questioned in the legal sense, we are free to seek and use whatever help we need in reaching decisions. No matter how critical the decision or how little, we truly understand its implications. If we have the capacity to seek support and are able to express our decision, there is no obligation to declare the support we had in reaching it. This is not so for people whose cognitive capacity is questioned, whether because of the inherent limitations of ID, as in my son's case, or because of diminishing cognitive capacity often brought on by age. Unlike the rest of us, to maintain equality, those two groups are required to prove that they can make decisions independently. If they cannot, they are typically headed for the legal oblivion of incapacity and guardianship...I believe that to be discriminatory" (Cole, 2019).

Hence, increasingly, persons with disabilities and their caregivers are regarding legal capacity as fundamental to their being recognized as equal to others in society. They view societal structures or practices that are predicated on individuals having to demonstrate that they possess certain cognitive abilities in order to enjoy the legal recognition of decision-making capacity as manifesting an ableist bias that is often not adverted to or acknowledged. These set up norms and standards that discriminate against persons who lack such abilities and position them at a relative disadvantage to others in society. What philosophical and ethical arguments underpin supported decision making? These often are not as well articulated as legal ones. In the next section, we attempt a coherent philosophical and ethical framework for supported decision making.

Philosophical and Ethical Underpinnings

What Is Moral Deliberating and Deciding?

Bernard Lonergan (1972; 1988) has described human knowing as a composite of four distinct sets of cognitive operations that interact and are self-correcting: The first can be characterized broadly as *experiencing* data perceived by the senses or being aware of internal states and activities of consciousness; the second, as *understanding* or making sense of these data; the third, as *judging*, when we weigh the adequacy of evidence and check our understanding; and the fourth, as *deliberating and deciding,* when we aim to know what is good or worthwhile, and commit to act or refrain from acting in certain ways to attain it concretely (Lonergan, 1972; 1988, pp. 18–21). It is this fourth set of operations, which though related to and dependent on the others, constitutes us as *moral subjects* who encounter our world and act on it, for good or ill. In so doing, our decisions and actions shape the unique person that we each are and will become (Lonergan, 1996).

Deliberating and deciding, like the other cognitive operations, are *intentional* operations. Elizabeth Anscombe (1958), in her influential work *Intention*, distinguishes nonvoluntary actions (such as physiological responses and instincts) from voluntary decisions to act or refrain from acting in certain ways. These latter always have some purpose, aim, or goal for the one who is deciding. It is something that this person either wishes to continue to enjoy or avoid in the present or hopes to realize or avoid in the future. The intention (or intentions) behind a decision, and not merely the outcome of this decision, has moral relevance.

Lonergan adds to this analysis by proposing that there is a fundamental, ultimate orientation behind all deliberating and deciding: All human beings, in virtue of being human, aim to enjoy and realize in their lives what is truly good, worthwhile, or valuable. Lonergan calls this the transcendental intending of value. We could understand this in terms of a spontaneous desire to know what will bring about overall well-being for and among persons. Lonergan's view aligns with what diverse thinkers, among others, have described as striving to attain harmony (e.g., Confucius and Mencius); happiness understood as a flourishing human life (e.g., Aristotle, Aquinas); natural goods (e.g., Finnis (1980, 2011)); the fulfillment of basic human needs (e.g., Abraham Maslow (1943)); coherence with volitionally necessary commitments (e.g., Frankfurt (1999)); or the cultivation of universally shared conceptions of human strengths (e.g., Dahlsgaard et al. (2005)). Some examples of contributors to overall human well-being might be health, security, having positive connections with others, having a sense of belonging, engaging in meaningful work or leisure, contributing to the common good, encountering beauty, and fulfilling spiritual longings.

Deliberating and deciding, for Lonergan, consists in judging whether certain concrete goals (e.g., enjoying this activity with this friend) align with my transcendental intending of value (does it truly bring about the overall well-being that I desire?). Alignment of the two is discovered or known when I experience an abiding complacency, a resting of one's heart, as it were. Lack of alignment can result in my becoming uneasy, agitated, or concerned (Sullivan, 2005). We all become better over time at making good decisions through a self-correcting process, with the support of close and trusted persons on whom we rely for advice, guidance, or modeling (e.g., parents, teachers, mentors, friends, members of one's cultural, and religious communities). Hence, deliberating and deciding are similar in structure to other cognitive operations, such as understanding and judging, because they are *intentional* (purposeful) and they can be *self-transcending* (i.e., our apprehending or discovering values can develop as we learn which goods in life genuinely bring about the overall well-being or fulfillment toward which all humans are oriented).

While deliberating and deciding are *cognitive* operations (they result in some discovery), they are distinct from other cognitive operations in that the pivotal insights on which they hinge, namely, *apprehensions or discoveries of value*, are *given in feelings* (Lonergan, 1972; 1998, p. 38). The insight that a certain concrete good is worth pursuing or refraining from pursuing, as explained above, is felt. It is an affirmation that is accompanied by experiences of complacency or concern. This cannot be reduced to any strictly *logical* deduction. It is an instance of an affective rather than an intellectual cognition.

It is important to specify, however, that, drawing on the work of Dietrich von Hildebrand, Lonergan distinguishes moral feelings, which are intentional responses to what is genuinely good, worthwhile, or valuable (i.e., responses to the transcendental notion of value, as explained above) from both non-intentional states or urges (e.g., fatigue, anxiety, hunger, thirst) and momentary or transient intentional responses to what is *merely* agreeable or disagreeable. Moral feelings can lead often to decisions to maintain or pursue what is agreeable or avoid what is disagreeable, but they can equally entail decisions to maintain or pursue what is disagreeable or arduous. Nonetheless, for Lonergan, moral feelings give "intentional consciousness its mass, momentum, drive, power. Without these feelings our knowing and deciding would be paper thin. Because of our feelings, our desires and our fears, our hope or despair, our joys and sorrows, our enthusiasm and indignation, our esteem and contempt, our trust and distrust, our love and hatred, our tenderness and wrath, our admiration, veneration, reverence, our dread, horror, terror, we are oriented massively and dynamically in a world mediated by meaning" (Lonergan, 1972; 1988, p. 32). Moral feelings are thus *existentially significant* for each person. They are "so deep and strong, especially when deliberately reinforced, that they channel attention, shape one's horizon, direct one's life" (Lonergan, 1972; 1988, p. 33).

In summary, deliberating and deciding entail apprehending or discovering values. Values can be authentic in both of these ways: (1) they are notions of what is good and worthwhile for enjoying overall well-being, which is what all humans intend or strive for, and (2) they are what truly matter to a person. They form the core of a person's identity and are the source of goals or of what a person hopes for in life.

Can Persons with Intellectual and Developmental Disabilities Hold and Express Authentic Values?

From the above, it can be concluded that persons with IDD or progressive neurocognitive impairments can hold authentic values. Like all other human beings, they are oriented toward what is genuinely good for them, worthwhile, and valuable, which contribute or will contribute to their overall well-being. Through moral feelings, they are able to discover among concrete goods what aligns with this intention and, with support as needed, can become increasingly proficient at this, even when their cognitive impairments are profound or are progressing. Such persons might articulate values in very concrete terms (e.g., "singing makes me happy" or "I like you") or indicate them by drawings, gestures, or other nonverbal behaviors (such as when a person with dementia recognizes and enjoys a favorite song from youth or a young adult with an ID claps joyfully to a tune). Those close to such persons, who know them well, can learn to interpret their indications of value reliably.

It is important to clarify that many such indications, namely, those that stem from apprehending values, are not simply *arbitrary choices* without any grounding or context. They are different in kind from whims and fleeting desires, or certain preferences based on these, which every person also has. A contemporary philosopher who elaborates on this morally significant distinction is Michael Bratman (2018). For Bratman, the human will is not just a function of exercising absolute freedom to act or not act (what Lonergan calls the "essential freedom" of human beings). It entails *planning* based on specific ends or values to which a person is committed over time (what Bratman calls "temporally extended intentions") and is embedded in relating to what others are doing (*shared intentions*). These intentions constitute the ground and context for this person's decisions and actions and manifests this person's self-governance or agency.

Can persons with profound IDD have such intentions and manifest such agency? Many caregivers affirm that they can. For instance, Eva Feder Kittay, a philosopher who has a child, Sesha, with a profound ID, writes this of Sesha's capacity to apprehend beauty and to share her enjoyment by occasionally turning her head toward her mother with a twinkle in the eye when she likes a particular part of a musical piece: "even with all that Sesha cannot do and seems not to be able to comprehend, her response to music and her sensitivity to people are remarkably intact. Perhaps her responsiveness to music is more than remarkably intact; it is quite simply remarkable" (Kittay, 2008 p. 153).

Interpreting these indications of value in another person, such as Kittay's interpreting Sesha's turning toward her as an intentional activity expressing the value of musical beauty and of company, partly involves being able to discern *the ground or context for that person* of her or his decisions. This is why reliable interpretations will depend on how close and how well the interpreter knows the person with profound or progressively deteriorating cognitive impairments. The reliability of such interpretations also will develop over time. Moreover, a reliable interpreter can usually discern when a person's decision is not coherent or is misaligned with the authentic values of that person. This type of familiar knowledge of a person resembles the tacit knowledge that Michael Polanyi (1962, p. 4) described. This sort of knowledge cannot be comprehensively explained or easily transferred to others who are unfamiliar with the context of the close relationship.

A further relevant point relates to Bratman's notion of *shared* intentions, which opens up the possibility of shared agency between or among persons. This brings us to the notion of relational autonomy.

Autonomy as Relational

Discussions of decision making in healthcare often refer to the principle of respect for patient autonomy in ethics and law. *Autonomy* has various meanings in these disciplines and in the public's understanding of this term. For instance, some understand autonomy in a way that emphasizes rationality. By *autonomy* they mean an individual's capacity for moral self-governance based on reason and will (Byers, 2016). For others, autonomy is understood in a way that focuses on individual freedom. By *autonomy* they mean freedom from being interfered with or coerced in matters that are deemed to be private (Rajczi, 2016). The former understanding is sometimes referred to as the *positive* sense of autonomy (or the concept of moral agency) and the latter as the *negative* sense of autonomy (or the concept of liberty of action). Still others accept aspects of both meanings. According to either sense of autonomy, however, certain persons whose reasoning or liberty is compromised due to cognitive impairments would be regarded as unable to make autonomous decisions.

These prevailing notions of autonomy in society inadequately attend to *relational* aspects of decision making in persons with and without disabilities. The equation of autonomy with *independence* is not the only way to understand autonomy. When we advert to our experiences of exercising autonomy to make decisions in health care and other areas of life, we *all* depend on others *to some degree* (e.g., to communicate effectively, help with understanding, give guidance or advice, etc.), and in regard to certain decisions more than others. A relational understanding of autonomy most adequately describes our experiences of making decisions.

If we understand autonomy as relational, we can re-conceive decision making as a *capability*, or better, a set of capabilities along the lines of other human capabilities discussed by proponents of the so-called capability approach in economics, political theory, and social ethics (Sen, 1979, May 22; Nussbaum, 2006). This approach to distributive justice focuses, not on fairness in distributing goods or making opportunities generally available, but on establishing the social conditions that make it *feasible* for each person to access goods and avail themselves of opportunities in society. A just society and a just world should promote human capabilities. *Capability* is a person's functioning in order to enjoy well-being. This includes personal abilities *and* environmental and social conditions to facilitate and support that functioning. When we conceive of making decisions as a set of capabilities, we open up the possibility that each of us might need various types and degrees of assistance from others to *exercise* our autonomy.

A clarification, however, should be made. One criticism of the capabilities approach is that it does not go far enough. While it attends to the importance of support from others to exercise capabilities, it does not account for those who might never become self-reliant, even with such support. Reflecting on the reality of human interdependency, Eva Feder Kittay comments on the capabilities approach to justice as follows: "[A] theory of justice based on capabilities…does not address dependency head-on and therefore I would argue still needs supplementation with an ethic of care" (Kittay, 2011, p. 51). We add also that it needs to be supplemented by an understanding of autonomy as relational, as discussed above. It could be that some persons might need decision-making supporters, not simply to help them to interpret their authentic values but also to *decide* (e.g., for their supporters to select from among treatment options those that most align with these values). From an understanding of autonomy as relational and interdependent, such persons can be recognized as exercising relational autonomy.

Respect for Autonomy and Other Ethical Principles in Healthcare

In contemporary bioethics, health law, and health care, there is a tendency to regard respect for a person's autonomy as the sole or primary consideration. This trend calls into question, what is referred to as *paternalism* in healthcare decision making, e.g., when authority figures (e.g., healthcare professionals or parents of adult patients with IDD) decide on treatments, either without involving the person in making the decision or despite the person's refusal of a treatment. Some bioethicists consider as "soft paternalism" instances in which healthcare providers persuade ("nudge") persons to accept treatments that they need (Blumenthal-Barby & Burroughs, 2012; Sunstein & Thaler, 2003; Verweij & van den Hoven, 2012).

Criticisms of the so-called paternalistic approaches in healthcare decision making can be valid; sometimes, however, they are based on an inadequate understanding of autonomy that equates autonomy with absolute independence. Also, those who reject paternalism in healthcare decision making often set up a false opposition between respecting a person's autonomy and other ethical principles such as beneficence, non-maleficence, and solidarity. It is beyond the scope of this paper to examine in detail the issue of involuntary treatments in health care, especially mental health care. For our purposes, it suffices to say that there can be approaches to providing treatments that would be both beneficial to persons in need of health care (or would reduce the risk of them being harmed) *and* support and promote the decision-making capabilities of such persons and their exercise of autonomy. Indeed, no treatment can be truly beneficent, non-maleficent, or promote solidarity with a person if it is simply imposed without effort to include the person in the healthcare planning process (Sullivan, et al., 2020a, b, recommendation 3(a), p. 364). What practical implications for healthcare does this philosophical and ethical framework for supported decision making have?

Applying Theory to Practice in Healthcare

Assessing Decision-Making Capacity

Most criteria for assessing decision-making capacity in health care establish *thresholds* for determining whether this capacity is present or absent in a person for a certain decision. These criteria set up a binary distinction: either the person can authorize the decision (because she or he is capable or competent) or the person cannot. If a supported decision-making approach is accepted, however, the focus of such capacity assessments will need to change.

First, we would have to acknowledge that a person's decision-making capabilities exist on a continuum for each of those capabilities (e.g., the operations in human knowing described by Bernard Lonergan).

Second, in assessing decision-making capacity, the focus is often only on those cognitive operations thought necessary for giving *informed* consent. Typically, the person must demonstrate being able to understand information relevant to proposed treatments and appreciate the consequences of being treated or not treated (i.e., what Lonergan described as the second and third operations in human knowing). In a supported decision-making approach, however, the pivotal step in healthcare decision making is to discern a person with IDD's values or what matters fundamentally to her or him. Here, the capacity to be assessed is the person's ability to hold and express such values in a way that at least one other person can validly understand. This is the basis for someone to help the person with IDD to select, from among offered healthcare options, the one most able to realize those values (the fourth operation in human knowing for Lonergan).

Third, a supported decision-making approach would recognize and affirm relational autonomy. This would have two implications: the focus of assessments of decision-making capabilities would include identifying supports that patients might need for any of the capabilities involved in making healthcare decisions. If it is determined that persons with IDD are capable of expressing their authentic values, even if they might sometimes require help from close and trusted decision-making supporters who know them well for other aspects of the healthcare decision-making process, their decisions with such support should be recognized as authoritative.

Making such assessments will not always be straightforward and will entail some modifications of usual practice and skills. However, some aspects of a supported decision-making approach to assessing capacity in health care are often already informally practiced in person- and family-centered health care. The central principles underlying good assessments of patients should hold in any context:

strive to know the patient and caregivers well, and build good relationships based on effective communication and trust.

Since applying supported decision making in health care is recent, there are few tools to guide healthcare providers in assessing adults with IDD using this approach. "Decision making in health care of adults with intellectual and developmental disabilities: promoting capabilities" (Sullivan et al., 2020a), to our knowledge, is the only such tool that has been developed. This tool is designed mainly for use in the province of Ontario, Canada, where only legally recognized substitute decision-makers can provide consent to treatment for persons assessed to lack decision-making capacity. In this regard, Ontario has a legal regime that is not yet aligned with what is required of Canada and other states that have ratified the CRPD (2014); however, the help of decision-making supporters to enable adults with IDD to meet the criteria for legal decision-making capacity in Ontario may be considered a reasonable accommodation under the province's human rights laws. Substitute decision-makers who do not know the person who is assessed to lack decision-making capacity well can also be encouraged to consult with persons who can reliably interpret the person's health-related goals and values. Hence, there can be legal recognition of supported decision making within these limits.

The Decision Making in Healthcare of Adults with IDD: Promoting Capabilities. Tool guides healthcare providers to determine which decision-making approach is appropriate to take with an adult with IDD: an independent, interdependent (i.e., supported), or substituted approach. Including the interdependent or supported approach broadens the spectrum of adults with IDD who can authorize healthcare decisions. The collaboration between the adult with IDD, with supports as needed, and the healthcare provider to reach agreed-upon goals of care is what we understand by shared decision making (Fig. 3.1).

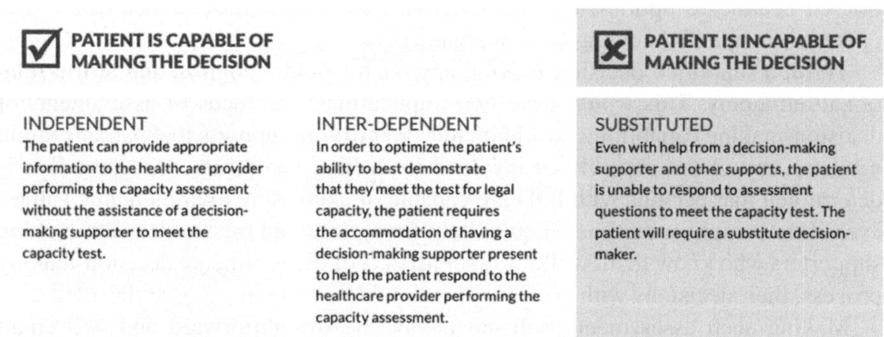

Fig. 3.1 Decision-making approaches across a spectrum of capabilities of adults with intellectual and developmental disabilities (Sullivan et al., 2020a, b)

Promoting Conditions to Optimize Communication with the Person in Need of Healthcare

The abovementioned tool prompts healthcare providers to explore various means to optimize communication with the person with IDD in need of health care. It is important that the healthcare provider adapts communication to the person's preferred communication method and asks the person whether she or he wishes to involve caregivers or others to whom the person is close and whom the person can trust to be a decision-making supporter.

Certain accommodations that are not usually considered should be offered to adults with IDD, for instance, scheduling appointments at an optimal time of day for the person, booking sufficient clinic time, and making the environment appropriate to put the person at ease (e.g., by accommodating any noise or light sensitivities the person might have).

Healthcare providers should be aware that some patients with IDD have difficulty expressing emotional distress related to their illness, past traumatic experience with health care, and other negative life events that could unduly influence their decisions to refuse certain proposed interventions. When being assessed, the patient's distress might be manifested by resistance or lack of engagement. The underlying causes of such behavior need to be identified and assiduously explored.

For instance, consider the circumstances of a young woman with IDD in the severe range, Miriam, who is prone to developing pneumonia. Miriam has a small range of single words that she can use, but she prefers to communicate mostly by gesturing and pointing. She also is very sensitive to intense lights and noise. Miriam lives in a group home where she is susceptible to a high rate of influenza (flu) transmission. Prior to that, between the ages of 5 and 25 years old, she had lived in an institution for persons with IDD. Her family has had little involvement in her life since she lived in that institution or the group home; however, her older brother, under the law, would be her substitute decision-maker if Miriam was assessed to lack decision-making capacity for a certain health decision. Miriam is usually accompanied to the primary care office by her group home worker, Ingrid, who has known her well for 10 years. This year, because of the COVID-19 pandemic, public health authorities have strongly advised that people receive flu and pneumococcal vaccines if they have a high risk of developing pneumonias. Ingrid explains this to Miriam, and Miriam agrees to visit her new family physician for these vaccinations. Before the visit, the physician consults with Ingrid about needed accommodations for Miriam and schedules an appointment early in the day when Miriam is usually calm and there are few people in the clinic. After developing rapport with Miriam, the physician asks whether she would like Ingrid to remain in the room. She indicates *yes* with a nod. When the physician brings up the issue of the vaccinations with Miriam during the visit, she seems initially confused. Ingrid helps to remind Miriam of what they had previously discussed regarding the vaccinations, and she seems to understand; however, when the physician brings out the needles, Miriam manifests agitation. The physician demonstrates the injection procedure on herself.

Miriam remains distressed and shakes her head repeatedly when the physician asks if she may do the same to her. The physician asks Ingrid about this distress. Ingrid maintains that Miriam had agreed to come to the office to be vaccinated when they had discussed this previously. Upon further exploration with Ingrid, the physician discerns that Miriam had some traumatic healthcare experiences while she lived in the institution. The physician manages to calm Miriam and explain to her that these injections will help her to keep healthy and enable her to do the activities she enjoys. Ingrid repeats this message to Miriam. The physician then offers Miriam the option of flu vaccination by a nasal spray, which she accepts, and schedules another visit for the pneumococcal vaccine injection, which can be administered by injection following desensitization training.

In this scenario, the physician is assessing Miriam's capacity to make a certain healthcare decision through optimizing communication and involving a decision-making supporter who knows Miriam well and is trusted by her. She avoids assuming that Miriam is incapable of making any healthcare decision because of the severity of her intellectual and communicative impairments and contacting her substitute decision-maker for consent for the vaccinations, which in Miriam's case would be her uninvolved older brother. She also avoids assuming that Miriam is capable of making this healthcare decision independently. She asks Miriam whether she would like Ingrid to support her in making this decision. Ingrid helps Miriam to communicate with the physician, and vice versa. With the help of Ingrid, the physician is able to discern that Miriam's initial resistance to the injections was likely motivated by strong responses to the trauma of her past healthcare experiences and that Miriam also wished to avoid the risk of developing a severe illness that would undermine what she enjoys in life. This involves the physician distinguishing between Miriam's spontaneous response to a dissatisfaction and her responses to values, i.e., those things Miriam truly hopes for and values.

Supported Decision Making and Shared Decision Making

In the above scenario, it is true the physician is offering the sort of shared decision making that should be practiced with *any* person in need of health care. Shared decision making, however, can sometimes be construed according to a consumer model of health care in which persons are assumed to be independent agents capable of negotiating health care with physicians as equals. A supported decision-making approach emphasizes relational autonomy and the importance of offering accommodations and supports in healthcare decision making to adults with IDD as needed, such as by offering an alternative means of flu vaccination in Miriam's circumstance. Miriam, however, would not have been able to assess independently the benefits and risks of vaccination via nasal spray versus injection. She still needed the support of Ingrid and her physician to help her interpret their relative risks, benefits, and burdens to her in terms of her health goals (to stay healthy), values (what she enjoys in life), and preferences (benefits without being traumatized by an injection).

From a supported decision-making approach, the healthcare provider should always endeavor to engage patients in the decision-making process regarding their care and include any decision-making supporters whose help a patient needs and wants. They should ensure that patients understand that there are options for care, that they can draw on help from their decision-making supporters at any time, and that a decision does not have to be made during a single visit. Miriam's physician, for example, defers the decision regarding the pneumococcal vaccine injection until a scheduled follow-up visit.

The scenario involving Miriam illustrates that healthcare providers also have some role in supporting decision-making capabilities of patients and the patient's decision-making supporters. It is ethically appropriate for healthcare providers not only to give factual information regarding possible interventions and their potential benefits, burdens, and risks. They should also be prepared to elicit and discuss the authentic values of patients, with the help of decision-making supporters as needed, and relate these values to health-related goals underlying various treatment options (e.g., prolonging life, improving function, alleviating distress, fostering important relationships, minimizing burdens for the patient and patient's family).

The healthcare provider should also be aware of and accommodate any cultural or religious perspectives regarding decision making in healthcare. For example, in certain cultural or religious traditions, respect for those in authority in the family or in health care might be accepted and expected by the person with IDD. In such situations, a greater role can be given to the decision-making support of family members whom the person values and trusts.

Above all, it is important in shared decision making to strive for common goals of care among patients, the caregivers on whom they depend, and healthcare providers. Prudent judgments by healthcare providers can often facilitate agreements. For example, when a treatment option involves uncertainty or ambiguity, a trial period for intervention might be an acceptable option to all. At other times, the healthcare provider might need to address concerns or reservations of patients by accepting a less medically effective option. For example, in the circumstance of Miriam above, flu vaccination via a nasal spray, while possibly less effective, can often be an acceptable alternative, along with other therapies in the long term to address Miriam's fears concerning needles.

In summary, shared decision making is important for healthcare providers to offer as part of patient-centered care, regardless of whether the adult with IDD is assessed to be capable independently or interdependently (i.e., with the help of one or more decision-making supporters) of reaching a certain healthcare decision or requires a substitute decision-maker.

Conclusion

Patients with IDD are capable of apprehending, holding, and expressing authentic values. As with any patient, these should guide decision making regarding their health care. Many such patients might also require decision-making supporters to help with some part of the decision-making process, e.g., understanding, judging, interpreting their authentic values, or deciding with them on acceptable options for treatment. Autonomy is relational. We all need various types or degrees of help from others, for certain kinds of decisions more than others. When needed, help from decision-making supporters and healthcare providers can promote an adult with IDD's decision-making capabilities and exercising relational autonomy. Ways should be found to respect the goals and authentic values of persons while providing them with treatments that are beneficial to them or would reduce the risk of them being harmed.

There remain further questions regarding supported decision making in healthcare. For example, what can and should be done to promote the decision-making capabilities of those patients who cannot decide independently but have no close persons who know them well enough to interpret their authentic values? What impact would supported decision making have on caregivers and other decision-making supporters who often have inadequate guidance, coaching, and other resources for their role? What safeguards would have to be in place to ensure that decision-making supporters undertake their role responsibly and ethically? Do healthcare providers have an ethical responsibility to advocate for policy and social changes, to legally recognize supported decision-making approaches in healthcare, and to foster the kinds of close relationships that socially isolated persons with intellectual and developmental disabilities need to support them in making healthcare decisions?

While continuing to address questions such as these, we can nonetheless conclude that supported decision-making approaches affirm the inherent dignity of patients with IDD, will enhance their participating in making healthcare decisions with supports as needed, and can enable their authority to make those decisions to be ethically and legally recognized.

References

Anderson, J., & Honneth, A. (2005). Autonomy, vulnerability, recognition, and justice. In J. Christman & J. Anderson (Eds.), *Autonomy and the challenges to liberalism: New essays* (pp. 127–149). Cambridge University Press.

Anscombe, G. E. M. (1958). *Intention*. Oxford University Press.

Bach, M., & Kerczner, L. (2010). *A new paradigm for protecting autonomy and the right to legal capacity: Advancing substantive equality for persons with disabilities through law, policy and practice*. Law Commission of Ontario. https://www.lco-cdo.org/wp-content/uploads/2010/11/disabilities-commissioned-paper-bach-kerzner.pdf

Blumenthal-Barby, J. S., & Burroughs, H. (2012). Seeking better health care outcomes: The ethics of using the "nudge.". *American Journal of Bioethics, 12*(2), 1–10.
Bratman, M. E. (2018). Agency, time, and sociality. In M. E. Bratman (Ed.), *Planning, time, and self-governance: Essays in practical rationality* (pp. 110–130). Oxford University Press.
Brucker, D. L., & Nord, D. (2016). Food insecurity among young adults with intellectual and developmental disabilities in the United States: Evidence from the National Health Interview Survey. *American Journal on Intellectual and Developmental Disabilities, 121*(6), 520–532.
Byers, P. (2016). Dependence and a Kantian conception of dignity as a value. *Theoretical Medicine and Bioethics, 37*, 61–69.
Byrne, G. (2018). Prevalence and psychological sequelae of sexual abuse among individuals with an intellectual disability: A review of the recent literature. *Journal of Intellectual Disabilities, 22*(3), 294–310.
Cole, A. (2019, September 4). *Presentation to PROBUS*, Perth, Ontario, Canada.
Dahlsgaard, K., Peterson, C., & Seligman, M. E. P. (2005). Shared virtue: The convergence of shared human strengths across cultures and history. *Review of General Psychology, 9*(3), 203–213.
Davidson, G., Kelly, B., MacDonald, G., et al. (2015). Supported decision making: A review of the international literature. *International Journal of Law and Psychiatry, 2015*(38), 61–67.
Dinerstein, R. D. (2012). Implementing legal capacity under article 12 of the UN convention on the rights of persons with disabilities: The difficult road from guardianship to supported decision making. *Human Rights Brief, 19*(2), 8–12. http://www.supporteddecisionmaking.org/sites/default/files/implementing_legal_capacity_article_12_un_convention.pdf
Finnis, J. (1980/2011). *Natural law and natural rights*. Oxford University Press.
Frankfurt, H. (1999). *Necessity, volition and love*. Cambridge University Press, essays 9, 11 & 14.
Honneth, A. (2007). *Disrespect: The normative foundations of critical theory*. Polity Press.
Hoole, L., & Morgan, S. (2011). 'It's only right that we get involved': Service-user perspectives on involvement in learning disability services. *British Journal of Learning Disabilities, 39*(1), 5–10.
Hughes, K., Bellis, M. A., Jones, L., et al. (2012). Prevalence and risk of violence against adults with disabilities: A systematic review and meta-analysis of observational studies. *Lancet, 379*(9826), 1621–1629.
Hughes-McCormack, L. A., Rydzewska, E., Henderson, A., et al. (2018). Prevalence and general health status of people with intellectual disabilities in Scotland: A total population study. *Journal of Epidemiology and Community Health, 72*, 78–85.
Hurd, C., Evans, C., & Renwick, R. (2018). "Having friends is like having marshmallows": Perspectives of transition-aged youths with intellectual and developmental disabilities on friendship. *Journal of Applied Research in Intellectual Disabilities, 31*(6), 1186–1196.
Kittay, E. F. (2008). At the margins of moral personhood. *Bioethical Inquiry, 5*, 137–156.
Kittay, E. F. (2011). The ethics of care, dependence, and disability. *Ratio Juris, 24*(1), 49–58.
Lonergan, B. (1972/1988). *Method in theology. Collected works of Lonergan, volume 14*. R. M. Doran & J. D. Dadovsky (Eds.), University of Toronto Press.
Lonergan, B. (1996). The subject. In W. F. J. Ryan & B. J. Tyrrell (Eds.), *A second collection: Papers by Bernard J.F. Lonergan. 2nd ed. Collected works of Lonergan, volume 11* (pp. 69–78). University of Toronto Press.
Maslow, A. H. (1943). A theory of human motivation. *Psychological Review, 50*(1), 370–396.
Nussbaum, M. (2006). *Frontiers of justice: Disability, nationality, species membership*. Harvard University Press/Belknap Press.
Polanyi, M. (1962). *The tacit dimension*. Anchor Books.
Potvin, L. A., Fulford, C., Ouellette-Kuntz, H., et al. (2019). What adults with intellectual and developmental disabilities say they need to access annual health examinations: System navigation and person-centred care. *Canadian Family Physician, 65*(Suppl. 1), S47–S52.
Rajczi, A. (2016). Liberalism and public health ethics. *Bioethics, 30*(2), 96–108.

Sen, A. (1979, May 22). *Equality of what? The Stanford University Tanner lecture on human values*, May 22, 1979. https://tannerlectures.utah.edu/_documents/a-to-z/s/sen80.pdf.
Simmons, M., & Gooding, P. (2017). Spot the difference: Shared decision making and supported decision making in mental health. *Irish Journal of Psychological Medicine, 34*(4), 275–286.
Sullivan, W. F. (2005). *Eye of the heart: Knowing the human good in the euthanasia debate.* University of Toronto Press.
Sullivan, W. F., & Heng, J. (2018). Supporting adults with intellectual and developmental disabilities to participate in health care decision making. *Canadian Family Physician, 64*(Suppl 2), S32–S36.
Sullivan, W. F., Diepstra, H., Heng, J., et al. (2018). Primary care of adults with intellectual and developmental disabilities: 2018 Canadian consensus guidelines. *Canadian Family Physician, 64*(4), 254–279.
Sullivan, W. F., Heng, J., Bach, M., et al. (2020a). *Decision making in health care of adults with intellectual and developmental disabilities: Promoting capabilities*. Developmental Disabilities Primary Care Program, Surrey Place. https://ddprimarycare.surreyplace.ca/tools-2/general-health/capacity-for-decision making/
Sullivan, W. F., Heng, J., DeBono, C., et al. (2020b). Promoting capabilities to make healthcare decisions. Consensus statement of the 9th IACB international colloquium. *National Catholic Bioethics Quarterly, 20*(2), 617–633.
Sulmasy, D. P., & Snyder, L. (2010). Substituted interests and best judgments: An integrated model of surrogate decision making. *Journal of the American Medical Association (JAMA), 304*(12), 1946–1947.
Sunstein, C., & Thaler, R. (2003). Libertarian paternalism is not an oxymoron. *University of Chicago Law Review, 70*, 1159–1202.
The Mental Capacity Act (U. K.). (2005). https://www.legislation.gov.uk/ukpga/2005/9/section/4
United Nations. (2006). *Convention on the rights of people with disabilities*. United Nations, articles 12(1) and 12(3). http://www.un.org/disabilities/documents/convention/convoptprot-e.pdf
United Nations Committee on the Rights of Persons with Disabilities. (2014). *General comment no. 1*. United Nations. https://documents-dds-ny.un.org/doc/UNDOC/GEN/G14/031/20/PDF/G1403120.pdf
Verweij, M., & van den Hoven, M. (2012). Nudges in public health: Paternalism is paramount. *American Journal of Bioethics, 12*(2), 15–17.
Williamson, H. J., Contreras, G. M., Rodriguez, E. S., et al. (2017). Health care access for adults with intellectual and developmental disabilities: A scoping review. *Occupation, Participation and Health, 37*(4), 227–236.

Chapter 4
Examining the Complexities of Support for Decision-Making Practice

Christine Bigby and Jacinta Douglas

Introduction

By asserting the equal recognition before the law of people with disabilities and their right to support to exercise legal capacity, Article 12 of the United Nations Convention on the Rights of Persons with Disabilities (CRPD) created new imperatives for thinking about decision making and people with cognitive disabilities. Scholars and disability activists have generated significant debate about the meaning of Article 12 and the implications of de-linking legal and mental capacity. Much of their attention focused on the inherent problems of existing guardianship provisions and finding alternative mechanisms to support decision making and protect people with cognitive disabilities from abuse in this sphere of their lives. The concept of *supported decision making* has been propelled to the fore as one such mechanism. It provides a means of recognising the role of supporters and distinguishing between decision support that puts the 'will and preferences' of the person at the centre, aligned with the intent of the CRPD, and formal substitute decision making or informal support based on perceptions of the person's 'best interests' (Committee on the Rights of Persons with Disabilities, 2014). The actual practice of supported decision making, however, has attracted much less attention than 'what ought to be' and its foundational legal and ethical frameworks.

This chapter considers the emerging evidence base about decision support and the practice of supported decision making developed from our programme of

C. Bigby (✉)
Living with Disability Research Centre, La Trobe University, Melbourne, VIC, Australia
e-mail: c.bigby@latrobe.edu.au

J. Douglas
Living with Disability Research Centre, La Trobe University, Melbourne, VIC, Australia

Summer Foundation, Melbourne, VIC, Australia
e-mail: j.douglas@latrobe.edu.au

I. Khemka, L. Hickson (eds.), *Decision Making by Individuals with Intellectual and Developmental Disabilities*, Positive Psychology and Disability Series,
https://doi.org/10.1007/978-3-030-74675-9_4

research at La Trobe University in Australia. For over 10 years, we have explored supporters' experiences of supporting people with cognitive disabilities (those with either intellectual disabilities or acquired brain injury) to make decisions and in turn their experiences of receiving support. We have compared and contrasted the experiences of these two groups and their family or paid supporters. This chapter is focused on decision supporters of people with intellectual disabilities. To avoid confusion, we use the term *support for decision making* when referring to the practice of support and *supported decision making* when referring to formal schemes that recognise the standing and role of supporters in providing support for decision making.

The overarching aim of our research programme has been to understand what constitutes good decision support practice, how to guide the practice of supporters, how the quality of practice might be judged and how to protect people with disabilities from abuse by holding supporters accountable for their practice. In researching these questions, we have also learned much about the issues supporters grapple with and how they resolve them as they try to provide support that respects the will and preferences of people with cognitive disabilities and their right to participate in decision making about their own lives. This chapter is based on our published empirical research and unpublished data from work in progress. These data represent a subset of parental supporters from a study that is examining the practice of decision support and the influence of a training and mentoring intervention. The sample comprised 37 parents who provided regular decision support to an adult with intellectual disabilities. Parents were between 47 and 74 years and mostly were mothers, and the adults they supported were between 18 and 39 years old and mostly lived at home with their families. The severity of their intellectual disability, reported by parents ranged from mild to severe to profound. The sample was recruited through the researchers' networks of advocacy, parent and disability support organisations and lived in three eastern states in Australia, Victoria, New South Wales and Queensland. Supporters were interviewed at multiple time points during the course of the study about the way they provided support, using a semi-structured schedule and measures of confidence and decision support strategies. After the first interview, 18 supporters participated in a 1-day training workshop about the La Trobe Support for Decision-Making Practice Framework and received up to 6 sessions of individual mentoring to assist in applying the Framework to their individual situation. More detailed information about the study methods are available from the authors. All the names are pseudonyms. The chapter is divided into five sections: (1) background and context; (2) experiences and processes of decision support and the significance of the decision, the context, and the relationship; (3) the development and application of the La Trobe Support for Decision-Making Practice Framework (framework); (4) the impact of training and mentoring in the Framework on supporters and people with intellectual disabilities and (5) conclusions and directions for future research.

Background and Context

Research on Involvement of People with Intellectual Disabilities in Decision Making

Our review of the literature from 2000 to 2014 found a limited evidence base on the practice of support for decision making. Predominantly, we identified research describing the limited participation of people with intellectual disabilities in decision making (Bigby et al., 2015a). Their lack of participation stemmed from external structural constraints, attitudes and skills of supporters, as well as the individual characteristics of people themselves (Dunn et al., 2010; Hawkins et al., 2011; Kjellberg, 2002; Antaki et al., 2009; Rossow-Kimball & Goodwin, 2009). For example, the institutional logic of services too often restricts opportunities for choice and decision making by attending to groups rather than individuals, applying organisational policies and risk averse management practices which focus on protection rather than potential opportunities of risk taking and primary supporters too frequently having a paternalistic rather than empowering orientation. Limited experiences of decision making as a result of structural factors and the low expectations of supporters further compound the individual difficulties of people with intellectual disabilities that hinder decision-making capacity, such as poor communication or problems understanding consequences or abstract concepts. In this early review up to 2014, research relevant to support for decision-making practice primarily focused on assessing individuals' capacity to make decisions and methods to enhance decision- and choice-making skills by drawing on basic cognitive science (Davies et al., 2003; Fisher et al., 2012; Khemka, 2000; Wennberg & Kjellberg, 2010). Too often though, strategies to enhance choice and decision making had been tested in controlled environments and stopped short of being about the real-world decision making.

Addressing Structural Constraints

The shift to a rights paradigm marked by the CRPD together with the growth of neoliberalism since the 1990s has changed policy visions by emphasising the exercise of choice and control by people with intellectual disabilities. New policies of individualisation of services and delivery through competitive market mechanisms were designed to remove some of the structural constraints and widen opportunities for decision making. Most evident have been the change in funding mechanisms whereby individualised funding has increasingly replaced block funding of disability support services in Europe, Scandinavia and North America (Fleming et al., 2019). A primary example of this is the Australian National Disability Insurance Scheme (NDIS) introduced progressively between 2013 and 2020. Its objectives included giving effect to government obligations under the CRPD and to "enable people with disability to exercise choice and control in the pursuit of their goals and the planning and delivery of their supports" (NDIS, 2013, sect 3 e). However, this type of reform was predicted to be less advantageous for people with intellectual disabilities compared to other disability groups as it worked best for "particular kinds of consumers in a contemporary landscape which privileges competence, capacity and individual independence" (Dowse, 2009, p. 573). A growing body of evidence suggests this is the case (Carey et al., 2018; Needham, 2013; Williams & Dickinson, 2016). For example, qualitative studies of NDIS planning reviewed by Bigby (2020) show that people with intellectual disabilities are frequently not included in decision making about funding or services planning. Recent ethnographic research from the UK suggests little has changed in culture or delivery of services in terms of participation in decisions despite major policy shifts. For example, Devi et al. (2020) demonstrate how multiple layers of organisational procedures and paperwork to assess decision capacity or risk generate a shift towards substitute away from supported decision making. From another perspective, in a case study of an accommodation service, McKearney (2020a) vividly describes the different strategies for persuasion that dominate staff practices.

Substitute and Supported Decision Making

Despite policy aims of increasing choice and control, reform of funding systems does not appear to have significantly addressed the limited participation in the decision making of people with intellectual disabilities. Indeed, in Australia since the introduction of the NDIS, there has been an increase in the number of people with intellectual disabilities who, through the appointment of a guardian as substitute decision-maker, have lost their legal capacity to make decisions (Bigby, 2020). Lack of formal mechanisms to provide support for decision making has clearly contributed to this loss of legal capacity at a time of increased opportunities for making decisions about support services.

Reflecting some of the intent of the CRPD, the NDIS assumed the right of people with intellectual disabilities to make decisions about their own lives and acknowledged the role of family members in providing support. However, the absence of a supported decision-making scheme and thus the lack of provisions to guide supporters or help them to be accountable for their support were belatedly acknowledged as a major issue (Bigby, 2020; Tune, 2019). To date the NDIS has relied on the long-standing option of appointing substitute decision-makers to act on a person's behalf either through nominee clauses or recourse to guardianship provisions in each state or territory (Cukalevski, 2019).

Appointing substitute decision-makers to act in a person's best interests reflects the doctrine of parens patriae stretching back to the Middle Ages (Carney, 1982). Although mechanisms have changed over time and differ across jurisdictions, appointment of substitute decision-makers remains the dominant approach in many jurisdictions. Supported decision making was developed in the 1990s as an alternative and first introduced in Sweden and some Canadian provinces. Essentially, supported decision making aims to support participation in decision making without taking away a person's right to make decisions. It is premised on enhancing individual autonomy and self-determination and the expression of preferences in the context of supportive and trusting relationships (Gordon, 2000). British Columbia's Representation Agreements, for example, recognise the shared nature of decision making and give formal standing to supporters whose roles vary from explaining issues and exploring options for a decision to interpreting communication and co-constructing preferences (Carney & Beaupert, 2013; Series, 2015).

The concept of supported decision making holds significant theoretical promise in realising the intent of Article 12 of the CRPD, ensuring retention of the rights and decisions driven by a person's own preferences. The term 'supported decision making' has become common parlance among advocates and disability support organisations and various models for organising support. In Australia, for example, over the last decade, various short-term pilot schemes have been developed, although characteristically without according legal standing to supporters or evidence-based training for supporters (Bigby et al., 2017). Governments, however, have been slow to make legal reforms and institute formal supported decision-making schemes. Where they have done so, new schemes more usually complement rather than replace guardianship (Then et al., 2018). Reasons for this include the scant empirical evidence about fidelity of the purpose of supported decision-making – whether it can deliver on promises of both empowerment and safeguarding.

Supported decision making places a lot of confidence and invests much power in supporters' actions. As Carney (2017, p. 48) suggests, it shifts the 'focus from the capacity of the person being assisted to the adequacy or otherwise of the capacity of those providing assistance'. To be true to purpose, supported decision making relies on the skills and integrity of supporters to explore options and interpret preferences. There is significant potential for undue influence and conflicts of interest as supporters are challenged to put aside their own values and avoid their interests interceding in practice, especially where they have a stake in decision outcomes. Indeed, some commentators regard this task to be so challenging that they suggest that supported decision making 'might actually have the opposite effect, disempowering such individuals or making them more vulnerable to manipulation, coercion or abuse' (Kohn et al., 2014, p.1114). Much more understanding of the processes of decision support and the constituents of good practice are required to build both the case for supported decision making and the capacity of decision supporters. Building this understanding has been the focus of our research.

Experiences and Process of Decision Support

The significance of relationships with supporters and the centrality of trust, equity and closeness have featured as pivotal in some writing about supported decision making (Arstein-Kerslake, 2017; Watson, 2016). However, as empirical research explored processes of decision support further, its complexity and the significance of other factors; the decision, the context of support and attributes of the person and their supporters, are becoming clearer (Browning, 2018). Each instance of decision support brings together a unique combination of factors that influence the actions of supporters and the outcome. No two decisions are alike, and the nature and intensity of support and the investment of supporters in outcomes vary with each decision. Curryer's Australian study showed, for example, that the support of mothers for their adult sons or daughters could range from empowering to paternalistic depending on the context and the decision (Curryer et al., 2020). Similarly, a Canadian study illustrated the continuum of influence that parents might exert over decisions spanned from providing suggestions, giving instructions to deciding what was possible (Taylor et al., 2019). Unpublished data from our recent fieldwork showed a very similar range of parental strategies and demonstrated how their actions ranged from controlling and directing to enabling and exploratory and were heavily influenced by the decision and its context. This contrast is illustrated by the following comments from parents talking about decision support:

Typically it's in a restaurant and we're eating and he'll decide something that we know it's going to be way too spicy for him and say, "That's going to be very hot for you". He says, "I'm happy" "Okay fine" and we'll let him do that, make the decision and he'll endure it. He will regret but endure it and we'll make sure we have something extra just in case he needs something more but we don't actually stop him from making decisions.

I provide the options, yes, I would say "this, this and this is available". And generally if – the downside about that is that if I think it's not appropriate, then I'm not going to provide her with that opportunity to make that decision.

These data illustrate the differing degrees of energy supporters invest in providing support for what may appear to be similar types of decisions. They show, too, how the supporter's stake in the decision outcome and the potentially negative impact on the person's well-being influences their support. For example, the father of one young man said, 'there's some decisions without consequence and I'll let him do it. So making a decision on a meal, it's up to him totally'. In contrast, for a similar decision also in relation to meals, a mother said she had given considerable thought to supporting her daughter to make decisions about what she ate and instigated a family culture that effectively restricted options to those she thought were healthy. She said,

...where health and food are concerned, we're going to have fruit for breakfast and we're going to have salad with our main meal...in a way, I'm not allowing too much independent decision making where those high level goals are so important.

As these few examples illustrate, tensions between supporting rights to make decisions and safeguarding the person's well-being can permeate the decision support process. A major task of supporters is to support people who have limited intellectual capacity and experiences to appreciate the breadth of options available to them and the consequences of their preferences to their immediate or longer-term well-being. Our data suggest that although parents and many paid supporters are deeply committed to the well-being of the people they support and their right to participate in decision making, they nevertheless want to ensure the 'right' decision is made. As one mother said about her son's decision making, 'He's pretty good. They are not always the right ones, that's the only thing'. A comment from another mother illustrates the tensions she experienced between prioritising her daughter's preferences whilst safeguarding what she perceived as her longer-term interests that she felt her daughter could not fully appreciate. She said,

> *So her idea of a plan would be that she'd like to work there five days a week. I don't see that as the best plan for her because I think there are now possibilities that weren't there before that she should explore. I don't want her to be too locked in on that and if she spends too much time there, she won't be able to do anything else.*

Support for decision making has been likened to emotional labour in light of its time-consuming and emotionally demanding nature (Wiesel et al., 2020). In our fieldwork, supporters talked about the extent of focused attention they needed to give to decision support and how much easier it was sometimes to make the decision themselves. Earlier studies (Bigby et al., 2015b) and our recent fieldwork point to the importance not only of equipping supporters with skills to grapple with the challenges they confront but also of providing peer or other formal emotional support. As one parent said reflecting on how she provided support, 'having some more support and training as to what else I could be doing. But I don't think I'm at the absolute top of where I could be. So, support and training in continuity, consistency.'

Process Models of Decision Support

Browning's (2018) model identified a similar high-level process that occurs for each instance of support – the identification of a decision opportunity and the dynamic interaction between the expression of the person's will and preferences about a decision and their supporter's response. This process is illustrated in Fig. 4.1.

The model helps to understand why each instance of decision support unfolds differently, by illustrating how it is influenced by the same combination of factors,

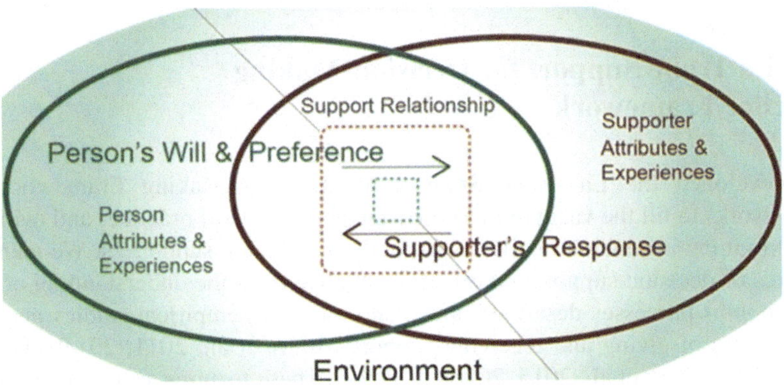

Fig. 4.1 Process model of decision support (from Browning et al., 2020, p.5)

but for every instance each factor has its own unique features – the attributes and experiences both the person and the supporter bring to the decision, the nature of the decision, the environment in which it occurs and the nature of the support relationship. Browing et al. (2020) describe, for example, the different attributes, experiences, values and priorities that supporters and the person they support brought to an instance of support about a similar type of decision, whether or not to go swimming, and how these shaped the way the process unfolded. Both supporters had a similar long-term trusting and mutually respectful relationship with the person they supported yet worked in very different contexts and brought different perspectives to their support. They write,

> *Lisa believed her role was to "help guide [Cecily] in a more healthy direction", perceiving Cecily's will and preferences as something to be shaped or "tweak[ed]". Whereas Annie believed her role was to "always try and remain neutral" and perceived Natalie's will and preferences as vulnerable to influence. Different environmental pressures, in these examples of service and funding contexts, helped to shape the process. Annie knew that Natalie's individualised funding might be threatened if she stopped participating in rehabilitation activities, and as a group home employee Lisa needed to ensure Cecily's heart condition and weight were being managed in line with medical recommendations. (Browing et al., 2020, p.8)*

Whilst not a practice model that identifies features of good practice, Browning's process model highlights the breadth of factors supporters need awareness of when they provide support. By illustrating the uniqueness of each instance of decision support, it also helps to alert supporters to the potential influence of their own values and context. This model also suggests the limited value to support practice of decision typologies, such as labelling decisions big or little, important or every day, by emphasising that each decision must be considered in relation to the person and the context. Importantly, what this work on decision support processes does is demonstrate empirically why decision support practice needs to be person-centred, tailored on every occasion to the individual, the decision and the context, and why supporters need to be reflective about their own skills and deliberative in the attributes and values they bring to the support process.

The La Trobe Support for Decision-Making Practice Framework

We developed the La Trobe Support for Decision-Making Framework (the Framework) to fill the vacuum of knowledge about support practice- and evidence-based training available to build the capacity of decision supporters. We identified features of decision support and effective practice from the understanding of decision support processes described above and our early empirical studies about the experiences of giving and receiving support (Bigby et al., 2011, 2015b; Douglas et al., 2015; Knox et al., 2015, 2016a, b, 2017). These features, or propositions, are summarised in Table 4.1.

Table 4.1 Features of decision support and effective practice

Features of decision support	Features of effective decision support practice
Has discernible components	Occurs in context of trusting relationships
Is iterative rather than linear	Supporters create opportunities for involvement
Involves multiple players	Supporters committed to the right to participate
Changes with every decision	Supporters understand the decision
Is shaped by the context of the decision	Supporters use individual-tailored strategies
Implementation of decisions may not rest with decision supporters	Supporters are aware of their own values
Support can be onerous	Supporters know the person well
Supporters require support for their role	Supporters explore options and understanding of preferences
	Supporters understand constraints on the decision

We used these propositions to develop the Framework and associated training materials (Bigby et al., 2019; Douglas & Bigby, 2018). The four-step process for developing and evaluating complex interventions (Craig et al., 2018) which we used is described in more detail by Douglas and Bigby (2018).

The Framework sets out the seven components or steps that all decision support practice involves, the three principles that should inform support and the broad strategies used across steps that are tailored to the individual and the step. The Framework is intended as a guide for supporters to decision support practice aligned with the intent of the CRPD, i.e. practice that priorities participation in decision making by a person with intellectual disabilities and puts respect for their will, rights and preferences at its centre. It also provides a set of criteria to serve as a means of reflective accountability for supporters. The Framework is applicable to paid or informal supporters of people with either intellectual disabilities or acquired brain injury and can be adapted to different policy and service contexts. For example, to date it has been used by family members, direct support workers in a variety of settings, support coordinators in a statutory insurance scheme for road traffic accident victims, an advocacy organisation to pilot a supported decision-making programme utilising paid supporters for people without informal support and an organisation responsible for substitute financial decision-making support of people with cognitive disabilities. Training in the Framework and follow-up mentoring to help supporters apply it to their own context have been shown to positively impact on the practice of supporters, by increasing their self-awareness and strategies that further participation (see later section). Figure 4.2 illustrates the components of the Framework, each of which is described with examples from our fieldwork in the following sections.

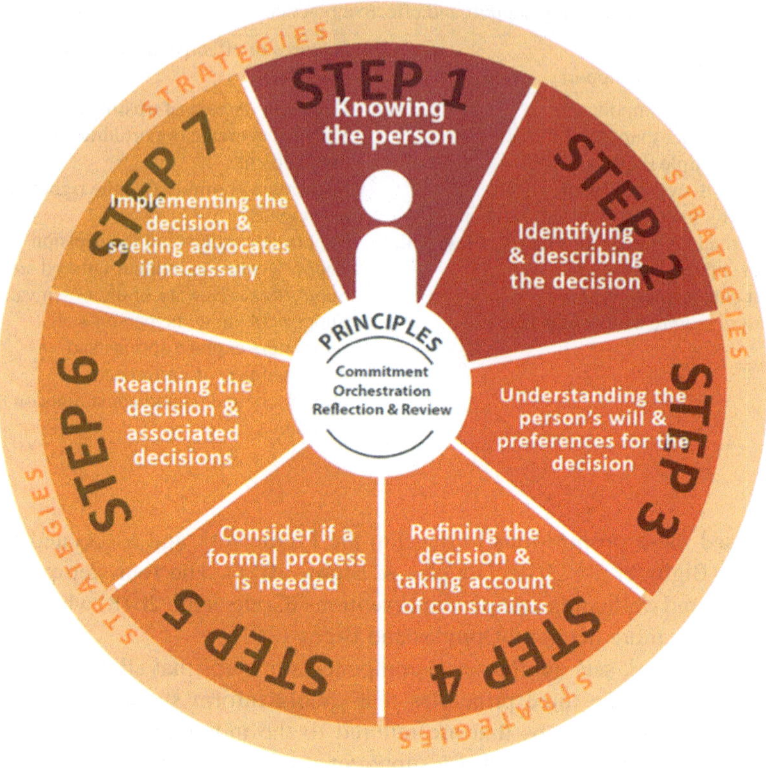

Fig. 4.2 La Trobe Support for Decision-making Practice Framework

Components or Steps

Decision-making support requires seven components regardless of its context or the nature of the decision. These components do not necessarily follow each other in a linear fashion, and supporters may need to revisit earlier steps as a decision unfolds. The amount of time each step requires and the challenges it poses for supporters depend on the multiple factors that influence the support process described earlier.

Step 1. Knowing the Person

Knowing the person is fundamental to support that is person-centred and individualised. Knowledge about the person is needed for all other steps and informing strategies such as adjusting communication or scaffolding information. Piecing together knowledge can be a lengthy process for new supporters and is a continuous task for long-term supporters as the person they support develops and broadens their experiences. Every person has many facets, their personality, strengths, weaknesses, skills, likes, dislikes, friends, family, acquaintances, long-term ambitions and short-term preferences. Members of a person's network know different things about them and see them from different perspectives. For example, parents or siblings are likely to know a person's family history, their experiences and major events in their life, whilst family doctors should know about their health and service providers about the services they use and the professionals who may also hold expert knowledge about things such as communication or social networks.

Some decisions may require seeking out particular types of knowledge about a person or their context: biographical, social, medical or systems. Seeking knowledge from other others, who might include long-term staff, friends or family, as well as drawing on one's own observations and experience help to build the type of rich and detailed picture about a person necessary for a good decision support. Many of the parents in our fieldwork talked about knowing their son or daughter better than anyone else, but nevertheless actively sought out information from others to complement their own for particular perspectives. For example, to help in getting to know activities that his daughter enjoyed, as she could not easily communicate this to him, one father talked about deliberately engaging several of the other women who attended the day programme and remarked how their information was preferable to that of staff who tended to put a positive spin on it. He said:

> *whenever I get a chance, I ask one of Heather's new friends what kind of day they think Heather had...Bree said Heather loved the bus, she hated the beach. It's very clear, no filtering or particular way of framing it. She just said what she saw. It's quite helpful in that sense.*

This step is often the one most challenging for paid supporters who come into someone's life for a short period or who are allocated little time to get to know the person. It is particularly difficult for those who support people with more severe and profound intellectual disabilities who have little verbal communication and simply cannot tell supporters about themselves. New supporters need to know where to find and take time to read existing information compiled about the person such as person-centred plans or support plans, seek out others who know the person and make time to be with the person as they move through different parts of their daily life. Generic skills such as active listening, observation, giving time to respond, adapting communication as well as adopting a curious not knowing stance are all useful for this step.

Step 2. Identifying and Describing the Decision

This step focuses attention on the core issues of the decision. As already highlighted each decision is unique and influenced by a wide range of factors. Being clear what these are helps supporters to be aware of their own values or stake in the decision. Describing the decision helps supporters have a sense of the parameters likely to impact on their practice, such as the time frame for making the decision, who else needs to be involved, the formal systems or services that might have to be engaged and the scope in terms of potential consequences for the person's life or others and other associated decisions that might be generated. This step, too, helps supporters understand resource or other types of constraints that may shape the options or be taken for granted rather than questioned by some of those involved.

Identifying and describing the decision establish a starting point for decision support and help to disentangle decisions from each other. For example, one of the paid supporters in our fieldwork talked about spending several meetings with a middle-aged man with intellectual disability who had sought help with decisions about how to use the funding allocated in different parts of his NDIS plan which had been prepared without his involvement. She said:

> *...the first thing we did was went through the whole plan, but really briefly, so talked about what this budget was for and this budget was for and this budget was for, and I did ask him quite an open question – what area in his life did he feel that he most needed help with, and he came back with a very definite answer of dealing with the separation of his wife. And so from there, I was like "okay" what budget could help with that in his NDIS plan and then focused on that.*

By understanding the service systems he was involved with, she was then able to describe the first decision in terms of whether he wanted to use his NDIS budget to continue to see the psychologist he had seen previously but who was no longer funded by the health system.

This step also helps to uncover hidden assumptions or decisions already made in which the person may not have participated, which may need to be revisited. In our fieldwork this was illustrated commonly in decisions about finding accommodation, where a parent had already made a decision that the person needed to move out of the family home. For example, as Margo said about her son, 'so, everything we do is to work towards independence. Yeah, independence…well, I did know I didn't want him living with me when I was 80'.

Finally, this step is helpful for supporters in identifying decision-making opportunities which may be overlooked. This can be important given that many of the parents in our fieldwork talked about the increasing confidence of their adult son or daughter as they gained more decision-making experience. One mother, for example, talked about how she looked for opportunities for her daughter to practice decision making, saying,

> *...if we're talking about "What do you want to have for your lunch today? Do you want a sandwich or a roll – because they're both in the freezer?" – I think making a conscious plan, on my part, to give her those options, and not just for me to pull the bread out of the freezer and make a sandwich.*

Step 3. Understanding the Person's Will and Preferences for the Decision

This is the 'blue sky' step when supporters assist in exploring all possible options, help the person to express their preferences and to understand the potential consequences. This step may be as straightforward as identifying and discussing the pros and cons of options to enable the person to express their preferences. For example,

Joy, a support coordinator, talked about how she had supported Kerrie to think about her preferences for different activity options by first listing her previous experiences and talking about those which she had enjoyed most. She said,

> *...a list of different activities that she's done over the years, and also what are the things that she wants to do. So we kind of made a priority list. And then from that list, okay, so you've only got certain hours in each day, so you can't do everything on your list. What's your first priority? What's your second priority? If you had a choice of anything you could do, what would it be? And that's kind of how we made our list, our initial list...Because it's got heaps and heaps of activities on it then we kind of went through and discussed each of them for their pros and cons. And how they would apply in Kerrie's life. And how she felt whether they were important or not.*

Similarly, a father talked about how his son tended to live in the moment and be happy with whoever was supporting him, so he had supported him to work out his preferences for which of the support workers should be employed regularly, saying,

> *If you ask him which is better – let's say for instance he's worked with several people...and you say the names and he'll say "Oh they're great", and "That's great", that's great". So there's no grading. So you have to say "Okay if A is great, it's a six out of ten, what's B?", and then he'll tell you. If you put a numerical figure he can easily do it. And every once in a while he stuffs it and says "Oh that's a nine. Hang on, that's a nine and a half". But it's clear enough...If you give him the tools. If you don't give him the tools, he won't because it's immediacy. Whatever is in today is good.*

For many people with intellectual disabilities, just talking through options is insufficient to enable them to understand what they might be or to express their preferences. Supporters need creativity to help a person who has had limited experience to explore options, understand the implications of each one and decide which they prefer. One strategy is giving the person an opportunity to directly experience options and use observation to learn about their preferences. For example, referring to decisions about leaving the family home, several parents had insisted their adult children experienced nights away from home in other places. As Kate said about her son using a respite service,

> *...he needs to experience what it's like to be out there with people who he knows and gets on with and seeing what that's like. Because without experiencing it he actually isn't going to be able to make a decision. So it's about getting him to understand what that means.*

Similarly, another parent talked about concrete strategies she had used to help her daughter understand what it might be like to go to a residential college. She said,

> *...taking her to college to look at the facility, meet people; we did that twice...They had a very nice prospectus so we went through that with her and tried to put it in simpler language where we could. We took a pros and cons approach. She's giving up her employment; I think that was a con. Leaving friends for a year, we talked about how we might be able to support that so she stayed in touch with them; she put together an address book and we practiced FaceTime. We said you could feel homesick; we talked about the example of her sister who is studying in the UK we said, "Jasmine's been through that, so if you were homesick you could maybe chat to her, maybe she'd be a good person." And we practiced FaceTiming with her sister.*

Similarly, it is important to help the person experience the potentially negative consequences of their preferences. One strategy is a risk-enabling approach which involves trying out the preference but minimising potential harm that may arise (Bigby et al., 2018a; Department of Health, 2010). A common example in our data involved decisions about travel and adults' preferences to use public transport independently. Supporters described various ways of helping the person experience using public transport alone and to minimise the distress or danger of getting lost. For example, one parent taught her daughter to phone her if she felt lost but before ringing to select a friendly looking female stranger whom she could ask to tell her mother exactly where she was. Another mother had taught her son to take pictures of station signs and send them to her using his phone, if he was unsure where he was.

Thinking about multiple options can be difficult for people who experience anxiety. One parent gave an example of her strategy of slowing down the pace and stretching consideration of options over a longer period. She said,

> *...I'll give him some options. If he goes, "Oh, I don't know", sometimes he gets a bit upset, so I'll just slow him down, because he's getting a bit anxious, so I'll just say, "Well, if you go here, this is what could happen, is that okay with you? Or, we go here?"... "Remember, it's not important, so whatever decision you make is fine." So I just reassure him. Just really calming him down. Most of the time you can walk away and he'll come back and talk.*

It is important at this stage to take time to listen carefully and avoid exerting undue influence. Many parents in our fieldwork talked about strategies such as deep listening and or 'sitting on the fence'. Joanne, for example, talked about taking a step back to really listen to what her son was saying about his activity preferences,

> *...taking on board what he's saying and trying to go deeper and deeper, and peel off the layers, and trying to discover what he's actually saying...Trying to ask him more questions, and questions that he'll understand better, in a different format, and really trying to get into the root of it basically...I'm more patient.*

Several parents talked about how, as they spent more time observing or listening to their son or daughter, they got to know them better and recognised that they did have preferences. For example, one mother said,

> *The more you sit back and let him have a go at things, the more you realise he does think. And I have to learn to be patient enough to respect that thinking and let it come out.*

The significance of this step is very apparent when it is not done well and supporters, for example, exercise their power to undermine participation by selectively deciding which options to consider and which to ignore, present options in biased ways, persuade the person towards particular options or simply fail to respect preference to favour their own. One of the many examples of such strategies in our data was the comment of one mother who said,

> *I have to basically do all the work [of exploring options] and then go, "Okay these are your choices, bang, bang, bang," and exclude all the ones that are just not suitable...I'm still controlling his choices in many ways because I am the one who is giving him the options. It's a bit of a sleight of hand in terms of if he was able to go online and Google stuff himself it would be a different conversation. But he can't do that.*

There are many reasons why supporters act in this way, some of which were considered in the earlier section on processes. Commonly supporters curtail exploration of options too soon because they are too focused on risks and practicalities and concerned about making the 'right decision' for a person's well-being. It is clear however that considering constraints on a decision (step 4) before step 3 poses the danger of unnecessarily restricting or disregarding options and ignoring preferences as too risky or impractical. Allowing sufficient time at step 3 provides a sound basis for prioritising preferences and understanding constraints at step 4.

Step 4. Refining the Decision and Taking Account of Constraints

A decision must be implementable, so it can be acted upon either by the person themselves, supporters or advocates. This step prioritises preferences and identifies constraints on the decision that may mean preferences have to be refined. The primary challenge for practice at this step is to avoid compromising preferences whilst also taking account of constraints. Common constraints that influence the realisation of preferences are the availability of resources such as money, time or support; impact of the decision on others; attitudes of others involved in implementation; organisational policies and procedures and risk to safety or longer-term well-being.

Finances constrain many different types of decisions. Refining preferences to take account of finances may mean helping the person select a purchase within their budget or postpone a purchase until they have sufficient funds. For example, one parent talked about how she helped her son to refine his preferences about a doll he wished to purchase, saying,

> *So I might guide him price wise, and say, "Jaxon, you've only $10 to spend, so these are the $10 Barbies that you can pick from." So that narrows the field down. And then I might offer him little advice like, the main thing that he likes to do with them is brush their hair, so I might steer him towards the ones with long hair, because often they get little haircuts at home too, and he likes to do them up in weird sort of beehives and all sorts of really intricate hair-do's. So then I might steer him in that direction too, like, "These ones have got long hair," so that I've narrowed the field down again, and then probably just let him mull it over.*

Time is another important resource that constrains decisions. This was illustrated by one parent who talked about how she helped her son to work through the time constraints on his preference to go and see a friend. She said,

> *I'd say, "I understand that, and I know that's important for you, but today we can't do that. Can we look at Saturday or something?" So again, options, looking at, "Well, if we go today, we could only probably stay an hour, but if we go on Saturday, it might be we could stay all morning."...Asking him what does he want to get out of it. What does he want, what is he expecting. And trying to, again, looking at, well, it depends on what this other person's availability is too. So we've got to negotiate and go back and forth. It's not just black and white, cut and dry, "This is what I want and that's it."*

The potential for harm to immediate safety or longer-term well-being are also frequent constraints. Similar strategies of enabling risk can be used in this step as the previous one, such as equipping a person to be confident to seek help if they get lost on public transport. The aim at this step, however, rather than helping someone to understand or experience consequences of risky options to help determine their preferences, is finding ways of carrying through preferences into decisions by minimising any potential associated harm. This was illustrated by a father who talked about how he identified and considered all the risks that might constrain his son from going to a comedy performance and intended to minimise any harm by unobtrusively tagging alone. He said,

> *...he's going to go to three sessions at the comedy theatre and I think that will be a bit tricky. But you know, at the end of the day, there will be some mature adults there, and they tend not to be younger adults in there. So, yes, he'll be a bit odd, he'll be a bit strange – I went to one with him last year, and he was a bit ridiculed, but I don't think he noticed. Nothing you can do about it, because he's really not mature enough...I may tag along to the point where I'll sit downstairs and have a drink and wait for him...The main concern would be he would laugh inappropriately and therefore be–well, worst case, booted out, which is fine. But he could get into a violent situation. But having assessed the people going there, they are quite mature and quite adult. They're not kids. They're not drunk. So, the risk of him being injured or getting in a fight would be much, much lower. It's like him going into a concert hall, as opposed to a rock concert.*

Thinking about a person's *will* as well as preferences can be useful at this step if practical constraints impede implementing a decision solely based on preferences. A person's will can be understood as the long-term values that underpin the direction of their life, whereas preferences are their more immediate wishes or desires (Szmukler, 2019). Supporters can use their understanding of the person's will to think more deeply about resolving tensions and refine what they might judge to be unrealisable preferences. For example, the desire to learn to drive may be so constrained by a person's poor literacy skills that it is unlikely to be realised. A supporter might refine that preference to one of using taxi's independently by drawing on their understanding of the person's long-term values around being independent. An example in our fieldwork was of a young woman whose expressed preferences were to move to the USA and become a singer, which supporters regarded as unrealistic and interpreted in light of her long-term held value of being creative and sought out options accordingly. Such strategies for refining preferences illustrate the power of supporters at this step in making judgements about constraints and reinterpreting preferences. They highlight too the ease with which supporters can compromise rights by overriding preference and the importance of supporter's self-awareness of factors that influence their interpretations.

At this step it is clear that supporters may have to judge whether the constraints on a preferred decision are so significant or the risks to a person's safety or longer-term well-being so great as to warrant overriding preferences. In doing so they may rely on their understanding of a person's will or chose to explore alternatives that are least restrictive of their rights. We argue, however, such actions must be deeply considered and may require recourse to formal processes.

Step 5. Consider if a Formal Process Is Needed

This step brings together the knowledge gained in the earlier steps about the decision, preferences, consequences, priorities and constraints. It is greatly influenced by context, role and formal standing of supporters. Most decision support for people with intellectual disabilities happens informally, provided by supporters who have no formal legal standing in respect of decision making. The widespread provision of informal support is situated in a context founded on the principle of the least restrictive alternative, common to many jurisdictions. Very simply this means informal support is preferred, considered to be the least restrictive of rights and should prevail unless there is reason to seek out formal alternatives, such as guardianship and appointment of a substitute decision-maker (e.g. see the 1986 Victorian Guardianship and Administration Act). The judgements and actions of informal decision supporters are guided by their understanding of disability policies, their own principles or the policies of their employer and reliant on their skills and support they derive from supervisors or peers. However, there may be circumstances when supporters consider more formal guidance and authority is required in following through or affirming their judgements.

Recourse to a formal process may be indicated by conflict among supporters or supporters' intentions to override preferences. Conflict arises between supporters about things such as whether identified or expressed preferences of a person are realistic, reflect their actual desires or are the result of undue influence or coercion. In our earlier studies, we found examples of conflict between family members and accommodation support staff about things such as holiday destinations, meals and dress codes which had gone unresolved in the absence of processes for mediation and led to disregard for expressed preferences (Bigby et al., 2018a, b).

Supporters make judgements about options, interpretations of preferences and determine constraints on decisions based on practicalities or risks to safety and well-being. In doing so they will sometimes infringe on the rights of the person to make decisions or shift their preferences from the centre of a decision. If they feel this is necessary or there is conflict among supporters, it is important to consider whether it is necessary to invoke a more formal process of decision support in order to safeguard the rights of the person being supported. Whether, for example, it may be necessary to seek the appointment of a guardian to act as a substitute decision-maker and bring to the process of decision support the procedures and safeguards associated with such appointments. The legislation relevant to the jurisdiction and policies of employers will determine the options available and how supporters may take steps to instigate a formal process. If a substitute decision-maker is appointed, the knowledge of informal supporters will be important in helping to inform their decision-making process.

Step 6. Reaching the Decision and Associated Decisions

This step is about reaching a decision that reflects prioritised preferences as closely as possible. For some people and contexts, it may also mean formally recording the decision and communicating it to others, such as a support coordinator or advocate who will take a leading role in getting it implemented. At this step too, it is important to return to the consequences of the decision being made, its likely implications for other parts of the person's life and the associated decisions that the person may also need support to reach.

We have likened decisions to Russian Babushka dolls that sit within each other, because as one decision is made another may appear as a result. One of the parents in our fieldwork captured this well when he talked about what he said to his daughter after she had showed him the recipe she wanted to make. He said, 'well, who do you want to do it with or where do you want to do it?' In thinking about associated decisions, the recursive nature of decision support practice becomes clear as the supporter loops back to step 2, identifying and describing other decisions.

Step 7. Implementing the Decision and Seeking Advocates if Necessary

Implementation of decisions does not necessarily rest with decision supporters. This step helps to illustrate differences between the role of advocates, whose work entails working with a person to get a decision that has already been made to be implemented and decision supporters whose role is in supporting the person to participate in making the decision.

If decision support has been done well, then issues of feasibility will have been addressed at step 4 when constraints were considered. Some empirical research suggests that decision making often falters at this step when, for example, paid staff accord low priority to implementing a decision or are opposed to it (Bigby et al., 2019; Burgen, 2016). In such cases an advocate may be needed to ensure a decision is taken seriously and oversee its implementation.

Whilst it may be feasible for a decision supporter to also implement a decision, it can also be time-consuming and demanding. This is quickly becoming apparent in individualised funding schemes where implementation rests with family members and many have likened it to a full-time job. For example, Henderson and Bigby (2021) explored the situation of a mother of a young man living in a group home who talked about how tiring it had been to get the service provider to implement decisions in her son's NDIS plan, and the only way she had managed this was to reduce her hours of work.

Principles of Decision Support Practice

Three principles underpin all aspects of decision support practice: (1) commitment to the person and their rights, (2) orchestration of others involved in the decision and the person's life and (3) reflection and review on one's own values, influence and decision support. Some of these have already been illustrated as we have described the seven steps.

Commitment to the Person and their Rights

Commitment requires the supporter to have a relationship with the person they support based on equity and respect for their right to 'make, communicate and participate in decisions that affect their lives and for their will, preferences and rights to direct decisions that affect their lives' (Australian Law Reform Commission, 2014). Explicit commitment by supporters to rights helps to avoid inadvertently using best interests as the touchstone for support. This may be particularly challenging for parents of adults who have had to gradually adjust their support from one of protection to rights as their son or daughter reaches adulthood. Reflecting on how difficult it was, one parent said,

> *I'm aware that it is difficult to transition from being the parent of a child to being the parent of an adult, because she's not capable of being fully adult-like, but legally, she is of age and people expect her to make her own decisions and be responsible for herself. So I think that sometimes I'm probably still very much acting in the mode of a parent of a child and possibly I'm quite controlling. It's very hard for me to back off there.*

Many parents in our fieldwork reflected on their conscious efforts to be committed to their adult child's rights to make their own decisions especially when they doubted the veracity of a decision or thought it to be foolhardy. As one mother said, 'I really want to be confident that she's making what I would regard as good choices'. Another said,

> *It has to be his choice. I'm learning that more and more. It has to be his choice. You've got to give the dignity of his choice. Even some of the things he does – he makes mistakes. If he blows it, whatever he's doing, well that's how we learn.*

Orchestration

Orchestration relates to the shared nature of decision support practice, bringing in others involved in the person's life or who have expertise about the decision at hand. It is premised on there being a primary supporter leading the process who draws in and coordinates other paid or unpaid supporters and mediates any differences between them. Parents in our fieldwork commented on the value of orchestration, in providing, for example, an additional person who was less emotionally involved than themselves or who brought a different perspective or expert knowledge to the decision. One parent talked about the value of having others involved with her daughter who could talk through options with her, she said,

> *So, on occasion, they've been able to raise things with Danielle that I would not be able to. So, I've become aware that that's a very useful technique because she's more likely to listen to other people than me on some matters...And I've sometimes said you might talk to your sister in law about those things. She can help you as well, go through the options and then you can make your decision independently.*

Applying this principle can also help to widen decision-making opportunities by encouraging others involved with the person to support their decision making. One parent, for example, talked about her conversation with a support worker along these lines when she had said,

> *We expect you to have a conversation with Brendon, don't make all his choices for him. Encourage him to speak to you, express himself, like speak in longer sentences, express ideas and things that he wouldn't necessarily talk to us about perhaps.*

Reflection and Review

This principle emphasises self-awareness and continuous reflection on practice as core to good decision support practice. Being self-reflective through all steps helps supporters to be conscious of their own influence, values or stake in the decision and better maintain a neutral approach that puts these aside. As we have already flagged, many parents talked about the difficulties of standing aside and respecting preferences. The process of reflection was useful in helping them realise how often they overstepped this mark and how influential they could be. One mother said, for example,

> *I reflected on the fact that virtually everything that Sally does has been decided by me. The reason why that sort of came to light is each Sunday night I do a whiteboard, which is on our fridge, and on the whiteboard I have seven days, and then on each of the days we write in what Sally's doing on that day...The fact that she's in work is because I have a goal for Sally that work should be part of her life. I also think, as far as possible, I'd like to think that she can be an independent traveller on public transport...When you think of just those couple of days and those key things, they were goals that I have set. I didn't really engage Sally in the decision-making process other than to say, you know, "Wouldn't it be a good idea if you went and got a job?"*

By reflecting on her relationship with her son, another mother had realised how easy it was for her to influence his preferences. She said,

> *Yes, because I'm his mother. And out of every person on the planet, he loves me the most. And therefore, he's acutely vulnerable to suggestions that I make and my view of him, and my attitude towards everything that he might do or might be. Therefore, it's very difficult to separate my intention from his.*

Being self-reflective enables supporters to articulate the steps and rationale behind the judgements and interpretations they have made. This helps them to be more accountable for their practice. This was illustrated by a parent talking about her reflection and review of the support she had provided about service options.

I probably spoke about the good things with the new service. But I probably only spoke about the bad things with the old one, which that alerts me...I could have changed that...I should have probably talked to him about the good times he's had at the old service, the times with Jaxon and Greg, the times that he's enjoyed the service, the beautiful environment it is, the way he's been there now for three years and it has been good...I could have reinforced all those things.

The Framework provides the basis for a simple checklist to help supporters review and then describe their actions around each step and principle for an instance of decision support and unpack the types of strategies they have used. The free online training resources developed as part of the Framework include this checklist and other tools that can be downloaded. These can be accessed at http://www.supportfordecisionmakingresource.com.au. Supporters can use this checklist not only for self-review but also to demonstrate accountability of their practice. Alluding to the importance of review of their practice as a safeguard for the rights of the person they supported, several supporters in our fieldwork drew attention to the few incentives to this in the current Australian context. Talking about her interactions with planners from the NDIS one parent said, for example,

...they don't really ask us whether the person that we are making the decision for has been consulted. There isn't any form that you get saying, "Did you speak to the participant about this? Do they agree?" It's nowhere. So they're perpetuating the old system, which is that parents act for children, and that doesn't matter how old the children are.

Mechanisms to assist supporters to reflect and review their practice will improve the quality of decision support as well as help in safeguarding rights. These are largely absent when decision support rests in the informal sphere but could include communities of practice, peer support groups or regular training or mentoring to supporters.

Strategies

Supporters must draw on a very broad range of strategies as they move through the Framework's steps and apply the principles to their practice. The examples and suggestions embedded throughout earlier sections illustrate the individually tailored decision and context-specific nature of strategies. Some of the different types of strategies are summarised in Table 4.2, and others are detailed in the training resources (Bigby et al., 2019).

Many of these strategies will already be familiar, as they reflect some of the core skills needed for effectively supporting people with intellectual disabilities to participate in planning their goals, engaging in meaningful activities and social relationships or participating in their communities. Some of these skills are illustrated in free online training programmes developed from research at the Living with Disability Research Centre. See, for example, resources about active support (Every Moment Has Potential, 2015), supporting inclusion (Bigby & Wiesel, 2015), enabling risk (Bigby et al., 2018a, b), supporting decision making (Bigby et al.,

Table 4.2 Broad strategies for decision support practice

Strategies
Attention to communication
Listening and engaging
Breaking things down
Networking
Researching options Educating about options, consequences and constraints
Creating opportunities to experience options and preferences
Enabling risk

2019) and those about communication from organisations such as Scope's Communication Resource Centre.

Impact of Training and Mentoring in the Framework

Since the Framework was developed, we have evaluated its impact on supporters' practice in a range of different settings and with different types of supporters. Much of the analysis remains work in progress and is not yet published, and some relates more specifically to professional support coordinators who work with people with acquired brain injury (Douglas et al., 2020). However, there is clear evidence emerging from these studies that when supporters are trained in the Framework and mentored to apply it, their confidence increases and they use more strategies that reflect supported decision-making principles (Bigby et al., 2021). Also, as a consequence of their changed practice, supporters describe more participation in the decision making of the people they support.

Support Strategies More Reflective of Supported Decision-Making Principles

In order to evaluate the impact of the Framework on the practice of supporters, Douglas and Bigby developed the Decision Support Questionnaire (DSQ) in 2015 (see Bigby et al., 2015c). The DSQ measures changes in the strategies that supporters used following a 1-day training workshop and then after a series of 1-hour mentoring sessions (more details of this and other measures used to evaluate training in the Framework are available from the authors and will be published in forthcoming papers). Our early analysis of data from a group of 18 supporter parents of adults with intellectual disabilities showed the increased use of strategies including checking the person who wants support to make the decision, working through options with the person, exploring new experiences and considering their own potential for influence. Results showed a reduced use of strategies that contradicted supported decision-making principles, such as reliance on thinking about the best interests of the person rather than their will and preferences. Some of these data are shown in Table 4.3.

The qualitative data illustrated the types of changes found in the quantitative data as well as the broader impact of the training on parents. The training had acted as a catalyst for parents to reflect on their approach to support. As one said, 'it made me stop and think'. Parents thought they had become more reflective and deliberative about providing decision support. As one mother said, 'I'm consciously doing it. Probably when we talked about it initially I just wasn't thinking as consciously as I perhaps am now'. The Framework gave parents a structure for thinking about support which they had not had previously, and many used the Framework diagram (see Fig. 4.2) as an aide-memoire. Being more reflective raised awareness of their influence, which some resolved to address by strengthening a commitment to the rights of their son or daughter. One father, for example, talked about shifting his approach from 'no risk to some risk', recognising that he didn't need to intervene as much as he had been doing. He explained a more reflective and deliberate approach to supporting his son, saying,

> *I think I'm more aware, personally, of making sure I ask him what he wants. I'm also more aware of if I'm guiding him or I'm just giving him some options and let him think about it.*

Slowing down to give the person more time to think and express their views was a new strategy that some supporters used after the training which they felt helped them to see better the person's perspective. One mother said, for example,

Table 4.3 Change to supporter's strategies for decision support following training in the La Trobe Support for Decision-Making Practice Framework

DSQ item	Pre-training mean	Post-training mean	Hypothesised change direction	Detected change direction	*p*
Focus on easy options	2.43	2.0	Decrease	Decrease	0.02*
Weigh up the advantages and disadvantages of the options with the person	3.09	3.40	Increase	Increase	0.05*
Consult other people who know the person in different situations	2.39	2.7	Increase	Increase	0.06^
Seek advice from a professional/expert	2.09	2.50	Increase	Increase	0.05*
Point out a range of options for the person	3.17	3.50	Increase	Increase	0.05*
Rely on what you think is best for the person	3.48	3.2	Decrease	Decrease	0.06^
Shift attention away from the decision to something else that needs to be achieved	1.77	1.50	Decrease	Decrease	0.07^

^ $p < 0.075$; * $p \leq 0.05$

I can see it more from his perspective now and I'm giving him more time and more options to sort of choose for himself. I want him to make choices for himself and I feel that I think I have more insight into it.

Changed Skills of People with Intellectual Disabilities and Confidence of Supporters

The qualitative data suggested that parents thought some of the people supported had also gained skills and confidence, as a result of their own changed support practice. Some parents reported, for example, that their son or daughter was making more decisions, being firmer in expressing their preferences and using better strategies to decide between options. One father said about his son,

He does now actually, sometimes bring up and say, "That would be nice to go and watch but then I've got this, this and this on, so I won't do that." So he does consider. He may not know all the constraints and in a way sometimes you help him. But when he knows the constraints he's quite good at it.

Another talked about the changes to her son's ability to express himself, since she had begun to slow down and pay more attention to listening. She said,

He's becoming more creative in how he communicates things and how he likes to express himself. And so he's continuing to progress in terms of his capacity to speak up, speak for himself, self-advocate.

Parents also drew attention to the positive effect of such changes to well-being, highlighting the benefits for the quality of life of exercising rights to decision making. One parent said, for example,

More and more lately, he gets to make decisions for himself. They're not massive, sometimes it's what are you going to wear today or what are you going to eat for breakfast. I think he feels good when he has the opportunity to choose for himself what he wants to do.

Not only did supporters comment on increased skills of those who they supported, they also reported significant increase in their confidence to provide support for decision making following participation in training. As a group, supporters' confidence in their ability to provide support for decision making increased significantly ($p = 0.03$) over the period of time from pretraining to post-training.

Conclusions and Directions for Future Research

The naysayers about supported decision making draw attention to the lack of evidence about what constitutes good practice and potential for supporters to abuse their power. In our empirical data, we saw many instances of decision support where there were conflicting interests, and that potentially looked like the exertion of undue influence or even coercion. These data help to identify the difficulties of judging the quality of support and supporter accountability. They demonstrate not only problems of taking supporters actions at face value but how these are compounded by the fact that every instance of decision support is person-centred and deeply embedded in the context and the decision.

As we have illustrated in the foregoing sections, support for decision making involves significant supporter interpretations of options, preferences and constraints of which close scrutiny opens up alternative interpretation. For example, there were instances of clear disregard for preferences by supporters who insisted the person explore alternative options first or foisted new experiences on them. Digging deeper, the beneficence of such actions might be explained in terms of expanding a person's horizons or helping them understand consequences of their preferences. When might this be justified and at what cost to the individual? McKearney (2020b) described, for example, similar actions by supporters in his ethnographic case study, demonstrating the anguish one man experienced when his immediate preferences were consistently ignored in favour of expanding his world. There were instances too when clearly expressed preferences were reinterpreted by supporters by reference to long-term values or goals (will) based on the rationale of the unrealistic nature of preferences. How accurate were such judgements and to what extent should they be tested? We found many examples where supporters rationalised favouring long-term goals over short-term preferences by invoking the person's inability to understand consequences of their preferences. This raises questions about the time and resources supporters should spend in finding ways to help understanding? It also brings to the surface issues about how supporters reach an understanding about the long-term goals they use as proxies for a person's will. Both in our data and other studies, there is evidence to suggest that some supporters settle on the long-term goals of the people they support in line with their own visions for their future (Curryer et al., 2020; Taylor et al., 2019). As McKearney (2020b) suggests, going beneath the surface of support reveals the 'inherently vulnerable, political, and interactive nature of claims to know what good care is', or in our case, what good support practice is. Such questions warrant further exploration before the hasty development of guidance to judge support practice.

One alternative approach might be to explore further the perspective of the people with intellectual disabilities who receive decision support, about their satisfaction with outcomes or process. Although our research included interviews with people with intellectual disabilities about their support, this approach is also fraught with the difficulties associated with taking things at face value. For example, there were many examples in our fieldwork where supporters talked about persuading the person they supported to their own point of view. In one instance a supporter described herself as a 'good salesperson', and in another a father described how he had won his son around to his own preferred option and been able to convince him this preference was his own rather than that of his fathers. There were also examples of satisfied decision-makers when it was apparent they had been given only scant information about their options or consequences. In such instance how should expressed satisfaction with the decision be judged or do the processes of support need to be excavated further before taking satisfaction at face value? Relying on the perspectives of the people supported to judge the quality of support also raises many of the well-rehearsed issues about inclusive research, particularly the exclusion of people with more severe and profound intellectual disabilities and the resources necessary to gain the trust of more able people and support their communication about complex issues (Bigby et al., 2014). Certainly, a more intensive approach is necessary than the intermittent interviews we conducted with the people with intellectual disabilities, in order to gain sufficient understanding of the decision, its context and their perceptions of support.

In our research we compared supporters' practice against an evidence-based Framework that reflects the components of effective practice aligned with the intent of supported decision making and the CRPD. This served as a proxy for judging the quality of practice and whether, after training, it shifted to align more strongly with supported decision making principles. We relied on the perspectives of parents or other supporters and their reflections on their practice – the process they had used, rationale for actions, how they described their practice, the changes that occurred following training and the knowledge and skills they had gained. We asked questions such as had they gone through the steps in the Framework, applied the principles and what strategies had they used, how did these reflect knowledge about the person, could they explain the rationale behind their actions and how had they dealt with competing imperatives of respecting preferences and taken account of constraints and risks to well-being. Essentially we focused on supporters' reflective capacity and ability to articulate the process they had used to provide support.

Carney et al. (2021) illustrates further the difficulties of the judging the quality of decision support and the absences of a bright line that distinguishes empowering from paternalistic support. Understanding further the nuances of decision support, and the hallmarks of good support which can be used to hold supporters to account, poses a significant new challenge for researchers. As does understanding support from the perspective of people with intellectual disabilities. We argue, however, that increasing supporters' understanding about support for decision-making practice, formally raising expectations about the supporters' practice in policy, disability support organisations and schemes such as the NDIS and embedding evidence-based training for supporters that promotes greater *participation* by people they support and enables supporters to be more *reflective* and *deliberative* is an important first step.

Acknowledgements These unpublished data are drawn from a study funded by the Australian Research Council and 12 Industry Partners. We acknowledge the other members of the research team involved in this study, Terry Carney, Shih-Ning Then, Ilan Wiesel, and Elizabeth Smith and also Lincoln Humphries and Lisa Hamilton who delivered training and mentoring. We would also like to thank the supporters and the people with intellectual disabilities they support who participated in the study.

References

Antaki, C., Finlay, W., & Walton, C. (2009). Choices for people with intellectual disabilities: Official discourse and everyday practice. *Journal of Policy and Practice in Intellectual Disabilities, 6*, 260–266. https://doi.org/10.1111/j.1741-1130.2009.00230.x

Arstein-Kerslake, A. (2017). *Restoring voice to people with cognitive disabilities: Realizing the right to equal recognition before the law*. Cambridge University Press.

Australian Law Reform Commission. (2014). *Equality, capacity and disability in commonwealth laws: Final report*. Australian Law Reform Commission.

Bigby, C. (2020). Dedifferentiation and people with intellectual disabilities in the Australian National Disability Insurance Scheme: Bringing research, politics and policy together. *Journal of Intellectual and Developmental Disabilities*. https://doi.org/10.3109/13668250.2020.1776852

Bigby, C., & Douglas, J. (2020). Supported decision making. In R. Stancliffe, P. Wehmeyer, & K. Shrogran (Eds.), *Choice, preference, and disability: Promoting self-determination across the lifespan* (pp. 45–66). Springer.

Bigby, C., & Wiesel, I. (2015). *Supporting inclusion – On-line training resources*. Translation of findings from ARC Linkage grant. http://supportinginclusion.weebly.com/

Bigby, C., Bowers, B., & Webber, R. (2011). Planning and decision making about the future care of older group home residents and transition to residential aged care. *Journal of Intellectual Disability Research, 55*, 777–789. https://doi.org/10.1111/j.1365-2788.2010.01297.x

Bigby, C., Frawley, P., & Ramcharan, P. (2014). Conceptualising inclusive research with people with intellectual disability. *Journal of Applied Research in Intellectual Disability, 27*(1), 3–12. https://doi.org/10.1111/jar.12083. http://hdl.handle.net/1959.9/513608

Bigby, C., Webber, R., & Bower, B. (2015a). Sibling roles in the lives of older group home residents with intellectual disability: Working with staff to safeguard wellbeing. *Australian Social Work, 68*, 453–468. https://doi.org/10.1080/0312407X.2014.950678

Bigby, C., Whiteside, M., & Douglas, M. (2015b). *Supporting people with cognitive disabilities in decision making – Processes and dilemmas*. Living with Disability Research Centre. http://hdl.handle.net/1959.9/319333

Bigby, C., Douglas, J., Carney, T., Wiesel, I., & Then, S. (2015c). *Effective decision making support for people with cognitive disability*. Australian Research Council, Linkage Projects, Proposal for funding.

Bigby, C., Douglas, J., Carney, T., Then, S., Wiesel, I., & Smith, E. (2017). Delivering decision making support to people with cognitive disability– What has been learned from pilot programs in Australia from 2010-2015. *Australian Journal of Social Issues, 52*(3), 222–240. http://onlinelibrary.wiley.com/doi/10.1002/ajs4.19/epdf

Bigby, C., Douglas, J., & Hamilton, L. (2018a). *Overview of literature about enabling risk for people with cognitive disabilities in context of disability support services*. La Trobe University. Retrieved from: http://hdl.handle.net/1959.9/563495

Bigby, C., Douglas, J., & Vassallo, S. (2018b). *Enabling risk: Putting positives first. An online learning resource for disability support workers*. Retrieved from: www.enablingriskresource.com.au

Bigby, C., Whiteside, M., & Douglas, J. (2019). Providing support for decision making to adults with intellectual disability: Perspectives of family members and workers in disability support services. *Journal of Intellectual and Developmental Disability*. Advance online publication. https://doi.org/10.3109/13668250.2017.1378873

Bigby, C., Douglas, J., Carney, T., Then, S., Wiesel, I., & Smith, E. (2021). Parents' reflections on using the La Trobe support for decision making practice framework. *Disability and Rehabilitation*.

Browing, M., Bigby, C., & Douglas, J. (2020). A process of decision making support: Exploring supported decision making practice in Canada. *Journal of Intellectual and Developmental Disabilities*. https://doi.org/10.3109/13668250.2020.1789269

Browning, M. J. (2018). *Developing an understanding of supported decision making practice in Canada: The experiences of people with intellectual disabilities and their supporters*. Doctoral thesis, La Trobe University, Melbourne, Australia.

Burgen, B. (2016). Reflections on the Victorian Office of the Public Advocate supported decision making pilot project. *Research and Practice in Intellectual and Developmental Disabilities, 3*(2), 165–181.

Carey, G., Malbon, E., Olney, S., & Reeders, D. (2018). The personalisation agenda: The case of the Australian National Disability Insurance Scheme. *International Review of Sociology, 28*(1), 20–34. https://doi.org/10.1080/03906701.2018.1425084

Carney, T. (1982). Civil and social guardianship for intellectually handicapped people. *Monash University Law Review, 8*(4), 199–232.

Carney, T. (2017). Supported decision making in Australia: Meeting the challenge of moving from capacity to capacity-building? *Law in Context, 35*(2), 44–63.

Carney, T., & Beaupert, F. (2013). Public and private bricolage-challenges balancing law, services & civil society in advancing CRPD supported decision making. *The University of New South Wales Law Journal, 36*, 175–201.

Carney, T., Bigby, C., Then, S., Wiesel, I., Douglas, J., & Smith, E. (2021). Paternalism to empowerment: All in the eye of the beholder? Disability and Society https://doi.org/10.1080/09687599.2021.1941781

Committee on the Rights of Persons with Disabilities. (2014). *General comment no. 1, article 12: Equal recognition before the law*. Author.

Craig, P., Dieppe, P., Macintyre, S., Michie, S., Nazareth, I., & Petticrew, M. (2018). Developing and evaluating complex interventions: The new medical research council guidance. *British Medical Journal 2008, 337*, a1655.

Cukalevski, E. (2019). Supporting choice and control—An analysis of the approach taken to legal capacity in Australia's National Disability Insurance Scheme. *Laws, 8*(2), 8. https://doi.org/10.3390/laws8020008

Curryer, B., Stancliffe, R., Wiese, M., & Dew, A. (2020). The experience of mothers supporting self-determination of adult sons and daughters with intellectual disability. *Journal of Applied Research in Intellectual Disability, 33*, 373–385. https://doi.org/10.1111/jar.12680

Davies, D. K., Stock, S. E., & Wehmeyer, M. L. (2003). A palmtop computer-based intelligent aid for individuals with intellectual disabilities to increase independent decision making. *Research and Practice for Persons with Severe Disabilities, 28*(4), 182–193.

Department of Health, United Kingdom. (2010). *Nothing ventured, nothing gained: Risk guidance for people with dementia*. Department of Health.

Devi, N., Prodinger, B., Pennycott, A., Sooben, R., & Bickenbach, J. (2020). Investigating supported decision making for persons with mild to moderate intellectual disability using institutional ethnography. *Journal of Policy and Practice in Intellectual Disabilities, 17*(2), 143–156.

Douglas, J., & Bigby, C. (2018). Development of an evidence-based practice framework to guide support for decision making. *Disability and Rehabilitation*. https://doi.org/10.1080/09638288.2018.1498546

Douglas, J., Bigby, C., Knox, L., & Browning, M. (2015). Factors that underpin the delivery of effective decision making support for people with cognitive disability. *Research and Practice in Intellectual and Developmental Disabilities, 2*, 37–44. https://doi.org/10.1080/23297018.2015.1036769

Douglas, J., Bigby, C., & Smith, E. (2020). *Building capability to support client decision making* (TAC project no: T005). Living with Disability Research Centre, La Trobe University. https://doi.org/10.26181/5fa20ac4bba07

Dowse, L. (2009). 'Some people are never going to be able to do that'. Challenges for people with intellectual disability in the 21st century. *Disability & Society, 24*(5), 571–584. https://doi.org/10.1080/09687590903010933

Dunn, M., Clare, I., & Holland, A. (2010). Living 'a life like ours': Support workers' accounts of substitute decision making in residential care homes for adults with intellectual disabilities. *Journal of Intellectual Disability Research, 54*, 144–160. https://doi.org/10.1111/j.1365-2788.2009.01228.x

Every Moment has Potential. (2015). *On-line training resource on active support developed in collaboration with Greystanes disability services*. http://www.activesupportresource.net.au/

Fisher, Z., Bailey, R., & Willner, P. (2012). Practical aspects of a visual aid to decision making. *Journal of Intellectual Disability Research, 56*(6), 588–599.

Fleming, P., McGilloway, S., Hernon, M., Furlong, M., O'Doherty, S., Keogh, F., & Stainton, T. (2019). Individualised funding interventions to improve health and social care outcomes for people with a disability: a mixed-methods systematic review. *Campbell Systematic Reviews Oslo, 15*. https://doi.org/10.4073/csr.2019.3

Gordon, R. (2000). The emergence of assisted (supported) decision making in the Canadian law of adult guardianship and substitute decision making. *International Journal of Law and Psychiatry, 23*, 61–77. https://doi.org/10.1016/S0160-2527(99)00034-5

Hawkins, R., Redley, M., & Holland, A. (2011). Duty of care and autonomy: How support workers managed the tension between protecting service users from risk and promoting their independence in a specialist group home. *Journal of Intellectual Disability Research, 55*, 873–884. https://doi.org/10.1111/j.1365-2788.2011.01445.x

Henderson, D., & Bigby, C. (2021). Stories from the wild west frontier: The NDIS experiences of people with severe and profound intellectual disabilities. In M. Cowden & C. McCullagh (Eds.), *The national disability insurance scheme: An Australian public policy experiment*. Palgrave MacMillan.

Khemka, I. (2000). Increasing independent decision making skills of women with mental retardation in simulated interpersonal situations of abuse. *American Journal on Mental Retardation, 105*, 387–401. https://doi.org/10.1352/0895-8017(2000)105<0387:IIDSOW>2.0.CO;2

Kjellberg, A. (2002). More or less independent. *Disability & Rehabilitation, 24*, 828–840. https://doi.org/10.1080/09638280210131745

Knox, L., Douglas, J., & Bigby, C. (2015). "The biggest thing is trying to live for two people": Spousal experiences of supporting decision making participation for partners with TBI. *Brain Injury, 29*, 747–757. https://doi.org/10.3109/02699052.2015.1004753

Knox, L., Douglas, J., & Bigby, C. (2016a). Becoming a decision making supporter for someone with acquired cognitive disability following TBI. *Research and Practice in Intellectual and Developmental Disabilities, 3*, 12–21. https://doi.org/10.1080/23297018.2015.1077341

Knox, L., Douglas, J., & Bigby, C. (2016b). "I won't be around forever": Understanding the decision making experiences of adults with severe TBI and their parents. *Neuropsychological Rehabilitation, 26*, 236–260. https://doi.org/10.1080/09602011.2015.1019519

Knox, L., Douglas, J., & Bigby, C. (2017). "I've never been a yes person": Decision making participation and self-conceptualisation after severe traumatic brain injury. *Disability Rehabilitation, 39*, 2250–2260. https://doi.org/10.1080/09638288.2016.1219925

Kohn, N. A., Blumenthal, J. A., & Campbell, A. T. (2014). Supported decision making: A viable alternative to guardianship? *Penn State Law Review, 177*, 1111–1157.

McKearney, P. (2020a). Challenging care: Professionally not knowing what good care is. *Anthropology and Humanism*. https://doi.org/10.1111/anhu.12302

McKearney, P. (2020b). What escapes persuasion: Why intellectual disability troubles 'dependence' in Liberal societies. *Medical Anthropology*, 1–14. https://doi.org/10.1080/01459740.2020.1805741

National Disability Insurance Scheme. (2013). *National Disability Insurance Scheme Act*. https://www.legislation.gov.au/Details/C2018C00276

Needham, C. (2013). *The boundaries of budgets: Why should individuals make spending choices about their health and social care?* Centre for Health and the Public Interest. https://chpi.org.uk/wp-content/uploads/2013/07/Boundaries-of-Budgets-Catherine-Needham-July-31.pdf

Rossow-Kimball, B., & Goodwin, D. (2009). Self-determination and leisure experiences of women living in two group homes. *Adapted Physical Activity Quarterly, 26*, 1–20.

Series, L. (2015). The development of disability rights under international law: From charity to human rights. *Disability & Society, 30*, 1590–1593. https://doi.org/10.1080/09687599.2015.1066975

Szmukler, G. (2019). "Capacity", "best interests", "will and preferences" and the UN convention on the rights of persons with disabilities. *World Psychiatry, 18*(1), 34–41.

Taylor, W., Cobigo, V., & Ouelette-Kunz, H. (2019). A family systems perspective on supporting self-determination in young adults with intellectual and developmental disabilities. *Journal of Applied Research in Intellectual Disabilities*. https://doi.org/10.1111/jar.12601

Then, S., Carney, T., Bigby, C., & Douglas, J. (2018). Supporting decision making of adults with cognitive disabilities: The role of law reform agencies – recommendations, rationales and influence. *International Journal of Law and Psychiatry*. https://doi.org/10.1016/j.ijlp.2018.09.001

Tune, D. (2019). *Review of the National Disability Insurance Scheme Act 2013. Removing red tape and implementing the NDIS participant service guarantee*. Commonwealth of Australia.

Watson, J. (2016). *The right to supported decision making for people rarely heard*. Doctoral thesis, Deakin University, Melbourne, Australia. Retrieved from http://dro.deakin.edu.au/eserv/DU:30083812/watson-theright-2016A.pdf

Wennberg, B., & Kjellberg, A. (2010). Participation when using cognitive assistive devices–from the perspective of people with intellectual disabilities. *Occupational Therapy International, 17*(4), 168–176.

Wiesel, I., Bigby, C., Carney, T., Then, S., Douglas, J., & Smith, E. (2020). The temporalities of supported decision making by people with cognitive disability. *Social and Cultural Geography*, 1–9.

Williams, I., & Dickinson, H. (2016). Going it alone or playing to the crowd? A critique of individual budgets and the personalisation of health care in the English national health service. *Australian Journal of Public Administration, 75*(2), 149–158. https://doi.org/10.1111/1467-8500.12155

Chapter 5
Parent and Sibling Roles in Decision Making with Individuals with Intellectual and Developmental Disabilities

Meghan Burke, Chung eun Lee, Zach Rossetti, and Sarah Hall

As shown throughout this book, individuals with intellectual and developmental disabilities (IDD) are key players in their own decision making. As with any individual—regardless of whether a disability is present—one may seek support or advice from others when making decisions. Often, these individuals may be family members including parents and siblings. Most research about the *family* of the individual with IDD refers to parents (not siblings). Unfortunately, little research and practice has been directed at examining the role of sibling and parent dynamics (i.e., the interplay of roles and responsibilities between individuals with IDD and their family members) in decision making involving individuals with IDD.

In this chapter, we delve into the extant literature about sibling and parent dynamics when considering decision making of individuals with IDD. For context, we begin by describing emerging trends in the role of the family in decision making for individuals with IDD. To this end, we discuss person-family interdependent

M. Burke (✉)
Department of Special Education, University of Illinois at Urbana-Champaign, Champaign, IL, USA
e-mail: meghanbm@illinois.edu

C. e. Lee
Department of Special Education, Chonnam National University, Gwangju, Republic of Korea

Vanderbilt University Medical Center, Nashville, TN, USA
e-mail: celee@jnu.ac.kr

Z. Rossetti
Wheelock College of Education & Human Development, Boston University, Boston, MA, USA
e-mail: zsr@bu.edu

S. Hall
Institute on Community Integration, University of Minnesota, Minneapolis, MN, USA
e-mail: hall1964@umn.edu

I. Khemka, L. Hickson (eds.), *Decision Making by Individuals with Intellectual and Developmental Disabilities*, Positive Psychology and Disability Series,
https://doi.org/10.1007/978-3-030-74675-9_5

planning as well as the role of the family in supported decision making. Then, we summarize the extant literature about decision making and the role of sibling and parent dynamics. Specifically, we discuss the types of decisions (i.e., formal and informal). We then discuss considerations for decision making for individuals with IDD, including personal characteristics, environmental demands, and support needs. We conclude with recommendations for future research and supports involving individuals with IDD and their families in the context of decision making.

Emerging Trends in the Role of Parent and Sibling Dynamics in Decision Making

Supporting individuals with IDD to make their own decisions is a critical area for research, practice, and policy. Whereas, historically, individuals with IDD have been relegated to passive roles, the current expectation is for individuals with IDD to take the reins of their own decision making (Jameson et al., 2015). This is an important shift for individuals with IDD. It is also an important shift for their families. Indeed, until recently, much of the decision making for individuals with IDD fell to their family members, specifically their parents. Because the goal for individuals with IDD is to play more active roles in their own decision making with their families playing supportive roles, it is important to reconcile the new role of the family while supporting the individual with IDD to be more independent in their decision making. Thankfully, recent research illustrates how families, including parents and siblings, can support their family members with IDD in decision making. In this chapter, we highlight two of these emerging trends: person-family interdependent planning and the family role in supported decision making.

Person-Family Interdependent Planning

In a seminal article about family involvement in transition planning, Kim and Turnbull (2004) argued that person-family interdependent planning (a combination of person-centered planning and family-centered planning) can enhance the quality of life for both the individual and their family. Unlike person-centered planning—which focuses solely on the individual with a disability—person-family interdependent planning ensures that the individual and the family are both involved. Relatedly, within person-family interdependent planning, the individual, with family input, chooses their own services and goals, exercises their rights and has economic resources, and is satisfied with services and their life situations.

Although there may be many reasons for person-family interdependent planning, Kim and Turnbull (2004) identify two primary rationales. First, person-family interdependent planning acknowledges that transition planning—and decisions more broadly—impacts both the individual with IDD and the family. Consider the decision for an individual with IDD to move outside of the family home. At first glance, this may be seen as a decision that impacts only the individual. Indeed, when individuals with IDD live independently outside of the family home, they enjoy greater well-being (Heller, et al., 2002) and increased choice making (Bradley et al., 2015). However, this decision also impacts the family. For example, when individuals with IDD move outside of the family home, their families report greater family quality of life, increased freedom to pursue their own interests, and improved familial relationships (Werner et al., 2009). There may be a differential effect on the benefits of out-of-home placement for individuals with IDD and their family members; mothers of adult children with autism spectrum disorder (ASD) reported that moving out of the family home yielded the greatest benefit for the individual with ASD, while co-residing in the family home yielded the greatest positive benefit for the family as a whole (Krauss et al., 2005). Person-family interdependent planning is needed to address how decisions impact both the individual with IDD and the family.

Second, person-family interdependent planning is needed as it highlights that ongoing support from the family may be needed to facilitate self-determination in decision making (Jordan & Dunlap, 2001). Further, person-family interdependent planning may be more critical for individuals with greater support needs. For example, individuals with greater support needs related to activities of daily living (e.g., eating, walking, paying bills) are more likely to rely on their parents and siblings for support with respect to recreation, employment, and housing (Sanderson et al., 2017). The role of such family members is not to replace the voice and subsequent decision making of the individual with IDD. Rather, in person-family interdependent planning, the purpose is for the family to support and encourage the individual with IDD to make decisions while also considering the perspectives and needs of their family members. The latter distinguishes person-family interdependent planning from supported decision making.

Supported Decision Making

As discussed in other chapters, supported decision making is a process wherein, with the support of family and friends, the individual with IDD makes the final decisions. Thus, within its very definition, supported decision making acknowledges the role of the family. Supported decision making is especially salient with respect to families as it is a marked shift from guardianship. Since the United Nations Convention on the Rights of Persons with Disabilities (CRPD) (2006) recognized supported decision making as a critical approach to emphasize and help implement universal legal capacity for individuals with IDD, many countries are increasingly replacing guardianship (wherein an individual, often the parent, makes most decisions for the individual with IDD) with supported decision making. Thus, the transition to supported decision making sparks a change not only in the role of the individual with IDD but also in the roles of parents and siblings, changing their roles from *decision-maker* to *supporter*.

Unfortunately, the limited extant research suggests that families have little knowledge or awareness of supported decision making. In a study of parents of individuals with IDD, parents reported knowledge only of guardianship; they were unfamiliar with alternatives to guardianship (Jameson et al., 2015). Similarly, in a study of adult siblings of individuals with IDD, siblings reported knowledge of guardianship with very limited knowledge of alternatives to guardianship, including supported decision making (Brady et al., 2019). As supported decision making becomes more common and is legislated in more state policies as a formal alternative to guardianship, it is expected that families will become more familiar with supported decision making and it will become increasingly popular among individuals with IDD. To date, unfortunately, there are few studies about awareness of individuals with IDD and their families about supported decision making.

Research Findings About Decision Making

Individuals with IDD and their families support each other throughout their lives. Such support may be especially relevant among siblings as the sibling relationship is the longest-lasting familial relationship, regardless of whether a disability is present. Indeed, among siblings without IDD, relationships and support are often reciprocal (Howe & Recchia, 2005). Siblings with and without IDD may also engage in reciprocal relationships and support through shared activities and family experiences. Kramer et al. (2013) found that siblings with IDD built reciprocity by enacting family roles (e.g., aunt, uncle) and actively participating in shared family activities, while siblings without IDD helped their brothers and sisters with IDD access community resources and opportunities. These were not equal exchanges, but the asynchronous exchanges of support resulted in reciprocity among the siblings (Kramer et al., 2013).

Although reciprocity is present, there is often a hierarchical nature in family relationships when a disability is present. The most obvious example of a power dynamic may exist between parents and their adult children. Indeed, regardless of whether a disability is present, parent-child relationships often have a power dynamic resulting with more power lying with the parent than the child. Over time, this dynamic tends to dissipate except when the adult child has a disability (Burke et al., 2015b).

Similarly, relationships between siblings with and without IDD tend to take on a hierarchical nature in which the sibling without IDD provides more support to their brother or sister with IDD (Avieli et al., 2019). Adult siblings without IDD take on a number of roles in regard to their brother or sister with IDD including caregiver, friend, advocate, legal representative, leisure planner, and informal service coordinator (Hall & Rossetti, 2018). Thus, as part of the multiple supports they provide to their brothers and sisters with IDD, siblings without IDD will often be well situated to support their brothers and sisters with IDD to make decisions in their lives. Indeed, among the limited extant research on (or related to) decision making by individuals with IDD, siblings without IDD can play a pivotal role in the decision making opportunities and outcomes of their siblings with IDD. Below, we discuss the role of the family in formal and informal decisions.

Formal Decisions

Formal decisions include significant or long-term life decisions about employment, post-secondary education, community living, and relationships. Overall, individuals with IDD engage infrequently in decision making related to these formal areas (Shogren et al., 2017). One area of prior research has examined the quality of life outcomes for adults with IDD related to decisions about the type of services and service delivery models. For example, Gross et al. (2013) conducted a case study of parent/guardian experiences and decisions during participant direction using, "a service delivery model in which the consumer of public benefits, or his or her surrogate decision maker, exercises some level of choice and control over the consumer's supports and services" (p. 88). They found that participant direction allowed parents and guardians of individuals with significant IDD to have more control, access, and choice related to their adult children's services, which they reported they valued because they could more effectively work toward their desired outcomes for their children. Their study did not include individuals with IDD as participants. As such, it is unclear whether the parents' involvement and goals mirrored the desires of the individuals with IDD. Indeed, most service delivery systems look to parents (not individuals with IDD) to access services (Taylor et al., 2017). The focus on parents—who may be guardians—may limit the participation of adults with IDD in decision making. Formal decisions (e.g., guardianship issues) are discussed later in this chapter.

Other studies have similarly found that parents (versus individuals with IDD themselves) may be more active participants in formal decisions for individuals with IDD. In their study of parent and sibling perspectives of future planning for individuals with IDD, Lee et al. (2019b) found that most parents made the formal decisions for their adult children with IDD. Additionally, Bigby et al. (2017) examined the perspectives of individuals with IDD on their supported living arrangements, finding that while individuals with IDD reported they enjoyed more choice and control in supported living arrangements compared to congregate settings and group homes, their decisions were limited to everyday (i.e., informal) decisions. Further, the individuals with IDD reported that their families, while supportive and regularly involved in their lives, made formal decisions, thus limiting their decision making. Indeed, large-scale research has also found that individuals with IDD are often not involved in their own formal decisions. In a study of the quality of life outcomes of older adults with IDD that analyzed the Personal Outcome Measures® data from 800 people with IDD aged 55 years old and older, the majority did not choose where and with whom to live, where to work, the services they received, and the personal goals upon which they focused (Friedman, 2019). These examples suggest an important consideration related to decision making in that individuals with IDD themselves may not have been directly involved in making the decisions. However, with supported decision making, individuals with IDD may become more involved in formal decisions.

Research has also found that siblings, similar to parents, may play more active (versus supportive) roles in formal decision making for individuals with IDD. Although siblings often report that they (compared to their parents) are less overprotective of their brothers and sisters with IDD (Burke et al., 2015b), siblings may still make formal decisions on behalf of (versus with) their brothers and sisters with IDD. In their study using a dyadic interview procedure with nine dyads of siblings with and without IDD, Burke et al. (2019) found that siblings (and parents) primarily made most formal decisions for their family members with IDD. Siblings without IDD reported utilizing their knowledge of their brother or sister with IDD when making decisions. For example, a sibling in this study explained, "Yeah, me, my mom and my sister we all kind of really made decisions together.…[We then] presented it in a light where he [Aaron] would think it would be a good idea."

Alternatively, in two of this study's nine dyads, individuals with IDD experienced greater involvement in decision making for formal decisions. In these two dyads, the siblings without IDD emphasized the agency of siblings with IDD and consistently focused on supporting them to make their own decisions. For example, one sibling dyad and their mother would travel to appointments to meet with service providers and discuss their options together (Burke et al., 2019). The other sibling dyad spoke daily and worked together to achieve the personal goals of the individual with IDD, as the sibling without IDD described:

> *Well, you know, he talks to me. When he wants to make changes, he works with me for the changes.…I'm not the parent, right? And I don't make decisions like a parent would make a decision for their kid. You know, I let him go. If he's gonna fail, he's gonna fail. Just a part of living, right? You gotta know what failing is like and know what winning is like.*

Inherent to his description of their decision-making process are considerations related to risk taking, as well as a potential difference between making decisions and decision-making support from siblings compared to parents.

Other studies have found that siblings and parents may disagree with respect to future planning and decision making for formal decisions, noting a pattern in which parents may focus on safety and security of individuals with IDD, while siblings may be more willing to take chances based on the individual's interests or preferences (Burke et al., 2019; Lee et al., 2019b). For example, in response to anticipated employment for individuals with IDD, some parents reported that their adult children with IDD would work in a standard position (e.g., grocery store bagger), whereas siblings emphasized finding an interest-based job and not settling for less (Lee et al., 2019b). In a study of sibling advocacy on behalf of their brothers and sisters with IDD, siblings without IDD advocated for their siblings with IDD to receive additional adult services, kept an eye on their experiences in their residential settings, and included them in daily activities based on their interests and daily activity preferences (Burke et al., 2015a). While siblings primarily determined these formal decisions, they did so based on the interests and preferences of their brothers and sisters with IDD. Such a pattern is critical to decision making involving individuals with IDD because while parents are primary advocates and caregivers during the individual's school years, siblings are more frequently taking over these roles in the adult service system as parents and individuals with IDD age (Burke et al., 2015b; Burke et al., 2019). Indeed, siblings without IDD engaged in caregiving, advocacy, and future planning for their brothers and sisters with IDD (Lee et al., 2019a).

Informal Decisions

Compared to their experiences in formal decisions, prior research (e.g., Burke et al., 2019) indicates that individuals with IDD engaged more frequently in decision making for informal decisions about recreational and leisure activities (e.g., choosing a restaurant or movie) and daily living choices (e.g., what clothes to wear). These *everyday* decisions were based on the preferences of individuals with IDD and were distinguished from the larger, more formal decisions about services, employment, and community living. Siblings (and parents) reported providing individuals with IDD opportunities to make such decisions as well as supports to facilitate decision making (e.g., prompts, waiting for responses) (Burke et al., 2019; Rossetti et al., 2020). For example, in a study that examined sibling experiences with decision making, all of the participants with IDD made their own decisions about daily choices such as what clothes to wear, what food to eat, and what to do at home, as well as about recreational and leisure activities such as visiting their sibling, going to see a specific movie, and eating out at a favorite restaurant (Burke et al., 2019). *Choices* generally refers to informal decisions (e.g., where to eat, what food to order), whereas "decision making" includes formal decisions. Siblings with and without IDD reported similar shared social activities that were often based on decisions by individuals with IDD, including hanging out and talking, eating at restaurants, going to movies, shopping at the mall, and completing chores together (Rossetti et al., 2020). Other informal decisions about recreational and leisure activities included taking a trip and going on vacation together, as one of the roles in which siblings without IDD engaged was that of a leisure planner or facilitator related to their brothers and sisters with IDD (Hall & Rossetti, 2018).

Other studies have similarly found that individuals with IDD make their own informal decisions. In one study, individuals with IDD living in supported living arrangements reported making decisions about what clothes to wear, what to do during the day, and where and when to go shopping (Bigby et al. 2017). While their decision making was a feature of increased opportunities to make choices in their independent living arrangements, these individuals also reported being in regular contact with and supported informally by parents, siblings, and other family members, which facilitated their informal decisions. Similarly, though parents made formal decisions for individuals with IDD in their study, Gross et al. (2013) reported that one of the four individuals with IDD expressed his preferences more than he had in the past when he was provided with increased opportunities to make informal decisions. These findings collectively indicate that with increased opportunities to make decisions and minimal support from siblings or parents to do so, individuals with IDD will more frequently engage in decision making about informal decisions.

Considerations of Factors Impacting Decision Making

Based on the different types of decisions, individuals with IDD may engage in various levels of decision making with their family members (i.e., parents, siblings). Relatedly, decision-making capacities may be influenced by a range of potential factors. Drawing from a social-ecological model of disability, Shogren and Wehmeyer (2015) proposed a framework to understand how individualized decision making operates among individuals with IDD. Using their framework, contextual factors which may impact decision making include personal characteristics, environmental demands, and support needs.

Personal Characteristics

Decision-making ability (i.e., decision-making capacity) often refers to the personal characteristics which build one's ability to make decisions or choices (Shogren & Wehmeyer, 2015). Of course, there are many personal characteristics which may influence decision making. With respect to individuals with IDD, historically, decision making has primarily revolved around one characteristic: cognitive ability (Suto et al., 2005). Specifically, there is a presumed relation between cognitive impairment and a limited capacity to make decisions. Individuals with cognitive impairment may have more difficulties in understanding relevant information and reasoning to make decisions based on their level of intellectual functioning (Morris et al., 1993). However, cognitive impairment does not mean that individuals with IDD cannot make decisions. Although cognitive ability may influence decision-making ability, it does not imply incapability of one's decision making. Indeed, with individualized and appropriate supports, individuals with disabilities can make decisions (Shogren et al., 2017).

Even though the focus on *cognitive ability* has impacted decision making among individuals with IDD, the importance of individuals with IDD making their own decisions has also long been recognized as important. In a seminal article, Perske (1972) introduced the concept of *the dignity of risk* which is the right for individuals with disabilities to take risks when engaging in life experiences, including making decisions. Families may be overprotective (Burke et al., 2015b) and, accordingly, fear risk-taking among their family members with IDD. To this end, families may eliminate the opportunities for individuals with IDD to make decisions. However, risk-taking opportunities are crucial to developing decision-making skills. As such, families could encourage individuals with IDD to participate in decision-making opportunities and learn from their experiences.

Another personal characteristic includes self-determination skills of individuals with IDD. Self-determination is critical for individuals with IDD to achieve goals and make decisions. Self-determination skills involve understanding one's strengths and needs, setting and working toward self-selected goals, advocating for one's self, and self-assessing one's own progress and outcomes (Dew et al., 2019). Without learning these skills, individuals with IDD may not seek out engaging in self-determined behaviors (e.g., problem-solving, goal setting, or self-advocacy) when decision-making opportunities are available. Self-determination can involve support for someone to be the causal agent in their lives, rather than being characterized by independent performance of behaviors (in this case, decision making) or self-sufficiency (Shogren & Broussard, 2011; Wehmeyer, 2005).

Environmental Demands

Environmental demands are often defined as the demands and requirements of the context within which a person functions (Shogren & Wehmeyer, 2015). Such demands may facilitate or hinder one's decision making. Similar to personal characteristics, there are many environmental demands, and, correspondingly, their effects on decision making are vast. For brevity, we focus on three particularly salient environmental demands when considering decision making among individuals with IDD and their families: inadequate adult disability services, guardianship or conservatorship status, and natural supports.

Inadequate Adult Disability Services Although individuals with IDD may have abilities and skills to make decisions with support from their families, inadequate adult disability services may preclude effective decision making (Lakin & Burke, 2019). For example, an individual with IDD may want to live in the community with individualized supports, and the family may be in favor of this decision. However, the lack of available community supports may preclude this decision from becoming a reality. In the United States, 43 states have waiting lists for adult disability services (e.g., home- and community-based services); accordingly, an estimated 193,828 individuals with IDD are waiting for services and, likely, remain in the family home while doing so (Larson et al., 2018).

The lack of adult services cannot be attributed solely to waiting lists. Even if there is not a waiting list, there is a national shortage of disability professionals (e.g., direct support providers, case workers, and other paid caregivers) due to low levels of compensation and physically demanding work (Bogenschutz et al., 2014). Although an adult with IDD may decide to live with supports in the community, the lack of available direct support providers may occlude this decision from being realized. Because of the limited adult disability services and supports, most adults with IDD live with their families until a crisis occurs (Heller & Factor, 2008). Due to limited services for individuals with IDD, parents may feel forced into more restrictive choices for individuals with IDD (thus not supporting individuals with IDD to make their own decisions). Specifically, families may feel compelled to make choices which emphasize safety and security, settling for services they can obtain rather than individualized services based on the individual's preferences and needs (Rossetti et al., 2015).

The role of the family in relation to decision making and adult disability services is also problematic. It is widely recognized that adult disability service system is fragmented, tedious, and difficult to navigate (Timmons et al., 2004). Because of the highly regulated and constantly changing bureaucracies and rules dictating adult services, individuals with IDD often rely on their families to navigate the adult service delivery systems (Timmons et al., 2004). Such families may support their family members with IDD in making decisions but struggle to identify and access needed services to make such decisions come to fruition (Rossetti et al., 2015). To this end, researchers have begun developing interventions to educate parents about the adult service delivery system. Consider the Volunteer Advocacy Project for Transition (VAP-T). In this 30-hour training, parents of young adults with autism learn to navigate adult services to identify, access, and advocate for services for their offspring with autism (Taylor et al., 2017). In this training, the person with autism is involved in some sessions and person-centered planning is the crux of the curriculum. Thus, the family may play an integral role in overcoming the barrier of inadequate services.

Guardianship or Conservatorship Status In addition to limited adult disability services, guardianship or conservatorship status is an environmental demand related to decision making. Historically, guardianship and conservatorship have been heavily pursued for individuals with IDD (Millar, 2003). However, guardianship and conservatorship often serve to reduce decision-making abilities of individuals with IDD. Indeed, Stancliffe and colleagues (2000) reported that individuals without guardians/conservators (versus individuals with IDD who had guardians/conservators) exercised more personal control. Further, individuals with IDD who have guardians (versus individuals who do not have guardians) have significantly fewer opportunities to make decisions (e.g., residence, medical care, financial); instead, decisions are often made by the guardians themselves (American Bar Association, 2017).

The family plays an important role in considering guardianship/conservatorship and decision making. Indeed, family members often act as the guardians or conservators for their family members with IDD (Jameson et al., 2015). However, it is unclear whether families understand how the roles of guardian and conservator can restrict decision making among individuals with IDD. In a study of ten adult siblings of individuals with IDD, all siblings reported not being aware of alternatives (e.g., supported decision making) to guardianship (Brady et al., 2019). Nearly all siblings perceived their brothers and sisters to have limited decision-making ability and, accordingly, planned to become guardians or conservators for their brothers and sisters with IDD. However, sibling participants did indicate some willingness to consider less restrictive forms of decision making. Thus, the lack of knowledge about alternatives to guardianship or conservatorship may explain the choices of families in considering such restrictive forms of decision making.

Natural Supports With respect to decision making, formal adult services and roles (e.g., guardianship/conservatorship) are important to consider; in addition, it is important to examine natural supports. Defined as unpaid supports, natural supports are provided by people associated with the individual with disabilities (Sanderson et al., 2017). Because 75% of individuals with IDD do not receive formal adult services (Watson, 2016), natural supports are especially important. Natural supports are often primarily provided by family members but also can be provided by neighbors or friends (Sanderson et al., 2017). These natural support providers could provide opportunities for individuals with IDD to perform decision making (Carter et al., 2009).

For example, parents may limit the choices for individuals with IDD as they may be overly protective or restrictive (Dew et al., 2019). Some research suggests that parents are unsure of whether their adult offspring with IDD understand *right* and *wrong* decisions; accordingly, to avoid *wrong decisions*, parents report making decisions for their children with IDD (Saaltink et al., 2012). Unsurprisingly, because of this disjunct between parents and adults with IDD, they may have disagreements about where to live in the future and which adult disability service is needed. On the other hand, siblings—who also serve as natural supports—report being less overprotective of their brothers and sisters with IDD (Burke et al., 2019). From dyadic interviews with ten siblings and their brothers and sisters with IDD, siblings reported encouraging their brothers and sisters with IDD to make their own daily life decisions (e.g., choosing clothes, choosing meals). In addition, siblings tried to encourage their parents to include their family members with IDD in decision making. Depending on natural support providers, decision making could be influenced in different ways by parents and siblings.

Further, natural support providers often know how to respond to the preferences of individuals with IDD. Often individuals with severe or profound disabilities were perceived as individuals with limited abilities to engage in decision making (Jameson et al., 2015). However, individuals with severe or profound disabilities may express their preferences or opinions using different communication modalities; natural support providers often know how to respond to these expressions appropriately (Watson, 2016). This is called "supporter responsiveness." For example, when individuals with IDD make specific eye movements revealing their decisions, natural support providers will notice their expression, interpret and assign meaning to the expression, and act on the meaning (Watson, 2016). Thus, natural support providers may be especially instrumental in supporting the decisions of individuals with IDD. Unfortunately, there are no intervention studies, to our knowledge, that have helped persons with IDD and their natural support providers to improve communication and responsiveness.

Support Needs

Support needs refer to the pattern and intensity of supports a person requires to participate in decision making (Shogren & Wehmeyer, 2015). Given that the support needs of individuals with IDD vary, it is important to individually consider the role of support needs in relation to a given person with IDD and their family. Support needs may include the need to address individual behaviors including maladaptive and adaptive behaviors.

Although there is mixed research, some evidence indicates a negative correlation between maladaptive behavior, choice making and personal control. Specifically, an individual with IDD who has significant maladaptive behaviors (e.g., internalizing or externalizing behaviors) is often less likely to have opportunities to make choices and to be able to exert control over their choices (Carter et al., 2009; Jones et al., 2018; Schalock, 1994; Stancliffe, 2001). Further, when an individual with IDD demonstrates maladaptive behaviors, the individual is significantly more likely to live in more restrictive settings (e.g., institutions) (Bambara et al., 1998). Such settings, because of their large and impersonal nature, make decision making among individuals with IDD uncommon. The role of maladaptive behaviors extends to families of individuals with IDD. When individuals with IDD demonstrate significant maladaptive behaviors, their families often perceive the individual with IDD to have limited decision-making capacity (Carter et al., 2009; Jones et al., 2018).

Adaptive behavior (i.e., activities of daily living) may influence decision making among individuals with IDD and their families. Having fewer adaptive behavior skills often correlates with having fewer opportunities to make decisions among individuals with IDD (Antaki et al., 2009; Burke et al., 2019). Also similar to the case with maladaptive behaviors, family members (i.e., siblings of individuals with IDD) may be more likely to have lower expectations and question the ability of their family member with IDD to make decisions when the individual has fewer functional abilities (Burke et al., 2019).

Conversely, greater levels of adaptive behavior often lead to increased opportunities for decision making (Heller et al., 2000; Jones et al., 2018; Lancioni et al., 1996). Indeed, individual adaptive behavior skills may have a direct correlation with decision making. For example, individuals with IDD are more likely to make decisions when they are able to *lock their door* and *vote in elections* (Jones et al., 2018). Other specific activities may also positively correlate with decision making including high-quality work performance; consistent task engagement; assignment accuracy and appropriate social/communicative behavior; and consistent eye contact (Heller et al., 2000; Lancioni et al., 1996). Thus, increasing training and supports to improve adaptive behavior of individuals with IDD may facilitate more decision-making opportunities.

Implications for Research and Practice

This chapter reviews the empirical literature with respect to the role of parent and sibling dynamics in decision making among individuals with IDD. The extant literature posits a jumping off point for future research and practice. Below, we identify implications for research and practice to better support individuals with IDD and their families.

Implications for Research

The extant research suggests the need for a more holistic understanding of the individual with IDD and their family in decision making. First, of the limited literature, most research only showcases the perspective of one individual—often the parent or sibling of the individual with IDD (Lee et al., 2019b). Few studies included the perspective of the individual with IDD (Burke et al., 2019; Kramer et al., 2013; Rossetti et al., 2020). The near-exclusion of individuals with IDD is a tremendous limitation of prior research. Moving forward, it is critical to fulfill the call of *nothing about us, without us*; research needs to include individuals with IDD as participants. To this end, increasingly there are advances recognizing the emerging person-family interdependent planning and supported decision-making models.

However, research also needs to go beyond the inclusion of individuals with IDD. The extant research also suggests conflicting perspectives among individuals with IDD, their parents, and their siblings (e.g., Burke et al., 2019; Lee et al., 2019b). Yet, none of the prior studies have included the perspectives of the parent, the sibling, and the individual with IDD within a given family unit. To truly understand the nature and effect of sibling and parent roles in decision making among individuals with IDD, it is critical for research to explore the perspectives of multiple family members from the same households.

Second, research needs to include diverse measures and methodologies. At present, the bulk of research relies on self-report (e.g., Gross et al., 2013; Burke et al., 2015b). While important, self-report is at risk for bias, misperception, and subjectivity. Further, research demonstrates that self-report may not accurately reflect reality. In the study by Burke et al. (2019), for example, siblings reported being more encouraging than parents of their brothers and sisters with IDD in making decisions. However, the dyadic interviews revealed that siblings often still make the major, formal decisions. To truly understand the phenomenon of decision making among individuals with IDD and their families, it is important to go beyond self-report.

To do this, researchers may consider observational measures of decision making. Indeed, during the dyadic interviews, the researchers were able to observe the dynamic between the sibling and the individual with IDD (Burke et al., 2019; Rossetti et al., 2020). To this end, researchers were able to see the exchange of information between the siblings and how they reflected, individually and together, about decision making. Future research may consider coding such interactions to operationalize and measure the role of the family member and the individual with IDD in decision making. As a complement to the first implication for research (i.e., including individuals with IDD in research), dyadic interviewing is also a method to ensure accessibility and responsiveness when interviewing individuals with IDD (Caldwell, 2014).

Third, research is needed about perceived decision-making capacity among individuals with IDD. To some extent, this perception dovetails with the literature suggesting that personal characteristics and support needs are considerations in decision making (Shogren & Wehmeyer, 2015). Especially with respect to guardianship, perceived decision-making capacity is a critical construct. However, there are few valid or reliable measures for decision-making capacity (Applebaum, 2007). Without a valid measure, how can decision-making capacity be such a perceived critical factor in decision making? Research is needed to understand what families are referring to when they discuss *decision-making capacity*. In addition, research is needed to identify more rigorous and well-defined measures of decision-making capacity. Relatedly, studies on how best to provide information and education are needed to help families to understand that decision-making capacity can be addressed by services and supports.

Implications for Practice

Moving forward, it is critical to change practice such that family members can play more supportive roles in decision making among individuals with IDD. For example, information needs to be available for families about less restrictive forms of decision making. Parents and siblings alike are often more familiar with restrictive forms of decision making (e.g., guardianship, conservatorship) and less familiar with other forms of decision making that include individuals with IDD to a greater degree (e.g., supported decision making) (Brady et al., 2019; Jameson et al., 2015). Strikingly, although siblings report being less overprotective than their parents (Burke et al., 2015b), siblings still reported relying on more restrictive forms of decision making. However, studies indicate that families want to play more supportive (and less authoritative) roles with respect to decision making for their family members with IDD (e.g., Burke et al., 2019).

To this end, information about less restrictive forms of decision making needs to be available for families of individuals with IDD. Specifically, families need to learn about supported decision making (Shogren & Wehmeyer 2015) and person-family interdependent planning (Kim & Turnbull, 2004). By educating families about various types of involvement in decision making, parents and siblings can make informed choices about their roles and participation in decision making among individuals with IDD.

Second, research-based interventions which may increase decision making among individuals with IDD need to become more common in regular practice. Throughout this chapter, we identified several ways to increase decision making among individuals with IDD while still considering the support and input of their families. Such ways included increasing opportunities for the individual with IDD, embracing the dignity of risk, improving self-determination, accessing adult services, increasing one's activities of daily living, and providing positive behavioral supports to better understand and reduce maladaptive behavior. The literature offers targeted research-based interventions addressing a few of these approaches, but such interventions need to become common practice and more well known.

Consider the *Future Is Now* curriculum (Heller & Caldwell, 2006). In this curriculum, individuals with IDD and their family members meet in a group to plan for the future. The curriculum reflects the input of the individual with IDD and other family members. Participants learn about decision making, adult services, and family communication. As a result of participating in this training, individuals with IDD demonstrated significantly increased choice making—not necessarily more formal decision making. However, the *Future Is Now* curriculum is not widely used by practitioners. Such an intervention may be helpful in promoting the role of individuals with IDD in decision making and educating families about their supportive roles in formal decisions.

One may also consider the Volunteer Advocacy Project-Transition (VAP-T). In this 30-hour curriculum, parents of young adults with autism learn about available adult services and supports (Taylor et al., 2017). The curriculum is rooted in person-centered planning. As a result, families learn about the desires and concerns of their family member with autism. Then, families identify, apply, and access needed services to fulfill those desires and address the concerns. After participating in the VAP-T, parents demonstrate significantly improved knowledge, advocacy, and empowerment; their adult children with autism demonstrate reduced unmet service needs and greater post-school outcomes (Taylor et al., 2017). Currently, efforts are underway to examine the efficacy of the VAP-T. If efficacious, the VAP-T may help educate parents and siblings about adult disability services (Lee et al., 2019b).

Conclusion

There are certainly many research and practice implications with respect to parent and sibling dynamics in decision making among individuals with IDD. Moving forward, we encourage researchers and practitioners to think holistically about family involvement in decision making as well as to use a variety of methods, measures, and interventions to ensure families play supportive roles in decision making. Although there is much work to be done, it is encouraging that individuals with IDD are taking the reins of their own decision making with their families, parents, and siblings alike, playing supportive roles.

References

American Bar Association. (2017). *Capacity definition & initiation of guardianship proceedings.* Retrieved from https://www.americanbar.org/groups/law_aging/resources/guardianship_law_practice.html

Antaki, C., Finlay, W. M. L., & Walton, C. (2009). Choices for people with intellectual disabilities: Official discourse and everyday practice. *Journal of Policy and Practice in Intellectual Disabilities, 6*(4), 260–266. https://doi.org/10.1111/j.1741-1130.2009.00230.x

Appelbaum, P. S. (2007). Assessment of patients' competence to consent to treatment. *The New England Journal of Medicine, 357*(18), 1834–1840. https://doi.org/10.1056/nejmcp074045

Avieli, H., Band-Winterstein, T., & Bergman, T. A. (2019). Sibling relationships over the life course: Growing up with a disability. *Qualitative Health Research, 29*, 1739–1750. https://doi.org/10.1177/1049732319837228

Bambara, L. M., Cole, C. L., & Koger, F. (1998). Translating self-determination concepts into support for adults with severe disabilities. *Journal of the Association for Persons with Severe Handicaps, 23*(1), 27–37. https://doi.org/10.2511/rpsd.23.1.27

Bigby, C., Boulda, E., & Beadle-Brown, J. (2017). Conundrums of supported living: The experiences of people with intellectual disability. *Journal of Intellectual and Developmental Disabilities, 42*(4), 309–319. https://doi.org/10.3109/13668250.2016.1253051

Bogenschutz, M. D., Hewitt, A., Nord, D., & Hepperlen, R. (2014). Direct support workforce supporting individuals with IDD: Current wages, benefits, and stability. *Intellectual and Developmental Disabilities, 52*(5), 317–329. https://doi.org/10.1352/1934-9556-52.5.317

Bradley, V., Fay, M. L., Giordano, S., Melda, K., Smith, D., Ficker-Terrill, C., & Tichá, R. (2015). *25 years after the ADA: What story does the data tell? Impact: Feature issue on the ADA and people with intellectual and developmental disabilities*. Institute on Community Integration.

Brady, A., Burke, M. M., Trent, L., & Oertle, K. (2019). Siblings of adults with intellectual and developmental disabilities: Their knowledge and perspectives on guardianship and the alternatives. *Journal of Applied Research in Intellectual Disabilities, 32*, 1078–1087. https://doi.org/10.1111/jar.12597

Burke, M. M., Arnold, C. K., & Owen, A. L. (2015a). Sibling advocacy: Perspectives about advocacy from siblings of individuals with intellectual and developmental disabilities. *Inclusion, 3*(3), 162–175. https://doi.org/10.1352/2326-6988-3.3.162

Burke, M. M., Fish, T., & Lawton, K. (2015b). A comparative analysis of adult siblings' perceptions towards caregiving. *Intellectual and Developmental Disabilities, 53*, 143–157. https://doi.org/10.1352/1934-9556-53.2.143

Burke, M. M., Lee, C., Hall, S., & Rossetti, Z. (2019). Understanding decision making among individuals with intellectual and developmental disabilities (IDD) and their siblings. *Intellectual and Developmental Disabilities, 57*, 26–41. https://doi.org/10.1352/1934-9556-57.1.26

Caldwell, K. (2014). Dyadic interviewing: A technique valuing interdependence in interviews with individuals with intellectual disabilities. *Qualitative Research, 14*, 488–507. https://doi.org/10.1177/1468794113490718

Carter, E. W., Owens, L., Trainor, A. A., Sun, Y., & Swedeen, B. (2009). Self-determination skills and opportunities of adolescents with severe intellectual and developmental disabilities. *American Journal on Intellectual and Developmental Disabilities, 114*(3), 179–192. https://doi.org/10.1352/1944-7558-114.3.179

Convention on the Rights of Persons With Disabilities. (2006). *G.A. Res. 61/106, U.N. Doc. A/RES/6/106 (2007).* Retrieved from http://www.un.org/disabilities/default.asp?id

Dew, A., Collings, S., Dillon Savage, I., Gentle, E., & Dowse, L. (2019). "Living the life I want": A framework for planning engagement with people with intellectual disability and complex support needs. *Journal of Applied Research in Intellectual Disabilities, 32*(2), 401–412. https://doi.org/10.1111/jar.12538

Friedman, C. (2019). Quality-of-life outcomes of older adults with severe disabilities. *Research and Practice for Persons with Severe Disabilities, 44*(4), 237–250. https://doi.org/10.1177/1540796919879227

Gross, J. M. S., Wallace, L., Blue-Banning, M., Summers, J. A., & Turnbull, A. (2013). Examining the experiences and decisions of parents/guardians: Participant directing the supports and services of adults with significant intellectual and developmental disabilities. *Journal of Disability Policy Studies, 24*(2), 88–101. https://doi.org/10.1177/1044207312439102

Hall, S. A., & Rossetti, Z. (2018). The roles of adult siblings in the lives of people with severe intellectual and developmental disabilities. *Journal of Applied Research in Intellectual Disabilities, 31*(3), 423–434. https://doi.org/10.1111/jar.12421

Heller, T., & Factor, A. (2008). Family support and intergenerational caregiving: Report from the State of the Science in Aging with Developmental Disabilities Conference. *Disability and Health Journal, 1*, 131–135. https://doi.org/10.1016/j.dhjo.2008.04.004

Heller, T., Miller, A. B., Hsieh, K., & Sterns, H. (2000). Later-life planning: Promoting knowledge of options and choice-making. *Mental Retardation, 38*, 395–406. https://doi.org/10.1352/0047-6765(2000)038<0395:lppkoo>2.0.co;2

Heller, T., Miller, A. B., & Hsieh, K. (2002). Eight-year follow-up of the impact of environmental characteristics on well-being of adults with developmental disabilities. *Mental Retardation, 40*, 366–378. https://doi.org/10.1352/0047-6765(2002)040<0366:eyfuot>2.0.co;2

Heller, T., & Caldwell, J. (2006). Supporting aging caregivers and adults with developmental disabilities in future planning. *Mental retardation, 44*(3), 189–202.

Howe, N., & Recchia, H. (2005). Playmates and teachers: Reciprocal and complementary interactions between siblings. *Journal of Family Psychology, 19*(4), 497–502. https://doi.org/10.1037/0893-3200.19.4.497

Jameson, J. M., Riesen, T., Polychronis, S., Trader, B., Mizner, S., Martinis, J., & Hoyle, D. (2015). Guardianship and the potential of supported decision making with individuals with disabilities. *Research and Practice for Persons with Severe Disabilities, 40*, 36–51. https://doi.org/10.1177/1540796915586189

Jones, J. L., Shogren, K. A., Grandfield, E. M., Vierling, K. L., Gallus, K. L., & Shaw, L. A. (2018). Examining predictors of self-determination in adults with intellectual and developmental disabilities. *Journal of Developmental and Physical Disabilities, 30*(5), 601–614. https://doi.org/10.1007/s10882-018-9607-z

Jordan, B., & Dunlap, G. (2001). Construction of adulthood and disability. *Mental Retardation, 39*, 286–296. https://doi.org/10.1352/0047-6765(2001)039<0286:coaad>2.0.co;2

Kim, K., & Turnbull, A. P. (2004). Transition to adulthood for students with severe intellectual disabilities: Shifting toward person-family interdependent planning. *Research And Practice for Persons with Severe Disabilities, 29*, 53–57.

Kramer, J., Hall, A., & Heller, T. (2013). Reciprocity and social capital in sibling relationships of people with disabilities. *Intellectual and Developmental Disabilities, 51*(6), 482–495. https://doi.org/10.1352/1934-9556-51.6.482

Krauss, M. W., Seltzer, M. M., & Jacobson, H. T. (2005). Adults with autism living at home or in non-family settings: Positive and negative aspects of residential status. *Journal of Intellectual Disability Research, 49*, 111–124. https://doi.org/10.1111/j.1365-2788.2004.00599.x

Lakin, K. C., & Burke, M. M. (2019). Looking forward: Research to respond to a rapidly aging population. *Research and Practice for Persons with Severe Disabilities, 44*(4), 280–292. https://doi.org/10.1177/1540796919882356

Lancioni, G. E., O'Reilly, M. F., & Emerson, E. (1996). A review of choice research with people with severe and profound developmental disabilities. *Research in Developmental Disabilities, 17*, 391–411. https://doi.org/10.1016/0891-4222(96)00025-x

Larson, S. A., Eschenbacher, H. J., Anderson, L. L., Taylor, B., Pettingell, S., Hewitt, A., Sowers, M., & Bourne, M. L. (2018). *In-home and residential long-term supports and services for persons with intellectual or developmental disabilities: Status and trends through 2016.*

University of Minnesota, Research and Training Center on Community Living, Institute on Community Integration.

Lee, C., Burke, M. M., & Arnold, C. K. (2019a). Examining the relation between disability severity among older adults with disabilities and sibling caregiving. *Research and Practice for Persons with Severe Disabilities, 44*(4), 224–236. https://doi.org/10.1177/1540796919879102

Lee, C., Burke, M. M., & Stelter, C. R. (2019b). Exploring the perspectives of parents and siblings toward future planning for individuals with intellectual and developmental disabilities. *Intellectual and Developmental Disabilities, 57*(3), 198–211. https://doi.org/10.1352/1934-9556-57.3.198

Millar, D. S. (2003). Age of majority, transfer of rights and guardianship: Considerations for families and educators. *Education and Training in Developmental disabilities, 38*, 378–397.

Morris, C. D., Niederbuhl, J. M., & Mahr, J. M. (1993). Determining the capability of individuals with mental retardation to give informed consent. *American Journal of Mental Retardation, 98*, 263–272.

Perske, R. (1972). Dignity of risk and the mentally retarded. *Mental Retardation, 10*(1), 24–27.

Rossetti, Z., Lehr, D., Lederer, L., Pelerin, D., & Huang, S. (2015). Parent perceptions of time spent meaningfully by young adults with pervasive support needs. *Research and Practice for Persons with Severe Disabilities, 40*(1), 3–19. https://doi.org/10.1177/1540796914566714

Rossetti, Z., Lee, C., Burke, M., & Hall, S. (2020). Perspectives about adult sibling relationships: A dyadic analysis of siblings with and without intellectual and developmental disabilities. *Research in Developmental Disabilities, 96*, 1–6. https://doi.org/10.1016/j.ridd.2019.103538

Saaltink, R., MacKinnon, G., Owen, F., & Tardif-Williams, C. (2012). Protection, participation and protection through participation: Young people with intellectual disabilities and decision making in the family context. *Journal of Intellectual Disability Research, 56*(11), 1076–1086. https://doi.org/10.1111/j.1365-2788.2012.01649.x

Sanderson, K. A., Burke, M. M., Urbano, R. C., Arnold, C. K., & Hodapp, R. M. (2017). Who helps? Characteristics and correlates of informal supporters to adults with disabilities. *American Journal on Intellectual and Developmental Disabilities, 122*, 492–510. https://doi.org/10.1352/1944-7558-122.6.492

Schalock, R. L. (1994). The concept of quality of life and its current applications in the field of mental retardation/developmental disabilities. In D. Goode (Ed.), *Quality of life for persons with disabilities: International perspectives and issues* (pp. 266–284). Brookline Books.

Shogren, K. A., & Broussard, R. (2011). Exploring the perceptions of self-determination of individuals with intellectual disability. *Intellectual and Developmental Disabilities, 49*(2), 86–102. https://doi.org/10.1352/1934-9556-49.2.86

Shogren, K. A., & Wehmeyer, M. L. (2015). A framework for research and intervention design in supported decision making. *Inclusion*, 3(1), 17–23.

Shogren, K. A., Wehmeyer, M. L., Lassmann, H., & Forber-Pratt, A. J. (2017). Supported decision making: A synthesis of the literature across intellectual disability, mental health, and aging. *Education and Training in Autism and Developmental Disabilities, 52*, 144–157. https://doi.org/10.1352/2326-6988-3.1.17

Stancliffe, R. J. (2001). Living with support in the community: Predictors of choice and self-determination. *Mental Retardation and Developmental Disabilities Research Reviews, 7*(2), 91–98. https://doi.org/10.1002/mrdd.1013

Suto, W. M. I., Clare, I. C. H., Holland, A. J., & Watson, P. C. (2005). Capacity to make financial decisions among people with mild intellectual disabilities. *Journal of Intellectual Disability Research, 49*, 199–209. https://doi.org/10.1111/j.1365-2788.2005.00635.x

Taylor, J. L., Hodapp, R. M., Burke, M. M., Waitz-Kudla, S. N., & Rabideau, C. (2017). Training parents of youth with autism spectrum disorder to advocate for adult disability services: Results from a pilot randomized controlled trial. *Journal of Autism & Developmental Disorders, 47*, 846–857. https://doi.org/10.1007/s10803-016-2994-z

Timmons, J. C., Whitney-Thomas, J., McIntyre, J. P., Butterworth, J., & Allen, D. (2004). Managing service delivery systems and the role of parents during their children's transitions. *Journal of Rehabilitation, 70*, 19–26.

Watson, J. (2016). Assumptions of decision making capacity: The role supporter attitudes play in the realisation of article 12 for people with severe or profound intellectual disability. *Laws, 5*(1), 6. https://doi.org/10.3390/laws5010006

Wehmeyer, M. L. (2005). Self-determination and individuals with severe disabilities: Re-examining meanings and misinterpretations. *Research & Practice for Persons with Severe Disabilities, 30*(3), 113–120. https://doi.org/10.2511/rpsd.30.3.113

Werner, S., Edwards, M., & Baum, N. T. (2009). Family quality of life before and after out-of-home placement of a family member with an intellectual disability. *Journal of Policy and Practice in Intellectual Disabilities, 6*, 32–39. https://doi.org/10.1111/j.1741-1130.2008.00196.x

Part II
Influences and Considerations in Decision Making

Chapter 6
Theoretical Perspectives on Decision Making

Ishita Khemka

Introduction

Decision making is a complex human activity with profound impact on one's adjustment, well-being, and quality of life. It refers to behaviors by which a preferred option or a course of action is chosen from among a set of alternatives based on certain criteria. Decision theories have been widely applied in many disciplines (e.g., economics, sociology, psychology, consumer behavior) to study organizational, group, or individual decision-making behaviors. Over decades of research, various decision theories across application domains have expanded our understanding of decision making as a process and the myriad factors that inform its development. From formally described decision rules for maximizing rationality and subjective utility to more realistic explanations of human capacity for rational decision making and the interdependent roles of cognition, motivation, and emotion, we consider the theoretical perspectives on decision making, which have significantly expanded to recognize individual differences in both ability and style of decision processing.

The chapter presents an updated review of key theoretical frameworks and related concepts that shape our present day understanding of the multifaceted nature of the independent decision making of adolescents and adults in the general population (see Hickson & Khemka, 2013, 2014; Khemka & Hickson, 2017 for earlier reviews). New insights from recent advances in brain and neurobiological research that have increased our understanding of the neural mechanisms underlying decision making are discussed. This research has yielded important explanations for when decision making is incomplete or suboptimal, especially in characterizing decision-making patterns linked with various etiologies or pathologies. This

I. Khemka (✉)
St. John's University, Queens, NY, USA
e-mail: khemkai@stjohns.edu

I. Khemka, L. Hickson (eds.), *Decision Making by Individuals with Intellectual and Developmental Disabilities*, Positive Psychology and Disability Series,
https://doi.org/10.1007/978-3-030-74675-9_6

comprehensive review highlights key factors and conditions that play informative roles in the study of decision making, particularly in interpersonal situations, a key area of interest to us.

This general body of decision-making research informs our (refers to Khemka & Hickson) study of decision making in individuals with intellectual and developmental disabilities (IDD). Drawing upon this research with the general population, we are better able to find ways to improve the overall effectiveness of the independent decision making of individuals with IDD. To that end, an explanatory model for the study of decision making is proposed – *Pathways Model of Decision Processing (referred to as Pathways Model)*, updated here from previous versions presented in Hickson and Khemka (2013, 2014) and Khemka and Hickson (2017).

Theories of Decision Making

Bounded Rationality

Normative theories of decision making have been central to the study of decision making in specifying decision rules for maximizing the utility of decision outcomes as the *rational* standards to which people's decision behaviors are compared (see Stanovich, 2010). However, Simon's (1967, 1983) concept of *bounded rationality* challenged the notion that people's decision behaviors are derived in normative ways. The view that human *rationality* is bounded (hence limited) resulted in a paradigm shift in decision theories to acknowledge that decision making is not a given *ideal* and that decision-makers deviate from rationality due to inherent factors such as their limitations in cognitive capacity, lack of will power, or because of existing situational constraints (see Gigerenzer & Selten, 2001). Typically, in many real-life situations, decisions are made under conditions of incomplete knowledge, and the likely outcomes of all possible options for the decision at hand may not be clear. People's weightings of the different options may also change over time. In addition, emotions and motivational beliefs have been identified as important individual influences that shape decision behaviors. All these influences divert peoples' decision-making away from the principles set forth by normative models (see Galotti, 2002). This has led to the understanding that people make decisions in real life, within their own capacity limitations (Selten, 2001). This process entails a decision maker considering and evaluating available decision options so that their decision outcome serves their goals in a satisfactory way (also called *satisficing* (Simon, 1983)), even though the selected option may not be the most optimal one. The satisficed suboptimal decision suffices as it delivers *good enough* outcomes, even though it might not be the most fully evaluated best possible decision in a situation (Brown, 2004; Gigerenzer & Goldstein, 1996; Simon, 1956). Schwartz et al. (2011) labeled this way of arriving at satisfactory outcomes as *robust satisficing* wherein a decision maker asks, "what is a 'good enough' outcome, and then seeks the option that will produce such an outcome under the widest set of circumstances" (pp. 209).

As early as 1986, Kuhl pointed out that each of the three basic psychological processes – cognition, motivation, and emotion – provides unique inputs during decision making, as well as interact with each other in complex ways to produce decision outcomes. In exploring the nature of these interdependencies, the normative models, over time, began to be replaced by alternative explanations of what and how real-life decisions are made. Nonnormative theories or models not only enhanced our understanding of the decision-making process but also contributed to an appreciation of individual differences (and constraints) in our cognitive functioning and hence in our ability to expend time and effort for fully calculated rational decisions.

A summary of a few main decision-making theories is briefly presented, followed by a discussion on the role of the three basic psychological processes in decision making.

Prospect Theory

Prospect theory (Kahneman & Tversky, 1979) describes the thought process that occurs when people make decisions while anticipating future outcomes. Its premise is that decision makers value losses and gains differently. In general, people tend to place a greater value on avoiding losses due to the associated sharper emotional reaction to loss than gain. Therefore, when provided with the option of a guaranteed win of a smaller amount ($50) as opposed to a 50-50 chance of winning $100 or potentially winning nothing, most people will choose the first option of winning $50, preferring a certain gain. When gains and losses are equal in likelihood, people are more likely to pick certainty over uncertain outcomes, i.e., the risk. Thus, the framing of a decision problem (in terms of losses or gains) can significantly alter how people assign differential values when evaluating alternatives and the decision outcome (see Kahneman, 2011).

Stage-Based Theories

Stage-based models of decision making (e.g., Byrnes, 1998; Gumpel, 1994; Janis & Mann, 1977) describe a deliberative decision-making process in terms of well-defined, sequential steps that provide insight into how a decision-maker proceeds from beginning to end to arrive at a reasoned decision in a specific situation. These models underscore the sequenced stepwise processing as providing for an optimal, rational decision-making plan. Most stage-based models focus on the processes of generating and evaluating options to arrive at a reasoned decision, while a few include the additional steps of implementing a decision (action) and learning from the consequences and feedback (see Byrnes, 1998). Stage-based accounts have provided a useful framework for identifying and interpreting individual differences in decision making, by assessing the ability of the decision-maker to successfully perform each component step. Although stage-based models typically imply that decision making proceeds with the specified sequenced steps, Galotti (2002) suggested that a predetermined order of deliberative decision-making steps was not necessary and instead proposed to look at components of decision making as being *phase-based*. For instance, she described the phase of the process known as decision structuring as a process during which the decision-maker creates a *short list* of options and considers (implicitly or explicitly) the criteria (e.g., goals) for evaluating those options.

Dual-Process Theories

Researchers have long recognized dual modes of intuition-based (referred to as Type 1) and reasoning-based (referred to as Type 2) processing during decision making (see Stanovich, 2011). Type 1 processes are fast, automatic, and effortless and are most readily (and therefore intuitively) available to a decision-maker for acting upon. Unless interrupted or overridden by slower, effortful, and controlled Type 2 processes that are considered more analytical and deliberative, it is assumed that most decision making proceeds initially with Type 1 thinking. Reyna and Brainerd's (2011) *fuzzy-trace* theory provides insight into the decision-makers' mental representations during decision making. According to this theory, decision-makers construct two types of representations (verbatim vs. gist) of the decision-making situation based on their past knowledge and values. The first type, verbatim, is based on getting the precise information needed for deliberative processing. The second type is based on a gist representation of the decision problem and often incorporates emotion-related information. The gist-based decision processing consists of impressionistic processes of intuition and therefore is regarded as being fuzzy. The theory posits that decision-makers do equally well with either approach, although those inclined toward gist-based processing might be more prone to adverse biased or framing effects.

Use of Heuristics

Kahneman and Tversky's (1979) research provided a basis for uncovering decision *heuristics*, or mental shortcuts, as the kinds of mechanisms and representations that individuals tend to use to simplify their decision process. Heuristics support satisfactory decisions to be made quickly and with ease (Shah & Oppenheimer, 2008). According to Gigerenzer and Goldstein (1996), if cognitive mechanisms (e.g., heuristics) involving fast and frugal reasoning allow for successful reasoning in the real world, then those mechanisms do not necessarily need to meet the formal norms of rational inferencing. In the social domain, researchers argue that under conditions of high unpredictability, competing goals, or time pressure, simple heuristics for decision making can serve an adaptive function. They can reduce the cognitive demands of searching through all available alternatives by offering an approximation to the available options to effectively handle the social interaction (see Hertwig et al., 2013).

Basic Processes in Decision Making

Cognition

Several sources of cognitive influences are known to contribute to decision-making competence and accuracy. Higher-order mental abilities help identify a problem that requires a decision to be made, generate alternatives, and anticipate and evaluate consequences (Gottfredson, 1997; Neisser et al., 1996). These mental abilities, most importantly, relate to the level of cognitive ability defined as "mental capability that…involves the ability to reason, plan, solve problems, think abstractly, comprehend complex ideas, learn quickly, and learn from experience" (Gottfredson, 1997, p. 13). Risk awareness and the ability to anticipate and predict challenges, or to see events as opportunities, and to adjust behavior, accordingly, plays a key role in decision making. This has been described as the adaptive nature of intelligence through information processing and analytical thinking (Haier, 2016; Hunt, 2011; Sternberg & Kaufman, 2011). Other cognition-related abilities that characterize decision functioning, for Type 2 reasoning-based processing, are associated with *fluid abilities*, such as speed of processing, memory, and executive functioning (Schaie & Willis, 2002).

Executive function, comprised of a group of related subprocesses (e.g., working memory, inhibitory control, and cognitive flexibility) (Miyake et al., 2000) has a prepotent effect on higher-order cognitive thought, action, and emotion (e.g., Zelazo et al., 2008) and, therefore, on decision making. Executive functioning abilities are psychological processes that underlie conscious, goal-directed thought and behavior. Emotions can play a heavy role on how individuals access their executive functioning and decision-making skills in fast-paced, high-pressured situations. Referred to as *hot cognitions* in decision making, these psychological processes driven by emotions are largely automatic, activated almost immediately in response to high-intensity emotion-based stimuli and high reactivity in the reward system (Steinberg, 2010). These decisions tend to be unplanned and impulsive, often subject to bias, and thus prone to error. In contrast, decisions based on *cool cognitions* are more reflective and focused on planning and slow deliberation (Prencipe et al., 2011).

The use of hot cognitions in decision making might be prompted by the nature of the decision task, such as in emergency situations (e.g., by first responders); reward or sensation seeking (e.g., dangerous driving, substance use); or in high-risk situations that bring intense emotions (e.g., sexual encounters). Hot cognitions are also more pervasive in situations of immature executive functioning, as commonly seen during the developmental period of adolescence, when executive functioning and the ability to modulate emotional experiences may lag the development of other cognitive skills, leading adolescents to engage more often in impulsive acts and to make errors in judgment (Diamond, 2002).

Motivation

Decision making has been defined as an intentional, goal-directed behavior. Motivation, the activation or energization of goal-oriented behavior is therefore an essential mechanism by which decision making proceeds in response to a specific decision problem. Motivation serves an important underlying function in decision making by providing the *motive* and the *effort* for engaging in and successfully completing a decision process. Motivation is seen in terms of internal (intrinsic motivation) and external (extrinsic motivation) forces that influence how we act or why we behave in a certain way. Motivation is tied to positive reinforcement when a decision results in a satisfactory outcome, or to a negative effect when the decision outcome is not successful. When successful actions are repeated often enough, those decisions become habits, being selected almost intuitively or automatically over time, with little to no planning.

Motivations can be linked to emotional processes during decision making or be generated in response to demands of specific situations. Not all situations are equally motivating or requiring the same level of motivational resources. Two general ways of how motivation might impact decision making have been studied: (a) the goal-driven intent of motivational inputs and (b) the motivational environment of the decision-maker as shaped by personal attributions, agencies of control, and emotion-laden beliefs that determine the level of engagement. Therefore, in the context of decision making, motivation pertains to both goal-related processes and to a person's self-beliefs about her/his capabilities as they relate to the extent to which a goal is attainable and worth pursuing. Motivational factors that are effective in one individual might not be effective in others. Therefore, motivation for decision making and the allocation of *self*-resources varies significantly based on individual differences and overlap with cognitive and emotional functions.

Goals and Intentions

Byrnes (1998) defined decision making as the process of choosing a course of action from among two or more alternatives while in the pursuit of one's goals. Hockey (2011) suggested that goal selection and motivational effort allocation are arbitrated by higher-order cognitions and control functions. From a biological perspective, several brain areas must work in coordination for the regulation of the different internal cost-benefit mechanisms relating to effort monitoring (e.g., anterior insula), reward (e.g., nucleus accumbens), and action outcomes (e.g., anterior cingulate cortex) (see Boksem & Tops, 2008). Individuals may be exposed to a full spectrum of attitudes, values, and cultural influences that interact to shape their motivational tendencies. Goals can be positive or negative. Positive goals are the ones that an individual decision-maker tries to attain, such as staying healthy, having friends, getting a tangible product, etc., while negative goals are the ones that the decision-maker tries to avoid or escape from, such as escaping from an embarrassing situation, coercion, or abuse.

Decisions range in complexity and level of goal evaluation. People are motivated by more than one goal and often experience conflicting pressures or motivations for and against certain goals. The impact of competing goals is dependent on their relative strength and ease of attainment. How we balance our goals and how we elect to pursue them depend on their motivational content, with certain types of goals receiving preference as specific needs arise (Kruglanski, 1996a, b, c).

Attitudes and Beliefs

Individuals often circumvent the use of detailed decision rules to simplify their decision making to a limited set of viewpoints or expectations based on their values, beliefs, or preferences. These viewpoints based on evaluations of past experiences and possible future events offer quick satisficing ways to reach a decision outcome. An individual's subjective reactions to decision making over time become predictable, simplified ways of approaching a decision situation.

Naturalistic decision-making models (see Greitzer et al., 2010; Klein, 1997, 2008) examine the kinds of decisions people make in challenging conditions of uncertainty, time pressure, and natural constraints, often in the course of their work (e.g., fire fighters, nurses). In contrast to structured approaches where the decision-making task and goal(s) are predefined and controllable, naturalistic views focus on how decision-makers use their experience (and expertise) to respond to ill-defined and often unplanned (as in emergency or crisis situations) situations by recognizing patterns and relying on subconscious, automatic matching of novel situations to previously encountered situations without engaging in a deliberative evaluative process. Rapid decision making is primed by an individual's recall of experienced events and long-term memory of cues, actions, and expectations that over time build into nonconscious, instinctive knowing patterns of how best to make in-the-moment decisions (see Allen, 2011). Often, in time-critical situations, the goals that the decision-maker wants to commit to are not known at the outset but get clarified while they are being pursued in the decision-making context.

Emotion

Emotion and Affect

Emotions play an integral role in shaping our decision behaviors. Research on emotion, affect (i.e., unspecified feelings, mood, or emotion-related traits), and decision making has increased rapidly, underscoring that "emotions constitute potent, pervasive, predictable, sometimes harmful and sometimes beneficial drivers of decision making" (Lerner et al., 2015) (p. 799). The role of emotions has been widely studied in the field of neuroscience, where decision making is linked with activity in different brain regions such as the amygdala-striatal-based emotional system or the ventromedial prefrontal cortex that underlie deliberative processing involving both emotion and cognition (De Martino et al., 2006; Reyna & Zayas, 2014). The interrelatedness of these systems is evident when impairment in these areas of the brain results in reduced ability to feel emotions and to make optimal decisions that cannot be explained simply by the resulting cognitive changes (Bechara et al., 1999). Reyna and Zayas (2014) drew attention to a third interconnected brain system, the insula, which can modulate emotional-cognitive activity to alter the perceived reward value of potential decision outcomes.

Extending our understanding of the role of emotion and its interaction with cognition increases our ability to clearly view the dichotomous systems of rational and emotional thinking involved in decision making and to account for the variation in decision behaviors that we commonly see among individuals. Evaluation of a situation in terms of the emotion and the affect it generates sets the intensity of decision-making engagement for the decision-maker and has consequences for her/his behaviors by compelling the actions required for decision making in that situation. Although emotions are primarily regarded as having a negative influence on a person's decision-making process (De Martino et al., 2006; Martin and Delgado, 2011), research reveals that whether a specific emotion ultimately improves or weakens a specific decision outcome depends on interactions among the cognitive and motivational mechanisms triggered by each emotion, within the overall framework of decision making. In a study by Zhao (2006) on the impact of induced positive vs. negative emotions on risky decision making, positive emotions were associated with risk-averse behaviors as a motivation to protect positive feeling states in contrast with negative emotions prompting risk-seeking behaviors as a way for elevating one's negative affective state.

Whether emotions have negative or positive valence, they strongly and routinely influence and shape behavior. Knowing about the different types of emotions that can impact decisions can offer cues to the decision-maker for how best to regulate the emotions toward her or his best interest. According to George and Dane (2016) and Lerner et al. (2015), emotions can operate at an integral level, arising primarily because of the specific decision task or situation at hand or they can be incidental, triggered by remembering a previously charged experience, but unrelated to the decision at hand. Effects of integral emotions can operate at both conscious (deliberatively) and nonconscious (intuitively) levels, although incidental emotions tend to operate largely automatically without much self-awareness.

Another mechanism by which emotions may influence decisions involves providing the implicit motivation, or goal, to pursue during decision making. According to Izard (2009), the motivational activation of emotions can come from either a momentary emotion-cognition interaction in deciding how to respond (on the spur of the moment) or via a more enduring impact of personality traits, shaped by emotions, that evolve developmentally over time. Either way, emotions play a critical role in the unconscious operation of the mental processes by providing the motivational value for a decision. The action tendencies (motivated behavior) stem from the quality of consciousness that characterizes the felt emotions and thereby informs decision making. Similarly, Zeelenberg et al. (2008) endorse this view as *feeling is for doing*, suggesting that discrete emotions trigger discrete implicit goals. Focusing on the motivational aspect of emotions allows for a better understanding of the role of emotion in decision making.

According to Nelissen et al. (2007), "an emotional state is characterized by a motivational tendency to the attainment or maintenance of a particular, emotion-specific end-state" (p. 902). Motivational response to the activated emotions sets the goal state which serves to anchor the decision-maker's evaluation of possible outcomes in the specific situation (e.g., anger provides the motivation to respond to injustice; anticipation of regret provides a reason to avoid excessive risk-taking) (Loomes & Sugden, 1982; Solomon, 1993). This indicates the involvement of a goal-activation mechanism in the decision process derived from the evaluation of the consequences of the experienced emotion. Since different emotions are associated with different goals, emotions are recognized for providing control precedence during the decision process because they direct decision-makers to different courses of action (Frijda, 1986). In other words, the experience of a particular emotion brings forth a specific goal that may overrule other goals.

Biases

Strong emotions or past impressions can remarkably bias decision making even when cognitive information would suggest otherwise rational, alternative courses of action (see Loewenstein, 1996; Loewenstein et al., 2001). *Biases* tend to attach themselves to decision targets, changing to strong preferences or habits over time that are then difficult to dislodge (Rozin & Royzman, 2001). An example of such a bias is when an individual who fears flying chooses to drive a long distance rather than taking a flight, well knowing that the base rates for death by driving are much higher than those for flying for the equivalent mileage (Gigerenzer, 2004). Similarly, decisions anchored by people's superstitions or long-term commitments can lead to bias in the decision-making process as only a limited set of possible alternatives might be canvassed during the decision search (Etzioni, 1988; Gilovich et al., 1985). For instance, people might select higher-priced items to match their favorite brand or lucky color even when the information of other lower cost options of similar quality is readily available. A well-studied bias in decision making is the impact of framing effects that can subjectively affect the weighting of the different attributes of the problem.

Emotional Arousal and Stress

The impact of emotional arousal and stress on decision making is well studied in the literature. In earlier understandings, the impact of arousal and stress on decision making was regarded largely as negative. For instance, it was assumed that under conditions of high stress or levels of anxiety, decision making would be less than optimal (e.g., Janis & Mann, 1977; Toda, 1980). However, research now suggests that stress and decision making are intricately connected, and whether the resulting effect is positive or negative depends not only on the type of decision-making situation but also on any confounding effects of individual factors such as personality and demographics. Gutnik et al. (2006) described the relationship between emotional arousal and decision making as one that resembles an inverted U-shape. Under high levels of arousal, important information may be overlooked as the focus of attention is too narrow; under low levels of arousal, the focus of attention might be too broad and include unnecessary irrelevant information in the decision process. On the other hand, moderate levels of arousal would be most conducive to a balanced allocation of attention and use of relevant information. In a review by Starcke and Brand (2012), both laboratory-induced stress and naturally occurring stress were found to mediate underlying mechanisms of decision making, including strategy use, feedback processing, and reward and punishment sensitivity. The effects were most likely to be detrimental (negative impact of stress) in situations involving risk avoidance, strategy use, or other higher-level systems. In other situations, where stress was perceived as a challenge rather than a threat, stress appeared to have a positive impact bringing about an increase in cognitions toward more adaptive decision making. According to them, "If stress confers an advantage or disadvantage in terms of outcome depends on the specific task or situation" (p.1228). In a recent study by Byrne et al. (2020), researchers found that induced acute stress allowed for greater tolerance of uncertainty during decision making, enhancing preference for high-uncertainty options that were associated with highly variable rewards but with long-term advantage over low-uncertainty options with smaller yet more consistent rewards. This indicates that stress can be manipulated in an adaptive way to make decision-makers physiologically cross over the initial threshold of uncertainty in favor of longer-term outcomes that lead to learning for maximizing rewards.

Suggested patterns of response to the levels of emotional arousal might vary not only by situation but also by the level of decisional control exercised, such as when an individual is known to have extensive experience with a decision domain (Klein, 2009). This was evident in a study by Leprohon and Patel (1995) who found the decision making of nurses in emergency telephone triage situations to be 100% accurate even though decisions were being made extremely rapidly under the highest levels of urgency. Therefore, evaluation of a situation, in terms of the intensity of affect and arousal it generates for the decision-maker, sets the mechanisms for the type and level of decision engagement and the consequences of the actions required for decision making in that situation. This explains how the experience of stress might influence decision making differently for different people, depending on an individual's ability to handle the impact of stress.

Although we commonly relegate the primary function of emotion to be a direct cause of behavior (e.g., as in fear causing fleeing), emotion also influences behavior indirectly via a feedback system (see Baumeister et al., 2007). Emotional states promote (or discourage) future behavior by providing feedback and appraisal of actions. Rapid, automatic affective responses, particularly those due to the intensity of the emotional experience, inform cognition and behavioral choices to create a strong memory to guide future decisions. The automatic affective responses in a decision situation might remind the person of past emotional outcomes and provide useful guides as to what emotional outcomes may be anticipated in the present, thereby creating a feedback cycle. Drawing from such emotional inputs, motivational efforts are impacted, fostering strong interdependence between these factors during decision making.

Interdependence of Cognitive, Motivational, and Emotional Processes in Decision Making

The decision-making parameters do not remain static across multiple decision-making contexts. Rather, they change as a function of shifting goals and priorities and emotional states. Decision theorists have drawn upon the work of Bandura (1997) and Weiner (1986) to illustrate how cognitive thinking during decision making can be influenced by individual perceptions of self-efficacy or control and the decision-maker's feelings of agency about their ability to carry out a particular action. Shah and Kruglanski (2000) proposed that the interaction of cognition with motivation during decision processing produces different types of motivational input that affect the extent and direction of information processing.

There is ample evidence to suggest that reasoned cognitive decision processing might depend on prior accurate emotional processing (Bechara & Damasio, 2005). The influence of factors related to emotional processing may occur at multiple levels of the decision-making process, both conscious and unconscious, including physiological reactions (e.g., Bechara et al., 1997, ten Brinke et al., 2019). As noted earlier, executive function is an umbrella term for a series of cognitive skills, including selective attention and inhibition (Diamond, 2013). These two skills are considered mechanisms of attentional control, as selective attention supports the focusing of attention on task-relevant cues, while inhibitory control relates to the inhibition of task-irrelevant cues (Eysenck et al., 2007). Attentional control theory posits that anxiety reduces top-down attentional control. That is, rather than attending to goal-directed information, individuals high in trait anxiety would be biased toward the processing of stimuli related to a perceived threat.

Optimal decision making in affective contexts relies on both the skills needed in cognitive decision making and an individual's affective control and reward sensitivity. Affective control refers to goal-directed, future-oriented processes triggered in contexts where emotion and motivation play a role and include balancing immediate gratification with long-term rewards (Zelazo & Carlson, 2012). Whereas cognitive control is related to cognitive functioning, affective control is related to emotional functioning (Poon, 2018). Affective control is necessary during decision making in affective contexts to downregulate emotional activation.

The interrelatedness of cognition and emotion in decision making is further reinforced when there is a high possibility of emotional bias leading to asymmetric processing of emotional stimuli. As recent research in laboratory settings has shown, the level of emotional bias (i.e., focus on negativity vs. positivity) in decision making can be impacted by several factors such as the level of arousal, type of task, or cultural background (see Yuan et al., 2019). Additionally, influential cognitive processes related to attention and interpretation of salient information during decision making might precede the impact of emotional bias. For instance, attention bias reflects how much attention is allocated to high-valence (high emotional value) information. After the initial perception and evaluation of the valence, an interpretation bias resolves how ambiguous information in decision situations might be interpreted, based on one's previous learning and memory. Such study of emotional bias, and related cognitive mechanisms, is particularly relevant in conditions of neurodevelopmental disorders related to anxiety or mental disorders where affective regulation (or dysregulation) underpins the psychopathologies. This promotes greater understanding of the intricately linked processes of emotion and cognition and reveals important mechanisms to target for intervention support. Lam et al. (2020) propose that "they should be studied synergistically rather than in isolation. Perhaps a productive way to move forward is to address whether different aspects of cognitive processing mediate/moderate specific aspects of emotional biases in healthy and mentally unwell individuals" (p 157).

Individual Determinants of Decision Making

Personal Factors

Understanding the factors that influence decision-making processes is essential to understanding how decisions are made. Several personal factors influence the process of decision making in ways that have an impact on outcomes. These factors include gender (Weller et al., 2018; Toplak et al., 2017), experience (Juliusson et al., 2005), age, and other individual differences (Bruine de Bruin et al., 2007).

Gender

Several studies have measured gender differences in decision-making styles or other personality traits that inform self-regulatory inputs during decision making and overall decision-making competence. Reported gender differences have been linked to both neuropsychological and biological differences (e.g., van den Bos et al. 2013) and socialization processes (Eagly, 1995), causing behavioral and motivational differences in the ways that decision goals are selected and acted upon by women and men. Findings of measured gender effects have not been consistent. For instance, Stanovich et al. (2017) reported nonsignificant gender differences on decision subcomponents, such as framing, risk knowledge, and decision confidence. In another study, Toplak et al. (2017) found men to perform better than women on the use of a composite heuristic and bias index that involved cognitive reflection in the process of decision making. Most studies have suggested that the observed gender differences occur in the context of one or more of other influential factors (e.g., age, personality, culture) and require further study.

In a study by Cross et al. (2011) that explored gender differences in impulsivity traits on behavioral risk-taking tasks, researchers found that the differences were related to motivational inputs rather than effortful behavioral control. Men showed significantly higher sensation seeking in comparison to women, with no significant differences on executive function tasks measuring effortful control. Delaney et al. (2015) found age- and gender-based individual differences in decision-making styles, with older people showing a greater preference for an independent/self-controlled decision-making style and women, relative to men, reporting a higher interpersonal-oriented dependent profile, in comparison to an affective/experiential profile when making important decisions. In a study by Weller et al. (2018), involving a large online community adult sample (N = 804), the researchers tested associations between specific personality dimensions and individual differences on indicators of decision-making competency (applying decision rules, resistance to framing, recognizing social norms, and consistency in risk perception). Among the subset of competencies examined, only the application-of-decision-rules scale showed evidence of differential functioning across groups, suggesting that gender differences may be confined to specific subcomponents, rather than reflecting global differences in decision making between women and men. Higher levels of conscientiousness, honesty/humility, and openness were associated with higher decision-making competencies, with two traits demonstrating gender-specific associations. Specifically, higher decision-making competency scores were associated with higher levels of emotionality (related to tendency to create emotional bonds with others) in women only, while low extraversion (related to approach motivations and greater potential reward sensitivity) was associated with higher decision-making competency scores for men. Although gender is an important personal factor in decision making, research on the relationship between gender and information processing styles (typically characterized as intuitive vs. analytical) used during decision making has been somewhat inconclusive. A few studies have found women to be higher in their intuitive thinking than men, who tend to engage more in deliberative thinking (e.g., Pacini & Epstein, 1999; Gigerenzer et al., 2013). In contrast, no gender differences in thinking styles have been found elsewhere (Delaney et al., 2015). Findings of these studies suggest that regardless of the underlying etiology of gender influences, personality traits related to self-regulatory cognitive and behavioral control (and cognitive flexibility) require further study to determine how these aspects might operate differently for men and women to shape increased rational thinking during decision making.

Experience

Decision making is impacted by experience as decision-makers learn how to respond optimally (or near optimally) based on prior observations; and motivations to engage in decision processing may change because of past interactions in a decision-making setting. Juliusson et al. (2005) indicate that past decisions influence the decisions people make in the future. Given a similar decision situation, people are more likely to decide in the same way as in the past, especially if the past decision had been a positive experience. To the extent that an experience did not lead to a successful outcome, people will tend to avoid making the mistake again and choose differently (Sagi & Friedland, 2007). However, using experience favorably for future decision making requires the decision-maker to examine outcomes of past decisions carefully and to be selective to the extent that past decisions are beneficial to future decision making.

Individual Differences

Research has indicated that individual differences related to age, socioeconomic status (SES), etc. influence decision making (Finucane et al., 2005). The influences of age on decision making have been most extensively studied during the developmental period of adolescence when reward-seeking and risk-taking behaviors significantly shape decision making along with situational effects such as influence of peers, social expectations, etc. Research studies (e.g., Cauffman et al., 2010) show that response inhibition, emotion regulation, calibration of reward seeking, and the capacity to anticipate and learn from negative consequences, all increase with maturation. Therefore, adolescents will adjust their decision-making behaviors over time with age and learning experience (see Hickson & Khemka (2014) for a more detailed discussion on the impact of age on decision making). Finucane et al. (2005) reported a significant difference in decision making across age, with a focus on older people experiencing declines in decision-making performance because of declining cognitive functioning and other age-related influences.

Cultural and Social Influences

Sociocultural attitudes and group norms solicit strong preferences during decision making, creating cultural influences that can have a significant impact on individual and group decisions. Cross-cultural variation is observed in the decision modes people use, their preferences, negotiation styles, creativity, and more. These differences among people can be attributed to culturalized norms or preferences. Culture refers to the "myriad ways of living exhibited by a particular group of people, ways that are transmitted from one generation to the next and which distinguish that group from others" (cf. Smith, 1997) (cited in Yates & de Oliveira, 2016, p. 106). Many other factors have been studied as a proxy for culture including nationality, religion, and social class as giving rise to distinctly identifiable *cultures* (Cohen, 2009). The cultural motivations for decision making stem from not only a person's cultural background but also from their values and ethics (Adler, 1991; Brake et al., 1995). These culturally guided attributions provide commitment to a course of action that is intended to serve the interests and values of groups of people within those cultures.

Much of the existing work on culture and decision making has focused on cultural dimensions such as comparing opposing cultural traits such as individualism-collectivism; independence-interdependence (Hofstede, 2011; Markus & Kitayama, 1991; Oyserman et al., 2002; Triandis, 2004); and recently, tightness-looseness as a measure of the strength of expected strictness to cultural norms (Gelfand et al., 2011; Yates & de Oliveira, 2016). These dimensions are associated with different conceptualizations of the *self*. Individualistic cultures bestow greater autonomy on the self and tend to value self-expression and personal goal pursuit independent of others' goals. By contrast, collectivistic cultures view the self as part of a whole, with the expectation to fit in working with his or her group's goals, i.e., to vary one's personal behavior according to the dominant social context.

Cognitive style has also been a popular framework with which to study cultural decision-making differences (Choi et al., 2007). Cognitive style has been theoretically linked to social orientations ascribed by different cultural groups, such that individualists are relatively analytic (attention to logic and details) and collectivists are relatively holistic (attention to relationships and context) (Varnum et al., 2010).

Although normative cultural differences hold strong influences on individual decision-making patterns as discussed above, research also indicates that broad social and cognitive patterns serve as drivers of more specific cultural differences among individuals. The emerging perspective is to acknowledge other relevant influences such as individual's processing styles, the use of strategies, attention, etc. as shaping decision behaviors instead of only the dominant cultural traits. Structures in social environments also guide decision behaviors by activating schemata that represent the ways in which the characteristics of certain events or objects are recalled (e.g., stereotypes, preferred worldviews, commitment along political party lines). In addition, the influence of culture is subjective, and individuals from one cultural group could very well import another culture's decision-making practices to improve their own decision-making effectiveness. In sum, decision-making practices based on culture may have many inherent influences that ultimately reinforce individual decision-making patterns over time.

Environmental and Situational Factors

The role of situational or contextual factors in decision making has been well discussed. The type of decision task (impersonal vs. personal), level of task complexity (dictated by number of evaluation criteria to be considered), or the presence of others in a decision situation can all have a sizeable impact on the process and outcome of decision making (Loewenstein et al., 1989). For instance, in an interpersonal context, the outcomes anticipated for other(s) in the situation may emerge as a potentially salient reference point during decision-making. The social motives might differ in interpersonal decision contexts depending on the nature of the relationship (e.g., in terms of familiarity, status) of the decision maker with others. The level of ambiguity or uncertainty in a decision situation might require a decision-maker to search for more information or adapt to imperfect information. Therefore, the situational effects of risk and uncertainty, often present in social situations, can have a significant impact on the decision-making outcome, independent of one's ability and style of information processing. Janis and Mann (1977) and Hoch and Loewenstein (1991) explain that in situations of high emotional intensity or time pressure, decision-makers might bypass logical and effortful consideration of options and their consequences. Loewenstein (1996) suggested that heightened drive/physical states (e.g., extreme hunger, thirst) or intense moods and emotions, which he labeled visceral factors, can influence an individual's decision process in ways that make them act contrary to their self-interest. "I don't know what got into me" or "I must have been crazy when…" (p 289) are common observed reactions to such out-of-control (irrational) decision making.

During the adolescent years, the influence of peers (O'Brien et al., 2011) can be an influential social contextual variable that impacts decision making by creating a strong situational bias toward greater reward sensitivity for engaging in risky behavior and gaining peer acceptance. Knowing the nature of the decision task is an important consideration in improving decision quality. Ayal et al. (2015) indicate that if individuals can flexibly adjust their information processing style to the situation on hand, they are more likely to make optimal decisions. That is, an intuitive style of information processing is likely to lead to better performance on an intuitive task, and reversely, higher analytical processing is more optimal for analytical-based decision tasks.

Decision-making processes are also influenced by environmental factors associated with the personal histories of individuals and the cultural context within which an individual operates as a decision-maker. Therefore, people's reasoning is guided by their goals, but within prevailing environmental constraints. Over and above the situational variants, the personal relevance of a decision to a person determines the level of effort and commitment to making the decision.

Extending the Research from the General Population to Individuals with IDD

Research on general decision-making theories as reviewed above provides a strong theoretical basis for understanding the components of decision making and the many factors that might impact individual decision making and its effectiveness. These theories provide insight into the complex nature of decision making and the multitude of factors that underpin the level of engagement and success on a decision-making task. Although many of the traditional decision-making research studies have involved laboratory-based experimental tasks with an emphasis on isolating important variables to study, the studies offer insights that can be extended to less controlled, more complex environments. They identify mechanisms and tools that can be used to predict key components of decision making in other domains such as social or interpersonal contexts. Familiarity with the decision-making literature for the general population also helps to evaluate strengths and weaknesses in the decision making of specific groups of individuals, such as those with IDD, and to predict key components of decision making that might help explain differences within and between people with and without IDD.

Most of the decisions we make in everyday life take place in the context of our environmental surroundings, taking into consideration a multitude of factors. Primarily, real-life decisions are made in a social context and might involve an interaction with another person, what we have referred to in our work as interpersonal decision making. Interpersonal decision making might involve a dyadic interaction where two people jointly decide, or it could involve an interpersonal interaction with another person that requires a decision to be made in a situation that represents some level of ambiguity or uncertainty. Most of our decision-making research has focused on this type of interpersonal decision making in situations with a threat of abuse, peer influence, peer aggression, etc. In social situations, the person making the decision will need to invoke a series of basic processes in the cognitive, emotional, and/or motivational domains to integrate the social information into the decision-making process. The information received from the social environment could be implicit (e.g., facial expressions, status of the other person), unconscious (e.g., interpretation of social norms based on experience), or overt and direct, such as a perpetrator using an overt lure or threat to coerce someone in a situation that is physically dangerous.

With the above theoretical underpinnings of decision making in mind, we have been conducting decision-making research for the past two decades with individuals with IDD, to see to what extent research findings based on the general population can inform the study of special populations, particularly individuals with IDD, for whom expected cognitive limitations are likely to interfere with effective decision making. The goal of our research has been to explore the separate as well as combined contributions that these cognitive, emotional, and motivational processes have on social, interpersonal decision making of adolescents and adults with IDD. Based on our current research perspectives on decision making among individuals with IDD, we have developed a model of decision processing that presents a comprehensive view of the important contributors to effective decision making for this group and predicts probable pathways for decision processing in response to a particular decision task in each situation.

Pathways Model of Decision Processing

The *Pathways Model of Decision Processing*, first presented by Hickson and Khemka in 2013 as a framework for decision-making research, is an explanatory model that can be used to assess and interpret decision-making behaviors of individuals with IDD across a range of interpersonal decision-making situations. The model draws from the vast theoretical advances in the study of decision making (e.g., Janis & Mann, 1977; Kahneman, 2011; Newell et al., 2015; Stanovich, 2011) and our own past work (e.g., Hickson & Khemka, 1999, 2001, 2013, 2014; Khemka & Hickson, 2006, 2017) focusing on the decision-making experiences of individuals with IDD. Since interpersonal decisions are made in the context of social situations and in relation to others, these decisions by nature tend to be open-ended and have outcomes that are unknown at the start of the process. Therefore, these decisions are highly sensitive not only to differences in decision-making ability and individual preferences but also to a host of outside influences relating to the social, cultural, and situational profile of the decision-making situation.

All in all, there is no one suggested way for an individual to approach an interpersonal decision task, but rather one must select an approach that matches inputs in a specific decision situation. The model, presented in Fig. 6.1, depicts four alternative pathways of decision processing that set forth the approach to information seeking and decision processing that distinctly defines each of the four pathways. Ideally, effective decision-makers should have a repertoire of decision-making tools that enables the flexible deployment of all four possible pathways calibrated to the demands of a particular situation. Further, the demands for decision making may vary substantially by the type of decision, as indicated by the situational/environmental bar at the base of the model. For example, financial/medical treatment decisions might require involvement of others and usually are made over a relatively longer period in contrast to decisions relating to personal relationships or social interactions, which are typically more immediate and self-determined. The model also highlights the myriad factors that might influence decision processing along the pathways and the interdependencies among them.

Moving upward, the next bar in the model highlights the neurodevelopmental and personal characteristics of the decision-maker as key determinants of which decision pathway will be selected and how it will be applied in a specific environment or situation. The personal characteristics which impact decision making might include age, gender, experience, and social and cultural influences. The neurodevelopmental factors cover etiology and a range of developmental attributes that relate to a person's biological and neurological makeup and determine an individual's

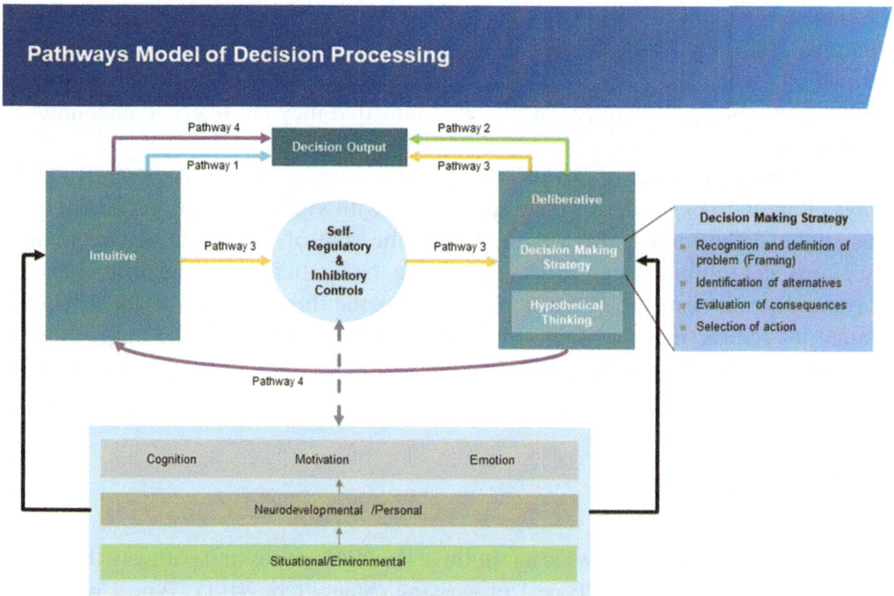

Fig. 6.1 The Pathways Model of Decision Processing

response capabilities in areas of motor functioning, learning, language, communication, neuropsychiatric problems, etc.

The topmost of the three bars in the model takes into consideration each of the basic mental processes of cognition, motivation, and emotion and illustrates how the interplay of these processes might influence an individual's approach to a decision-making situation and her/his ability to adaptively satisfy personal goals to the extent possible. In the domain of cognition, control of cognitive thought and action is determined by aspects of memory, attention, executive functioning, comprehension, consequential thinking, and communication. Motivational determinants in interpersonal decision making are twofold and include those that define individual goal identification and prioritization and those that invoke personal agency beliefs related to one's self efficacy as a decision-maker. As discussed in this chapter, the emotion-based components that likely impact decision behaviors include a range of affective factors (e.g., stress, arousal, feelings), including values and beliefs, and are shaped by personal experience.

In our research studies, we have intended to expand our understanding of what constitutes an effective decision. In general, the decisional intent in healthy relationship situations or positive peer interactions is to be prosocial, while in situations involving abuse or coercive peer interactions, the decisional intent is to be self-protective. Our research methodology has been to use tasks, such as brief descriptive vignettes, to portray interpersonal decision problems based on real-life situations. Respondents are typically asked what the key decision-maker, the protagonist, should do in each of the vignette situations. An overarching goal of Khemka and Hickson's research is to identify interventions and supports to assist individuals with IDD to calibrate their individual selection of decision-processing pathways to the specific needs of the situations that they are likely to encounter in their lives.

The *Pathways Model of Decision Processing* presents the four alternative pathways along with examples of their application with vignette situations drawn from our studies. The model also acknowledges the role of individual self-regulatory functions and inhibitory controls in upholding the application of the different pathways and, in the case of *Pathway 3*, illustrates the direct impact of these mechanisms in redirecting the type of processing to pursue considering implied consequences for the ultimate decision outcome.

Pathway 1: Intuitive

Pathway 1 (follow blue arrow) is an intuitive decision processing pathway that relies almost entirely on intuitive Type 1 processing (Stanovich, 2011). Type 1 processes are deployed rapidly and automatically with little or no effort. This type of responding can apply to a range of decisions, with the possibility of both adaptive and nonadaptive outcomes. In its simplest form, the pathway refers to most everyday decisions that are routine or have minor or nonconsequential outcomes. It might involve selecting, based on familiarity, and associated with emotions, the decision action that represents a strong distinct emotion (such as choosing one's breakfast cereal). In interpersonal situations, an intuitive style of decision making might be optimal when alternatives are limited within a controlled set of conditions and decisions that have minor payoffs or losses. For example, in a healthy coworker relationship situation, Anna's coworker, Marvin, may say to Anna, "It is a really nice day. Would you like to go out for lunch today?" Most likely, Anna would be able to use *Pathway 1* to make a quick decision based on whether she is free and/or whether she would like to go out to lunch with Marvin. This type of decision process might also be called upon if a person is offered a ride home by a friend.

In highly emotionally charged situations, when the decision-maker is guided mostly by strong affective feelings or emotions, the decision could bear high emotional intensity (inducing hot cognition), and the result could be a more rushed, hypervigilant, intuitive decision outcome. These types of decision have a high probability of being nonadaptive as they are made quickly with no evaluation of alternative options. For instance, in situations of peer pressure to engage in sexual activity or substance abuse or other activities that might offer sudden high payoffs, the decision-maker might respond to the possibility of a thrill or reward with a rapid sensation-seeking, risk-taking decision to participate.

In highly time-pressured situations, intuitive decision making might be the most practical way to respond. When time is limited for decision processing, decision-makers responding affectively (based on emotion, general mood state) may be prone to making unwanted errors in their decision making. They may also rely on simplified rules or heuristics in lieu of more extensive processing. The use of heuristics can sometimes outperform more deliberative, analytic decision processing (Payne et al., 1993), but that typically requires careful integration of decision information for success. Finally, decision-makers who act on the spur of the moment by relying on memory, experience, or a quick representation of the available information may follow *Pathway 1* intuitively. There is agreement that the primary source of such intuitive decisions is largely from memory (Kahneman & Klein, 2009) and they are typically based upon a gist-based representation of the situation, as also described in *fuzzy-trace theory* (Reyna, & Brainerd, 1995). The intuitive processes here might relate to associations based on simple learning and retrieval of feelings of liking or disliking or of previous similar successful actions in the interpersonal/social domain. Alternatively, intuitions could be based on accumulating information proceeding through a quick, automatic process involving integration of memory traces with currently available information (Glockner & Witteman, 2010). The extent to which these processes result in adaptive decisions depends on how well the meaning (gist) is derived from memory and whether attention is given to select details of the situation. Undue memory interference, bias in developing meaning based on experience, or excessive affective input might disrupt the balance needed to proceed successfully for this intuitive pathway. Since these decisions are based on intuitive, impressions that are not modified by conscious deliberation, the quality of the decisions made depends on the adaptive nature of those impressions.

Pathway 2: Deliberative

Pathway 2 (follow green arrow) involves reasoned, deliberative (Type 2) processing based on a detailed, verbatim representation of a situation. Type 2 processing is slow and analytic (Stanovich, 2011) and typically involves multiple components or the use of a stepwise strategy for the generation and evaluation of alternative courses of action. Complex interpersonal situations with serious long-term consequences, such as when a person is faced with a decision about what to do regarding a long-term, abusive relationship with a family member or caregiver, often require deliberative *Pathway 2* processing. Individuals with ID tend to encounter difficulty with applying multiple strategy components and with the hypothetical thinking needed to project possible short-term and long-term consequences of potential decision options. For example, in the following abuse situation, whenever Jayde's uncle comes to family parties, he forces Jayde into a bedroom and touches her private parts. Jayde's uncle tells Jayde not to tell anyone. This situation is complex with multiple possible solutions and serious potential consequences. Jayde may need to apply a deliberative strategy to consider her options and their possible consequences and to find the best way to stop the abuse in the immediate situation (a short-term decision) as well as in the future (a long-term decision). Other examples of such reasoned, calculated decisions might pertain to complex life decisions relating to living, employment, medical treatment, or consent to a sexual relationship. In applying this well-reasoned deliberative approach to decision processing for highly analytical decisions, there is likely to be a discrete stage of cognitive processing after initial perception and attention, leading to an iterative cognitive cycle of acquiring and evaluating information needed for decision making.

Pathway 3: Intuitive to Deliberative

Pathway 3 (follow yellow arrow) involves a shift from an initial intuitively based (Type 1) decision process to a deliberative (Type 2) process after self-regulatory and inhibitory mechanisms detect that the situation requires a more deliberative process and overrides the initial Type 1 process. Starting from an intuitive, affect-based decision process, the decision-maker regulates to a more cognitively based reflective (and deliberative) process of generating and evaluating options to arrive at a decision that addresses the goal that becomes relevant to the situation. This pathway accounts for relying on one's emotions or on intuition but prompts the decision-maker to assess the situation more accurately by thinking systematically. Lerner et al. (2015) note this approach to be more adaptive to decision making in highly emotional situations.

The use of this pathway might occur in interpersonal situations that at first glance appear benign, but later reveal themselves to involve threats or coercion, initiating consideration of options and a deliberative decision to satisfy the goal of being safe. For example, in a seemingly benign cyber situation, Jamal has been visiting an online forum for guitar players who share songs that they like. Then, Ian, another forum user, posts, "Jamal must be retarded if he likes that song." Jamal will likely need to employ a *Pathway 3* process, involving a shift from an initial intuitive automatic process with respect to his use of the online forum to a reasoned reflective decision-making process to figure out how best to handle the cyber abuse consisting of a humiliating verbal taunt by Ian. This involves the interplay of the two systems as a quick automatic decision is interrupted to allow for more thoughtful control and a deliberative decision to be made.

Cognitive processing functions related to attention and interpretation may play a significant role in mediating the impact of emotional bias as decision making proceeds through initial automatic and subsequent strategic processing (Yuan et al., 2019). In some cases, the level of affective regulation (or dysregulation) reached might be limited by underlying psychopathological conditions or neurodevelopmental constraints (e.g., IDD, ADHD, anxiety, mental disorders). The self-regulatory and inhibitory controls that help navigate between the reasoned and intuitive systems are not only shaped by such individual differences, but over time provide feedback to inform the further development of those characteristics.

Pathway 4: Deliberative to Intuitive

Finally, a *fourth pathway, Pathway 4,* (follow magenta arrow) develops over time because of practice from repeated decision-making experiences in a specific domain. Repeated deliberative decision processing can build familiarity with alternatives and their likely consequences in a specific domain such that, over time, a new pathway of decision processing evolves to create a decision process in the form of a learned response to be activated from memory in a particular type of situation. Since the rehearsed decision is recalled readily from memory, it resembles an intuitive Type 1 response that proceeds almost automatically. This pathway is most often associated with the type of decisions that emergency workers rely on when having to make skilled, quick decisions in emergency situations (Klein, 1998, 2009). It is also a plausible pathway for those who have built up a repertoire of effective decision actions that they can draw upon and implement rapidly in emergency situations. Klein (2008) described this as a recognition-primed decision that allows an individual to draw upon their experience and arrive at a course of action without having to deliberatively weigh all the alternatives. When primed by experience, or training, with a previous pattern, that has been repeated often, such decision responses could be considered intuitive in the sense they have come to be easily recognized and applied without the use of reasoning or a detailed logical routine. Intuitions, in these situations, become highly accurate, reliable, and credible sources of information for making decisions. These decisions can be emotionally charged and have major consequences. They are simply made through a quick selection of an action and can result in a habit in the long term. For instance, when approached by a group of unfamiliar peers as opposed to one peer to go to a neighborhood park, the immediate response might be to decline the invitation and get to a safe place quickly. When prompted by a telemarketer to share personal financial information, a *Pathway 4* response learned through experience would be to deny the request and to quickly hang up.

Conclusion

The literature reviewed in this chapter articulates the patterns, processes, and variations in decision-making behavior that have been considered across the typical range of human functioning and provides the basis for a comparative view of the decision-making capacities and processes utilized by individuals who are discrepant from the norm, such as individuals with IDD.

Decision making is a largely cognitive psychological process, shaped by substantial inputs from our emotional and motivational systems. Decision making is construed as a goal-oriented process of arriving at a decision action that best meets (or satisfices) an end goal. Decision making can proceed as a deliberative, step-by-step process (*Pathway 2*) or be based upon the use of a rapid, intuitive process

(*Pathway 1*). Regulatory processes can trigger a shift from an intuitive to a deliberative process (*Pathway 3*), or determinants derived from experience or training can result in a shift from a deliberative to a recognition-based intuitive process (*Pathway 4*). In adverse situations, the entire planful process of decision making can be subverted, in favor of affect-based, intuitive responding. Decision-making outcomes are not always optimal.

Given these various ways in which decision making can proceed, it is useful to outline the different pathways, as summarized above, that the decision-maker can take in the decision process. The *Pathways Model of Decision Processing* provides a guide for looking at the various potential determinants of independent decision making for individuals with IDD and describes four different pathways to decision making, Aspects of decision making can be targeted through specific training methods to improve the decision-making capabilities and experience of individuals with IDD. See Chap. 11 in this volume for an application of the *Pathways Model* in identifying key processes that constitute loci of difficulties in decision making for individuals with IDD.

References

Adler, N. J. (1991). *International dimensions of organization behavior* (2nd ed.). PWS-Kent Publishing.

Allen, D. (2011). Information behavior and decision making in time-constrained practice: A dual-processing perspective. *Journal of the American Society for Information Science and Technology, 62*, 2165–2181.

Ayal, S., Rusou, Z., Zakay, D., & Hochman, G. (2015). Determinants of judgment and decision-making quality: The interplay between information processing style and situational factors. *Frontiers in Psychology, 6*, 1088. https://doi.org/10.3389/fpsyg.2015.01088

Bandura, A. (1997). *Self-efficacy: The exercise of control*. Freeman.

Baumeister, R. F., Vohs, K. D., Nathan DeWall, C., & Zhang, L. (2007). How emotion shapes behavior: Feedback, anticipation, and reflection, rather than direct causation. *Personality and Social Psychology Review, 11*(2), 167–203. https://doi.org/10.1177/1088868307301033

Bechara, A., & Damasio, A. R. (2005). The somatic marker hypothesis: A neural theory of economic decision. *Games and Economic Behavior, 52*(2), 336–372. https://doi.org/10.1016/j.geb.2004.06.010

Bechara, A., Damasio, H., Damasio, A., & R. (1997). Deciding advantageously before knowing the advantageous strategy. *Science, 275*(5303), 1293–1295.

Bechara, A., Damasio, H., Damasio, A. R., & Lee, G. P. (1999). Different contributions of the human amygdala and ventromedial prefrontal cortex to decision making. *Journal of Neuroscience, 19*(13), 5473–5481. https://doi.org/10.1523/JNEUROSCI.19-13-05473.1999

Boksem, M. A. S., & Tops, M. (2008). Mental fatigue: Costs and benefits. *Brain Research Reviews, 59*(1), 125–139. https://doi.org/10.1016/j.brainresrev.2008.07.001

Brake, T., Walker, D. M., & Walker, T. (1995). *Doing business internationally: The guide to cross-cultural success*. Irwin.

Brown, R. (2004). Consideration of the origin of Herbert Simon's theory of 'Satisficing'' (1933-1947). *Management Decision, 42*(10), 1240–1256. https://doi.org/10.1108/00251740410568944

Bruine de Bruin, W., Parker, A., & Fischhoff, B. (2007). Individual differences in adult decision making competence. *Journal of Personality and Social Psychology, 92*, 938–956. https://doi.org/10.1037/0022-3514.92.5.938

Byrne, K. A., Peters, C., Hunter, C., Willis, H. C., Phan, D., Cornwall, A., & Worthy, D. A. (2020). Acute stress enhances tolerance of uncertainty during decision making. *Cognition, 205*, 104448. https://doi.org/10.1016/j.cognition.2020.104448

Byrnes, J. P. (1998). *The nature and development of decision making: A self-regulation model*. Erlbaum.

Cauffman, E., Shulman, E., Steinberg, L., Claus, E., Banich, M., Graham, S., et al. (2010). Age differences in affective decision making as indexed by performance on the Iowa Gambling Task. *Developmental Psychology, 46*, 193–207.

Choi, I., Koo, M., & Choi, J. (2007). Individual differences in analytic versus holistic thinking. *Personality & Social Psychology Bulletin, 33*, 691–705. https://doi.org/10.1177/0146167206298568

Cohen, A. B. (2009). Many forms of culture. *The American Psychologist, 64*(3), 194–204. https://doi.org/10.1037/a0015308

Cross, C. P., Copping, L. T., & Campbell, A. (2011). Sex differences in impulsivity: A meta-analysis. *Psychological Bulletin, 137*(1), 97–130. https://doi.org/10.1037/a0021591

De Martino, B., Kumaran, D., Seymour, B., & Dolan, R. J. (2006). Frames, biases, and rational decision making in the human brain. *Science, 313*, 684–687.

Delaney, R., Strough, J., Parker, A. M., & de Bruin, W. B. (2015). Variations in decision-making profiles by age and gender: A cluster-analytic approach. *Personality and Individual Differences, 85*, 19–24. https://doi.org/10.1016/j.paid.2015.04.034

Diamond, A. (2002). Normal development of prefrontal cortex from birth to young adulthood: Cognitive functions, anatomy, and biochemistry. In D. Stuss & R. Knight (Eds.), *Principles of frontal lobe function*. Oxford University Press. https://doi.org/10.1093/acprof:oso/9780195134971.003.0029

Diamond, A. (2013). Executive functions. *Annual Review of Psychology, 64*, 135–168. https://doi.org/10.1146/annurev-psych-113011-143750

Eagly, A. H. (1995). The science and politics of comparing women and men. *American Psychologist, 50*(3), 145–158. https://doi.org/10.1037/0003-066X.50.3.145

Etzioni, A. (1988). Normative-affective factors: Toward a new decision-making model. *Journal of Economic Psychology, 9*, 125–150.

Eysenck, M. W., Derakshan, N., Santos, R., & Calvo, M. G. (2007). Anxiety and cognitive performance: Attentional control theory. *Emotion (Washington, DC), 7*(2), 336–353. https://doi.org/10.1037/1528-3542.7.2.336

Finucane, M., Mertz, C., Slovic, P., & Schmidt, E. (2005). Task complexity and older adults' decision-making competence. *Psychology and Aging, 20*, 71–84. https://doi.org/10.1037/0882-7974.20.1.71

Frijda, N. H. (1986). *Studies in emotion and social interaction. The emotions*. Cambridge University Press; Editions de la Maison des Sciences de l'Homme.

Galotti, K. M. (2002). *Making decisions that matter: How people face important life choices*. Erlbaum.

Gelfand, M. J., Raver, J. L., Nishii, L., Leslie, L. M., Lun, J., Lim, B. C., … Yamaguchi, S. (2011). Differences between tight and loose cultures: A 33-nation study. *Science, 332*, 1100–1104.

George, J. M., & Dane, E. (2016). Affect, emotion, and decision making. *Organizational Behavior and Human Decision Processes, 136*, 47–55. https://doi.org/10.1016/j.obhdp.2016.06.004

Gigerenzer, G. (2004). Dread risk, September 11, and fatal traffic accidents. *Psychological Science, 15*(4), 286–287. https://doi.org/10.1111/j.0956-7976.2004.00668.x

Gigerenzer, G., & Goldstein, D. G. (1996). Reasoning the fast and frugal way: Models of bounded rationality. *Psychological Review, 103*(4), 650–669. https://doi.org/10.1037/0033-295X.103.4.650

Gigerenzer, G., & Selten, R. (2001). Rethinking rationality. In G. Gigerenzer & R. Selten (Eds.), *Bounded rationality: The adaptive toolbox* (pp. 1–12). The MIT Press.
Gigerenzer, G., Galesic, M., & Garcia-Retamero, R. (2013). Stereotypes about Men's and Women's intuitions: A study of two nations. *Journal of Cross-Cultural Psychology, 45*, 62–81. https://doi.org/10.1177/0022022113487074
Gilovich, T., Vallone, R., & Tversky, A. (1985). The hot hand in basketball: On the misperception of random sequences. *Cognitive Psychology, 17*, 295–314.
Glockner, A., & Witteman, C. (2010). Beyond dual-process models: A categorisation of processes underlying intuitive judgment and decision making. *Thinking and Reasoning, 16*(1), 1–25.
Gottfredson, L. (1997). Mainstream science on intelligence: An editorial with 52 signatories, history, and bibliography. *Intelligence, 24*(1), 13–23. https://doi.org/10.1016/S0160-2896(97)90011-8
Greitzer, F. L., Podmore, R., Robinson, M., & Ey, P. (2010). Naturalistic decision making for power system operators. *International Journal of Human-Computer Interaction, 26*, 278–291.
Gumpel, T. (1994). Social competence and social skills training for persons with mental retardation: An expansion of behavioral paradigm. *Educational and Training in Mental Retardation and Developmental Disabilities, 29*, 194–201.
Gutnik, L. A., Hakimzada, A. F., Yoskowitz, N. A., & Patel, V. L. (2006). The role of emotion in decision making: A cognitive neuroeconomic approach towards understanding sexual risk behavior. *Journal of Biomedical Informatics, 39*(6), 720–736. https://doi.org/10.1016/j.jbi.2006.03.002
Haier, R. J. (2016). *The neuroscience of intelligence*. Cambridge University Press.
Hertwig, R., Hoffrage, U., & the ABC Research Group. (2013). *Simple heuristics in a social world*. Oxford University Press.
Hickson, L., & Khemka, I. (1999). Decision making and mental retardation. In L. M. Glidden (Ed.), *International review of research in mental retardation* (Vol. 22, pp. 227–265). Academic Press.
Hickson, L., & Khemka, I. (2001). The role of motivation in the decision making of people with mental retardation. In H. N. Switzky (Ed.), *Personality and motivational differences in persons with mental retardation* (pp. 199–256). Erlbaum.
Hickson, L., & Khemka, I. (2013). Problem solving and decision making. In M. Wehmeyer (Ed.), *The Oxford handbook of positive psychology and disability*. Oxford University Press.
Hickson, L., & Khemka, I. (2014). The psychology of decision making. In R. Hodapp (Ed.), *International review of research in developmental disabilities* (Vol. 47, pp. 185–229). Elsevier.
Hoch, S. J., & Loewenstein, G. F. (1991). Time inconsistent preferences and consumer self-control. *Journal of Consumer Research, 17*, 492–507.
Hockey, G. R. J. (2011). A motivational control theory of cognitive fatigue. In P. L. Ackerman (Ed.), *Decade of behavior/science conference. Cognitive fatigue: Multidisciplinary perspectives on current research and future applications* (pp. 167–187). American Psychological Association. https://doi.org/10.1037/12343-008
Hofstede, G. (2011). Dimensionalizing cultures: The Hofstede model in context. *Online Readings in Psychology and Culture, 2*(1). https://doi.org/10.9707/2307-0919.1014
Hunt, E. (2011). *Human intelligence*. Cambridge University Press.
Izard, C. E. (2009). Emotion theory and research: Highlights, unanswered questions, and emerging issues. *Annual Review of Psychology, 60*, 1–25. https://doi.org/10.1146/annurev.psych.60.110707.163539
Janis, I. L., & Mann, L. (1977). *Decision making: A psychological analysis of conflict, choice, and commitment*. The Free Press.
Juliusson, E. A., Karlsson, N., & Gärling, T. (2005). Weighing the past and the future in decision making. *European Journal of Cognitive Psychology, 17*(4), 561–575. https://doi.org/10.1080/09541440440000159
Kahneman, D. (2011). *Thinking, fast and slow*. Farrar, Straus, and Giroux.
Kahneman, D., & Klein, G. (2009). Conditions for intuitive expertise: A failure to disagree. *American Psychologist, 64*(6), 515–526.

Kahneman, D., & Tversky, A. (1979). Prospect theory: An analysis of decision under risk. *Econometrica, 47*(2), 263–291. https://doi.org/10.2307/1914185

Khemka, I., & Hickson, L. (2006). The role of motivation in the decision making of adolescents with mental retardation. In L. M. Glidden & H. N. Switzky (Eds.), *International review of research in mental retardation, volume 31*. Elsevier.

Khemka, I., & Hickson, L. (2017). The role of positive psychology in interpersonal decision making. In K. A. Shogren, M. Wehmeyer, & N. N. Singh (Eds.), *Handbook of positive psychology in intellectual and developmental disabilities: Translating research into practice* (pp. 145–171). Springer.

Klein, G. (1997). The recognition-primed decision (RPD) model: Looking back, looking forward. In C. E. Zsambok & G. A. Klein (Eds.), *Naturalistic decision making* (pp. 131–140). Lawrence Erlbaum Associates.

Klein, G. (1998). *Sources of power: How people make decisions*. MIT Press.

Klein, G. (2008). Naturalistic decision making. *Human Factors, 50*(3), 456–460. https://doi.org/10.1518/001872008X288385

Klein, G. (2009). *Streetlights and shadows: Searching for the keys to adaptive decision making*. MIT Press.

Kruglanski, A. W. (1996a). Motivated social cognition: Principles of the interface. In E. T. Higgins & A. W. Kruglanski (Eds.), *Social psychology: Handbook of basic principles* (pp. 493–520). The Guilford Press.

Kruglanski, A. W. (1996b). A motivated gatekeeper of our minds. In R. M. Sorrentino & E. T. Higgins (Eds.), *Handbook of motivation and cognition*. Guilford.

Kruglanski, A. W. (1996c). Goals as knowledge structures. In P. M. Gollwitzer & J. A. Bargh (Eds.), *The psychology of action: Linking cognition and motivation to behavior* (pp. 599–618). Guilford Press.

Kuhl, J. (1986). Motivation and information processing: A new look at decision making, dynamic change, and action control. In M. Sorrentino & E. T. Higgins (Eds.), *Handbook of motivation and cognition: Foundations of social behavior*. Guilford.

Lam, C. L. M., Leung, C. J., Yiend, J., & Lee, T. M. C. (2020). The implication of cognitive processes in emotional bias. *Neuroscience and Biobehavioral Reviews, 114*, 156–157.

Leprohon, J., & Patel, V. L. (1995). Decision making strategies for telephone triage in emergency medical services. *Medical Decision making, 15*, 240–253.

Lerner, J. S., Li, Y., Valdesolo, P., & Kassam, K. (2015). Emotion and decision making. *Annual Review of Psychology, 66*(1), 799–823.

Loewenstein, G. (1996). Out of control: Visceral influences on behavior. *Organizational Behavior and Human Decision Processes, 65*, 272–292. https://doi.org/10.1006/obhd.1996.0028

Loewenstein, G. F., Thompson, L., & Bazerman, M. H. (1989). Social utility and decision making in interpersonal contexts. *Journal of Personality and Social Psychology, 57*(3), 426–441. https://doi.org/10.1037/0022-3514.57.3.426

Loewenstein, G. F., Weber, E. U., Hsee, C. K., & Welch, N. (2001). Risk as feelings. *Psychological Bulletin, 127*(2), 267–286. https://doi.org/10.1037/0033-2909.127.2.267

Loomes, G., & Sugden, R. (1982). Regret theory: An alternative theory of rational choice under uncertainty. *The Economic Journal, 92*, 805–824.

Markus, H. R., & Kitayama, S. (1991). Culture and the self: Implications for cognition, emotion, and motivation. *Psychological Review, 98*(2), 224–253. https://doi.org/10.1037/0033-295X.98.2.224

Martin, L., & Delgado, M. R. (2011). The influence of emotion regulation on decision making under risk. *Journal of Cognitive Neuroscience, 23*(9), 2569–2581.

Miyake, A., Friedman, N. P., Emerson, M. J., Witzki, A. H., Howerter, A., & Wager, T. D. (2000). The unity and diversity of executive functions and their contributions to complex "Frontal Lobe" tasks: A latent variable analysis. *Cognitive Psychology, 41*(1), 49–100. https://doi.org/10.1006/cogp.1999.0734

Neisser, U., Boodoo, G., Bouchard, T. J., Boykin, A. W., Brody, N., Ceci, S. J., et al. (1996). Intelligence: Knowns and unknowns. *American Psychologist, 51*, 77–101.

Nelissen, R. M. A., Dijker, A. J. M., & De Vries, N. K. (2007). Emotions and goals: Assessing relations between values and emotions. *Cognition and Emotion, 21*(4), 902–911. https://doi.org/10.1080/02699930600861330

Newell, B. R., Lagnado, D. A., & Shanks, D. R. (2015). *Straight choices: The psychology of decision making*. Psychology Press.

O'Brien, L., Albert, D., Chein, J., & Steinberg, L. (2011). Adolescents prefer more immediate rewards when in the presence of their peers. *Journal of Research on Adolescence, 21*, 747–753.

Oyserman, D., Coon, H. M., & Kemmelmeier, M. (2002). Rethinking individualism and collectivism: Evaluation of theoretical assumptions and meta-analyses. *Psychological Bulletin, 128*(1), 3–72. https://doi.org/10.1037/0033-2909.128.1.3

Pacini, R., & Epstein, S. (1999). The relation of rational and experiential information processing styles to personality, basic beliefs, and the ratio-bias phenomenon. *Journal of Personality and Social Psychology, 76*(6), 972–987. https://doi.org/10.1037/0022-3514.76.6.972

Payne, J. W., Bettman, J. R., & Johnson, E. L. (1993). *The adaptive decision maker*. Cambridge University Press.

Poon, K. (2018). Hot and cool executive functions in adolescence: Development and contributions to important developmental outcomes. *Frontiers in Psychology, 8*, 2311.

Prencipe, A., Kesek, A., Cohen, J., Lamm, C., Lewis, M., & David, Z. (2011). Development of hot and cool executive function during the transition to adolescence. *Journal of Experimental Child Psychology, 108*(3), 621–637.

Reyna, V. F., & Brainerd, C. J. (1995). Fuzzy-trace theory: An interim synthesis. *Learning and Individual Differences, 7*(1), 1–75. https://doi.org/10.1016/1041-6080(95)90031-4

Reyna, V. F., & Brainerd, C. J. (2011). Dual processes in decision making and developmental neuroscience: A fuzzy-trace model. *Developmental Review, 31*, 180–206.

Reyna, V. F., & Zayas, V. (2014). *The neuroscience of risky decision making*. American Psychological Association.

Rozin, P., & Royzman, E. B. (2001). Negativity bias, negativity dominance, and contagion. *Personality and Social Psychology Review, 5*(4), 296–320. https://doi.org/10.1207/S15327957PSPR0504_2.S2CID4987502

Sagi, A., & Friedland, N. (2007). The cost of richness: The effect of the size and diversity of decision sets on post-decision regret. *Journal of Personality and Social Psychology, 93*(4), 515–524. https://doi.org/10.1037/0022-3514.93.4.515

Schaie, K. W., & Willis, S. L. (2002). *Adult development and aging* (5th ed.). Prentice-Hall.

Schwartz, B., Ben-Haim, Y., & Dacso, C. (2011). What makes a good decision? Robust satisficing as a normative standard of rational decision making. *Journal for the Theory of Social Behaviour, 41*(2), 209–227.

Selten, R. (2001). What is bounded rationality? In G. Gigerenzer & R. Selten (Eds.), *Bounded rationality: The adaptive toolbox* (pp. 13–36). The MIT Press.

Shah, J. Y., & Kruglanski, A. W. (2000). Aspects of goal networks: Implications for self regulation. In M. Boekaerts, P. R. Pintrich, & M. Zeidner (Eds.), *Handbook of self regulation* (pp. 85–110). Academic.

Shah, A. K., & Oppenheimer, D. M. (2008). Heuristics made easy: An effort-reduction framework. *Psychological Bulletin, 134*(2), 207–222. https://doi.org/10.1037/0033-2909.134.2.207. Simon's (1967, 1983).

Simon, H. A. (1956). Rational choice and the structure of the environment. *Psychological Review, 63*(2), 129–138.

Simon, H. A. (1967). Motivational and emotional controls of cognition. *Psychological Review, 74*(1), 29–39. https://doi.org/10.1037/h0024127

Simon, H. A. (1983). *Reason in human affairs*. Stanford University Press.

Solomon, R. C. (1993). *The passions: Emotions and the meaning of life*. Hackett.

Stanovich, K. E. (2010). *Decision making and rationality in the modern world*. Oxford University Press.
Stanovich, K. E. (2011). *Rationality and the reflective mind*. Oxford University Press.
Stanovich, K. E., West, R. F., & Toplak, M. E. (2017). *The rationality quotient*. MIT Press. https://doi.org/10.7551/mitpress/9780262034845.001.0001
Starcke, K., & Brand, M. (2012). Decision making under stress: A selective review. *Neuroscience and Biobehavioral Reviews, 36*, 1228–1248.
Steinberg, L. (2010). A dual systems model of adolescent risk-taking. *Developmental Psychobiology, 52*, 216–224.
Sternberg, R. J., & Kaufman, S. B. (Eds.). (2011). *Cambridge handbook of intelligence*. Cambridge University Press.
ten Brinke, L., Lee, J. J., & Carney, D. (2019). Physiological reactions when observing lies and truths: Evidence and an intervention to enhance accuracy. *Journal of Personality and Social Psychology, 117*, 560–578. https://doi.org/10.1037/pspi0000175
Toda, M. (1980). Emotion in decision making. *Acta Psychologica, 45*, 133–155.
Toplak, M. E., West, R. F., & Stanovich, K. E. (2017). Real-world correlates of performance on heuristics and biases tasks in a community sample. *Journal of Behavioral Decision making, 30*, 541–554. https://doi.org/10.1002/bdm.1973
Triandis, H. C. (2004). The many dimensions of culture. *Academy of Management Executive, 18*, 88–93. https://doi.org/10.5465/AME.2004.12689599
van den Bos, R., Homberg, J., & de Visser, L. (2013). A critical review of sex differences in decision making tasks, focus on the Iowa Gambling Task. *Behavioral Brain Research, 238*, 95–108.
Varnum, M. E., Grossmann, I., Kitayama, S., & Nisbett, R. E. (2010). The origin of cultural differences in cognition: Evidence for the social orientation hypothesis. *Current Directions in Psychological Science, 19*(1), 9–13. https://doi.org/10.1177/0963721409359301
Weiner, B. (1986). *An attributional theory of motivation and emotion*. Springer-Verlag.
Weller, J., Ceschi, A., Hirsch, L., Sartori, R., & Costantini, A. (2018). Accounting for individual differences in decision-making competence: Personality and gender differences. *Frontiers in Psychology, 9*, 2258. https://doi.org/10.3389/fpsyg.2018.02258
Yates, J. F., & de Oliveira, S. (2016). Culture and decision making. *Organizational Behavior and Human Decision Processes, 136*, 106–118.
Yuan, J., Tian, Y., Huang, X., Fan, H., & Wei, X. (2019). Emotional bias varies with stimulus type, arousal and task setting: Meta-analytic evidences. *Neuroscience and Biobehavioral Reviews, 107*, 461–472. https://doi.org/10.1016/j.neubiorev.2019.09.035
Zeelenberg, M., Nelissen, R., Breugelmans, S., & Pieters, R. (2008). On emotion specificity in decision making: Why feeling is for doing. *Judgment and Decision making, 3*, 18–27.
Zelazo, P. D., & Carlson, S. M. (2012). Hot and cool executive function in childhood and adolescence: Development and plasticity. *Child Development Perspectives, 6*(4), 354–360.
Zelazo, P. D., Carlson, S. M., & Kesek, A. (2008). Development of executive function in childhood. In C. A. Nelson & M. Luciana (Eds.), *Handbook of developmental cognitive neuroscience* (2nd ed., pp. 553–574). MIT Press.
Zhao, J. (2006) The Effects of Induced Positive and Negative Emotions on Risky Decision making. Talk Presented at the 28th Annual Psychological Society of Ireland Student Congress, Maynooth, May 2006, 22 p.

Chapter 7
Etiology and Decision Making: Considerations and Implications

Brittney L. Goscicki, Cynde K. Josol, Marisa H. Fisher, Elisabeth M. Dykens, and Robert M. Hodapp

Introduction

As do those in the general population, individuals with intellectual and developmental disabilities (IDD) must make significant life decisions. These decisions involve a host of personal issues—for example, choosing to take a particular job or managing one's finances—as well as navigating interpersonal situations with family members, friends, and romantic partners. Indeed, the ability to manage personal and interpersonal affairs partly defines what it means to function as an adult in our society, and these issues can at times be challenging for any of us.

For individuals with IDD, however, managing personal and interpersonal affairs can become even more difficult. Individuals with IDD are prone to increased vulnerability to maltreatment and victimization, as well as to reduced agency over their

B. L. Goscicki (✉) · R. M. Hodapp
Department of Special Education, Peabody College, Vanderbilt University, Vanderbilt Kennedy Center's University Center for Excellence in Developmental Disabilities, Nashville, TN, USA
e-mail: brittney.l.goscicki@vanderbilt.edu; robert.hodapp@vanderbilt.edu

C. K. Josol · M. H. Fisher
Department of Counseling, Educational Psychology, and Special Education, Michigan State University, East Lansing, MI, USA
e-mail: josolcyn@msu.edu; fishermh@msu.edu

E. M. Dykens
Department of Psychology and Human Development, Peabody College, Vanderbilt University, Vanderbilt Kennedy Center's University Center for Excellence in Developmental Disabilities, Nashville, TN, USA
e-mail: elisabeth.m.dykens@vanderbilt.edu

I. Khemka, L. Hickson (eds.), *Decision Making by Individuals with Intellectual and Developmental Disabilities*, Positive Psychology and Disability Series,
https://doi.org/10.1007/978-3-030-74675-9_7

lives more generally (Fisher et al., 2016). But why these individuals more often undergo such negative experiences is only beginning to be understood. One important factor likely relates to decision making, the ability to make important, reasoned decisions from among several options. For individuals with IDD, it is essential to understand their decision-making patterns, so that interventions can be created to address key deficits and lead to improved life outcomes.

In considering such decision-making abilities, we acknowledge the difficulties that decisions can entail for anyone. Typically, everyday decisions are made with certain degrees of incompleteness or uncertainty, wherein a person has inadequate knowledge of all of the salient elements or may not appreciate that certain, seemingly important issues are irrelevant to the present choice (Hickson & Khemka, 2014). Even when one has identified all of the relevant factors, the processing activities involved in decision making place significant demands on a person's cognitive system. Rarely, for example, are we able to keep in mind (and appropriately weigh) every variable in making any major life decision, nor can we always determine which personal or environmental factors are important, which are unimportant, and which might change in importance in the years to come. For this reason, heuristics or *mental shortcuts* are often utilized to streamline decision making; oftentimes, these shortcuts are based on one's habits, values, and past experiences (Kahneman & Tversky, 1979). Given that heuristics are used to simplify the information-processing components of decision making, it is important to understand the specific decision patterns of those with IDD.

At the same time, persons with IDD are not homogeneous, and the population with IDD is actually comprised of individuals with different etiologies or causes. Individuals might have Williams syndrome, Prader-Willi syndrome, Down syndrome, or any of approximately 1000 different genetic IDD conditions (Hodapp & Dykens, 2012). Groups with many of these etiologies have their own phenotypic characteristics (Dykens et al., 2000). To date, however, decision-making studies have generally examined heterogeneous samples of persons with IDD (Hickson & Khemka, 2014). As a result, we know little about how etiology impacts specific decision-making patterns, including both adaptive and dysfunctional styles.

This chapter summarizes advances made in understanding the link between etiology and decision-making behaviors among individuals with IDD. For the purposes of this chapter, the decision-making processes discussed pertain to independent rather than supported decision-making. The terms *decision* and *choice* are used interchangeably, as the terms can be used synonymously in everyday contexts and situations. We begin with key background information, including principles used when examining etiology and issues of decision-making in general and for those with IDD. We then address etiological differences in decision-making patterns and their corresponding behavioral implications. We conclude with etiology-specific recommendations to improve the decision making of persons with different forms of IDD.

Etiology in IDD: Basic Issues

Over the past three decades in the field of IDD, important strides have been made in our understandings of etiology. We now know, for example, that approximately 1000 genetic anomalies cause IDD. Over the years, we have also advanced our understanding of these causes (Dykens, in press). Traditionally, 75% of cases of IDD were thought to arise from unknown causes (and went by a variety of names), with only 25% due to one or another identified genetic or nongenetic etiology (Zigler, 1967). With recent genetic advances, this ratio of unknown:known percentages has gradually changed, such that the ratio now more closely approximates a 50:50 split (Karam et al., 2015).

Beyond the increasing numbers of identified etiologies that account for higher percentages of the overall IDD population, the IDD field has increasingly appreciated that, when it comes to behavior, etiology matters. Although illustrated in many ways, one needs only to compare over the decades the numbers of articles on genetic conditions that have appeared in the "Journal of Intellectual Disability Research," the "American Journal on Intellectual and Developmental Disabilities," "Intellectual and Developmental Disabilities," and the "Journal of Policy and Practice in Intellectual Disabilities." Across these major IDD research journals, one sees large decade-by-decade increases in the number of etiology-related research articles. From relatively few articles—usually limited only to those with Down syndrome—during the 1980s (Hodapp & Dykens, 1994), the numbers of psychological research articles have increased exponentially during the 1990s and 2000s to include etiologies such as Williams syndrome, Prader-Willi syndrome, and fragile X syndrome (Hodapp & Dykens, 2012). Admittedly, this increased research focus remains centered on only a few of the 1000 known etiological conditions; still, these increases illustrate the importance of etiology in modern IDD research.

Genotype: Phenotype Approach

To understand the effects of genetic etiology on behaviors, one must go beyond sheer numbers and return to the classic discovery of genotypes and phenotypes. Geneticists have long distinguished between an individual's genotype—one's particular genetic inheritance—and one's phenotype, or the outward manifestation of that genotype. In most cases, phenotypes have been thought to consist of height, hair, skin, eye color, or other physical features. In addition, high-blood pressure, cancer, obesity, or other physical and health issues have been considered phenotypes.

In more recent years, IDD researchers have applied this genotype-phenotype approach to behavior (Dykens et al., 2000). First proposed by William Nyhan (1972) in relation to Lesch-Nyhan disease (a disorder characterized by severe self-injurious behavior), this concept was explored more fully beginning in the 1990s (Flynt & Yule, 1994; Hodapp & Dykens, 1994). The idea is that, like physical features or health problems, specific genetic anomalies can also have outcomes, or phenotypes, that relate to behavior. The term behavioral phenotypes implies this genotype: phenotype connection.

Behavioral Phenotypes: Three Principles

Before discussing how the behavioral phenotypes of certain genetic conditions relate to decision-making styles, we first provide three overarching principles. These principles show the strengths and limits to this approach.

Probabilistic Nature of Behavioral Phenotypes

When considering a chromosomal anomaly—for example, the presence of a third chromosome 21 in Down syndrome or the deletion on chromosome 7q11.23 in Williams syndrome—we generally see probabilistic, not deterministic, outcomes. In contrast to a deterministic outcome—for which a genetic disorder determines the outcome for every individual with the syndrome—probabilistic outcomes make it more likely that groups with a genetic condition will demonstrate the *usual* or *characteristic* behavioral feature. Behavioral phenotypes of specific genetic syndromes, then, involve "the heightened probability or likelihood that people with a given syndrome will exhibit a certain behavioral and developmental sequence relative to those without the syndrome" (Dykens, 1995, p. 523).

A good example involves Down syndrome. On average, groups with Down syndrome show difficulties in several aspects of language, with grammar as an area of special deficit relative to overall mental age. But not every person with Down syndrome shows such grammatical difficulties. In a book-length exploration, Rondal (1995) detailed the language abilities of a 32-year-old Frenchwoman named Francoise. Although by all measures Francoise had trisomy 21 (i.e., Down syndrome), she nevertheless uttered long and complex sentences. Rondal (1995, p. 117) reports her saying (translated), "And that does not surprise me because dogs are always too warm when they go outside" ("Et ca m'etonne pas parce que les chiens ont toujours trop chaud quand ils vont a la port"). While most individuals with Down syndrome do show grammatical weaknesses relative to their overall functioning levels (in cognition or in overall communication), this single counterexample shows that the effects of a genetic anomaly are probabilistic, not deterministic, in nature.

Few Behavioral Outcomes Are Unique to a Single Syndrome

A second important point relates to the uniqueness of a particular behavioral outcome. Referred to as *total vs. partial specificity* (Hodapp, 1997), most etiology-related behaviors are not specific to one and only one genetic etiology. Examples of total specificity include the extreme self-mutilation in most individuals with Lesch-Nyhan syndrome (Schretlen et al., 2005), extreme hyperphagia (overeating) in Prader-Willi syndrome (Dykens et al., 2007), or cat-cry in infants with 5p-syndrome (Sigafoos et al., 2009). In each instance, no other syndromes seem to show these behaviors to anywhere near the same extent.

In most cases, however, particular outcomes are *shared* (i.e., show partial specificity) with more than a single syndrome. Two or more syndromes might, for example, show a particular proneness to aggression, depressive episodes, obsessions, or other maladaptive behavior or demonstrate a specific profile of cognitive strengths and weaknesses. These shared behaviors likely arise because, in terms of genetic etiologies, there exist greater numbers of genetic etiologies (over 1000) than possible behavioral outcomes. Or, to quote the clinical geneticist John Opitz (1985), "The causes are many, but the final developmental pathways are few" (p. 9).

Phenotypes Involve Many Aspects of Health-Physical Characteristics and Behavior That Change Over Time

Just as, in the general population, phenotypic outcomes might include a person's height, hair color, and high blood pressure, so too do phenotypic outcomes relate to multiple aspects of physical-medical and behavioral characteristics. In many cases, physical-medical and behavioral characteristics even relate to one another.

To again use Down syndrome as an example, studies over many decades have shown that, on average, children with Down syndrome show particular linguistic-communicative deficits, especially in the areas of articulation and grammar. At the same time, from 35% to 70% of individuals with Down syndrome also show hearing problems (Porter & Tharpe, 2010), and most individuals show a characteristic physical structure of the tongue in terms of thickness, size, and placement within the mouth (Bunton & Leddy, 2011). In turn, these children's hearing problems and larger, thicker tongues have also been implicated as contributing to the weaknesses in articulation and inconsistent use of grammatical morphemes such as word endings like "s" or "-es" for plurals and "-ed" for regular past tense (Abbeduto et al., 2007).

Within behavior itself, specific etiologies show connections to various aspects of behavior. In several syndromes, individuals are more prone to maladaptive behavior psychopathology. Examples here include the propensity in Prader-Willi syndrome to show obsessions and compulsions (Dykens et al., 1996) and, in Williams syndrome, to show high rates of fears and phobias (Dykens, 2003) and generalized anxiety (Kennedy et al., 2006; Leyfer et al., 2009). In addition to psychopathology, several etiological groups also demonstrate specific patterns of cognitive-linguistic strengths and weaknesses (e.g., the relatively low levels of grammar in Down syndrome), as well as etiology-related patterns of strengths-weaknesses on certain IQ tests. The best example might be the pattern of "simultaneous processing over sequential processing" found on the Kaufman Assessment Battery for Children (or K-ABC) test, a pattern that has been noted among boys with fragile X syndrome (Dykens et al., 1987) and individuals with Prader-Willi syndrome (Dykens et al.,

1992). Importantly, phenotypes evolve and change with development throughout the entire life course (Dykens, 2013). Thus, hyperphagia in Prader-Willi syndrome begins in early childhood and waxes and wanes over time; the anxieties and fears in people with Williams syndrome may intensify with advancing age; and risks for Alzheimer's dementia are heightened in middle-aged adults with Down syndrome.

One important aspect of behavior involves personality and motivation. Individuals with Down syndrome, even from their toddler years onward, attend more to people than to objects (Kasari & Freeman, 2001). These individuals also use their social abilities to try to avoid performing difficult cognitive tasks, a pattern of behaviors that Pitcairn and Wishart (1994) referred to as engaging in *party tricks*. Similarly, children with Williams syndrome, even from very early ages, have been described as having social and outgoing personalities (Davies et al., 1998; Doyle et al., 2004), exhibiting high levels of social motivation. Differentiating among studies of temperament-personality, social attention, and hyper-sociability/approachability, Thurman and Fisher (2015) have noted how various aspects of the *sociable personality* of Williams syndrome relate to real-world outcomes (see below).

Decision Making Process

When considering how each of us makes decisions, we begin with the idea that decision making is a complex process. First, we describe decision making in the general population, before briefly describing a current framework, the *Pathways of Decision Processing model* (Hickson & Khemka, 2013, 2014), that synthesizes the roles of mental processes (i.e., cognition, motivation, and emotion) to describe different plausible decision-making pathways used by persons with various etiologies of IDD.

Decision Making in the General Population

A critical consideration in decision making is the complexity of the situation or task itself. Specifically, differences exist between relatively straightforward decisions as opposed to complex decisions—one example might be the difference between the decision of what to wear for a job interview (easy decision problem) vs. where to go to college (complex decision problem). Compared to decisions concerning wardrobe choices, complex problems like college selection typically involve some uncertainty and ambiguity, especially in the early stages.

When embarking on complex decision making, one may need to use strategies (Mintzberg et al., 1976; Powell et al., 2011), especially in cases of decision making under uncertainty. Making strategic decisions in uncertain circumstances differs from decision making in which the context is well known. For example, if one is unsure of what to wear for a job interview, one can perform a simple online search for recommendations. In contrast, a strategic approach may be more appropriate for deciding which college to attend, especially in the early stages, when many unknowns exist (e.g., waiting for notifications from other colleges). An individual's comprehension is strengthened only if they continuously work through the complexity of the problem (Mason & Mitroff, 1981; Polasky et al., 2011).

While the complexity of a problem may seem limitless, the human information processing capacity is not. At any one time, for example, humans can keep in mind only 7 ± 2 separate items (Miller, 1956). Similarly, humans are also notoriously inept at *weighing* different risks, which explains why we are incredibly fearful about the risks of very low-risk events such as crashing in airplanes or experiencing shark attacks, but less worried about higher-risk events such as experiencing a house fire or car accident. Each of these human inadequacies may affect strategic decision making. Therefore, for complex, situations, the decision-making process may rely heavily on strategies that can help simplify the complexity of the problem.

In the general population literature, various descriptive models exist of decision making, each involving a number of different stages (e.g., Halpern-Felsher & Cauffman, 2001; Mazzolini, 1981; Steptoe-Warren et al., 2011). These models, which share some common characteristics, can be organized into three distinct stages (Schwenk, 1984).

Stage 1: Problem Identification/Goal Formulation

In numerous decision-making models for complex situations, problem identification and goal formulation consist of two activities: recognition and diagnosis (Mintzberg et al., 1976; Steptoe-Warren et al., 2011). Recognition refers to identifying the problem and considering all possible information. Consider again the problem of deciding which college to attend. Prospective students need to consider all available information: housing options, financial aid services, whether the college has one's desired courses or majors, social life, and even location (in a city, suburban or rural area; near vs. far from home). Such information helps the prospective student make decisions on the college that will best fit their education and career goals.

In almost every case, however, information can change or be inaccurate. In this first stage, then, one needs to perform a diagnostic process, which involves determining which types of information a person needs to collect and how to attain it. If, for example, a prospective student wants to know about a particular college's social life, is it better to consult the college's official website, peruse a popular guidebook of US colleges, or talk to someone from their high school who attends that college? And what if different sources disagree?

Stage 2: Alternatives Generation

After identifying the problem and formulating possible solutions, one needs to next generate strategic alternatives. In doing so, individuals may search their memories or their environments (e.g., helpful heuristics) for alternative solutions to the problem (Halpern-Felsher & Cauffman, 2001). To solve the problem of a company operating at a loss, for instance, one might think about filing for bankruptcy, applying for federal or state loans, reducing the number of staff, or charging more for products or services. But if too rigid during this stage, a person might never consider potentially better alternatives. It is also possible to overfocus on a single solution. In the example above, before searching for alternative solutions, one might quickly decide to file for bankruptcy, defending bankruptcy as the only possible solution (i.e., single outcome calculation; Steinbruner, 1974). One might even rationalize this decision by highlighting the negative outcomes of all others—*If I receive a business loan, I will have to pay it back*, or *If I increase prices, then my customers will take their business elsewhere*. Failure to generate and consider all possible alternative solutions may then negatively affect the success of strategic decision making.

Stage 3: Evaluation and Selection

Regardless of which strategic alternatives are considered, a decision-maker must ultimately select a solution to the problem. In the company example, one must decide if filing for bankruptcy really is the better solution compared to applying for a loan, reducing the number of staff, or increasing prices. But certain cognitive biases can affect the evaluation and selection of viable solutions. For example, even if a problem presents high levels of uncertainty, one can become overly confident about dealing with the problem's complexity due to representative bias, or overestimating the extent to which a solution to a similar problem can be generalized to another problem (Chen et al., 2007).

These three general stages of strategic decision making—identifying the problem, generating alternative solutions, and evaluating and selecting among those solutions—are not mutually exclusive. Instead, Schwenk (1984) states that these stages are mutually dependent. The process of decision making is also iterative and cyclical, reflecting the ambiguity and uncertainty of complex problems.

Pathways Model of Decision Processing for IDD

In summarizing the findings about decision making for complex problems, our knowledge derives mostly from the general population. For the most part, decision making models are lacking for the IDD population. But one model, Hickson and Khemka's (2013, 2014) the *Pathways Model of Decision Processing*, synthesizes extant models and outlines four possible social decision-making pathways used by individuals in both the general and IDD populations. Based on both intuition and deliberative strategizing, these pathways capture the roles that cognitive ability, motivation, and emotion all play in Decision making, as well as the situation's contextual demands. To capture how individuals with IDD might exhibit either adaptive or aberrant decision-making patterns, we now delineate the different pathways in this model and provide examples for each.

The first pathway involves low-effort and rapid intuitive decision making. In Pathway 1, an individual makes a quick decision, with little hard thinking and using a gist-based representation of context. For example, if someone's friend invites them to dinner, the person could use this pathway to quickly decide whether they are free and would like to accept the invitation. Although mostly applicable to relatively simple problems, individuals with IDD may exhibit atypical Pathway 1 processing patterns if they display rigid thinking or might have difficulties regulating emotions and intentions.

In contrast, Pathway 2 features calculated, reasoned decision making. Like generating reasoned alternatives to complex problems, such decision making employs higher-order hypothetical thinking to generate and analyze alternatives. Examples might involve deciding if one should live independently or considering how to handle an abusive relationship. Such situations are complex, with several possible solutions and profound current and future consequences. But given the high cognitive demand that such deliberative and planful decisions imply, individuals with IDD may show significant deficits with this pathway.

Pathway 3 accounts for reasoned decision making, but only after an individual's regulatory mechanisms override their initial intuitive, automatic processing. For example, a person might originally decide to comply with an abuser's demand, but then learn to stop and think about how to escape the situation by formulating and evaluating a course of action. In another instance, a person might share a game they like on an online forum, but then be verbally taunted by another user. In response, the abused game player would need to shift from their original intuitive processing to more reasoned decision making in order to stop the abuse (e.g., continue to play the game even when taunted or play another game instead). Given that persons with IDD often lack these regulatory skills (Hickson & Khemka, 2014), they may be prone to not perceiving when an initially benign interpersonal situation later becomes coercive.

Finally, Pathway 4 features individual-specific decision feedback loops that, from repeated experiences, develop over time. For example, consider an individual who has a long learning history of telling another adult when someone is bothering them. This person may have relied on this particular response over and over again in similar circumstances. This response pattern might prove inadequate for persons with IDD, who may have limited past experiences with effective decision making (Hickson & Khemka, 2014). In such cases, individuals with IDD may require explicit instruction to develop a repertoire of different strategies to draw from.

As implied above, decision-making pathways share many similarities for those with and without IDD. Individuals may rely on similar strategies for making complex decisions. By providing a structure for targeting areas of need, the *Pathways Model of Decision Processing* (Hickson & Khemka, 2013, 2014) can be useful in addressing the unique circumstances of individuals with IDD.

Tying Etiology to Decision Making: Three Examples

In this section, we discuss three etiological groups, describing how etiology may impact decision making in relation to the *Pathways Model of Decision Processing* (Hickson & Khemka, 2013, 2014). Given the paucity of current research in this area, the ensuing examples and explanations are largely speculative; future research should yield more definitive conclusions.

Williams Syndrome

Overview

Williams syndrome is a complex neurodevelopmental disorder caused by a deletion of about 28 genes on chromosome 7 (at 7q11.23; Hillier et al., 2003). Williams syndrome occurs in approximately 1 in 7500 to 20,000 live births each year (Pober, 2010), affecting both genders equally (American Academy of Pediatrics Committee on Genetics, 2001). Individuals with Williams syndrome present a unique cognitive and behavioral profile consisting of mild to moderate intellectual disability (ID) (e.g., average IQ between 50 and 60 (Martens et al., 2008; Searcy et al., 2004)) and relative strengths in adaptive social skills compared to daily living and motor skills (Fisher et al., 2016; Mervis & Pitts, 2015). In fact, early characterizations of Williams syndrome described a group that exhibits excellent language abilities despite the presence of ID (Thurman & Fisher, 2015). Those with Williams syndrome also show heightened experiences of anxiety and attention/emotion regulation difficulties (Dykens, 2003; Leyfer et al., 2006).

Salient Behavioral Phenotype

While those with Williams syndrome show many interesting cognitive and behavioral characteristics, an especially salient trait involves hyper-sociability. Regardless of whether the person is familiar or unfamiliar (Thurman & Fisher, 2015), these individuals display an extreme interest in interacting with other people (Klein-Tasman et al., 2011; Mervis et al., 2003). Persons with Williams syndrome have been described as overly friendly (Gosch & Pankau, 1997), charming (Fryns et al., 1991), a friend to everyone (Beuren et al., 1962), and never going unnoticed in a group setting (Dykens & Rosner, 1999). Such recurrent descriptions have led researchers to hypothesize that hyper-sociability is a syndrome-specific trait—a part of the behavioral phenotype of Williams syndrome (Rothbart & Derryberry, 1981; Thomas et al., 1968).

Hyper-sociability can be observed across the life span. Using nine temperament characteristics outlined by Thomas and Chess (1977), Tomc et al. (1990) compared the temperament profiles of 1- to 12-year-old typically developing children to same-aged children with Williams syndrome. Compared to the typically developing children, those with Williams syndrome demonstrated higher ratings of approaching others. Klein-Tasman and Mervis (2003) extended the earlier study to compare parent reports of 8- to 10-year-old children with Williams syndrome to (IQ-matched) children with other neurodevelopmental disorders. Even compared to those with other disorders, children with Williams syndrome displayed more sociability, including greater excitement and positive anticipation for expected pleasurable events and orientation toward people, with less negative affect.

Hyper-sociability has also been observed in adults with Williams syndrome. Interviewing caregivers of 70 adults aged 19–39 years, Davies et al. (1998) found that the vast majority of parents/caregivers expressed concern about their adult child's hyper-sociability, which often expressed itself in the form of over-friendliness. Using the parent-report Salk Institute Sociability Questionnaire and comparing those with Williams syndrome to (mental- and chronological age-matched) groups with other disabilities and to typically developing age-mates, heightened sociability was again reported for individuals with Williams syndrome, both across the life span (Doyle et al., 2004; Haas et al., 2010) and across cultures (Zitzer-Comfort et al., 2007).

While a seemingly desirable trait, hyper-sociability may place individuals with Williams syndrome at risk for social victimization, as they consistently display a lack of inhibition toward approaching unfamiliar people (Fisher et al., 2014; Thurman & Fisher, 2015). Other characteristics may also compound these risks. For example, due to a heightened sense of empathy and a desire to please, individuals with Williams syndrome often give the wrong impression about their intentions (Davies et al., 1998) or misread the intentions of others (Jawaid et al., 2012). Together with hyper-sociability, these factors may make it more difficult to make appropriate decisions in social relationships.

Hyper-sociability and impaired decision making may place those with Williams syndrome at high risk for experiencing abuse and victimization. Specifically, these characteristics may place them at risk of engaging with a stranger who has ill intentions, or even walking away with a stranger who presents a lure (Fisher et al., 2014). Because they often display overly affectionate behaviors toward familiar individuals (e.g., touching, kissing, and hugging others), they are also at increased risk for sexual abuse or victimization (Davies et al., 1998). Indeed, in one study of 38 young adults with Williams syndrome, Fisher et al. (2013) found that 79% of caregivers reported that these individuals had experienced some form of victimization during their lifetime. In the same study, caregivers reported that 16% of these adults had experienced some form of physical or sexual abuse, and other studies found that 20% of individuals with Williams syndrome had reported a sexually abusive experience (Davies et al., 1998).

Behavioral Ties to Decision Making

Williams syndrome's behavioral traits present unique challenges regarding potential targets for interventions. Due to this group's hyper-sociability and heightened risks of victimization, it is critical to address decision making in the context of social relationships and interactions. In addition, IDD decision-making frameworks like the *Pathways Model of Decision Processing* (Hickson & Khemka, 2013) can help identify which approaches individuals with Williams syndrome can take to make the most appropriate decisions.

Consider a situation in which an adult with Williams syndrome is approached by a stranger at a park. The stranger states that their car has a flat tire and asks for help. This scenario thus involves a complex situation, mostly because the intent of the stranger is unknown. The individual with Williams syndrome may use Pathway 2, which centers around reasoned reflection to generate and evaluate possible alternative solutions. On the one hand, the adult with Williams syndrome might consider agreeing to help, especially if the stranger seems nice or if a companion can also come along. Conversely, the adult might decline to help, noting that the person is unfamiliar and one should be wary of strangers.

In the case of decision-making by adults with Williams syndrome, however, Pathway 2 may be inhibited if the adult also possesses a heightened sense of empathy and a desire to please others. To address the influence of such traits, the adult with Williams syndrome may need to shift to Pathway 3, interrupting and overriding initial intuitive decisions. Depending on the strength of hyper-sociability and the desire to please others, overcoming one's original (prosocial) intuition might require explicit training.

In some decision-making contexts, the hyper-sociability of individuals with Williams syndrome may actually serve as a strength. Due to their hyper-sociability, individuals with Williams syndrome are also highly attentive to the facial expressions of others (Thurman & Fisher, 2015). Attentiveness to facial features can automatically trigger quick judgments about the trustworthiness of a person, thereby affecting the decision-making process (Engell et al., 2007). Current research suggests that individuals with Williams syndrome are able to perceive and select relevant physical cues to accurately predict the trustworthiness of other people (Gomez et al., 2020). This strength of individuals with Williams syndrome can be particularly useful in "stranger-lure" situations like the one described above and can be incorporated into interventions.

Prader-Willi Syndrome

Overview

Prader-Willi syndrome is a genetic disorder caused by a disruption on the paternally inherited chromosome 15. In most cases (about 2/3), there exists a paternal deletion—or missing part—on chromosome 15q11-13; less commonly, the child receives two chromosome 15's from the mother (called "maternal uniparental disomy") (for a review see Cassidy et al., 2012). Prader-Willi syndrome occurs in ~1 in 15,000–25,000 births and equally impacts both males and females. Prader-Willi syndrome is characterized by mild-to-moderate ID, growth hormone deficiencies, sleep disturbances (apnea, excessive daytime sleepiness), psychiatric disorders, infantile hypotonia (i.e., weak muscles), hyperphagia (overeating), and high risks of obesity (Cassidy et al., 2012).

Salient Behavioral Phenotype

Hyperphagia and associated food-seeking behaviors are salient and life-threatening features of Prader-Willi syndrome. Hyperphagia begins in early childhood and is associated with aberrant neural networks involved in satiety and reward. The impaired satiety response results in a state in which individuals are habitually hungry yet rarely feel full, what Holland et al. (2003) describe as starvation that manifests as obesity in food-rich environments.

While the severity of hyperphagia varies across individuals and over time, people with Prader-Willi syndrome often engage in food-seeking behaviors, sometimes in very clever ways. They may sneak food, manipulate others or bargain for food, and repeatedly ask about meals and food (Dykens et al., 2007). Managing hyperphagia in the syndrome requires external controls and interventions that ensure food security or safety. These interventions include vigilant supervision around food; locking food sources (e.g., kitchen cabinets, pantries, refrigerators, garbage cans); a reduced calorie diet; providing meals at predictable times; and ensuring food security in school, work, and community settings (Dykens et al., 2007).

Individuals with Prader-Willi syndrome show behavioral challenges that negatively impact their ability to get along with others, relate to peers, and maintain friendships. They often exhibit aggression, irritability, compulsivity, insistence on sameness, and rigid thinking (Dykens et al., 2019). Typically, these individuals become upset at sudden or unexpected changes in their schedules and are prone to such compulsive behaviors as skin-picking, repetitive questioning, ordering and rearranging items, and hoarding of nonfood items (Dykens et al., 1996). In addition to these difficult behaviors, persons with Prader-Willi syndrome typically show impairments in accurately perceiving the emotions, intentions, and behaviors of both themselves and others. Specifically, they may misinterpret benign social mishaps (someone accidently bumping into them) as people being "mean" and also have trouble recognizing sadness, anger, or fear (Dykens et al., 2019).

Behavioral Ties to Decision Making

Although deficits in both cognition and social cognition may impede decision making, we suspect that being in a constant state of hunger serves as a dominant force that disrupts the ability of many individuals with Prader-Willi syndrome to make appropriate decisions. The cognitive deficits commonly present among these individuals likely limit their capacity to comprehend the factors necessary for making decisions. For example, if a decision involves several complex variables, an individual with Prader-Willi syndrome might struggle to understand the implications of each possible choice. In addition, cognitive deficits are likely to hinder the ability to generate and evaluate alternatives. Similar to those with other IDD subtypes, individuals with Prader-Willi syndrome also have difficulties with perceiving both their own emotions and those of others (Dykens et al., 2019).

Hyperphagia likely causes individuals with Prader-Willi syndrome to have atypical decision-making motivations. Specifically, such individuals might base their decisions around the attainability of food, which can impact goal selection and hinder effective decisions. In addition, given their inflexibility and social deficits, persons with Prader-Willi syndrome might also have impairments in decision making that involve a change of routine or conversing with other people (Hickson & Khemka, 2014). Persons with the syndrome might also be unable to integrate emotional contexts during ambiguous situations to make an informed decision. Finally, individuals with Prader-Willi syndrome might experience anxiety during decision making (Hickson & Khemka, 2013, 2014), thereby increasing the time needed to finalize a choice.

Given these cumulative aberrant characteristics, individuals with Prader-Willi syndrome are likely to show atypical decision-making patterns across all pathways in the *Pathways Model of Decision Processing* (Hickson & Khemka, 2013). Granted, few studies have focused on samples with Prader-Willi syndrome, making speculative all such hypotheses. Still, consider Pathway 1, which is marked by intuitive, automatic decision making. Since individuals with Prader-Willi syndrome show impairments in emotional processing, they might be precluded from having access to a useful heuristic. Similarly, the hyperphagia and limited cognitive capacity present in this population is equally likely to impact Pathway 2, which necessitates the ability to generate decision-alternatives. Pathway 3, which necessitates regulatory mechanisms overriding initial intuitive, automatic processing, might be impacted by both the inflexibility and hunger seen in people with this syndrome. Finally, these individuals might retain underdeveloped Pathway 4 patterns, which rely on learning from the accumulation of repeated experiences. Thus, individuals with Prader-Willi syndrome might require explicit instruction before they are able to make more appropriate and strategic decisions.

Despite these deficits, individuals with Prader-Willi syndrome possess certain skills that foster a strengths-based approach for formulating decision-making interventions. This population tends to have strong visual-spatial skills (e.g., shape recognition), which is manifested in an interest in jigsaw puzzles (Dykens, 2002). Interventionists might capitalize on this interest by using puzzles as a source of engagement and flow. For example, a decision-making intervention might entail a tangible component (e.g., worksheets comprised of puzzles) or the use of puzzles as a reinforcer. In addition, individuals with Prader-Willi syndrome show strengths in acquiring information visually (Chedd et al., 2006). Therefore, interventionists should consider utilizing illustrations, photographs, and/or videos in their decision-making curricula.

Down Syndrome

Overview

The most well-known and common genetic (chromosomal) disorder of ID, Down syndrome, features a 160-year history of research. Individuals with Down syndrome are best identified through characteristic facial features. Beyond these physical features, approximately 50% of all newborns experience heart defects, and high rates also have been noted for gastrointestinal problems (about 50%), hypothyroidism (50%–90%), and leukemia (cumulative risk of 2.1% over the first 5 years; Roizen, 2010). Among physical conditions, probably the most common are obesity (50%–60%), hearing problems (50%–80%), and vision problems (20%–70%), although atlantoaxial instability (i.e., excessive movement looseness at bone or ligament joints, 15%) and skin conditions (15%) have also been noted (Pueschel, 1996). Starting in their late 40s and throughout their 50s, many individuals with Down syndrome—up to 40% or more—show dementia related to Alzheimer's disease (Zigman, 2013).

Salient Behavioral Phenotype

When considering salient behavioral characteristics of Down syndrome, one must consider both cognitive-linguistic and personality features. Cognitively, individuals with Down syndrome range widely in levels of intelligence, with overall IQ levels generally in the "moderate" range (IQ's 50–60; Hodapp and Fidler, in press). Historically, they have been thought to have better "visual vs. auditory" processing, with short-term memory better in visual versus auditory modalities (Hodapp & Fidler, in press). More recently, studies have focused on deficits in several areas of executive functioning, including tasks involving switching attention, inhibitory control, and planning (Daunhauer et al., 2017). From early childhood into the adult years, persons with Down syndrome generally show particular weaknesses in language, with pronounced problems in speech articulation (possibly related to the size and placement of the tongue; Bunton and Leddy, 2011) and in grammar (Abbeduto et al., 2007).

Of particular concern here, however, may be a set of etiology-related personality-motivational characteristics. Long recognized as being socially oriented toward others, children with Down syndrome display social skill strengths mostly at the lowest levels of social abilities, not for higher-level social skills (Cebula et al., 2010). Thus, while children with Down syndrome may look and smile toward others, they have difficulties in identifying emotions and may even have "greater than mental age (MA)-level" problems in inferring from stories with negative emotions such as fear or sadness (Jahromi et al., 2008; Williams et al., 2005).

Recent attention has also been paid to the ways in which individuals with Down syndrome use their sociability. By often performing what Pitcairn and Wishart

(1994) dubbed *party tricks*, these individuals sometimes use their sociability to avoid performing difficult cognitive tasks (see also Kasari and Freeman, 2001). While such attempts to avoid performing difficult tasks may seem charming to observers, parents and teachers may feel otherwise, especially if they would like to promote independence for their child. In addition, compared to others with IDD and to a mental age-matched typical sample in a frustrating puzzle task, Jahromi et al. (2008) noted that children with Down syndrome more often looked to the experimenter but did not specifically ask for help. Given their propensities to have difficulties understanding negative emotions and to avoid performing challenging tasks, individuals with Down syndrome may have etiology-related personality characteristics that might interfere with making decisions and addressing other issues in everyday life.

Behavioral Ties to Decision Making

Like others with IDD, persons with Down syndrome often present cognitive deficits that affect their capacity to understand the requisite contextual information to make decisions. Individuals with Down syndrome are also similar to those with other IDD subtypes in that they have deficits with interpreting the emotions of others.

But certain etiology-related characteristics may also influence their decision making. Specifically, deficits in auditory processing might inhibit individuals with Down syndrome's abilities to comprehend relevant information to make an informed decision. In addition, their tendencies to use their sociability to avoid challenging tasks likely lessen their experiences in actually practicing effective decision making (Kasari & Freeman, 2001; Pitcairn & Wishart, 1994). Essentially, if accustomed to having someone else perform tasks for them, these individuals might lack a learning history of independently making decisions, or even knowing when or how to solicit help. Finally, given deficits with identifying negative emotions, persons with Down syndrome might erroneously assume the best intentions in others, with less awareness of the risks involved in dealing with strangers (Fisher et al., 2013).

Given these cumulative characteristics, individuals with Down syndrome likely exhibit atypical decision-making patterns across all four decision-making pathways (Hickson & Khemka, 2013). In Pathway 1, marked by intuitive automatic processing, individuals with Down syndrome may show impairments with emotional processing, particularly with negative emotions, thereby precluding them from accessing a useful decision action. If a situation evokes fear, these individuals might struggle to effectively process this information. A limited cognitive capacity might serve to impact Pathway 2, which necessitates the ability to conceptualize decision alternatives. Deficits with inhibitory control might further impact Pathway 3, which requires regulatory mechanisms to override initial intuitive, automatic processing. Finally, persons with Down syndrome might retain restricted Pathway 4 patterns, especially if they have had fewer learning experiences because, over the years, they have over-relied on others.

As with the other etiologies discussed previously, we advocate for a strengths-based approach in designing decision-making interventions for persons with Down syndrome. These individuals have competency in pragmatic skills, whereby they use stronger social relatedness skills to compensate for language deficits (Hodapp & Fidler, in press). Thus, interventions might capitalize on these social skills by including an interpersonal component (e.g., role-playing) to engage participants, possibly even performing all trainings in group training formats. Similarly, the general finding for "visual-over-auditory" processing might predispose more successful learning when interventions use posters, reminder boards, and other visual materials. Good social skills might also help these individuals relay their decision to others within social contexts.

Etiologic Implications for Decision Making: Specific and General

As noted above, each group presents particular phenotypic characteristics that likely influence their decision-making patterns. For this reason, we now describe potential considerations and recommendations by syndrome. We then discuss general implications.

Etiology-Related Decision-Making Considerations for Those with Each Syndrome

Williams Syndrome

Given the behavioral characteristics of individuals with Williams syndrome, interventions can potentially target decision making in terms of social relationships and interactions, with the ultimate goal of decreasing rates of sexual and other forms of victimization. In this regard, stranger safety training can be an effective means of improving the decision-making abilities of these individuals during a *stranger danger* situation. In 2014, Fisher developed a stranger safety training program to teach adults with Williams syndrome how to respond to a lure from a stranger. Following 3 days of stranger safety training, 62% of participants demonstrated the use of the learned safety skills at least once after the intervention, and only three participants (out of 21) agreed to leave with a stranger.

It is noteworthy, however, that not all persons with Williams syndrome can easily benefit from such training. Parent reports indicate that attempts to teach their child with Williams syndrome not to talk to strangers often come up empty-handed (Jones et al., 2000). Oftentimes, individuals with Williams syndrome can *tell* you how they should act around strangers, but they are then unable to actually *demonstrate* this behavior in real-world situations (Uher et al., this volume). In the victimization literature, most abuse is perpetrated by family members or individuals who are well-known to the victim (Stevens, 2013). However, the overwhelming desire to approach and engage with strangers in individuals with Williams syndrome makes them uniquely vulnerable to victimization by strangers. This attraction to strangers needs to be specifically (and repeatedly) addressed in decision-making interventions, possibly even supplemented with a stranger safety training component.

Prader-Willi Syndrome

To improve their intuition-based decision-making capabilities, individuals with Prader-Willi syndrome should receive explicit instruction in learning how to identify the emotions and intentions of others. We advise that instruction particularly targets the emotions of anger, sadness, and fear; individuals with Prader-Willi syndrome oftentimes present deficits in recognizing these emotions in others, thereby precluding them from accessing a useful heuristic in recognizing potential situations of vulnerability and/or abuse.

At the same time, individuals with Prader-Willi syndrome may also need to receive strategies for increasing their flexibility; such instruction would better enable them to make more adaptive decisions based on context. Providing strategies to reduce anxiety might enable persons with Prader-Willi syndrome to consider all alternatives. Finally, it should be emphasized that even though many with the syndrome understand the *right* choices that they *should* make, being in a state of starvation may override their ability to actually implement the *right* decision. As stated by a young woman with Prader-Willi syndrome, "At home we have to have food locked up and stuff like that but if something is left out, it's just calling to me…it kind of feels like I can't think of anything or do anything unless, until I eat it."

Down Syndrome

In considering etiology-informed interventions in Down syndrome, we first advise that, whenever feasible, these individuals receive visual (as opposed to auditory) supports. Interventionists might attempt to become creative in generating many different types of such supports, including (but not limited to) posters, graphic organizers, calendars, hands-on materials, and visual technology (e.g., video examples and pictures). With visual supports, individuals with Down syndrome might be better equipped to comprehend the relevant factors for making decisions. Further, such individuals might benefit from explicit instruction in identifying the emotions and intentions of others, particularly negative ones, in order to strengthen their intuition-based decision-making skills. Being able to recognize negative emotions such as anger might help counter potentially maladaptive abusive situations.

Finally, practitioners working with persons with Down syndrome might foster the use of independent decision-making strategies, especially when confronted with complex, difficult problems. Again, this independent decision making would run counter to the *usual* behaviors of many with Down syndrome, in which teens and adults might more often look to others for solutions. Along these lines, teachers, parents, and service providers might present decision-making opportunities repeatedly throughout the day, along with appropriate supports or instruction. Given such repeated opportunities for successful, independent decision making, persons with Down syndrome might develop a learning history of commonly occurring, independent decision making. In addition, parents and service providers might also explicitly teach when it is appropriate to request assistance, fading prompts over time during skill acquisition (e.g., shifting from physical guidance to gestural prompts).

General Considerations

As noted for individuals with Williams syndrome, Prader-Willi syndrome, and Down syndrome, each disorder comes with its own etiology-related set of strengths and weaknesses. As a result, individuals with each of these conditions experience difficulties that might affect their decision-making abilities. For example, because of their hyper-sociability and desire to please, individuals with Williams syndrome may struggle with making difficult interpersonal decisions. Those with Prader-Willi syndrome feature hyperphagia that may lead them to struggle with making appropriate health- or food-related decisions. Those with Down syndrome possess party tricks and a positive orientation that may impede making difficult decisions, especially when such decisions relate to fear- or anger-inducing emotions.

In linking etiology-related characteristics to decision making, then, it is also important to address the issue of hypothetical versus real-world decisions. Indeed, across different genetic ID conditions, individuals are often able to be taught how to make appropriate decisions, and they might be able to display appropriate decision-making skills "on paper" (Fisher et al., 2014). At the same time, these individuals are often unable to actually put these skills "into action."

Consider as an example the social approach behavior of individuals with Williams syndrome. In their study comparing self-reported approach behaviors to real-life social approach behaviors, Fisher et al. (2014) found that most participants with Williams syndrome reported that they would not approach strangers. But when placed in a natural social situation, nearly all approached strangers, with most participants approaching multiple strangers. A mismatch existed between knowing and doing. On the one hand, adults with Williams syndrome have internalized the message that they should not talk to strangers and are able to repeat this message when asked. But when placed in a real-world situation, they do not always act appropriately—indeed, most participants, even after indicating they would avoid strangers and seeming to have learned about stranger danger, nevertheless proceeded to interact with strangers (Fisher et al., 2014). Similar disconnects between knowledge and action have been reported for those with Prader-Willi syndrome. Although they have been taught that they should limit their food consumption and are even able to state which foods are most healthy for them, they are often unable to put these ideas into action (Dykens et al., 2007).

In situations in which etiology-related physiological tendencies might be hard to override, we speculate based on previous research on decision making and IDD etiology that it would be important to consider behavioral approaches when designing and implementing decision-making interventions. That is, individuals with certain forms of IDD might benefit from learning specific stages or rules that remove the cognitive or emotional components of the decision, instead leading the individual to simply choose the intuitive response option. More specifically, behavioral interventions may be appropriate for decision making regarding high-pressure situations such as scenarios of potential sexual victimization or abuse. As described in another chapter (Uher et al., this volume), through behavioral interventions, these intuitive responses can help individuals with IDD to recognize particular situations that require specific, quick, and effective responses.

This mismatch between knowing and doing also has implications for research and intervention. Given the discrepancy between what the person should do and what the person actually does, decision-making training requires an observational component. Especially for behaviors that are difficult to inhibit, behavioral interventions need to include assessments in the natural setting, thereby providing tangible evidence as to whether individuals with IDD can be taught and use decision-making skills. Knowing what to do is not enough; to make better decisions, individuals with these (and other) etiologies must override their *natural* tendencies.

Concluding Thoughts

For individuals with and without IDD, decision making can be quite complex. Although the ability to make complex decisions is important for all of us, improving the decision-making skills of individuals with IDD is critically important. But persons with IDD present several atypical characteristics that impact their decision making. Across the broader IDD population, cognitive deficits often hinder the generation and consideration of choice alternatives. In addition, interventions for individuals with IDD must generalize learned skills to actual, real-world situations. Moreover, there is a lack of research on the strengths and weaknesses of decision making for individuals with IDD, resulting in poor understanding of their abilities and lack of direction when designing and implementing interventions.

Interventions also need to consider behavioral phenotypes. Compared to other IDD groups, individuals with Williams syndrome struggle with making decisions regarding *stranger danger* situations, and those with Prader-Willi syndrome have difficulties overcoming preoccupations with food, rigid thinking, and accurately perceiving their own and others' emotions. By describing the different ways in which individuals with IDD might approach a decision-making process, the *Pathways Model of Decision Processing* (Hickson & Khemka, 2013) helps identify which etiology-related behavioral traits influence the decision-making process and how these traits might be addressed in interventions. Ultimately, we need to successfully address etiology-related characteristics and foster decision-making skills that generalize beyond the intervention setting. Only then will individuals with IDD, of whatever etiology, be empowered to make their own decisions and function independently in the modern society.

References

Abbeduto, L., Warren, S. F., & Conners, F. A. (2007). Language development in Down syndrome: From the prelinguistic period to the acquisition of literacy. *Mental Retardation and Developmental Disabilities Research Reviews, 13*, 247–261. https://doi.org/10.1002/mrdd.20158

American Academy of Pediatrics Committee on Genetics. (2001). Health care supervision for children with Williams syndrome. *Pediatrics, 107*, 1192–1204.

Beuren, A. J., Apitz, J., & Harmjanz, D. (1962). Supravalvular aortic stenosis in association with mental retardation and a certain facial appearance. *Circulation, 26*, 1235–1240. https://doi.org/10.1161/01.CIR.26.6.1235

Bunton, K., & Leddy, M. (2011). An evaluation of articulatory working space area in vowel production of adults with Down syndrome. *Clinical Linguistics and Phonetics, 25*, 321–334. https://doi.org/10.3109/02699206.2010.535647

Cassidy, S. B., Schwartz, S., Miller, J. L., & Driscoll, D. J. (2012). Prader-Willi syndrome. *Genetics in Medicine, 4*(1), 10–26. https://doi.org/10.1038/gim.0b013e31822bead0

Cebula, K. R., Moore, D. G., & Wishart, J. G. (2010). Social cognition in children with Down's syndrome: Challenges to research and theory building. *Journal of Intellectual Disability Research, 54*, 113–134. https://doi.org/10.1111/j.1365-2788.2009.01215.x

Chedd, N., Levine, K., & Wharton, R. (2006). Educational considerations for children with Prader-Willi Syndrome. In *Management of Prader-Willi Syndrome* (3rd ed., pp. 302–316). Springer. https://doi.org/10.1007/978-0-387-33536-0_11

Chen, G., Kim, K. A., Nofsinger, J. R., & Rui, O. M. (2007). Trading performance, disposition effect, overconfidence, representativeness bias, and experience of emerging market investors. *Journal of Behavioral Decision Making, 20*(4), 425–451.

Daunhauer, L. A., Gerlach-McDonald, B., Will, E., & Fiddler, D. J. (2017). Performance and ratings based measures of executive function in school-aged children with Down syndrome. *Developmental Neuropsychology, 42*, 351–368. https://doi.org/10.1080/87565641.2017.1360303

Davies, M., Udwin, O., & Howlin, P. (1998). Adults with Williams syndrome: Preliminary study of social, emotional and behavioural difficulties. *The British Journal of Psychiatry, 172*, 273–276.

Doyle, T. F., Bellugi, U., Korenberg, J. R., & Graham, J. (2004). "Everybody in the world is my friend". Hypersociability in young children with Williams syndrome. *American Journal of Medical Genetics, 124A*, 263–273. https://doi.org/10.1002/ajmg.a.20416

Dykens, E. M. (1995). Measuring behavioral phenotypes: Provocations from the "new genetics". *American Journal on Mental Retardation, 99*, 522–532.

Dykens, E. M. (2002). Are jigsaw puzzle skills "spared" in persons with Prader-Willi syndrome? *Journal of Child Psychology and Psychiatry, 43*(3), 343–352. https://doi.org/10.1111/1469-7610.00025

Dykens, E. M. (2003). Anxiety, fears, and phobias in persons with Williams syndrome. *Developmental Neuropsychology, 23*, 291–316.

Dykens, E. M. (2013). Aging in rare intellectual disability syndromes. *Developmental Disabilities Research Reviews, 18*, 75–83. https://doi.org/10.1002/ddrr.1130

Dykens, E. M. (in press). Behavioral phenotypes and genomic testing: Scientific and societal implications. In L. M. Glidden (Ed.), *APA handbook of intellectual and developmental disabilities*. APA.

Dykens, E. M., & Rosner, B. A. (1999). Refining behavioral phenotypes: personality-motivation in Williams and Prader-Willi syndromes. *American Journal on Mental Retardation, 104*(2), 158–169. https://doi.org/10.1352/0895-8017(1999)104<0158:RBPPIW>2.0.CO;2

Dykens, E. M., Hodapp, R. M., & Leckman, J. F. (1987). Strengths and weaknesses in intellectual functioning of males with fragile X syndrome. *American Journal of Mental Deficiency, 92*, 234–236.

Dykens, E. M., Hodapp, R. M., Walsh, K. K., & Nash, L. (1992). Profiles, correlates, and trajectories of intelligence in Prader-Willi syndrome. *Journal of the Academy of Child and Adolescent Psychiatry, 31*, 1125–1130. https://doi.org/10.1097/00004583-199211000-00022

Dykens, E. M., Leckman, J. F., & Cassidy, S. B. (1996). Obsessions and compulsions in Prader-Willi syndrome. *Journal of Child Psychology and Psychiatry, 37*, 995–1002. https://doi.org/10.1111/j.1469-7610.1996.tb01496.x

Dykens, E. M., Hodapp, R. M., & Finucane, B. (2000). *Genetics and mental retardation syndromes: A new look at behavior and treatments*. Paul H. Brookes Publishing Company.

Dykens, E. M., Maxwell, M., Pantino, E., Kossler, R., & Roof, E. (2007). Assessment of hyperphagia in Prader-Willi syndrome. *Obesity (Silver Spring, Md.), 15*(7), 1816–1826. https://doi.org/10.1038/oby.2007.216

Dykens, E. M., Roof, E., Hunt-Hawkins, H., Daniell, C., & Jurgensmeyer, S. (2019). Profiles and trajectories of impaired social cognition in people with Prader-Willi syndrome. *PLOS One, 14*(10), e0223162. https://doi.org/10.1371/journal.pone.0223162

Engell, A. A., Haxby, J. V., & Todorov, A. (2007). Implicit trustworthiness decisions: Automatic coding of face properties in the human amygdala. *Journal of Cognitive Neuroscience, 19*(9), 1508–1519. https://doi.org/10.1162/jocn.2007.19.9.1508

Fisher, M. H. (2014). Evaluation of a stranger safety training program for young adults with Williams syndrome. *Journal of Intellectual Disability Research, 58*, 903–914. https://doi.org/10.1111/jir.12108

Fisher, M. H., Moskowitz, A. L., & Hodapp, R. M. (2013). Differences in social vulnerability among individuals with autism spectrum disorder, Williams syndrome, and Down syndrome. *Research in Autism Spectrum Disorders, 7*, 931–937. https://doi.org/10.1016/j.rasd.2013.04.009

Fisher, M. H., Mello, M. P., & Dykens, E. M. (2014). Who reports it best? A comparison between parent-report, self-report, and the real-life social behaviors of adults with Williams syndrome. *Research in Developmental Disabilities, 35*, 3276–3284. https://doi.org/10.1016/j.ridd.2014.08.011

Fisher, M. H., Lense, M. D., & Dykens, E. M. (2016). Longitudinal trajectories of intellectual and adaptive functioning in adolescents and adults with Williams syndrome. *Journal of Intellectual Disability Research, 60*, 920–932. https://doi.org/10.1111/jir.12303

Flynt, J., & Yule, W. (1994). Behavioural phenotypes. In M. Rutter, E. Taylor, & L. Hersov (Eds.), *Child and adolescent psychiatry: Modern approaches* (3rd ed., pp. 666–687). Blackwell Scientific.

Fryns, J. P., Borghgraef, M., Volcke, P., & Van den Berghe, H. (1991). Adults with Williams syndrome. *American Journal of Medical Genetics, 40*(2), 253.

Gomez, A., Costa, M., Lio, G., Sirigu, A., & Demily, C. (2020). Face first impression of trustworthiness in Williams syndrome: Dissociating automatic vs decision based perception. *Cortex, 132*, 99–112. https://doi.org/10.1016/j.cortex.2020.07.015

Gosch, A., & Pankau, R. (1997). Personality characteristics and behaviour problems in individuals of different ages with Williams syndrome. *Developmental Medicine & Child Neurology, 39*(8), 527–533. https://doi.org/10.1111/j.1469-8749.1997.tb07481.x

Haas, B. W., Hoeft, F., Searcy, Y. M., Mills, D., Bellugi, U., & Reiss, A. (2010). Individual differences in social behavior predict amygdala response to fearful facial expressions in Williams syndrome. *Neuropsychologia, 48*(5), 1283–1288. https://doi.org/10.1016/j.neuropsychologia.2009.12.030

Halpern-Felsher, B., & Cauffman, E. (2001). Costs and benefits of a decision: Decision making competence in adolescents and adults. *Journal of Applied Developmental Psychology, 22*, 257–273.

Hickson, L., & Khemka, I. (2013). Problem solving and decision making. In M. Wehmeyer (Ed.), *The Oxford handbook of positive psychology and disability* (pp. 198–225). Oxford University Press.

Hickson, L., & Khemka, I. (2014). The psychology of decision making. *In International Review of Research in Developmental Disabilities, 47*, 185–229. https://doi.org/10.1016/B978-0-12-800278-0.00006-3

Hillier, L. W., Fulton, R. S., Fulton, L. A., Graves, T. A., Pepin, K. H., Wagner-McPherson, C., Layman, D., Maas, J., Jaeger, S., Walker, R., Wylie, K., Sekhon, M., Becker, M. C., O'Laughlin, M. D., Schaller, M. E., Fewell, G. A., Delehaunty, K. D., Miner, T. L., Nash, W. E., Cordes, M., … Wilson, R. K. (2003). The DNA sequence of human chromosome 7. *Nature, 424*(6945), 157–164. https://doi.org/10.1038/nature01782

Hodapp, R. M. (1997). Direct and indirect behavioral effects of different genetic disorders of mental retardation. *American Journal on Mental Retardation, 102*, 67–79. https://doi.org/10.1352/0895-8017(1997)102<0067:DAIBEO>2.0.CO;2

Hodapp, R. M., & Dykens, E. M. (1994). Mental retardation's two cultures of behavioral research. *American Journal on Mental Retardation, 98*, 675–687.

Hodapp, R. M., & Dykens, E. M. (2012). Genetic disorders of intellectual disability: Expanding our concepts of phenotypes and of family outcomes. *Journal of Genetic Counseling, 21*, 761–769. https://doi.org/10.1007/s10897-012-9536-4

Hodapp, R. M., & Fidler, D. J. (in press). Down syndrome. In L. M. Glidden (Ed.), *APA handbook of intellectual and developmental disabilities*. American Psychological Association Press.

Holland, A. J., Whittington, J. E., Butler, J., Webb, T., Boer, H., & Clarke, D. (2003). Behavioural phenotypes associated with specific genetic disorders: Evidence from a population-based study of people with Prader-Willi syndrome. *Psychological Medicine, 33*, 141–154. https://doi.org/10.1017/s0033291702006736

Jahromi, L. B., Gulsrud, A., Kasari, C., & Dykens, E. (2008). Emotional competence in children with Down syndrome: Negativity and regulation. *American Journal of Mental Retardation, 113*, 32–43. https://doi.org/10.1352/0895-8017(2008)113[32:ECICWD]2.0.CO;2

Jawaid, A., Riby, D. M., Owens, J., White, S. W., Tarar, T., & Schulz, P. E. (2012). 'Too withdrawn' or 'too friendly': Considering social vulnerability in two neurodevelopmental disorders. *Journal of Intellectual Disability Research, 56*, 335–350. https://doi.org/10.1111/j.1365-2788.2011.01452.x

Jones, W., Bellugi, U., Lai, Z., Chiles, M., Reilly, J., Lincoln, A., & Adolphs, R. (2000). II. Hypersociability in Williams syndrome. *Journal of Cognitive Neuroscience, 12*(Suppl. 1), 30–46.

Kahneman, D., & Tversky, A. (1979). Prospect theory: An analysis of decision under risk. *Econometrica, 47*, 263–291. https://doi.org/10.2307/1914185

Karam, S. M., Riegel, M., Segal, S. L., Félix, T. M., Barros, A. J. D., Santos, I. S., Matijasevich, A., Giugliani, R., & Black, M. (2015). Genetic causes of intellectual disability in a birth cohort: A population-based study. *American Journal of Medical Genetics Part A, 167*(6), 1204–1214. https://doi.org/10.1002/ajmg.a.37011

Kasari, C., & Freeman, S. F. N. (2001). Task-related social behavior in children with Down syndrome. *American Journal on Mental Retardation, 106*, 253–264. https://doi.org/10.1352/0895-8017(2001)106<0253:TRSBIC>2.0.CO;2

Kennedy, J. C., Kaye, D. L., & Sadler, L. S. (2006). Psychiatric diagnoses in patients with Williams syndrome and their families. *Jefferson Journal of Psychiatry, 20*, 22–31.

Klein-Tasman, B. P., & Mervis, C. B. (2003). Distinctive personality characteristics of 8-, 9-, and 10-year-olds with Williams syndrome. *Developmental neuropsychology, 23*(1-2), 269–290. https://doi.org/10.1080/87565641.2003.9651895

Klein-Tasman, B. P., Li-Barber, K. T., & Magargee, E. T. (2011). Honing in on the social phenotype in Williams syndrome using multiple measures and multiple raters. *Journal of Autism and Developmental Disorders, 41*, 341–351. https://doi.org/10.1007/s10803-010-1060-5

Leyfer, O. T., Woodruff-Borden, J., Klein-Tasman, B., Fricke, J., & Mervis, C. B. (2006). Prevalence of psychiatric disorders in 4–16- year olds with Williams syndrome. *American Journal of Medical Genetics Part B (Neuropsychiatric Genetics), 114B*, 615–622. https://doi.org/10.1002/ajmg.b.30344

Leyfer, O., Woodruff-Borden, J., & Mervis, C. B. (2009). Anxiety disorders in children with Williams syndrome, their mothers, and their siblings: Implications for the etiology of anxiety disorders. *Journal of Neurodevelopmental Disorders, 1*, 4–14. https://doi.org/10.1007/s11689-009-9003-1

Martens, M. A., Wilson, S. J., & Reutens, D. C. (2008). Williams syndrome: A critical review of the cognitive, behavioral, and neuroanatomical phenotype. *Journal of Child Psychology and Psychiatry, 49*, 576–608. https://doi.org/10.1111/j.1469-7610.2008.01887.x

Mason, R. O., & Mitroff, I. I. (1981). *Challenging strategic planning assumptions*. Wiley.

Mazzolini, R. (1981). How strategic decisions are made. *Long Range Planning, 14*, 85–96.

Mervis, C. B., & Pitts, C. H. (2015). Children with Williams syndrome: Developmental trajectories for intellectual abilities, vocabulary abilities, and adaptive behaviour. *American Journal of Medical Genetics Part C, 9999*, 1–14. https://doi.org/10.1002/ajmg.c.31436

Mervis, C. B., Morris, C. A., Klein-Tasman, B. P., Bertrand, J., Kwitny, S., Appelbaum, L. G., & Rice, C. E. (2003). Attentional characteristics of infants and toddlers with Williams syndrome during triadic interactions. *Developmental Neuropsychology, 23*, 243–268.

Miller, G. A. (1956). The magical number seven, plus or minus two: Some limits on our capacity for processing information. *Psychological Review, 63*, 81–97. https://doi.org/10.1037/0033-295X.101.2.343

Mintzberg, H., Raisinghani, P., & Theoret, A. (1976). The structure of "unstructured" decision processes. *Administrative Science Quarterly, 21*, 246–275. https://doi.org/10.2307/2392045

Nyhan, W. L. (1972). Behavioral phenotypes in organic genetic diseases: Presidential address to the Society of Pediatric Research (May 1, 1971). *Pediatric Research, 6*, 1–9.

Opitz, J. M. (1985). Editorial comment: The developmental field concept. *American Journal of Medical Genetics, 21*, 1–11.

Pitcairn, T. K., & Wishart, J. G. (1994). Reactions of young children with Down's syndrome to an impossible task. *British Journal of Developmental Psychology, 12*, 485–489. https://doi.org/10.1111/j.2044-835X.1994.tb00649.x

Pober, B. (2010). Williams-Beuren syndrome. *New England Journal of Medicine, 362*, 239–252. https://doi.org/10.1056/NEJMra0903074

Polasky, S., Carpenter, S. R., Folke, C., & Keeler, B. (2011). Decision making under great uncertainty: Environmental management in an era of global change. *Trends in Ecology and Evolution, 26*(8), 398–404. https://doi.org/10.1016/j.tree.2011.04.007

Porter, H., & Tharpe, A. M. (2010). Hearing loss among persons with Down syndrome. *International Review of Research in Mental Retardation, 39*, 195–220. https://doi.org/10.1016/S0074-7750(10)39007-0

Powell, T. C., Lovallo, D., & Fox, C. R. (2011). Behavioral strategy. *Strategic Management Journal, 32*(13), 1369–1386. https://doi.org/10.1002/smj.968

Pueschel, S. M. (1996). Young people with Down syndrome: Transition from childhood to adulthood. *Mental Retardation and Developmental Disabilities Research Reviews, 2*, 90–95. https://doi.org/10.1002/(SICI)1098-2779(1996)2:2<90::AID-MRDD6>3.0.CO;2-0

Roizen, N. (2010). Chapter one- Overview of health issues among persons with Down syndrome. *International Review of Research in Mental Retardation, 39*, 2–23. https://doi.org/10.1016/S0074-7750(10)39001-X

Rondal, J. (1995). *Exceptional language development in Down syndrome*. Cambridge University Press.

Rothbart, M. K., & Derryberry, D. (1981). Development of individual differences in temperament. *Advances in Developmental Psychology, 1*, 37–86.

Schretlen, D. J., Ward, J., Meyer, S. M., Yun, J., Puig, J. G., Nyhan, W. L., Jinnah, H. A., & Harris, J. C. (2005). Behavioral aspects of Lesch-Nyhan disease and its variants. *Developmental Medicine and Child Neurology, 47*, 673–677.

Schwenk, C. R. (1984). Cognitive simplification processes in strategic decision making. *Strategic Management Journal, 5*, 111–128. https://doi.org/10.1002/smj.4250050203

Searcy, Y. M., Lincoln, A., Rose, F., Klima, E., Bavar, N., & Korenberg, J. R. (2004). The relationship between age and IQ in adults with Williams syndrome. *American Journal on Mental Retardation, 109*, 231–236. https://doi.org/10.1352/0895-8017(2004)1092.0.CO;2

Sigafoos, J., O'Reilly, M. F., & Lancioni, G. E. (2009). Editorial: Cri-du-chat. *Developmental Neurorehabilitation, 12*, 119–121.

Steinbruner, J. D. (1974). *The cybernetic theory of decision*. Princeton University Press.

Steptoe-Warren, G., Howat, D., & Hume, I. (2011). Strategic thinking and decision making: Literature Review. *Journal of Strategy and Management, 4*, 238–250. https://doi.org/10.1108/17554251111152261

Stevens, E. (2013). Safeguarding vulnerable adults: Exploring the challenges to best practice across multi-agency settings. *The Journal of Adult Protection, 15*(2), 85–95. https://doi.org/10.1108/14668201311313596

Thomas, A., & Chess, S. (1977). *Temperament and development*. Brunner/Mazel.

Thomas, A., Chess, S., & Birch, H. G. (1968). *Temperament and behaviour disorders in children*. New York University Press.

Thurman, A. J., & Fisher, M. H. (2015). The Williams syndrome social phenotype: Disentangling the contributions of social interest and social difficulties. *International Review of Research in Developmental Disabilities, 49*, 191–227. https://doi.org/10.1016/bs.irrdd.2015.06.002

Tomc, S. A., Williamson, N. K., & Pauli, R. M. (1990). Temperament in Williams syndrome. *American Journal of Medical Genetics, 36*(3), 345–352.

Uher A., Josol, C. K., & Fisher, M. H. (this volume). Behavioral approaches to teaching decision making. In I. Khemka & L. Hickson (Eds.), *Decision making by individuals with IDD: Research and practice* (Positive psychology and disabilities series). Springer.

Williams, K. R., Wishart, J. G., Pitcairn, T. K., & Willis, D. (2005). Emotion recognition by children with Down syndrome: Investigation of specific impairments and error patterns. *American Journal on Mental Retardation, 110*, 378–392.

Zigler, E. (1967). Familial mental retardation: A continuing dilemma. *Science, 155*, 292–298. https://doi.org/10.1126/science.155.3760.292

Zigman, W. B. (2013). Atypical aging in Down syndrome. *Developmental Disabilities Research Reviews, 18*, 51–67. https://doi.org/10.1002/ddrr.1128

Zitzer-Comfort, C., Doyle, T., Masataka, N., Korenberg, J., & Bellugi, U. (2007). Nature and nurture: Williams syndrome across cultures. *Developmental Science, 10*(6), 755–762. https://doi.org/10.1111/j.1467-7687.2007.00626.x

[illegible]

Chapter 8
The Impact of Anxiety on Decision Making in Individuals with Intellectual and Developmental Disabilities or a Diagnosis on the Autism Spectrum

Dawn Adams and Stephanie Malone

The past decade has seen a growing research and clinical interest in the prevalence and presentation of mental health disorders in individuals with intellectual and developmental disabilities (IDD) and in those with a diagnosis on the autism spectrum. One of the most frequently researched mental health disorders is that of anxiety. Anxiety is associated with a feeling of the uneasy anticipation of a threatening event. It is a typical reaction to stress which can be beneficial in some situations (Beck, 1972). Anxiety disorders differ from normal feelings of nervousness or anxiousness in that they are often disabling. Although viewed as interchangeable, anxiety differs from fear as anxiety involves the anticipation of a future concern or event, whilst fear is an emotional response to an immediate threat. Individuals with anxiety may experience physiological and/or cognitive symptoms. The *Diagnostic and Statistical Manual of Mental Disorders, Fifth Edition* (*DSM-5*; American Psychiatric Association, 2013) lists anxiety disorders as separation anxiety disorder, selective mutism, specific phobia, social phobia, panic disorder, agoraphobia and generalised anxiety disorder.

This chapter focuses upon the impact of anxiety on the decision-making process in individuals with IDD and those on the autism spectrum, a topic that has not (to date) been explicitly researched. Although autism is often classified as a developmental disability, there is a large amount of autism-specific research both in the area of mental health and decision making, so within this chapter, the two literature bases (IDD and autism-specific) are discussed independently. We therefore open with a brief overview of anxiety, its prevalence, presentation and assessment in individuals with IDD and those on the autism spectrum. Next, the influence and impact of anxiety on the decision-making process in the neurotypical population is reviewed and

D. Adams (✉) · S. Malone
Autism Centre of Excellence, Griffith University, Brisbane, QLD, Australia

Griffith Institute for Educational Research, Griffith University, Brisbane, QLD, Australia
e-mail: Dawn.Adams@Griffith.edu.au; Stephanie.Malone@Griffith.edu.au

I. Khemka, L. Hickson (eds.), *Decision Making by Individuals with Intellectual and Developmental Disabilities*, Positive Psychology and Disability Series,
https://doi.org/10.1007/978-3-030-74675-9_8

discussed, allowing us to finally consider the relevance of these typically developing models to IDD or autism. This literature is drawn upon to hypothesise whether there may be any differences in the direction or strength of relations described within these neurotypical models. This highlights the importance of considering both diagnosis (autism/IDD) and anxiety as key variables in the process of decision making. Finally, as the majority of work that has been done in this specific area of decision making has been experimental and laboratory-based, the chapter ends on possible implications for real-life personal and interpersonal decision making.

Anxiety in IDD and Autism

Mental Health and Anxiety in IDD

Individuals with IDD are at increased risk compared with those without IDD of experiencing mental health problems (Smiley et al., 2007). Systematic reviews report that 30–50% of individuals with IDD have a co-occurring mental health disorder (Einfeld et al., 2011) compared to only 3–6% of those without an IDD (Green et al., 2005; Polanczyk et al., 2015). Using national data sets, children and adolescents with IDD are reported to be 4–5 times more likely to show symptoms of diagnosable mental health problems than neurotypical children (Emerson & Hatton, 2007).

To date, the majority of reviews have reported mental health disorders as a single construct rather than focussing upon a specific diagnosis such as anxiety. A systematic review of anxiety disorders in individuals with IDD reported prevalence rates of 2–22% (Reardon et al., 2015) with a meta-analysis reporting pooled prevalence estimates of 4.7% (Maiano et al., 2018). When these were divided into specific subtypes of anxiety, Maiano et al. (2018) reported pooled prevalence estimates of 11.5% for specific phobia, 5% for separation anxiety disorder, 2.7% for social phobia, 2.4% for obsessive-compulsive disorder (OCD) and 2.2% for generalised anxiety disorder. As is noted for neurotypical youth (Costello et al., 2011), significantly more children were reported to experience separation anxiety disorder, and significantly more adolescents were reported to experience specific phobia, OCD and social phobia.

The assessment of anxiety in those with IDD is reviewed in Reardon et al. (2015), and the assessment of anxiety, specifically for individuals with severe or profound IDD, is reviewed in Flynn et al. (2017). Both reviews highlight that whilst some measures show promise, the sparse work around the reliability and validity of these measurements means that further work is needed in this area. Flynn et al. concluded that the Diagnostic Assessment for the Severely Handicapped Scale-II (DASH-II; Matson, 1995) is a strong, reliable measure of mental health problems and can be used to effectively assess mood disorders (depression and mania), anxiety and schizophrenia in adults with severe to profound IDD. However, no reliable and valid measures of anxiety or mental health disorders were found specifically for use with children or adolescents with severe or profound IDD.

Anxiety has been well documented to impact the quality of life (Olatunji et al., 2007) and academic and social performance (de Lijster et al., 2018) in neurotypical children. However, the impact of anxiety in individuals with IDD is less researched. Children with IDD who experience anxiety are more likely than typically developing children with anxiety to have additional co-occurring diagnoses, with studies suggesting that they are 4–10 times more likely to have attention-deficit hyperactivity disorder (ADHD) or externalising problems (Green et al., 2015). Children with IDD and elevated anxiety have also been shown to have higher levels of sleep problems as well as higher levels of behavioural challenges (Rzepecka et al., 2011), which aligns with the suggestion that behavioural challenges may be indicators of anxiety issues in both children and adults with IDD (Hemmings et al., 2006). Anxiety in individuals with IDD has been shown to negatively impact parent physical and mental health (Gray et al., 2011).

Models of Anxiety in IDD

The neural circuitry of anxiety and depression is better understood in the typically developing population than in those with an IDD. Within the typically developing population, there is accumulating evidence to suggest an association between executive dysfunction and heightened levels of both anxiety and depression (Hosenbocus & Chahal, 2012; Micco et al., 2009; Ottowitz et al., 2002), with suggestions that disruptions of "frontally located" executive functions may reduce the inhibition of posterior emotion systems, resulting in elevated anxiety and depressive symptoms (Galynker et al., 1998; Ohta et al., 2008; Ray & Zald, 2012). However, to date, such relationships have not been explored in individuals with IDD. Even though those with an IDD will inevitably have some challenges in cognitive functioning, there can still be considerable variability within and between individuals in skills such as working memory, verbal memory, planning, inhibition and other executive functioning skills. The lack of consideration of the impact of such variability and the dearth of research in this area has resulted in a call for further studies to better understand the interaction between IDD, relative cognitive impairments/variability in cognitive profiles and psychiatric disorders in both children and adults (McCarthy & Barbot, 2016).

There has been limited research examining the cognitive processes related to mental health in individuals with IDD. Those studies that have explored this area have found that the same cognitive processes mediate mental health for individuals with IDD as for typically developing individuals (Dagnan & Sandhu, 1999; Dagnan & Waring, 2004). Cognitive theories suggest that anxiety is a result of cognitions which centre around themes of threat, danger, unpredictability and uncertainty (e.g. Beck & Weishaar, 1989). Threat perception theory of anxiety (e.g. Beck, 1985) hypothesises that individuals with anxiety have anxiety-related schemata that direct processing resources towards threat-relevant information resulting in cognitive biases related to attention, interpretation and memory (e.g. Clark & Wells, 1995; Rapee & Heimberg, 1997). Threat perception was evaluated in one of the only studies to explore cognitions relating to anxiety in those with IDD: Houtkamp et al. (2017) found that negative interpretations of social situations were associated with higher levels of social anxiety in adolescents with mild IDD. This, coupled with research demonstrating that individuals with IDD have similar coping strategies to individuals without IDD (e.g. Da Costa et al., 2000; Kurtek, 2016), suggests that the same cognitive biases are implicated in the experience of anxiety for both groups. This is further supported by a systematic review of cognitive behavioural therapy for anxiety in adults with IDD (Dagnan et al., 2018) which found that all but 1 of the 19 studies reviewed reported a positive outcome of cognitive behavioural therapy, highlighting the role of cognitions in anxiety in individuals with IDD. However, further, longitudinal work is needed in order to understand trajectories and pathways between these cognitive processes and anxiety symptomatology.

Anxiety in Autism

Anxiety is now recognised as the most common co-occurring diagnosis for individuals on the autism spectrum. Meta-analyses suggest prevalence rates of anxiety to be 40% in children (van Steensel et al., 2011) and 27% in adults (Hollocks et al., 2019), with prevalence rates of around 11–12% for depression across adolescents and adults (e.g. (Hudson et al., 2019). A recent meta-analysis (Lai et al., 2019) combined data from 68 studies and 169,829 participants with a diagnosis on the autism spectrum and found a pooled prevalence of 20% for anxiety disorders across the lifespan, a prevalence rate that is approximately twice as high as those reported in typically developing population meta-analyses (Steel et al., 2014).

Reviews of the research show a tenfold increase in the number of publications focussing upon anxiety in individuals on the autism spectrum in the last 10 years (Vasa et al., 2018). This recent work has led to the recognition that the identification of anxiety in individuals on the autism spectrum can be complicated by two main factors. The first is diagnostic overshadowing: the misattribution of symptoms (i.e. mental health symptoms) to another diagnosis that the person has (i.e. autism). For example, it may be difficult to identify social anxiety symptoms in an individual on the spectrum as these symptoms may be interpreted as a consequence of a social communication challenge related to the profile of autism characteristics (Adams et al., 2019). The second factor is the suggestion of the presence of both *typical* (i.e. consistent with *DSM-5* anxiety symptomatology) and *atypical* (i.e. those that are specific to anxiety in individuals on the autism spectrum) anxiety symptoms which can occur in isolation or, most commonly, in combination with each other (Kerns et al., 2014). The suggestion of autism-specific anxiety characteristics has led to the development of autism-specific anxiety measures including the Anxiety Scale for Children-Autism Spectrum Disorders (ASC-ASD self-report and parent-report versions; Rodgers et al., 2016), Anxiety Scale for Autism-Adults (ASA-A; Rodgers et al., 2020) and the Parent-Rated Anxiety Scale for Youth with Autism Spectrum Disorder (Scahill et al., 2019).

The impact of anxiety on individuals on the autism spectrum has only recently been explored. Within a few years, published work has highlighted that anxiety symptomatology has a more significant impact on child's quality of life than autism characteristics (Adams et al., 2019; Adams et al., 2020). Parents report that their child's anxiety impacts their child's educational experience (Simpson et al., 2020) more than other areas of their daily life (den Houting et al., 2020). Child anxiety levels are also positively correlated with parent anxiety levels and specific aspects of the child's anxiety profile, namely intolerance of uncertainty, influence parenting and family behaviour (Adams & Emerson, 2020).

Models of Anxiety in Autism

Alexithymia and Interoception Anxiety in individuals on the autism spectrum needs to be understood within the context of the frequently reported challenges or differences in interoception and alexithymia. Interoception is the process of identifying, understanding and responding appropriately to the patterns of internal sensory signals related to internal body experiences such as pain, temperature, itch, sensual touch and hunger (Craig, 2003). Recognising these signals is an essential part of recognising emotions and their associated physiological sensations. Although in individuals on the autism spectrum the brain structures and possible hypo- and hyper-functional connections responsible for interoception have not yet been confirmed (DuBois et al., 2016), studies have shown that within neurotypical populations, interoceptive awareness (e.g. heart rate) is integrated with emotion-related signals received from other related neural structures (e.g. amygdala) within the insular cortex (Gogolla, 2017).

There are suggestions that interoception, or some of the three aspects of interoception (interoceptive awareness, interoceptive sensitivity/accuracy and interoceptive sensibility), may be attenuated in individuals on the autism spectrum (Mul et al., 2018). However, further work is needed to understand how interoceptive differences link to the behavioural and cognitive characteristics of autism (DuBois et al., 2016) and the neural circuitry of anxiety.

Alexithymia is the process of identifying and then describing one's own emotions. Elevated alexithymia is reported to be present in approximately 12% of the typically developing population (McGillivray et al., 2017) and in 50% of those on the autism spectrum (Kinnaird et al., 2019). In the general population, alexithymia mediates the relationship between interoceptive sensibility and anxiety (Palser et al., 2018). In adolescents on the autism spectrum, it has been shown to mediate the relationship between autism traits and anxiety (Pickard et al., 2020), suggesting that a greater subjective, self-reported awareness of one's own bodily states may exacerbate anxiety.

Executive Function and Attentional Control Executive function is an umbrella term for a series of cognitive skills, including selective attention and inhibition

(Diamond, 2013). These two skills are considered mechanisms of attentional control, as selective attention supports the focusing of attention on task-relevant cues, whilst inhibitory control relates to the inhibition of task-irrelevant cues (Eysenck et al., 2007). Attentional control theory (based upon neurotypical data and models) posits that anxiety reduces top-down attentional control by increasing the degree to which attention is stimulus driven. That is, rather than attending to goal-directed information, individuals high in trait anxiety would be biased towards the processing of stimuli related to a perceived threat.

Multiple studies have demonstrated an association between heightened anxiety levels in individuals on the autism spectrum and challenges with the executive functioning skill of set-shifting using both direct assessments (Hollocks et al., 2014; Zimmerman et al., 2017) and informant-based measures (Wallace et al., 2016). As all of these studies were cross-sectional, causality cannot be inferred, but it could be hypothesised that reduced cognitive flexibility may make it difficult for individuals to approach and adjust to new settings and/or to recognise, act and shift their emotional state, both of which would result in increased levels of anxiety.

Cognitive Processes Intolerance of uncertainty (IU) was discussed as a way of understanding anxiety in autism by Rodgers et al. (2012) and expanded upon by Boulter et al. (2014). They report a "causal mediational model" in which IU almost completely mediated the relationship between autism and anxiety. Wigham et al. (2015) further highlighted the role that IU has on the interaction between anxiety and autism traits, and Hodgson et al. (2017) shared parental perspectives on how IU is a recognisable and important construct associated with anxiety distinguishable from, but related to, features of autism. IU and sensory hypersensitivity have been shown to mediate the relationship between autism characteristics and anxiety in adolescents on the spectrum, even when additional mediators (alexithymia, parent intolerance of uncertainty) were adjusted (Pickard et al., 2020). Research specifically focussing upon anxiety in adults with autism has also shown that IU is related to general anxiety (Maisel et al., 2016) and specifically to social anxiety (Pickard et al., 2020).

The main alternative cognitive model of anxiety focuses upon threat expectancy. This refers to an increased expectation that one will experience a threatening event and that the outcome of such an event will be negative (Aue & Okon-Singer, 2015). Compared to IU, threat expectancy as a risk factor for anxiety has been less researched in individuals on the autism spectrum. The typical paradigms used to assess this construct involve the use of threatening faces; however, these techniques may be impacted by differences in facial processing in individuals on the spectrum. This limits the sensitivity and specificity of such methods, as both aversion to social stimuli and bias towards threat cues may be noted in individuals on the spectrum with and without anxiety (see White et al., 2014, for a discussion). In contrast to the results associated with anxiety in neurotypical populations, both Hollocks et al. (2013) and May et al. (2015) did not find an association between anxiety and attentional bias to threat in individuals on the autism spectrum. When exploring cognitive and physiological factors in the same model, Hollocks et al. (2016) found that both cognitive and physiological factors were related to the degree of anxiety symptoms, but only the physiological parameters were significant predictors of current anxiety disorders. Further work is needed to explore if this was due to the attentional bias task or anxiety measure used, or whether the threat perception cognitive processing bias differs between individuals on the autism spectrum and neurotypical populations.

Summary

A growing body of research has highlighted the prevalence of anxiety in both the autism and IDD populations (Lai et al., 2019; Maiano et al., 2018; van Steensel et al., 2011). This is particularly pronounced for individuals on the autism spectrum who are twice as likely as their neurotypical counterparts to be diagnosed with an anxiety disorder across the lifespan. Anxiety has been shown to have a wide-reaching impact on behaviours and life skills for those without autism and IDD (e.g. academic performance; de Lijster et al., 2018). Research, however, is yet to examine fully the impact of anxiety in individuals with IDD or those on the autism spectrum, and there is very limited work exploring the interplay between anxiety and decision making within these populations. Consideration of this topic is vital as decision making plays a critical role in our lives: each day we encounter a range of situations which can be as simple as deciding what to cook for dinner and as complex as making a life-changing decision such as a deciding upon who to live with or who should provide any support needed. As such, given the prevalence of anxiety in these populations, it is important to increase our understanding of how anxiety may impact decision making. To begin furthering our understanding within the IDD and autism populations, it is first useful to explore the impact of anxiety on decision making in neurotypical individuals. This may then provide a framework from which to consider the interplay between these two factors in IDD and autism.

Anxiety and Decision Making in Neurotypical Individuals

There is a wealth of literature which suggests that sound and rational decision making depends on prior accurate emotional processing (Bechara & Damasio, 2005). The influence of factors related to emotional processing, such as emotions and anxiety, may occur at multiple levels of the decision making process, both conscious and unconscious. This section will review the cognitive and somatic/physiological components of anxiety that may underpin the relation between anxiety and decision making in neurotypical samples who are either diagnosed with an anxiety disorder or are high in trait anxiety (e.g. Bensi & Giusberti, 2007; Miu et al., 2008).

Role of Emotions in Decision Making

Emotions, particularly anxiety, are characterised by physiological arousal, including changes in skin conductance responses (SCR; Barrett, 2012) and heart rate (Wang et al., 2018). The somatic marker hypothesis (Bechara et al., 1997; Damasio, 1994) posits that these physiological changes serve as critical (conscious or unconscious) indicators in decision making (especially risky decisions), guiding an individual in their selection of the safer option (Bechara & Damasio, 2005; Schwarz & Clore, 2007; Yip & Schweitzer, 2016). Over time, emotions and their corresponding somatic markers become associated with particular situations and the previous outcomes and inform future decisions. Reactivation of the previous somatic markers and the outcomes can either be via the "body loop" (e.g. changes in the body which are projected into the brain) or the "as-if body loop" (cognitive representations of experiences causing bodily changes). Empirical support for the somatic marker hypothesis has been obtained using the Iowa gambling task (IGT; Bechara et al., 1997). The object of this task is to win as much money as possible by drawing cards from a series of four decks. Two of these decks (A and B) contain cards that represent high rewards ($100) but large losses resulting in an overall loss; the remaining two decks (C and D) contain cards of lower rewards ($50) but smaller losses, leading to an overall gain. As such, the most advantageous strategy is to select cards from either deck C or D. The early trials of the IGT are thought to represent decision making under ambiguity, and the later trials (once the participant has had an opportunity to work out the outcomes of the deck) involve a more conscious decision-making process and decision making under risk.

The somatic marker hypothesis (Bechara et al., 1997) has been explored in a series of studies, including those comparing the performance of adults with lesions to their prefrontal cortex to that of neurotypical adults without lesions (e.g. Bechara et al., 1997; Bechara & Damasio, 2002; Reimann & Bechara, 2010). These lesions are of particular interest as the prefrontal cortex is critical to the generation and regulation of emotion (Dixon et al., 2017). Consequently, when contemplating the selection of a card from a risky deck (i.e. A or B), the anticipatory SCRs of neurotypical individuals increase, thus providing a somatic marker which guides them away from making a risky choice. In comparison, individuals with prefrontal cortex lesions did not experience this increase in SCR and therefore were more likely to select a card from the risky pile than were the healthy adults.

In order for emotions to impact decision making, it is necessary for the individual to interpret the somatic markers and determine the valance of the emotion. This can occur either consciously or unconsciously (e.g. Bechara et al., 1997; ten Brinke et al., 2019). Individuals with anxiety disorders have a heightened sensitivity to the physiological characteristics of emotion and therefore demonstrate enhanced interoceptive awareness (e.g. Ehlers & Breuer, 1996; Pineles & Mineka, 2005). Moreover, anxiety disorders are often associated with increased physiological activity (e.g. Monk et al., 2001; Pauli et al., 1997). Taken together, these suggest that both baseline and reactionary somatic markers may differ in individuals with anxiety and highlight potential ways in which anxiety may impact upon decision making and the need to explore this experimentally.

Using the IGT, Miu et al. (2008) examined how somatic markers relate to the complex decision making of individuals high in anxiety. Prior to an advantageous card selection (i.e. from deck C or D), adults high in trait anxiety experienced an increase in heart deceleration and SCR amplitude. Heart rate was also found to be sensitive to the behavioural outcome: for adults high in trait anxiety, greater heart deceleration was observed following punishment trials (i.e. cards representing loss) than reward trials (i.e. cards representing gains). However, in order for these somatic markers to be effective in guiding a person's decision making, they need to be interpreted correctly. Miu et al. found that despite the potentially adaptive somatic markers experienced prior to an advantageous decision (i.e. reduction in heart rate), adults high in anxiety performed at a lower level on the IGT than those with low trait anxiety. This is in contrast to what we would expect given the greater interoceptive awareness experienced by individuals with anxiety which could be utilised to make an advantageous decision. One possible explanation for this is the tendency for individuals high in anxiety to focus on fewer cues (Leon & Revelle, 1985); therefore, the participants may have disregarded the somatic markers in favour of utilising the more easily identified cue of reward size (leading to the selection of the disadvantageous pile giving the greater reward). Regardless of performance on the task, this study supported the somatic marker hypothesis in that these crucial somatic markers varied in relation to decision making and, if attended to and interpreted correctly, could guide the individual to the safer decision.

The Role of Cognitive Biases in Decision Making

The impact of cognitive biases on the experience of anxiety has long been observed. Early research highlighted the tendency of individuals with anxiety to interpret ambiguous material as threatening (Butler & Mathews, 1983) and to overestimate both the likelihood that an unpleasant event will occur and that the outcomes of these events will be negative (Butler & Mathews, 1983; McManus et al., 2000). This is interpreted as demonstrating an inherent threat expectancy. Anxiety is also associated with increased selective attention towards threatening information in that individuals with anxiety often detect threatening stimuli more quickly than nonthreatening stimuli (see MacLeod et al., 2019). An intolerance of uncertainty is also frequently associated with anxiety, with this intolerance underpinned by a fear of the unknown and a belief that one lacks the ability to overcome or avoid future negative events (Carleton, 2012). Each of these biases has been implicated in the decision-making strategies of individuals with anxiety.

In a clear demonstration of how threat expectancy can influence the decision making of individuals with anxiety, Mitte (2007) presented a series of scenarios to undergraduate students. Each scenario depicted an event with two potential choices: one risky and one safe. Compared to their low-anxiety peers, individuals high in trait anxiety (a) were more likely to select the safe option and (b) identified a greater cost to the risky decision. Moreover, high-anxiety individuals believed it was more likely that a negative event would happen to themselves than to other people. This heightened threat expectancy in individuals with anxiety was therefore associated with a tendency to avoid risky behaviour, a finding that has been replicated across many decision-making paradigms, including the IGT (Mueller et al., 2010, but see Zhang et al., 2015; the Balloon Analog Risk Task [BART], Maner et al., 2007; and questionnaire-based research, Maner et al., 2007).

Intolerance of uncertainty is also associated with impaired decision making (e.g. Ladouceur et al., 1997). When confronted with a decision for which the potential outcomes are uncertain, individuals with anxiety arguably find this situation discomforting due to an increased intolerance for the unknown. In order to assuage this discomfort, those with anxiety (and therefore likely to have higher levels of intolerance of uncertainty) would be motivated to reach a decision as quickly as possible. This is reflected by a "jump-to-conclusions" reasoning style (Bensi & Giusberti, 2007). Bensi and Giusberti (2007) contrasted the performance of individuals high and low in trait anxiety across three tasks, each with uncertain outcomes. For all tasks, participants were able to request as much information as they required in order to make their final decision. Consistently, individuals high in trait anxiety requested less information before reaching a decision. This arguably impacted on their overall accuracy as they often reached the incorrect conclusion. Importantly, this lower performance level was not reflective of a lack of interest or understanding of the task rules, and therefore represents a trade-off between accuracy and a desire to reduce the discomfort associated with uncertainty by ending the task quickly.

When making decisions, the amount of attention paid towards an aversive choice is predictive of its avoidance (Armel et al., 2008). Therefore, if a person with anxiety already has an attentional bias towards a threat, this may increase the likelihood that they would select the safer alternative (Hartley & Phelps, 2012). This attentional bias may also influence the information gathered during the active information search process (Huber et al., 1997). Using a search-and-rescue paradigm, participants (with varying levels of trait anxiety) considered the costs and benefits of each potential route that could be taken to reach the lost explorers (Matthews et al., 2011). Although not consistent across conditions, trait anxiety was associated with a bias during the information search process. That is, individuals high in anxiety spent more time considering information relating to possible losses, thus highlighting a potential attentional bias towards threat. This offers some support for the importance of attentional control when making judgements.

Summary

Research examining the influence of anxiety on decision making in neurotypical individuals has highlighted a range of anxiety-related characteristics that influence the decision-making process. These include cognitive biases and an increased interoceptive awareness. These characteristics often result in differences in the decision-making strategies used by individuals high in anxiety compared to those used by their peers with low anxiety. For example, high anxiety is associated with a "jump-to-conclusion" style of reasoning (Bensi & Giusberti, 2007) and, in some contexts, a tendency to focus on sampling information relating to the riskier choice (Matthews et al., 2011). Consequently, these strategies often result in individuals with anxiety reaching a final decision that differs (and not always positively) from that reached by their low-anxiety peers. Taken together, these findings emphasise the importance of considering the impact that anxiety can have on decision making, particularly for populations in which experiences of anxiety are prevalent.

Anxiety and Decision Making in Individuals with IDD or Autism

Research into the decision making of neurotypical individuals has identified a range of anxiety-related factors that can influence decision making. Whilst there has been some work exploring the presence of these cognitive biases and interoceptive awareness in autism, there is extremely limited research into these areas in individuals with IDD. Therefore, although we are able to draw upon the neurotypical research to theorise about the impact of anxiety on decision making for individuals on the autism spectrum, there is not enough literature at present to hypothesise about this relation for individuals with IDD. This emphasises a clear need for research to identify the mechanisms underpinning anxiety in individuals with IDD; this is even more pressing given the prevalence of anxiety in IDD. With this knowledge, we can then begin to understand how anxiety influences decision making in this population. To date, research examining factors affecting the decision making of individuals with IDD has focused on constructs such as learned helplessness (e.g. Jenkinson, 1999) and self-determination and supported decision making (e.g. Blanck & Martinis, 2015). This is clearly an important focus for research; however, given the prevalence of anxiety in IDD, it is important to enhance our understanding of its impact on decision making (e.g. Reardon et al., 2015).

The focus of this final section will therefore be on hypothesising and modelling the specific impact of anxiety on decision making in autism. To date, this is not an area that has been researched or explored in any depth despite the finding that more than half of individuals on the autism spectrum self-report that their autism characteristics interfere with the decision-making process *often* or *always* and that the perceived level of interference increases as anxiety increases (Luke et al., 2012). As the knowledge base around anxiety in IDD and autism increases, we are hopeful that researchers will also consider exploring the impact that the neurological and psychological processes associated with anxiety have on critical tasks or processes, such as decision making.

The Potential Influences of Anxiety on Decision Making in Individuals on the Autism Spectrum

Anxiety May Increase the Role of Emotions in Decision Making in Individuals on the Autism Spectrum

A number of differences have been observed in the decision-making process of individuals on the autism spectrum, compared to neurotypical controls; some of these are summarised in Table 8.1. Using well-known paradigms, Vella et al. (2018) found that those on the autism spectrum sampled more information and took longer to make their decisions. This difference is consistent with the dual process theory which posits that individuals on the autism spectrum utilise a more deliberative, effortful style of reasoning rather than an automatic, intuitive style (Brosnan et al., 2014). This aligns with self-reported difficulties in making decisions quickly (Luke et al., 2012), supporting the possibility that individuals on the autism spectrum have difficulties in rapid reasoning characteristic of an intuitive approach. This additional time may reflect a bias towards a slower, more deliberative (rather than intuitive) reasoning style. This tendency to make decisions based upon deliberate reasoning rather than intuitively may be associated with the challenges or differences with interoception or alexithymia discussed earlier. Given that both neurotypical individuals and individuals on the autism spectrum with anxiety disorders have a heightened sensitivity to the physiological characteristics of emotion, and therefore demonstrate enhanced interoceptive awareness (e.g. Ehlers & Breuer, 1996; Pickard et al., 2020; Pineles & Mineka, 2005), one would hypothesise that individuals on the autism spectrum with anxiety would show increased intuitive decision making processes. Shah et al.'s (2016) study of *framing effects* allowed for a test of this hypothesis. Framing effects describe the impact of the way a decision is framed (i.e. the same outcome can be framed as a loss, e.g. *bet $50; lose $20*, or a gain, e.g. *bet $50; keep $30*) on the outcome. In line with previous research, Shah et al.'s study confirmed that the framing effect was significantly smaller in individuals on the autism spectrum than in the neurotypical group. Importantly, after accounting for autism and alexithymia, increased anxiety was associated with larger framing effects, representing an increased role of emotion in the decision-making process. This initial work provides preliminary evidence for the hypothesis, suggesting that the presence of anxiety in individuals on the autism spectrum may increase the tendency to make intuitive or emotionally driven decisions over deliberate ones (see Table 8.1). However, more work is needed to explore this possibility and the broader interrelationship of autism, anxiety, alexithymia and interoception on decision making.

Table 8.1 Potential impact of autism characteristics and anxiety on decision making in individuals with autism

Decision-making stage	Impact of autism characteristics	Impact of anxiety in autism	Possible, hypothesised outcomes or clinical presentations[a]
Decision-point recognition	Challenges in quickly interpreting social and emotional cues (Allman et al., 2005; Robic et al., 2015) Preference for routine and predictability (Minassian et al., 2007)	Anxiety is associated with an increased intolerance of uncertainty (Rodgers et al., 2012) which may lead to avoidance of decision points	Missing or late identification of decision points. Anxiety may further increase a preference for routine and avoidance of decisions
Option generation	Challenges in executive functioning and imagination (Hill, 2004) reducing options generated	Anxiety further increases set-shifting difficulties (Hollocks et al., 2014; Zimmerman et al., 2017)	Difficulties considering multiple options which increase with anxiety, especially under time pressure. Tendency to stick to "known" or "safe" options
Option evaluation and elimination of options based on somatic markers	Reduced interoception and alexithymia makes it harder to read and integrate somatic cues into decision making (De Martino et al., 2008). Reduced likelihood of somatic-based decisions which increases decision making and/or response time (Vella et al., 2018)	Anxiety increases "framing effects" (Shah et al., 2016), meaning emotions are more involved in decisions. Relative increase in interoception makes somatic cues relatively prominent	Slower decision making in general, but relatively more emotionally driven (and potentially not fully evaluated) or "jump to" decisions when anxious
Cognitive evaluation and elimination of remaining options	Motivated by a fear of loss rather than the gaining of rewards (South et al., 2014). Overthinking and being overwhelmed by options (Luke et al., 2012). Tendency towards rule-based decisions (Shah et al., 2016)	Anxiety increases risk-taking but also further increases the need to succeed due to increased fear of failure (South et al., 2011)	Avoidance of failure at all costs. When anxious, there may be a further increase in avoidance of starting tasks or trying new things. Anxiety changes the approach towards "risky" decisions
Decision executed	Challenges in planning and executing and monitoring plan (Hill, 2004) Avoidant decision making	Anxiety may possibly increase the fear of negative evaluation or judgements of decision (Luke et al., 2012)	Challenges developing and implementing plan once decision is made. Increased anxiety may reduce ease of discussing (or adjusting) options and strategies due to fear of negative evaluation

[a]The lack of research means that these are tentative hypotheses rather than evidence-based statements

Anxiety May Change the Desired or Feared Outcome of the Decision in Individuals on the Autism Spectrum

A series of studies exploring a specific type of decision making (risk-based decision) have suggested that individuals on the autism spectrum may have differing motivation driving the decision made. Instead of being motivated by the value of previous rewards as per typically developing children on the IGT, individuals on the autism spectrum are more likely to be motivated by a fear of loss (South et al., 2014). Whilst this risk aversion may in some contexts (including gambling) prove advantageous, in many everyday situations, avoidance of situations with unknown outcomes would result in the avoidance of many social and other activities.

The literature on the influence of anxiety on decision making in neurotypical individuals would suggest that the presence of anxiety increases risk aversion, so it could be hypothesised that individuals on the autism spectrum with co-occurring anxiety would show an even more prominent fear of loss and show even more significant risk-avoidance behaviours. To date, there have not been any studies using the IGT in individuals on the autism spectrum with varying levels of anxiety. South et al. (2011) used an alternative decision-making task, the BART (Lejuez et al., 2002), to explore the impact of IQ and trait anxiety on risk-based decision making in children on the autism spectrum compared to neurotypical controls. In contrast to the typically developing group, South et al. found that increased anxiety was associated with increased risk-taking and longer decision-making times in individuals on the autism spectrum. This may be linked to the reduced threat-related attentional bias in individuals on the autism spectrum with anxiety discussed earlier in this chapter (Hollocks et al., 2013; May et al., 2015). A reduced bias towards threatening stimuli or outcomes may mean that the anxiety does not cause functioning to be limited, resulting in risk avoidance. Instead, the authors hypothesise that the increased anxiety in the autism group leads to an increased motivation to succeed due to a fear of failure. This hypothesis could have potentially been explored by the authors by looking at participants' responses to wins and losses on the BART and the impact each win or loss had on the next decision (i.e. how many pumps to put into the balloon), although this was not within the scope of the original article. However, it may also be that individuals on the autism spectrum who experience increased trait anxiety are experiencing increased arousal, which makes it difficult to process relevant and discard irrelevant external and internal cues, leading to a longer decision-making process and a poorer (i.e. riskier) decision outcome. This would align with the suggestion of the somatic marker hypothesis and the work of Bechara and Damasio (2005) in neurotypical individuals that, in the absence of somatic markers, one cannot estimate the adverse consequences of a decision and therefore make a decision focussed on short-term reward. Regardless of the cause, the relationship between the risk-based decision making and anxiety differs between individuals on the spectrum and neurotypical individuals, highlighting the need for more work in the area.

Whilst the IGT and BART are both risk-based decision-making tasks, they differ in the decision to be made. The initial 40 trials of the IGT (when the person does not know much about the relative risks and benefits of each deck) have been termed decision making under ambiguity and the final 60 described as decision making under risk, as participants have gained enough experience with the decks to learn their relative risks and benefits (Brand et al., 2007). The process for the BART can vary (see Lauriola et al., 2014, for a meta-analysis of BART studies), but generally there is no strategy or cost/benefit ratio to learn, so the ambiguous decision is solely on the option or winning or losing money. In South et al.'s (2011) study, participants selected the number of pumps for the balloon at the beginning of the trial, instead of having to select whether or not to add an additional pump each time. In neurotypical individuals, the impact of anxiety on decision making differs dependent upon whether the decision is made under ambiguity or risk (Buelow & Barnhart, 2017) and the time between which the decision is made and the outcome (Buelow et al., 2013), neither of which has been systematically explored in individuals on the autism spectrum. Given the increased challenges around managing uncertainty in individuals on the spectrum with anxiety (e.g. Rodgers et al., 2012), it could be hypothesised that there would be a significant shift from decision making under ambiguity to decision making under risk, although this is yet to be explored within the research literature.

Anxiety May Increase Challenges in Set-Shifting and Generalisation, Making It More Difficult for Individuals on the Autism Spectrum to Make Informed and Flexible Decisions

Many individuals on the autism spectrum have challenges in executive functioning, and as discussed at the start of this chapter, multiple studies have demonstrated an association between heightened anxiety levels in individuals on the autism spectrum and challenges with the executive functioning skill of set-shifting using both direct assessments (Hollocks et al., 2014; Zimmerman et al., 2017) and informant-based measures (Wallace et al., 2016). There is some evidence for a relationship between inhibition, set-shifting and performance on decision-making tasks in neurotypical individuals (e.g. Campbell-Meiklejohn et al., 2008). This has not yet been explored systematically in individuals on the autism spectrum, but it is hypothesised that decision making (or at least the decision-making experiments such as the IGT) that requires the review, synthesis and reflection of performance in order to inform and potentially change approach (as per the bottom line of Table 8.1) will be more challenging for those individuals on the autism spectrum with elevated anxiety than for those who are not experiencing anxiety.

Summary and Areas for Future Exploration

The possibility for decision-making tasks to disentangle and accommodate for the impact of anxiety and the impact of IDD/autism characteristics is an area that requires further development. A different route to both decision making and anxiety may mean that standard tests may not be measuring what they would typically measure. For example, Top Jr. et al. (2016) reported that individuals on the autism spectrum showed less differentiation in amygdala activity between the neutral and startle conditions of their experimental paradigm. If such experimental paradigms are being used, it is essential they have sufficient sensitivity and specificity to inform researchers about the neural circuitry of anxiety and/or decision making in individuals on the autism spectrum.

As the experimental paradigms lead to a more sophisticated understanding and to models of anxiety and its impact on individuals on the IDD and/or autism spectrum (including their decision making), it is important that anxiety itself is considered a spectrum rather than a dichotomous variable (Mazefsky & Herrington, 2014). Whilst the decision making in autism literature has predominantly focussed upon experimental paradigms, it is important to recognise that these may not immediately translate into the real-world decisions that are being made every second by individuals with IDD and/or autism who are experiencing anxiety each day. Such decisions are likely to require the incorporation of multiple sources of information, for example, social-decision making, which Luke et al. (2012) noted was significantly more difficult for adults on the autism spectrum than neurotypical controls, requires the incorporation of the available external social and environmental cues as well as the internal processes of theory of mind, executive functioning and empathy, many of which are not required within a laboratory-based task. The need for further work to explore the impact of anxiety in those with IDD and those on the autism spectrum on different types of decision tasks is clear.

Table 8.1 brings together our hypothesised outcomes and impact of anxiety on the stages of the decision making in autism (as discussed in this chapter). Anxiety may increase a person's tendency to wish to stick to perceived "safe" or "known" decisions, as they increase certainty and are perceived to decrease the risk of failure. Anxiety may also increase the likelihood of an individual on the autism spectrum making an emotionally driven decision which has not necessarily been fully evaluated. As this is due to a relative (rather than absolute) increase in the somatic cues within an individual, it is therefore critical that anxiety is considered prior to all individuals with autism (and/or IDD) making large or life-changing decisions, not only in those with recognised anxiety disorders. Finally, anxiety may decrease the likelihood of an individual to be able to recognise or admit to a problem relating to the decision due to fear of negative evaluation. This highlights the need for genuine and available support not only during but after significant decisions are made. Due to the limited research in the area, it has not been possible to hypothesise for the impact of anxiety on decision making in IDD, although many of the above recommendations could be considered good practice to support decision making in individuals with IDD and anxiety until further research is available. The limited research literature has also meant that it has not been possible to consider the differing types of anxiety (e.g. social anxiety, performance anxiety, generalised anxiety). It may be that more can be learnt from their similarities and differences on the decision-making process.

Anxiety and its impact on decision making in IDD and autism is under-researched, complex and multifaceted. Models will likely stem from multidisciplinary teams who are able to understand the interaction between cognitive and behavioural, neurobiological and, of course, life experiences. Such models allow for a new opportunity to explore the intersection between cognition and emotion in IDD and in autism (South et al., 2011) and may enhance and further our understanding on one of the many impacts of anxiety on the daily life of those with IDD or on the autism spectrum.

References

Adams, D., Clark, M., & Keen, D. (2019). Using self-report to explore the relationship between anxiety and quality of life in children on the autism spectrum. *Autism Research, 12*(10), 1505–1515. https://doi.org/10.1002/aur.2155

Adams, D., Clark, M., & Simpson, K. (2020). The relationship between child anxiety and the quality of life of children, and parents of children, on the autism spectrum. *Journal of Autism and Developmental Disorders, 50*(5), 1756–1769. https://doi.org/10.1007/s10803-019-03932-2

Adams, D., & Emerson, L. M. (2020). Family accommodation of anxiety in a community sample of children on the autism spectrum. *Journal of Anxiety Disorders, 70*, ARTN 102192. https://doi.org/10.1016/j.janxdis.2020.102192

Adams, D., MacDonald, L., & Keen, D. (2019). Teacher responses to anxiety-related behaviours in students on the autism spectrum. *Research in Developmental Disabilities, 86*, 11–19. https://doi.org/10.1016/j.ridd.2018.12.009

Allman, J. M., Watson, K. K., Tetreault, N. A., & Hakeem, A. Y. (2005). Intuition and autism: A possible role for Von Economo neurons. *Trends in Cognitive Sciences, 9*(8), 367–373. https://doi.org/10.1016/j.tics.2005.06.008

American Psychiatric Association. (2013). Diagnostic and statistical manual of mental disorders (5th ed.). https://doi.org/10.1176/appi.books.9780890425596

Armel, K. C., Beaumel, A., & Rangel, A. (2008). Biasing simple choices by manipulating relative visual attention. *Judgment and Decision making, 3*(5), 396–403.

Aue, T., & Okon-Singer, H. (2015). Expectancy biases in fear and anxiety and their link to biases in attention. *Clinical Psychology Review, 42*, 83–95. https://doi.org/10.1016/j.cpr.2015.08.005

Barrett, L. F. (2012). Emotions are real. *Emotion, 12*(3), 413–429. https://doi.org/10.1037/a0027555

Bechara, A., & Damasio, A. R. (2005). The somatic marker hypothesis: A neural theory of economic decision. *Games and Economic Behavior, 52*(2), 336–372. https://doi.org/10.1016/j.geb.2004.06.010

Bechara, A., & Damasio, H. (2002). Decision-making and addiction (Part I): Impaired activation of somatic states in substance dependent individuals when pondering decisions with negative future consequences. *Neuropsychologia, 40*(10), 1675–1689. https://doi.org/10.1016/S0028-3932(02)00015-5

Bechara, A., Damasio, H., Tranel, D., & Damasio, A. R. (1997). Deciding advantageously before knowing the advantageous strategy. *Science, 275*(5304), 1293–1295. https://doi.org/10.1126/science.275.5304.1293

Beck, A. T. (1972). Cognition, anxiety, and psychophysiological disorders. In C. D. Spielberger (Ed.), *Anxiety: Current trends in theory and research* (pp. 343–357). Academic Press.

Beck, A. T. (1985). Cognitive therapy. In H. I. Kaplan & B. J. Sadock (Eds.), *Comprehensive textbook of psychiatry/IV: Vol. 2* (pp. 1432–1438). Williams and Wilkins.

Beck, A. T., & Weishaar, M. (1989). Cognitive therapy. In A. Freeman, K. M. Simon, L. E. Beutler, & H. Arkowitz (Eds.), *Comprehensive handbook of cognitive therapy* (pp. 21–36). Plenum Press.

Bensi, L., & Giusberti, F. (2007). Trait anxiety and reasoning under uncertainty. *Personality and Individual Differences, 43*(4), 827–838. https://doi.org/10.1016/j.paid.2007.02.007

Blanck, P., & Martinis, J. G. (2015). "The right to make choices": The national resource center for supported decision making. *Inc, 3*(1), 24–33. https://doi.org/10.1352/2326-6988-3.1.24

Boulter, C., Freeston, M., South, M., & Rodgers, J. (2014). Intolerance of uncertainty as a framework for understanding anxiety in children and adolescents with autism spectrum disorders. *Journal of Autism and Developmental Disorders, 44*(6), 1391–1402. https://doi.org/10.1007/s10803-013-2001-x

Brand, M., Recknor, E. C., Grabenhorst, F., & Bechara, A. (2007). Decisions under ambiguity and decisions under risk: Correlations with executive functions and comparisons of two different gambling tasks with implicit and explicit rules. *Journal of Clinical and Experimental Neuropsychology, 29*(1), 86–99. https://doi.org/10.1080/13803390500507196

Brosnan, M., Chapman, E., & Ashwin, C. (2014). Adolescents with autism spectrum disorder show a circumspect reasoning bias rather than 'jumping-to-conclusions'. *Journal of Autism and Developmental Disorders, 44*(3), 513–520. https://doi.org/10.1007/s10803-013-1897-5

Buelow, M. T., & Barnhart, W. R. (2017). The influence of math anxiety, math performance, worry, and test anxiety on the Iowa gambling task and balloon analogue risk task. *Assessment, 24*(1), 127–137. https://doi.org/10.1177/1073191115602554

Buelow, M. T., Okdie, B. M., & Blaine, A. L. (2013). Seeing the forest through the trees: Improving decision making on the Iowa gambling task by shifting focus from short- to long-term outcomes. *Frontiers in Psychology, 4*. https://doi.org/10.3389/fpsyg.2013.00773

Butler, G., & Mathews, A. (1983). Cognitive-processes in anxiety neurosis. *Bulletin of the British Psychological Society, 36*(May), A49.

Campbell-Meiklejohn, D. K., Woolrich, M. W., Passingham, R. E., & Rogers, R. D. (2008). Knowing when to stop: The brain mechanisms of chasing losses. *Biological Psychiatry, 63*(3), 293–300. https://doi.org/10.1016/j.biopsych.2007.05.014

Carleton, R. N. (2012). The intolerance of uncertainty construct in the context of anxiety disorders: Theoretical and practical perspectives. *Expert Review of Neurotherapeutics, 12*(8), 937–947. https://doi.org/10.1586/Ern.12.82

Clark, D. M., & Wells, A. (1995). A cognitive model of social phobia. In R. G. Heimberg, M. R. Liebowitz, D. A. Hope, & F. R. Schneier (Eds.), *Social phobia: Diagnosis, assessment, and treatment* (pp. 69–93). The Guilford Press.

Costello, E. J., Egger, H. J., Copeland, W., Erklani, A., & Angold, A. (2011). The developmental epidemiology of anxiety disorders: Phenomenology, prevalence, and comorbidity. In W. K. Silverman & A. P. Field (Eds.), *Anxiety disorders in children and adolescents* (2nd ed., pp. 56–75). Cambridge University Press.

Craig, A. D. (2003). Interoception: The sense of the physiological condition of the body. *Current Opinion in Neurobiology, 13*(4), 500–505. https://doi.org/10.1016/S0959-4388(03)00090-4

Da Costa, D., Larouche, J., Dritsa, M., & Brender, W. (2000). Psychosocial correlates of prepartum and postpartum depressed mood. *Journal of Affective Disorders, 59*(1), 31–40. https://doi.org/10.1016/S0165-0327(99)00128-7

Dagnan, D., Jackson, I., & Eastlake, L. (2018). A systematic review of cognitive behavioural therapy for anxiety in adults with intellectual disabilities. *Journal of Intellectual Disability Research, 62*(11), 974–991. https://doi.org/10.1111/jir.12548

Dagnan, D., & Sandhu, S. (1999). Social comparison, self-esteem and depression in people with intellectual disability. *Journal of Intellectual Disability Research, 43*, 372–379. https://doi.org/10.1046/j.1365-2788.1999.043005372.x

Dagnan, D., & Waring, M. (2004). Linking stigma to psychological distress: Testing a social-cognitive model of the experience of people with intellectual disabilities. *Clinical Psychology & Psychotherapy, 11*(4), 247–254. https://doi.org/10.1002/cpp.413

Damasio, A. R. (1994). *Descartes' error: Emotion, rationality and the human brain*. Penguin.

de Lijster, J. M., Dieleman, G. C., Utens, E. M. W. J., Dierckx, B., Wierenga, M., Verhulst, F. C., & Legerstee, J. S. (2018). Social and academic functioning in adolescents with anxiety disorders: A systematic review. *Journal of Affective Disorders, 230*, 108–117. https://doi.org/10.1016/j.jad.2018.01.008

den Houting, J., Adams, D., Roberts, J., & Keen, D. (2020). Brief report: Investigating the impact of anxious symptomatology in autistic children. *International Journal of Disability Development and Education*. https://doi.org/10.1080/1034912x.2020.1727420

Diamond, A. (2013). Executive functions. *Annual Review of Psychology, 64*, 135–168. https://doi.org/10.1146/annurev-psych-113011-143750

Dixon, M. L., Thiruchselvam, R., Todd, R., & Christoff, K. (2017). Emotion and the prefrontal cortex: An integrative review. *Psychological Bulletin, 143*(10), 1033–1081. https://doi.org/10.1037/bul0000096

DuBois, D., Ameis, S. H., Lai, M. C., Casanova, M. F., & Desarkar, P. (2016). Interoception in autism spectrum disorder: A review. *International Journal of Developmental Neuroscience, 52*, 104–111. https://doi.org/10.1016/j.ijdevneu.2016.05.001

Ehlers, A., & Breuer, P. (1996). How good are patients with panic disorder at perceiving their heartbeats? *Biological Psychology, 42*(1-2), 165–182. https://doi.org/10.1016/0301-0511(95)05153-8

Einfeld, S. L., Ellis, L. A., & Emerson, E. (2011). Comorbidity of intellectual disability and mental disorder in children and adolescents: A systematic review. *Journal of Intellectual & Developmental Disability, 36*(2), 137–143. https://doi.org/10.1080/13668250.2011.572548

Emerson, E., & Hatton, C. (2007). Mental health of children and adolescents with intellectual disabilities in Britain. *British Journal of Psychiatry, 191*, 493–499. https://doi.org/10.1192/bjp.bp.107.038729

Eysenck, M. W., Derakshan, N., Santos, R., & Calvo, M. G. (2007). Anxiety and cognitive performance: Attentional control theory. *Emotion, 7*(2), 336–353. https://doi.org/10.1037/1528-3542.7.2.336

Flynn, S., Vereenooghe, L., Hastings, R. P., Adams, D., Cooper, S. A., Gore, N., Hatton, C., Hood, K., Jahoda, A., Langdon, P. E., McNamara, R., Oliver, C., Roy, A., Totsika, V., & Waite, J. (2017). Measurement tools for mental health problems and mental well-being in people with severe or profound intellectual disabilities: A systematic review. *Clinical Psychology Review, 57*, 32–44. https://doi.org/10.1016/j.cpr.2017.08.006

Galynker, I. I., Cai, J., Ongseng, F., Finestone, H., Dutta, E., & Serseni, D. (1998). Hypofrontality and negative symptoms in major depressive disorder. *Journal of Nuclear Medicine, 39*(4), 608–612. https://www.ncbi.nlm.nih.gov/pubmed/9544664

Gogolla, N. (2017). The insular cortex. *Current Biology, 27*(12), R580–R586. https://doi.org/10.1016/j.cub.2017.05.010

Gray, K. M., Piccinin, A. M., Hofer, S. M., Mackinnon, A., Bontempo, D. E., Einfeld, S. L., Parmenter, T., & Tonge, B. J. (2011). The longitudinal relationship between behavior and emotional disturbance in young people with intellectual disability and maternal mental health. *Research in Developmental Disabilities, 32*(3), 1194–1204. https://doi.org/10.1016/j.ridd.2010.12.044

Green, H., McGinnity, A., Meltzer, H., Ford, T., & Goodman, R. H. (2005). *Mental health of children and young people in Great Britain 2004*. Palgrave Macmillan.

Green, S. A., Berkovits, L. D., & Baker, B. L. (2015). Symptoms and development of anxiety in children with or without intellectual disability. *Journal of Clinical Child and Adolescent Psychology, 44*(1), 137–144. https://doi.org/10.1080/15374416.2013.873979

Hartley, C. A., & Phelps, E. A. (2012). Anxiety and decision making. *Biological Psychiatry, 72*(2), 113–118. https://doi.org/10.1016/j.biopsych.2011.12.027

Hemmings, C. P., Gravestock, S., Pickard, M., & Bouras, N. (2006). Psychiatric symptoms and problem behaviours in people with intellectual disabilities. *Journal of Intellectual Disability Research, 50*, 269–276. https://doi.org/10.1111/j.1365-2788.2006.00827.x

Hill, E. L. (2004). Executive dysfunction in autism. *Trends in Cognitive Sciences, 8*(1), 26–32. https://doi.org/10.1016/j.tics.2003.11.003

Hodgson, A. R., Freeston, M. H., Honey, E., & Rodgers, J. (2017). Facing the unknown: Intolerance of uncertainty in children with autism spectrum disorder. *Journal of Applied Research in Intellectual Disabilities, 30*(2), 336–344. https://doi.org/10.1111/jar.12245

Hollocks, M. J., Jones, C. R. G., Pickles, A., Baird, G., Happe, F., Charman, T., & Simonoff, E. (2014). The association between social cognition and executive functioning and symptoms of anxiety and depression in adolescents with autism spectrum disorders. *Autism Research, 7*(2), 216–228. https://doi.org/10.1002/aur.1361

Hollocks, M. J., Lerh, J. W., Magiati, I., Meiser-Stedman, R., & Brugha, T. S. (2019). Anxiety and depression in adults with autism spectrum disorder: A systematic review and meta-analysis. *Psychological Medicine, 49*(4), 559–572. https://doi.org/10.1017/S0033291718002283

Hollocks, M. J., Ozsivadjian, A., Matthews, C. E., Howlin, P., & Simonoff, E. (2013). The relationship between attentional bias and anxiety in children and adolescents with autism spectrum disorders. *Autism Research, 6*(4), 237–247. https://doi.org/10.1002/aur.1285

Hollocks, M. J., Pickles, A., Howlin, P., & Simonoff, E. (2016). Dual cognitive and biological correlates of anxiety in autism spectrum disorders. *Journal of Autism and Developmental Disorders, 46*(10), 3295–3307. https://doi.org/10.1007/s10803-016-2878-2

Hosenbocus, S., & Chahal, R. (2012). A review of executive function deficits and pharmacological management in children and adolescents. *Journal of the Canadian Academy of Child and Adolescent Psychiatry, 21*(3), 223–229. https://www.ncbi.nlm.nih.gov/pubmed/22876270

Houtkamp, E. O., van der Molen, M. J., de Voogd, E. L., Salemink, E., & Klein, A. M. (2017). The relation between social anxiety and biased interpretations in adolescents with mild intellectual disabilities. *Research in Developmental Disabilities, 67*, 94–98. https://doi.org/10.1016/j.ridd.2017.06.003

Huber, O., Wider, R., & Huber, O. W. (1997). Active information search and complete information presentation in naturalistic risk decision tasks. *Acta Psychologica, 95*, 15–29. https://doi.org/10.1016/S0001-6918(96)00028-5

Hudson, C. C., Hall, L., & Harkness, K. L. (2019). Prevalence of depressive disorders in individuals with autism spectrum disorder: A meta-analysis. *Journal of Abnormal Child Psychology, 47*(1), 165–175. https://doi.org/10.1007/s10802-018-0402-1

Jenkinson, J. C. (1999). Factors affecting decision making by young adults with intellectual disabilities. *American Journal on Mental Retardation, 104*(4), 320–329.

Kerns, C. M., Kendall, P. C., Berry, L., Souders, M. C., Franklin, M. E., Schultz, R. T., Miller, J., & Herrington, J. (2014). Traditional and atypical presentations of anxiety in youth with autism spectrum disorder. *Journal of Autism and Developmental Disorders, 44*(11), 2851–2861. https://doi.org/10.1007/s10803-014-2141-7

Kinnaird, E., Stewart, C., & Tchanturia, K. (2019). Investigating alexithymia in autism: A systematic review and meta-analysis. *European Psychiatry, 55*, 80–89. https://doi.org/10.1016/j.eurpsy.2018.09.004

Kurtek, P. (2016). Role of anxiety as a trait and state in youth with mild intellectual disability: Coping with difficult situations. *Journal of Policy and Practice in Intellectual Disabilities, 13*(3), 236–245. https://doi.org/10.1111/jppi.12150

Ladouceur, R., Talbot, F., & Dugas, M. J. (1997). Behavioral expressions of intolerance of uncertainty in worry – Experimental findings. *Behavior Modification, 21*(3), 355–371. https://doi.org/10.1177/01454455970213006

Lai, M. C., Kassee, C., Besney, R., Bonato, S., Hull, L., Mandy, W., Szatmari, P., & Ameis, S. H. (2019). Prevalence of co-occurring mental health diagnoses in the autism population: A systematic review and meta-analysis. *Lancet Psychiatry, 6*(10), 819–829. https://doi.org/10.1016/S2215-0366(19)30289-5

Lauriola, M., Panno, A., Levin, I. P., & Lejuez, C. W. (2014). Individual differences in risky decision making: A meta-analysis of sensation seeking and impulsivity with the balloon analogue risk task. *Journal of Behavioral Decision making, 27*(1), 20–36. https://doi.org/10.1002/bdm.1784

Lejuez, C. W., Read, J. P., Kahler, C. W., Richards, J. B., Ramsey, S. E., Stuart, G. L., Strong, D. R., & Brown, R. A. (2002). Evaluation of a behavioral measure of risk taking: The Balloon Analogue Risk Task (BART). *Journal of Experimental Psychology-Applied, 8*(2), 75–84. https://doi.org/10.1037//1076-898x.8.2.75

Leon, M. R., & Revelle, W. (1985). Effects of anxiety on analogical reasoning: A test of three theoretical models. *Journal of Personality and Social Psychology, 49*(5), 1302–1315. https://doi.org/10.1037//0022-3514.49.5.1302

Luke, L., Clare, I. C., Ring, H., Redley, M., & Watson, P. (2012). Decision making difficulties experienced by adults with autism spectrum conditions. *Autism, 16*(6), 612–621. https://doi.org/10.1177/1362361311415876

MacLeod, C., Grafton, B., & Notebaert, L. (2019). Anxiety-linked attentional bias: Is it reliable? *Annual Review of Clinical Psychology, 15*, 529–554. https://doi.org/10.1146/annurev-clinpsy-050718-095505

Maiano, C., Coutu, S., Tracey, D., Bouchard, S., Lepage, G., Morin, A. J. S., & Moullec, G. (2018). Prevalence of anxiety and depressive disorders among youth with intellectual disabilities: A systematic review and meta-analysis. *Journal of Affective Disorders, 236*, 230–242. https://doi.org/10.1016/j.jad.2018.04.029

Maisel, M. E., Stephenson, K. G., South, M., Rodgers, J., Freeston, M. H., & Gaigg, S. B. (2016). Modeling the cognitive mechanisms linking autism symptoms and anxiety in adults. *Journal of Abnormal Psychology, 125*(5), 692–703. https://doi.org/10.1037/abn0000168

Maner, J. K., Richey, J. A., Cromer, K., Mallott, M., Lejuez, C. W., Joiner, T. E., & Schmidt, N. B. (2007). Dispositional anxiety and risk-avoidant decision making. *Personality and Individual Differences, 42*(4), 665–675. https://doi.org/10.1016/j.paid.2006.08.016

De Martino, B., Harrison, N. A., Knafo, S., Bird, G., & Dolan, R. J. (2008). Explaining enhanced logical consistency during decision making in autism. *Journal of Neuroscience, 28*(42), 10746–10750. https://doi.org/10.1523/JNEUROSCI.2895-08.2008

Matson, J. L. (1995). *The diagnostic assessment for the severely handicapped-revised (DASH-II)*. Disability Consultants, LLC.

Matthews, G., Panganiban, A. R., & Hudlicka, E. (2011). Anxiety and selective attention to threat in tactical Decision making. *Personality and Individual Differences, 50*(7), 949–954. https://doi.org/10.1016/j.paid.2010.09.005

May, T., Cornish, K., & Rinehart, N. J. (2015). Mechanisms of anxiety related attentional biases in children with autism spectrum disorder. *Journal of Autism and Developmental Disorders, 45*(10), 3339–3350. https://doi.org/10.1007/s10803-015-2500-z

Mazefsky, C. A., & Herrington, J. (2014). Autism and anxiety: Etiologic factors and transdiagnostic processes. In I. Davis, E. Thompson, S. W. White, & T. Ollendick (Eds.), *Handbook of autism and anxiety* (pp. 91–103). Springer.

McCarthy, J. B., & Barbot, B. (2016). The need for research on intellectual disabilities and severe psychiatric disorders in children and adolescents. *Journal of Mental Disorders and Treatment, 2*(112), 2.

McGillivray, L., Becerra, R., & Harms, C. (2017). Prevalence and demographic correlates of alexithymia: A comparison between Australian psychiatric and community samples. *Journal of Clinical Psychology, 73*(1), 76–87. https://doi.org/10.1002/jclp.22314

McManus, F., Clark, D. M., & Hackmann, A. (2000). Specificity of cognitive biases in social phobia and their role in recovery. *Behavioural and Cognitive Psychotherapy, 28*, 201–209. https://doi.org/10.1017/S1352465800003015

Micco, J. A., Henin, A., Biederman, J., Rosenbaum, J. F., Petty, C., Rindlaub, L. A., Murphy, M., & Hirshfeld-Becker, D. R. (2009). Executive functioning in offspring at risk for depression and anxiety. *Depression and Anxiety, 26*(9), 780–790. https://doi.org/10.1002/da.20573

Minassian, A., Paulus, M., Lincoln, A., & Perry, W. (2007). Adults with autism show increased sensitivity to outcomes at low error rates during decision making. *Journal of Autism and Developmental Disorders, 37*(7), 1279–1288. https://doi.org/10.1007/s10803-006-0278-8

Mitte, K. (2007). Anxiety and risky decision making: The role of subjective probability and subjective costs of negative events. *Personality and Individual Differences, 43*(2), 243–253. https://doi.org/10.1016/j.paid.2006.11.028

Miu, A. C., Heilman, R. M., & Houser, D. (2008). Anxiety impairs decision making: Psychophysiological evidence from an Iowa gambling task. *Biological Psychology, 77*(3), 353–358. https://doi.org/10.1016/j.biopsycho.2007.11.010

Monk, C., Kovelenko, P., Ellman, L. M., Sloan, R. P., Bagiella, E., Gorman, J. M., & Pine, D. S. (2001). Enhanced stress reactivity in paediatric anxiety disorders: Implications for future cardiovascular health. *International Journal of Neuropsychopharmacology, 4*(2), 199–206. https://doi.org/10.1017/S146114570100236X

Mueller, E. M., Nguyen, J., Ray, W. J., & Borkovec, T. D. (2010). Future-oriented decision making in generalized anxiety disorder is evident across different versions of the Iowa gambling task. *Journal of Behavior Therapy and Experimental Psychiatry, 41*(2), 165–171. https://doi.org/10.1016/j.jbtep.2009.12.002

Mul, C. L., Stagg, S. D., Herbelin, B., & Aspell, J. E. (2018). The feeling of me feeling for you: Interoception, alexithymia and empathy in autism. *Journal of Autism and Developmental Disorders, 48*(9), 2953–2967. https://doi.org/10.1007/s10803-018-3564-3

Ohta, H., Yamagata, B., Tomioka, H., Takahashi, T., Yano, M., Nakagome, K., & Mimura, M. (2008). Hypofrontality in panic disorder and major depressive disorder assessed by multichannel near-infrared spectroscopy. *Depression and Anxiety, 25*(12), 1053–1059. https://doi.org/10.1002/da.20463

Olatunji, B. O., Cisler, J. M., & Tolin, D. F. (2007). Quality of life in the anxiety disorders: A meta-analytic review. *Clinical Psychology Review, 27*(5), 572–581. https://doi.org/10.1016/j.cpr.2007.01.015

Ottowitz, W. E., Dougherty, D. D., & Savage, C. R. (2002). The neural network basis for abnormalities of attention and executive function in major depressive disorder: Implications for application of the medical disease model to psychiatric disorders. *Harvard Review of Psychiatry, 10*(2), 86–99. https://doi.org/10.1093/hrp/10.2.86

Palser, E. R., Fotopoulou, A., Pellicano, E., & Kilner, J. M. (2018). The link between interoceptive processing and anxiety in children diagnosed with autism spectrum disorder: Extending adult findings into a developmental sample. *Biological Psychology, 136*, 13–21. https://doi.org/10.1016/j.biopsycho.2018.05.003

Pauli, P., Dengler, W., Wiedemann, G., Montoya, P., Flor, H., Birbaumer, N., & Buchkremer, G. (1997). Behavioral and neurophysiological evidence for altered processing of anxiety-related words in panic disorder. *Journal of Abnormal Psychology, 106*(2), 213–220. https://doi.org/10.1037/0021-843x.106.2.213

Pickard, H., Hirsch, C., Simonoff, E., & Happe, F. (2020). Exploring the cognitive, emotional and sensory correlates of social anxiety in autistic and neurotypical adolescents. *Journal of Child Psychology and Psychiatry.* https://doi.org/10.1111/jcpp.13214

Pineles, S. L., & Mineka, S. (2005). Attention biases to internal and external sources of potential threat in social anxiety. *Journal of Abnormal Psychology, 11*(4), 314–318. https://doi.org/10.1037/0021-843X.114.2.314

Polanczyk, G. V., Salum, G. A., Sugaya, L. S., Caye, A., & Rohde, L. A. (2015). Annual research review: A meta-analysis of the worldwide prevalence of mental disorders in children and adolescents. *Journal of Child Psychology and Psychiatry, 56*(3), 345–365. https://doi.org/10.1111/jcpp.12381

Rapee, R. M., & Heimberg, R. G. (1997). A cognitive-behavioral model of anxiety in social phobia. *Behaviour Research and Therapy, 35*(8), 741–756. https://doi.org/10.1016/S0005-7967(97)00022-3

Ray, R. D., & Zald, D. H. (2012). Anatomical insights into the interaction of emotion and cognition in the prefrontal cortex. *Neuroscience and Biobehavioral Reviews, 36*(1), 479–501. https://doi.org/10.1016/j.neubiorev.2011.08.005

Reardon, T. C., Gray, K. M., & Melvin, G. A. (2015). Anxiety disorders in children and adolescents with intellectual disability: Prevalence and assessment. *Research in Developmental Disabilities, 36*, 175–190. https://doi.org/10.1016/j.ridd.2014.10.007

Reimann, M., & Bechara, A. (2010). The somatic marker framework as a neurological theory of decision making: Review, conceptual comparisons, and future neuroeconomics research. *Journal of Economic Psychology, 31*(5), 767–776. https://doi.org/10.1016/j.joep.2010.03.002

Robic, S., Sonié, S., Fonlupt, P., Henaff, M. A., Touil, N., Coricelli, G., ... Schmitz, C. (2015). decision making in a changing world: A study in autism spectrum disorders. *Journal of Autism and Developmental Disorders, 45*(6), 1603–1613. https://doi.org/10.1007/s10803-014-2311-7

Rodgers, J., Farquhar, K., Mason, D., Brice, S., Wigham, S., Ingham, B., Freeston, M., & Parr, J. (2020). Development and initial evaluation of the anxiety scale for autism - adults. *Autism in Adulthood, 2*(1), 24–33. https://doi.org/10.1089/aut.2019.0044

Rodgers, J., Glod, M., Connolly, B., & McConachie, H. (2012). The relationship between anxiety and repetitive behaviours in autism spectrum disorder. *Journal of Autism and Developmental Disorders, 42*(11), 2404–2409. https://doi.org/10.1007/s10803-012-1531-y

Rodgers, J., Wigham, S., McConachie, H., Freeston, M., Honey, E., & Parr, J. R. (2016). Development of the anxiety scale for children with autism spectrum disorder (ASC-ASD). *Autism Research, 9*(11), 1205–1215. https://doi.org/10.1002/aur.1603

Rzepecka, H., McKenzie, K., McClure, L., & Murphy, S. (2011). Sleep, anxiety and challenging behaviour in children with intellectual disability and/or autism spectrum disorder. *Research in Developmental Disabilities, 32*(6), 2758–2766. https://doi.org/10.1016/j.ridd.2011.05.034

Scahill, L., Lecavalier, L., Schultz, R. T., Evans, A. N., Maddox, B., Pritchett, J., Herrington, J., Gillespie, S., Miller, J., Amoss, R. T., Aman, M. G., Bearss, K., Gadow, K., & Edwards, M. C. (2019). Development of the parent-rated anxiety scale for youth with autism spectrum disorder. *Journal of the American Academy of Child and Adolescent Psychiatry, 58*(9), 887–896. e882. https://doi.org/10.1016/j.jaac.2018.10.016

Schwarz, N., & Clore, G. (2007). Feelings and phenomenal experiences. In A. W. Kruglanksi & E. T. Higgins (Eds.), *Social psychology: Handbook of basic principles* (pp. 385–407). Guildford Press.

Shah, P., Catmur, C., & Bird, G. (2016). Emotional decision making in autism spectrum disorder: The roles of interoception and alexithymia. *Molecular Autism, 7*, ARTN 43. https://doi.org/10.1186/s13229-016-0104-x

Simpson, K., Adams, D., Wheeley, E., & Keen, D. (2020). Parent perspectives on the presentation, triggers, impact, and support of anxiety in young children on the autism spectrum. *Journal of Child and Family Studies, 29*(2), 572–582. https://doi.org/10.1007/s10826-019-01576-5

Smiley, E., Cooper, S. A., Finlayson, J., Jackson, A., Allan, L., Mantry, D., McGrother, C., McConnachie, A., & Morrison, J. (2007). Incidence and predictors of mental ill-health in adults with intellectual disabilities: Prospective study. *British Journal of Psychiatry, 191*, 313–319. https://doi.org/10.1192/bjp.bp.106.031104

South, M., Chamberlain, P. D., Wigham, S., Newton, T., Le Couteur, A., McConachie, H., Gray, L., Freeston, M., Parr, J., Kirwan, C. B., & Rodgers, J. (2014). Enhanced decision making and risk avoidance in high-functioning autism spectrum disorder. *Neuropsychology, 28*(2), 222–228. https://doi.org/10.1037/neu0000016

South, M., Dana, J., White, S. E., & Crowley, M. J. (2011). Failure is not an option: Risk-taking is moderated by anxiety and also by cognitive ability in children and adolescents diagnosed with an autism spectrum disorder. *Journal of Autism and Developmental Disorders, 41*(1), 55–65.

Steel, Z., Marnane, C., Iranpour, C., Chey, T., Jackson, J., Patel, V., & Silove, D. (2014). The global prevalence of common mental disorders: A systematic review and meta-analysis 1980-2013. *International Journal of Epidemiology, 43*(2), 476–493. https://doi.org/10.1093/ije/dyu038

ten Brinke, L., Lee, J. J., & Carney, D. R. (2019). Different physiological reactions when observing lies versus truths: Initial evidence and an intervention to enhance accuracy. *Journal of Personality and Social Psychology, 117*(3), 560–578. https://doi.org/10.1037/pspi0000175

Top, D. N., Jr., Stephenson, K. G., Doxey, C. R., Crowley, M. J., Kirwan, C. B., & South, M. (2016). Atypical amygdala response to fear conditioning in autism spectrum disorder. *Biological Psychiatry: Cognitive Neuroscience and Neuroimaging, 1*(4), 308–315. https://doi.org/10.1016/j.bpsc.2016.01.008

van Steensel, F. J. A., Bogels, S. M., & Perrin, S. (2011). Anxiety disorders in children and adolescents with autistic spectrum disorders: A meta-analysis. *Clinical Child and Family Psychology Review, 14*(3), 302–317. https://doi.org/10.1007/s10567-011-0097-0

Vasa, R. A., Keefer, A., Reaven, J., South, M., & White, S. W. (2018). Priorities for advancing research on youth with autism spectrum disorder and co-occurring anxiety. *Journal of Autism and Developmental Disorders, 48*(3), 925–934. https://doi.org/10.1007/s10803-017-3320-0

Vella, L., Ring, H. A., Aitken, M. R., Watson, P. C., Presland, A., & Clare, I. C. (2018). Understanding self-reported difficulties in decision making by people with autism spectrum disorders. *Autism, 22*(5), 549–559. https://doi.org/10.1177/1362361316687988

Wallace, G. L., Kenworthy, L., Pugliese, C. E., Popal, H. S., White, E. I., Brodsky, E., & Martin, A. (2016). Real-world executive functions in adults with autism spectrum disorder: Profiles of impairment and associations with adaptive functioning and co-morbid anxiety and depression. *Journal of Autism and Developmental Disorders, 46*(3), 1071–1083. https://doi.org/10.1007/s10803-015-2655-7

Wang, C. A., Baird, T., Huang, J., Coutinho, J. D., Brien, D. C., & Munoz, D. P. (2018). Arousal effects on pupil size, heart rate, and skin conductance in an emotional face task. *Frontiers in Neurology, 9*, ARTN 1029. https://doi.org/10.3389/fneur.2018.01029

White, S. W., Mazefsky, C. A., Dichter, G. S., Chiu, P. H., Richey, J. A., & Endick, T. H. (2014). Social-cognitive, physiological, and neural mechanisms underlying emotion regulation impairments: Understanding anxiety in autism spectrum disorder. *International Journal of Developmental Neuroscience, 39*, 22–36. https://doi.org/10.1016/j.ijdevneu.2014.05.012

Wigham, S., Rodgers, J., South, M., McConachie, H., & Freeston, M. (2015). The interplay between sensory processing abnormalities, intolerance of uncertainty, anxiety and restricted and repetitive behaviours in autism spectrum disorder. *Journal of Autism and Developmental Disorders, 45*(4), 943–952. https://doi.org/10.1007/s10803-014-2248-x

Yip, J. A., & Schweitzer, M. E. (2016). Mad and misleading: Incidental anger promotes deception. *Organizational Behavior and Human Decision Processes, 137*, 207–217. https://doi.org/10.1016/j.obhdp.2016.09.006

Zhang, L., Wang, K., Zhu, C., Yu, F., & Chen, X. (2015). Trait anxiety has effect on decision making under ambiguity but not decision making under risk. *PLoS One, 10*(5), e0127189. https://doi.org/10.1371/journal.pone.0127189

Zimmerman, D., Ownsworth, T., O'Donovan, A., Roberts, J., & Gullo, M. J. (2017). Associations between executive functions and mental health outcomes for adults with autism spectrum disorder. *Psychiatry Research, 253*, 360–363. https://doi.org/10.1016/j.psychres.2017.04.023

Chapter 9
Affective Decision Making and Peer Influence in Youth with Intellectual Disability

Anika Bexkens and Christoph M. Müller

Introduction

Adolescents with intellectual disability (ID) are regularly referred to special services due to problems that result from suboptimal decision making, such as risky behavior, rule-breaking, or the subsequent consequences, such as physical or mental trauma. An example of this is the boy who was arrested by the police, but had not realized the packages he was transporting for his *friends* were filled with drugs. Another example is the girl who was referred for behavioral problems but was actually suffering from trauma after repeatedly getting into unsafe situations at the insistence of her peers. To help adolescents with ID engage in safe decision making, it is important to analyze their decision-making process and determine which skills are needed.

However, the skills required to make optimal decisions depend not only on the decision-making process itself but also on the decision-making context. Decisions can be made in both cognitive and affective (i.e., including social) contexts, which require different skills. A special affective context during adolescence is one formed by peers. Peers are same- age individuals who have either been voluntarily selected, such as friends, or who are members of involuntary groups, such as all students in a classroom. Through various processes of influence, peers have an impact on how adolescents make decisions in this affective context. Negative peer influence in the affective decision-making process can lead to short-term increases in individual risk-taking and the long-term socialization of problem behaviors. The challenges

A. Bexkens (✉)
Developmental and Educational Psychology, Leiden University, Leiden, The Netherlands
e-mail: a.bexkens@fsw.leidenuniv.nl

C. M. Müller
Department of Special Education, University of Freiburg, Freiburg im Breisgau, Germany
e-mail: Christoph.mueller2@unifr.ch

I. Khemka, L. Hickson (eds.), *Decision Making by Individuals with Intellectual and Developmental Disabilities*, Positive Psychology and Disability Series,
https://doi.org/10.1007/978-3-030-74675-9_9

related to affective decision making and susceptibility to peer influence may be particularly pronounced for youth with ID. Their difficulties in cognition and social competence may contribute to suboptimal decision making in affective contexts and increased susceptibility to peer influence in this regard.

This chapter discusses research on affective decision making and susceptibility to peer influence in youth with ID. ID is characterized by significant limitations in adaptive and cognitive functioning (typically about two standard deviations below the reference norm) that must originate in the developmental period (American Association on Intellectual and Developmental Disabilities, 2021). Some of the cited studies in this chapter also include adolescents with borderline intellectual functioning (i.e., having an IQ between 70 and 85). Given the dearth of research with adolescents with ID and affective decision making, we also present findings based on adults with ID when available or on typically developing adolescents when informative. We first consider the decision-making process in the affective context and its determinants, before turning to the role of the developmental period of adolescence and peer influence on affective decision making. A comprehensive review of studies on affective decision making and peer influence in adolescents with ID follows. This review on the one hand includes experimental studies, which focus on the immediate impact of affective context or of peers on decisions. On the other hand, correlational studies are reviewed, which examine daily behavior and longer-term developmental processes of peer influence that subsequently may impact affective decision making over time. In addition, an overview of affective decision-making support is provided. Based on our review, we then discuss implications and future research directions.

Decision Making in Cognitive Versus Affective Contexts

Decisions made in cognitive contexts lack pronounced emotional components at the time of the decision. Although there may be hypothetical rewards and losses, the decision-maker does not experience them as immediate consequences of the decision. This type of decision making is sometimes referred to as *cool* decision making. An example would be an adolescent sitting at home on her computer, trying to choose between two similar mobile phones. The options differ in some aspects (e.g., price, color, etc.), and this information can be used to choose between the options. After deciding, the adolescent is still at home on her computer and there are no immediate gains or losses. In cognitive contexts, decision making relies mainly on reasoning skills, including basic cognitive control functions such as inhibition and working memory, which are goal-directed, future-oriented processes. During decision making adolescents are required to mentally run through different scenarios and evaluate the outcomes of potential decisions, a process that relies heavily on working memory capacity and processing speed (Doidge et al., 2019; Stanovich, 2011).

In contrast, other decision-making situations involve emotionally charged contexts in which decisions come with immediate losses or rewards. This is referred to as affective or *hot* decision making. For example, an adolescent is at a mobile phone dealer and her friends are present. Her best friend is very excited and tells her the expensive blue mobile phone looks absolutely amazing. In this situation, the adolescent must still evaluate whether she should choose the expensive, blue phone or an affordable, plain one, which will have to be ordered and can be taken home only next week. To do so she will need to regulate the excitement she feels when imagining herself taking the blue phone home right away with the prospect of being admired by her best friend.

Thus, optimal decision making in affective contexts relies on both the skills needed in cognitive decision making and an individual's affective control and reward sensitivity. Affective control refers to goal-directed, future-oriented processes triggered in contexts where emotion and motivation play a role and includes balancing immediate gratification with long-term rewards (Zelazo & Carlson, 2012). Whereas cognitive control is related to cognitive functioning, such as academic performance, affective control is related to emotional functioning (Poon, 2018). Affective control is necessary during decision making in affective contexts to downregulate emotional activation, for instance, in response to reward, so that it does not interfere with the decision-making process. How much affective control is necessary in a given situation differs between persons, who may differ in the intensity of their reaction to reward or emotional stimuli. Generally, people who experience more emotional reactivity in relation to losses or gains tend to make more risky decisions (Charpentier et al., 2016; Penolazzi et al., 2012).

Neuroimaging studies suggest that cognitive and affective control operate from different neural systems. Cognitive control, which is also referred to as cool executive functioning, has consistently been linked to the prefrontal cortex, suggesting that cognitive decision making mainly relies on prefrontal functions. In contrast, affective control, which is also referred to as hot executive functioning, has been linked to the paralimbic cortex, limbic system, and connections between the prefrontal cortex and limbic system (Leshem et al., 2020).

Affective Decision Making in Adolescence

The impact of the affective context on decision making is present particularly in adolescence. Results of experimental tasks indicate that reward sensitivity resembles an inverted U-shape in relation to age that peaks in mid- to late adolescence (Braams et al., 2015; Poon, 2018). The implication is that the same situations elicit increased affective reactions in adolescents compared to children and adults and therefore require more affective control. However, cognitive control abilities, required for reasoning and regulating behavior, increase linearly with age and are still developing throughout adolescence (Casey et al., 2011; Cauffman et al., 2010; Luna et al., 2010; Poon, 2018). Research using performance-based cool vs hot executive function tasks shows that cool executive functioning, or cognitive control, matures earlier than hot executive functioning, or affective control, which matures later in adolescence (Prencipe et al., 2011, but for a review also see Zelazo & Carlson, 2012). Moreover, the connections between the cognitive control regions and the reward system necessary for affective control strengthen over time, allowing for more stability in control over emotions at a later age (Van Duijvenvoorde et al., 2016).

The differential timing in the development of cognitive and affective control processes has been linked to coinciding developmental changes in the structure and function of the control and reward networks of the brain that also parallel the peak in real-life risk-taking behaviors (Steinberg, 2008). This may be explained by a mismatch between a high need for control due to increased reward sensitivity, even as affective control is still maturing. An often-used metaphor describes adolescents as cars with a high tendency to drive fast, even as their brakes are not strong enough to handle this speed. Alternatively, some authors (Crone & Dahl, 2012) propose the brakes may be flexibly used by adolescents because they are motivated to take risks. Risk-taking behavior in this context may be socially accepted, such as engaging in extreme sports. However, often the outcome of inadequate cognitive and affective control in adolescence is behavior that contradicts normative expectations and is therefore considered problematic by the social environment (see also Jahromi et al., 2008). Examples of this are engaging in dangerous games (e.g., Internet challenges that involve risky behavior) and undercontrolled behavior problems, such as impulsive, risk-taking, and antisocial behaviors (Hobson et al., 2011; Van Gelder, 2013).

Peer Influence on Affective Decision Making in Adolescence

Adolescent decision making is rarely a solo expedition, but rather a collaborative effort between peers, who become increasingly important to adolescents in this developmental period (Larson et al., 1996; Steinberg, 2004). The need for social acceptance and the possibility of peer rejection become powerful motivators (Steinberg, 2008). For instance, adolescents may perceive that peers have certain expectations or will reinforce them for specific decisions. The presence of peers during decision making therefore represents an affective context, which is especially important during adolescence.

Several studies provide evidence on the influence of peers as an affective context in adolescence. Risky driving is one domain where this has often been demonstrated. Automobile accidents are much more likely to occur when adolescents drive with their friends than when they drive alone (Allen & Brown, 2008). Adolescents who were presented with a simple laboratory driving task take more risks overall compared to adults, but especially in the presence of peers (Chein et al., 2011; Gardner & Steinberg, 2005). In addition, peer presence increases neural activity in the reward system, and this enhanced activity was associated with more risky decisions (Albert et al., 2013; Foulkes & Blakemore, 2016). This suggests that the presence of peers creates a neurobiological vulnerability to risk-taking behavior, as it is an affective context that leads to a larger temporary imbalance between reward and control processes (see also Casey et al., 2011; Forbes & Dahl, 2010). In addition, in decision-making situations with peers present, evidence suggests that neural networks associated with social processing are activated (e.g., the temporoparietal junction). These neural networks have been linked to inferring others' mental states (Kilford et al., 2016). Optimal affective decision making amidst peers therefore may require not only proficient affective control but also the ability to account for the perspectives and intentions of peers.

Sensitivity to peers as an affective context is also reflected in longer-term peer influence processes that can impact future decision making in affective contexts. For instance, consider the ways in which peers influence the development of behavior problems. Many behavior problems are characterized by low affective control and suboptimal affective decision making (e.g., Hobson et al., 2011). Insights into the long-term influence of peers on problem behaviors may thus provide information on how peers shape affective decision-making styles over longer periods of time. In this regard, peer selection and peer socialization need to be differentiated. Peer selection is important in the context of affective decision making, as it refers to the tendency for adolescents to select peers who are similar to themselves as interaction partners. For example, adolescents who exhibit delinquent behaviors tend to have more delinquent friends than other adolescents (Warr, 2002). The implication for affective decision making is that peers tend to select a social, and therefore affective, context that may reinforce their own decision-making tendencies. Peer socialization, in contrast, means that adolescents change their behavior over time due to peer influence.

Peer socialization can take different forms (for overviews, see Brechwald & Prinstein, 2011; Dishion & Tipsord, 2011; Warr, 2002). One important mechanism is social learning, including processes such as imitation and peer reinforcement (e.g., Akers & Jensen, 2006; Bandura, 1986). Social learning can take place when adolescents observe peer behaviors associated with affective decision making and perceive the consequences that stem from these behaviors. For example, an adolescent may observe a student who has less control over his emotions during a conflict and who attacks another person. This behavior is followed by peer reinforcement (i.e., peers are laughing), which may lead the observing adolescent to also exert less control over his emotions and show this behavior in the future. When experiencing peer reinforcement himself, this will contribute to a generalization of the behavioral pattern and a tendency toward impulsive decisions. In such processes, adolescents learn affective decision-making tendencies from each other over time.

Another process underlying peer socialization of affective decision making is the tendency for adolescents to orient themselves toward group norms. Often, they try to avoid deviating from these norms in order to avoid social sanctions from peers. Group norms can relate to perceived attitudes (i.e., injunctive norms) or to behaviors (i.e., descriptive norms) that dominate a group of peers (e.g., Veenstra et al., 2019). With regard to affective decision making, an agreed-upon peer group norm in a classroom may be that undercontrolled disruptive behavior is undesirable as it disturbs others (i.e., injunctive norm), which is also evident in a low mean level of such behavior in class (i.e., descriptive norm). In this situation, adolescents may be motivated to use more affective control during decision making to avoid deviating from the peer group. The effects of descriptive and injunctive peer group norms on problem behaviors characterized by low affective control and impulsive behavior have been shown repeatedly (e.g., Megens & Weerman, 2012; Mercer et al., 2009; Müller et al., 2018).

Although openness to peer influence is a main characteristic of adolescence, individuals differ substantially in their peer influence susceptibility (Monahan et al., 2009). This has become evident both in terms of general openness to peer influence (e.g., Steinberg & Monahan, 2007) and peer influence on specific problem behaviors characterized by low affective control (e.g., Brechwald & Prinstein, 2011). In terms of peer influence on antisocial behaviors, for example, research suggests that less self-control and more risk tolerance are related to increased susceptibility to peer influence (for a systematic review, see Müller & Minger, 2013). One explanation may be that adolescents with such characteristics are more sensitive to social reward and have fewer capabilities to resist the strong emotional activation that results from social cues in the peer context. The relationship between peer influence susceptibility and other factors is less clear. For instance, family dysfunction, less advantageous parenting behavior, and lower social competence were all found to be associated with increased peer influence susceptibility to antisocial behavior (e.g., Allen et al., 2012; Gardner et al., 2008; Müller et al., 2017; Prinstein et al., 2001; Wright et al., 2001).

In summary, at least five important factors can be identified that impact adolescent affective decision making and susceptibility to peer influence. First, cognitive control is necessary for successfully following through with a planful decision-making process. Second, in adolescence affective control is important for counterbalancing increased reward sensitivity in response to (social) reward. Third, social cognition is required to consider the drives and intentions of peers and evaluate whether to go along with peers. Fourth, social learning, including feedback learning and imitation, and orientation toward group norms affect how peers influence each other's behavior and impact affective decision making over longer time periods. Finally, both individual and contextual factors appear to moderate peer influence susceptibility in adolescents.

Affective Decision Making and Peer Influence in Adolescents with ID

In light of the factors that determine affective decision making and peer influence susceptibility, adolescents with ID are likely to face specific challenges in both. First, consistent evidence demonstrates that adolescents with ID experience difficulties in cognitive control over thoughts and actions, such as inhibition and working memory (Bexkens et al., 2014; Van der Molen et al., 2010), and these difficulties may be even greater in adolescents with ID who also display externalizing behavior problems (Schuiringa et al., 2017; Van Nieuwenhuijzen et al., 2009). Cognitive control difficulties may render the cognitive decision-making process, which is also part of affective decision making, more vulnerable. In line with this, there is some evidence that cognitive decision making in adults with ID may be more focused on one-dimensional strategies: individuals focus on one aspect of a decision while ignoring other potentially important aspects, indicating difficulties balancing competing dimensions (Jenkinson & Nelms, 1994; Khemka et al., 2009; Willner et al., 2010). One study by Bexkens et al. (2016) replicated these findings in adolescents. These authors used a cool decision-making task, in which participants were asked to choose one of two gambling machines. As it was a paper-and-pencil task, there were no real consequences to the decision. The gambling machines were each characterized by three dimensions (i.e., each had a certain gain, chance of loss, and amount of loss), and this information could be used to make optimal decisions. Items were constructed in such a way that decision strategies could be analyzed. Results showed adolescents with ID had a much higher prevalence of employing guessing strategies and one-dimensional strategies than typically developing adolescents, whereas the prevalence of integrative strategies (i.e., using multidimensional strategies) was higher in the typically developing group. These findings suggest that adolescents with ID have difficulty regulating themselves during the cognitive decision-making process. These difficulties can be expected to be compounded when adolescents with ID must also exert emotional control to deal with an affective context.

Second, the role of the adolescence-specific peak in reward sensitivity should be considered in relation to ID. Pubertal hormonal changes are thought to drive a restructuring of the reward networks in the brain that coincide with this peak in reward sensitivity. Whether adolescents with ID experience the same peak in reward sensitivity has never been directly tested. However, indirect evidence suggests that pubertal development in adolescents with ID may be on par with typically developing adolescents. More specifically, research on menstrual development finds no evidence of differential timing of pubertal development in girls with ID compared to typically developing girls (Nazli & Chavan, 2017). If the same is found for boys, it is probable that the developmental trajectory of reward sensitivity in adolescents with ID parallels that of adolescents without ID. In turn, the difficulties they face in cognitive control would render them less able to deal with a heightened response to reward and therefore would be more likely to impact their decision making, especially during adolescence.

Third, impaired social cognition in adolescents with ID (Jacobs & Nader-Grosbois, 2020) may contribute to difficulties in peer influence situations. More specifically, limitations in the ability to take another's perspective can impair the ability of adolescents with ID to correctly interpret peer intentions. This may lead to increased vulnerability to social manipulation by peers and therefore to further increases in susceptibility to peer influence. Also, specific biases in social information processing in adolescents with ID make it more probable that they will experience increased affective states during decision making. Difficulties in basic cognitive control functions, such as focused attention and working memory, are related to a tendency of adolescents with ID to attribute more hostile and purposeful intent to peers in ambiguous or even clearly accidental situations (Van Rest et al., 2019, 2020). This attribution of hostile or purposeful intent was linked to aggression, suggesting an increased affective state that is difficult to downregulate (Van Rest et al., 2019). This is particularly the case when having inhibition problems as often observed in adolescents with ID (Van Rest et al., 2019). Concluding, different problems in social cognition of adolescents with ID can be expected to lead to specificities in their affective decision making.

Fourth, motivational and social-emotional factors may cause adolescents with ID to rely more on others' ideas than their own. Their external locus of motivation (Switzky, 2006) may lead adolescents with ID to exhibit decreased skepticism of peer advice (Greenspan, 2017). This is also related to increased outer-directedness in individuals with ID. Outer-directedness refers to a tendency to rely more on external than internal cues during problem solving (Bybee & Zigler, 1998; Tanaka et al., 2001). Outer-directedness was found in children, adolescents, and adults with ID, but it was greatest in adults. This suggests increasing outer-directedness with age, which may be related to the high degree of failure that adults with ID experience (Tanaka et al., 2001). Outer-directedness is compounded by feelings of lower self-efficacy (Khemka & Hickson, 2006), which may make adolescents with ID more prone to trusting others instead of relying on their own ideas (Greenspan, 2006). Problems in social competence that are often seen in adolescents with ID (Guralnick, 2017) may further reduce the ability to resist negative peer influence (e.g., Allen et al., 2012). Together, these specificities could contribute to less critical reflection of peers' behaviors in social learning processes, a strong orientation to explicit behavioral group norms, and lower resistance to peer influence.

Fifth, certain contextual factors known to moderate peer influence susceptibility in typical development are likely also important in affective decision making for adolescents with ID. For example, children and adolescents with ID are at higher risk for family dysfunction: Children with disabilities are three to four times as likely to experience child maltreatment or neglect (e.g., Corr & Santos, 2017). Given that research with adolescents without ID showed a link between family dysfunction and susceptibility to peer influence (e.g., Prinstein et al., 2001), adolescents with ID can also be expected to be at increased risk for susceptibility to peer influence. Attachment problems, which are more prevalent in individuals with ID compared to typically developing individuals (Hamadi & Fletcher, 2019; Schuengel et al., 2013), could increase openness to peer influence in affective decision-making situations (Allen et al., 2012; Buck et al., 2013).

Finally, other aspects of the social situation could impact the role peers play in affective decision making. For example, students with ID are at greater risk of rejection by their peers than typically developing students and those with other disabilities (De Laat et al., 2013; Zic & Igri, 2001). In this situation, adolescents with ID may be especially willing to conform to their peers and engage in risk-taking in order to gain popularity and feel they are part of a peer group (see findings by Peake et al., 2013, who investigated this question in a sample of typically developing adolescents). Additionally, peer rejection has been linked to increased stress responses in social situations in typically developing adolescents (Stroud et al., 2009), increasing the need for affective control. As adolescents with ID are likely to experience peer rejection, this may impact their subsequent affective decision making.

In sum, multiple factors suggest that adolescents with ID exhibit specificities in their affective decision making, including the role peers play in these processes. In order to see how far this expectation is supported by research that directly investigated affective decision making, we review such studies in the following section.

Evidence on Affective Decision Making in Adolescents with ID

Most studies of affective decision making in adolescents with ID make use of experimental paradigms during which decision outcomes are immediately experienced by participants. Some studies use gambling tasks that include consequences in terms of gains and losses of fictional (or real) incentives. Another example is delayed gratification, also called temporal discounting, namely, tasks where participants must choose between a small immediate reward and a large delayed reward.

With respect to affective decision making in nonsocial contexts (i.e., absent peers), little research has focused on adolescents with ID. Bexkens et al. (2019) used the Balloon Analogue Risk Task where participants are asked to pump a balloon with the click of a mouse. For each pump they win (fictional) money, but if they pump the balloon too far it explodes and the money for that balloon is lost. The task has 30 trials that are set to different explosion points. Typically developing adolescents' performance on the Balloon Analogue Risk Task has been linked to real-life behaviors associated with suboptimal affective decision making such as substance use, sexual risk-taking, and delinquency (Hopko et al., 2006; Hunt et al., 2005). In two studies that used the Balloon Analogue Risk Task in adolescents with mild ID or borderline intellectual functioning, Bexkens and colleagues found no evidence of increased affective decision-making problems compared to typically developing adolescents for either boys (Bexkens et al., 2019) or boys and girls (Wagemaker et al., 2020). If anything, there is some evidence that adolescents with ID were more careful and less tempted to take risks to increase rewards (Wagemaker et al., 2020). The only difference that emerged was that adolescents with ID were less behaviorally consistent than adolescents without ID, which is in line with a previous study on decision making in adults with ID (Willner et al., 2010). Low behavioral consistency in the Balloon Analogue Risk Task has previously been suggested to be related to lower adherence to decision rules (Rolison et al., 2012; Wallsten et al., 2005).

A study by Dymond et al. (2010) used the Iowa Gambling Task. This affective decision-making task consists of a computerized gambling game. Participants are presented with two card decks and are told that choosing cards can win them money, but losses are also possible. One deck has small rewards and small losses (advantageous deck) and the other has slightly bigger rewards but much larger losses (disadvantageous). Participants are instructed that one of the decks is more advantageous than the other. During the task, participants can learn which deck is which by selecting cards and gaining feedback from their gains and losses. The results by Dymond et al. (2010) indicate that adults with ID chose advantageously on the task, suggesting they were not overly tempted to only select from the larger reward, high loss deck. Compared to adults without ID, however, they did not learn as much from their experiences as the task proceeded, suggesting that implicit information on gains and losses was not sufficient to improve their decision making.

In delay of gratification paradigms, participants choose between a small immediate reward and a large delayed reward. The reward can either be hypothetical (i.e., without actual gains) or actual (i.e., in the sense that rewards are contingent on the choices either by earning currency in a computer game or by receiving real earnings). The more real the reward, the more the paradigm is considered to provide an affective decision-making context. Findings in children and adults with ID are mixed, and a meta-analysis showed no consistent overall difference in performance of individuals with and without ID (Bexkens et al., 2014). However, this result may be related to large differences between the paradigms used and study samples. Individual studies show that adults with ID tend to have a preference for smaller immediate rewards as opposed to larger delayed rewards (Willner et al., 2010) and children with ID have more difficulty waiting for a desired toy than typically developing children (McIntyre et al., 2006).

Taken together, studies that directly investigated affective decision making in ID do not clearly support the above derived expectation that adolescents with ID exhibit increased difficulties specific to this domain of decision making. In fact, most of the studies reviewed show no evidence of a difference between individuals with ID and typically developing individuals when solving affective decision-making tasks. If confirmed by further research, this could mean that differences in decision-making between adolescents with and without ID are mainly driven by differences in skills related to cognitive control. However, this is only a very tentative hypothesis, as studies on this topic are scarce, especially with regard to affective control and affective decision making in daily life.

Evidence on Peer Influence and Affective Decision Making in Adolescents with ID

Experimental Studies

With respect to peer influence on affective decision making, experimental studies conducted in laboratory settings suggest that susceptibility to peer influence is increased in adolescents with ID compared to typically developing adolescents. Khemka et al. (2016) used vignettes and showed that adolescents with ID made particularly suboptimal decisions in vignettes that contained some sort of social coercion (Khemka & Hickson, 2009; Khemka et al., 2016). In addition, the previously discussed Balloon Analogue Risk Task studies also included a peer influence condition (Bexkens et al., 2019; Wagemaker et al., 2020). In both studies a group of adolescents with mild ID or borderline intellectual functioning performed the task without information on peers, as discussed above. A second group performed the task in a peer influence condition. Peer influence was operationalized by displaying pictures of same-sex peers in the corner of the screen and playing audio files containing statements recorded by these peers. All statements were risk-encouraging and were formulated either in a positive (i.e., *you are really cool if you press more*) or negative (i.e., *you are a chicken if you quit*) manner. Comparing the non-peer and peer version of the task in adolescent boys with mild ID or borderline intellectual functioning and typically developing adolescents, the difference in performance between these conditions was larger in the ID group. This indicates a greater effect of peer influence on affective decision making in adolescents with mild ID or borderline intellectual functioning (Bexkens et al., 2019).

Model-based analysis of Balloon Analogue Risk Task performance showed that peer influence increased risk-taking propensity and decreased risk perception for adolescents with mild ID or borderline intellectual functioning, but also increased certainty about their decision (Bexkens et al., 2019). This finding is consistent with studies showing increased outer-directedness among individuals with ID (Tanaka et al., 2001), as it indicates that peer influence leads to increased certainty about their decisions for adolescents with ID. Some evidence suggests the susceptibility to peer influence on risk-taking behavior observed in the Balloon Analogue Risk Task may be specific to boys with ID. Wagemaker et al. (2020) conducted a follow-up study that used the same experiment that included both boys and girls with mild ID or borderline intellectual functioning. The study also distinguished between a positive and a negative peer condition. All statements made by the virtual peers in these conditions were still risk-encouraging. However, the positive condition statements were positive and inclusive to the group (e.g., *respect if you pump the balloon further*), whereas in the negative condition, peer statements were more belittling or exclusive to the group (e.g., *if you stop pumping the balloon you are a wimp*). Results showed that the peer influence susceptibility effect was specific to boys with ID. Additionally, the effect was only present in the negative feedback condition and not in the positive condition, providing evidence that the large effect of peers in adolescents with ID may be driven by a tendency to avoid negative peer feedback.

The degree of specificity of the effect of peers on adolescents' affective decision making remains an open question. A study by Egger et al. (2021) provides some insights. The study tested for the influence of both peers and nonsocial cues on social judgment making, by asking adolescents with ID and comparison groups to rate the coolness of adolescents who appeared in photographs. After study participants rated all adolescents on a scale that ranged from uncool to cool, they received information on ratings provided by same-age peers and were asked to repeat their assessments. These peer ratings were always presented in a standardized distance from participants' own ratings. Any changes in judgment from the original judgment to the judgment made upon seeing how peers had rated the photographs were used to measure peer influence susceptibility. In line with the pattern in the previously reported studies, results showed that adolescents with ID made significantly larger changes from the solo to the peer condition than typically developing adolescents. Interestingly, the same group difference was also seen in a study condition where participants were presented meaningless visual stimuli (i.e., blinking signals on the rating scale): Again, adolescents with ID made significantly larger shifts toward the presented stimuli than typically developing adolescents. Across all conditions, the way that adolescents with ID were influenced in their social judgments was similar to results in mental age-matched children. The results from this study confirm the increased peer influence susceptibility in adolescents with ID found in the previously reported experimental studies. However, the fact that adolescents with ID changed their ratings not only toward peer opinions but also toward nonsocial stimuli brings up an interesting alternative hypothesis for the findings. Further experimental research is needed to discern whether peers as an affective context have a unique role in influencing decisions of adolescents with ID or whether the peer impact is rather just one aspect of a general outer-directedness (which also includes influence from other social and nonsocial sources).

Correlational Studies

Additional insights on peer influence susceptibility in adolescents with ID come from correlational studies conducted in naturalistic settings. These studies shed light on the ways in which the peer context may impact behavior related to affective decision making in daily life. Furthermore, these studies often examine longer time periods than laboratory-based experimental studies to help better understand the peer socialization of affective decision making. Typically, these studies look at the behavioral outcomes of suboptimal affective decision making and focus on antisocial behaviors (including, e.g., disruptive, aggressive, oppositional, and delinquent behaviors). During adolescence, antisocial behaviors often take place when adolescents are among their peers (Warr, 2002), and in this context they also rely on individual affective decision-making competence.

Regarding antisocial behavior and peer influence susceptibility in adolescents with ID, interesting insights are provided by Steinberg and Monahan (2007). These researchers investigated the role of IQ in peer influence susceptibility among juvenile offenders (who did not necessarily have an ID). They developed and used the Resistance to Peer Influence Scale, a self-report measure that in samples of adolescents without ID was found to be valid and reliable. Increased susceptibility to peer influence measured with this scale was related to greater impulsivity, antisocial risk-taking, concern over others' opinions, and increased antisocial behavior (Steinberg & Monahan, 2007). In the study among adolescent offenders, the authors found that lower IQ was associated with more self-reported susceptibility to peer influence. Although strongly limited by the correlational nature of the analyses and a lack of detailed information on participants in the low IQ range, these findings suggest that peers could play a role in antisocial behavior exhibition and related affective decision-making processes for adolescents with ID. However, increased susceptibility to peer influence, as measured using the same scale, was also found in non-offending youth with ID (i.e., adolescents with presumably lower levels of antisocial behaviors; Dekkers et al., 2019). Increased openness to peer influence among adolescents with ID therefore may not be limited to the domain of antisocial behaviors and related affective decision-making processes.

Van der Put et al. (2014) provide additional insights into the role of peers in affective decision making in the context of antisocial behavior. The authors studied reports from probation officers who had conducted standardized interviews with 12- to 18-year-old offenders with and without ID. The two groups did not differ significantly in correlation between recidivism and reports on antisocial friends, gang membership, or resistance to influence from antisocial peers. Other analyses showed the prevalence of probation officer-reported "no resistance to influence from antisocial peers" was 51% for adolescent offenders with ID versus 49% for those without ID (difference not significant; Asscher et al., 2012). This result contradicts the study by Dekkers et al. (2019), who used self-reported data from adolescents with ID and found greater susceptibility to peer influence. These contradictory findings suggest this information may be sensitive to data collection method and information source.

Müller et al. (2021) used a longitudinal approach to examine peer influence on various types of problem behaviors in 1125 students with ID in Swiss special needs schools for students with ID. Such schools in Switzerland can only be attended by students with a clinical diagnosis of ID. About 67.7% of the study participants had adaptive behavior levels in the low or extremely low range according to the reference norms of the Adaptive Behavior Assessment System-3 (Harrison & Oakland, 2015). A total of 179 special needs classrooms were included in the study. Antisocial and disruptive behaviors were assessed using the Developmental Behaviour Checklist (Einfeld et al., 2002). The authors found no significant effect of baseline classroom levels of antisocial and disruptive behaviors on individual development of such behaviors over time, controlling for individual problem behaviors, and other factors at baseline. This finding stands in contrast to studies with typically developing adolescents that used a similar methodological approach and found that classmates influence antisocial behavior development (e.g., Müller et al., 2016). It also contradicts cross-sectional findings on increased self-reported peer influence susceptibility among adolescents with lower IQ and high levels of antisocial behaviors (i.e., juvenile offenders; Steinberg & Monahan, 2007). However, the sample in the Müller et al. (2021) study differed substantially from other peer influence studies, as it included many students with profound levels of ID. Although evidence suggests students with profound and severe multiple disabilities are also sensitive to their peer environment (Brady et al., 1991; Logan et al., 1998; Nijs & Maes, 2014; Scherler & Müller, 2018), peer influence on antisocial and disruptive behavior may nevertheless be more pronounced in students with mild or moderate levels of ID.

Interestingly, the study by Müller et al. (2021) nevertheless found classroom peer effects on anxiety as measured by the Developmental Behaviour Checklist (Einfeld et al., 2002). Anxiety is not an antisocial behavior but is also closely interrelated with emotional processing and peer influence. It is therefore worth considering peer effects on anxiety in students with ID in the context of affective decision making. Müller et al. (2021) reported that greater classroom levels of anxiety among students in special needs classrooms for students with ID at baseline predicted higher future levels of individual anxiety, controlling for students` baseline individual anxiety levels, and other covariates. Similar analyses on internalizing behaviors in classrooms of typically developing students did not show such classroom peer group effects (instead, only girls influenced girls; Müller et al., 2020). The results for students with ID might be explained by increased emotional contagion in groups of peers with ID. Emotional contagion refers to "the tendency to automatically mimic and synchronize facial expressions, vocalizations, postures, and movements with those of another person and, consequently, to converge emotionally" (Hatfield et al., 2009, p. 190). This process can involve both conscious and unconscious processes and explains how anxiety might spread in groups (Kelly & Barsade, 2001). Studies with typically developing individuals show that, among other factors, levels of susceptibility to emotional contagion depend on one's ability to regulate one's emotions (Papousek et al., 2008). Some support for this potential explanation is the finding that the peer effect was moderated by the degree of homogeneity in students' anxiety levels in the classroom: The more homogeneous students in class were regarding their anxiety levels, the greater the peer effect. It is possible that greater homogeneity in anxiety levels makes it easier for individuals to synchronize their emotional states. Pulling all this together, adolescents with ID might show increased susceptibility to emotional contagion in peer group contexts, which may in turn impact on their affective decision making. However, clearer conclusions on this issue require greater study on emotional contagion in ID and how this affects decision making by adolescents with ID.

When considered together, results from correlational studies in naturalistic settings provide suggestive yet mixed findings. The partially contradicting results on peer influence susceptibility may be due to the small number of studies, all of which investigated different aspects of the topic. The studies also differed in the samples investigated and in how they chose to assess peer influence susceptibility. While some studies addressed peer influence susceptibility as a general personal characteristic (e.g., using self-reported data), others examined longitudinal relations between peer characteristics and individual behavioral changes. Overall, the review of correlational studies is in line with results from experimental studies in that they confirm peer influence susceptibility in adolescents with ID regarding behaviors associated with suboptimal affective decision making. However, it is less clear whether this susceptibility is greater than in those without ID and how it relates to different behavioral domains and underlying affective decision-making processes.

Evidence on Supporting Adolescents with ID in Their Affective DecisionMmaking and Resistance to Peer Influence

Given the specificities and challenges in affective decision making faced by adolescents with ID, particularly in a peer context, the evidence on effective support measures must be considered. We identified two studies that investigated interventions to support decision making in affective decision-making tasks. Both studies relate to adults with ID but may also provide insights relevant for adolescents with ID. Both Bailey et al. (2011) and Fisher et al. (2012) used a visual decision making aid in delay of gratification paradigms to support cognitive decision-making processes in affective contexts. The visual aid helped participants to visualize the two choice options. For instance, when choosing between a small reward now and a larger reward in two weeks, each choice option was associated with two decision dimensions: reward size and waiting time. For each option, participants could visualize reward size with a green bar and waiting time with a red bar. This way they could more easily compare the decision dimensions of each choice option. Visual aid use led to substantial improvements in decision making. Participants made few errors in using the visual aid, and knowledge on how to use it during decision making was perfectly retained at two months follow-up. These results suggest that supporting cognitive decision making can be generally effective for improving decision making and in an affective context.

Another study, also conducted with adults with ID, used an adapted version of the Iowa Gambling Task. An intervention called symbol labeling was applied to support decision making. After each block of trials, participants were asked to put a happy smiley face next to the advantageous deck and an unhappy smiley face next to the disadvantageous deck. They also received feedback on the accuracy of their choices. If they labeled a deck incorrectly, the smiley faces disappeared and they were asked to perform the task again. This intervention resulted in a higher net gain for the ID group that received the intervention compared to a group that did not receive the intervention. The intervention group improved so much their scores were nearly comparable to those of non-ID participants (Dymond et al., 2010). This indicates that adults with ID may not pick up on implicit decision dimensions when using feedback from gains and losses, but an intervention that makes these decision dimensions explicit can help improve decision making.

We identified only one study that focused specifically on supporting the decision making of individuals with ID in peer contexts. Khemka et al. (2016) developed a curriculum, *PEER-DM*, to teach adolescents with ID to apply a set of decision-making steps when faced with different types of negative peer influence situations. Following the decision-making steps helped individuals make better and safer decisions by carefully considering different options and potential consequences. Using the curriculum, participants also practiced assertive decision making. The curriculum contains six lessons, during which adolescents are taught a four-step protocol for decision making. The contents focus on recognizing peer influence situations and empowering adolescents to stand up for themselves and their safety. Participants are taught to evaluate opposing goals and consequences, find language to resist peer influence, and build up other skills. When the curriculum was tested in a group of adolescents with a broad range of intellectual or developmental disabilities (mean IQ of disability subtypes ranged from 45 to 85), it showed a large positive effect, compared with a randomly assigned control group, in improving decision making in hypothetical negative peer influence situations (Khemka et al., 2016). As noted by Hickson and Khemka (2013), an important next step will be to assess whether improvements in decision making through such a curriculum will hold up in real-life affective contexts.

Conclusion

An important proportion of adolescents' daily activities and decisions occurs among peers. Peers provide an affective context so that additional decision-making skills are needed, compared to making decisions in cognitive contexts. Although adolescents with ID face the same demands in affective decision making as typically developing adolescents, they may have fewer skills to rely on to manage and control these processes. Our review highlighted the ways in which the main challenges appear to relate to difficulties in underlying cognitive decision making competencies and increased susceptibility to peer influence. However, our conclusions have to remain tentative because the state of knowledge is still very limited.

Generally, few studies have investigated directly the processes involved in affective decision making and peer influence in adolescents with ID. Our review therefore also included findings more broadly associated with these processes and studies in typically developing adolescents or adults with ID, where appropriate. Given the state of the literature, we consider it important to extend the types of experiments conducted to investigate the questions at hand. For example, to our knowledge no studies have directly tested affective control in adolescents with ID, although insights on this topic are much needed. Furthermore, the hypothesis that challenges in affective decision making may be mainly driven by limited cognitive decision making skills should be further investigated. Here, it may be beneficial to use a multi-method approach to determine how adolescents with ID make affective decisions across different tasks and contexts. Using several decision-making tasks, questionnaires, and observations in the same study would provide stronger evidence and would help to better understand the links between results from experimental research and adolescents' real-life behaviors.

With regard to the role of peers, most studies suggest adolescents with ID show increased susceptibility to peer influence during affective decision making. It is less clear, however, whether this susceptibility is specific to peer influence or whether it also relates to influences from other social groups. Moreover, the drivers of the peer influence effect remain unknown. For example, possible drivers include increased affective activation in response to peers, a greater impact of this activation on the cognitive decision-making process, or motivational factors such as a general outer-directedness. Future experiments should include task conditions that take general outer-directedness into account and compare peer effects to those of other social and nonsocial contexts. Psychophysiological measurements could further elucidate the affective impact of the context on adolescents with ID.

With regard to correlational studies in naturalistic settings, we see a need for more longitudinal studies on the peer socialization of affective decision-making styles in adolescents with ID. The insights from this review suggest such studies could focus on a broad set of behaviors associated with affective decision making, include different methodologies, and use various sources of information (e.g., direct observations and reports by different informant groups). Given the large variability in competencies among adolescents with ID and the fact that most studies investigated the situation of only adolescents with mild ID, there is also a need to further investigate how severity of ID and other individual and contextual factors moderate peer influence on affective decision making. Further integration of experimental and naturalistic approaches will be necessary if we are to gain additional insights into the ways in which underlying decision processes affect daily life decisions and behavior for adolescents with ID.

Finally, replicating, with adolescents, the interventions to support adults with ID in cognitive decision making would provide useful insights. Moreover, any decision-making aids developed for this purpose should be tested in affective contexts that also include peers. With a specific focus on increased peer influence susceptibility in affective decision making, future interventions for adolescents with ID may also benefit from considering insights derived from studies among typically developing adolescents in the school context. For example, evidence suggests teachers can impact the effect of peer influence on disruptive behaviors related to affective control. Two studies showed that teachers can buffer negative peer influence on disruptive behaviors in a classroom of typically developing early adolescents, through processes including emotional support, as well as interesting and academically supportive instruction (Müller et al., 2018; Shin & Ryan, 2017). Based on developmental research, Farmer et al. (2011, 2018) made several suggestions for how teachers may actively promote positive peer relations and peer influence processes, which could provide interesting perspectives for also supporting adolescents with ID. A focus on the positive aspects of peer influence may be of special importance for adolescents with ID, as the extant literature primarily focuses on the risky and negative side of affective decision making and peer influence. While stressing specific risks is important given the social vulnerability of adolescents with ID, it may also be beneficial to consider the opportunities provided by peer learning for optimal affective decision making.

As a whole, this chapter highlights the importance but also the complexity of affective decision-making processes and peer influence among adolescents with ID. Future research should seek to better understand the specificities of these processes and develop tools to support adolescents with ID, in order to support self-determination in real-life affective decisions to the greatest degree possible.

References

Akers, R. L., & Jensen, G. F. (2006). The empirical status of social learning theory of crime and deviance: The past, present, and future. In F. T. Cullen, J. P. Wright, & K. R. Blevins (Eds.), *Advances in criminological theory. Taking stock: The status of criminological theory* (pp. 37–76). Transaction Publishers.

Albert, D., Chein, J., & Steinberg, L. (2013). The teenage brain: Peer influences on adolescent decision making. *Current Directions in Psychological Science, 22*(2), 114–120. https://doi.org/10.1177/0963721412471347

Allen, J. P., & Brown, B. B. (2008). Adolescents, peers, and motor vehicles. The perfect storm? *American Journal of Preventive Medicine, 35*(3 SUPPL), 289–293. https://doi.org/10.1016/j.amepre.2008.06.017

Allen, J. P., Chango, J., Szwedo, D., Schad, M., & Marston, E. (2012). Predictors of susceptibility to peer influence regarding substance use in adolescence. *Child Development, 83*(1), 337–350. https://doi.org/10.1111/j.1467-8624.2011.01682.x

American Association on Intellectual and Developmental Disabilities (Ed.). (2021). *Intellectual disability: Definition, classification, and systems of support* (12th ed.). AAIDD.

Asscher, J. J., van der Put, C. E., & Stams, G. J. J. M. (2012). Differences between juvenile offenders with and without intellectual disability in offense type and risk factors. *Research in Developmental Disabilities, 33*(6), 1905–1913. https://doi.org/10.1016/j.ridd.2012.05.022

Bailey, R., Willner, P., & Dymond, S. (2011). A visual aid to decision making for people with intellectual disabilities. *Research in Developmental Disabilities, 32*(1), 37–46. https://doi.org/10.1016/j.ridd.2010.08.008

Bandura, A. (1986). *Social foundations of thought and action: A social cognitive theory*. Prentice Hall.

Bexkens, A., Huizenga, H. M., Neville, D. A., Collot d'Escury-Koenigs, A. L., Bredman, J. C., Wagemaker, E., & Van der Molen, M. W. (2019). Peer-influence on risk-taking in male adolescents with mild to borderline intellectual disabilities and/or behavior disorders. *Journal of Abnormal Child Psychology, 47*(3), 543–555. https://doi.org/10.1007/s10802-018-0448-0

Bexkens, A., Jansen, B. R. J., Van der Molen, M. W., & Huizenga, H. M. (2016). Cool decision making in adolescents with behavior disorder and/or mild-to-borderline intellectual disability. *Journal of Abnormal Child Psychology, 44*(2), 357–367. https://doi.org/10.1007/s10802-015-9996-8

Bexkens, A., Ruzzano, L., Collot d'Escury-Koenigs, A., Van der Molen, M. W., & Huizenga, H. M. (2014). Inhibition deficits in individuals with intellectual disability: A meta-regression analysis. *Journal of Intellectual Disability Research, 58*(1), 3–16. https://doi.org/10.1111/jir.12068

Braams, B. R., van Duijvenvoorde, A. C. K., Peper, J. S., & Crone, E. A. (2015). Longitudinal changes in adolescent risk-taking: A comprehensive study of neural responses to rewards, pubertal development, and risk-taking behavior. *Journal of Neuroscience, 35*(18), 7226–7238. https://doi.org/10.1523/JNEUROSCI.4764-14.2015

Brady, M. P., Martin, S., Williams, R. E., & Burta, M. (1991). The effects of fifth graders' socially directed behavior on motor and social responses of children with severe multiple handicaps. *Research in Developmental Disabilities, 12*(1), 1–16. https://doi.org/10.1016/0891-4222(91)90019-O

Brechwald, W. A., & Prinstein, M. J. (2011). Beyond homophily: A decade of advances in understanding peer influence processes. *Journal of Research on Adolescence, 21*(1), 166–179. https://doi.org/10.1111/j.1532-7795.2010.00721.x

Buck, K. A., Kretsch, N., & Harden, K. P. (2013). Positive attentional bias, attachment style, and susceptibility to peer influence. *Journal of Research on Adolescence, 23*(4), 605–613. https://doi.org/10.1111/jora.12026

Bybee, J., & Zigler, E. (1998). Outerdirectedness in individuals with and without mental retardation: A review. In J. A. Burack, R. M. Hodapp, & E. F. Zigler (Eds.), *Handbook of mental retardation and development* (pp. 434–461). Cambridge University Press.

Casey, B. J., Jones, R. M., & Somerville, L. H. (2011). Braking and accelerating of the adolescent brain. *Journal of Research on Adolescence, 21*(1), 21–33. https://doi.org/10.1111/j.1532-7795.2010.00712.x

Cauffman, E., Shulman, E. P., Steinberg, L., Claus, E., Banich, M. T., Graham, S., & Woolard, J. (2010). Age differences in affective decision making as indexed by performance on the Iowa Gambling Task. *Developmental Psychology, 46*(1), 193–207. https://doi.org/10.1037/a0016128

Charpentier, C. J., De Neve, J. E., Li, X., Roiser, J. P., & Sharot, T. (2016). Models of affective decision making: How do feelings predict choice? *Psychological Science, 27*(6), 763–775. https://doi.org/10.1177/0956797616634654

Chein, J. M., Albert, D., O'Brien, L., Uckert, K., & Steinberg, L. (2011). Peers increase adolescent risk taking by enhancing activity in the brain's reward circuitry. *Developmental Science, 14*(2), F1–F10. https://doi.org/10.1111/j.1467-7687.2010.01035.x

Corr, C., & Santos, R. M. (2017). Abuse and young children with disabilities: A review of the literature. *Journal of Early Intervention, 39*(1), 3–17. https://doi.org/10.1177/1053815116677823

Crone, E. A., & Dahl, R. E. (2012). Understanding adolescence as a period of social-affective engagement and goal flexibility. *Nature Reviews Neuroscience, 13*(9), 636–650. https://doi.org/10.1038/nrn3313

de Laat, S., Freriksen, E., & Vervloed, M. P. J. (2013). Attitudes of children and adolescents toward persons who are deaf, blind, paralyzed or intellectually disabled. *Research in Developmental Disabilities, 34*(2), 855–863. https://doi.org/10.1016/j.ridd.2012.11.004

Dekkers, L. M. S., Bexkens, A., Hofman, A. D., De Boeck, P., Collot d'Escury, A. L., & Huizenga, H. M. (2019). Formal modeling of the resistance to peer influence questionnaire: A comparison of adolescent boys and girls with and without mild-to-borderline intellectual disability. *Assessment, 26*(6), 1070–1083. https://doi.org/10.1177/1073191117698754

Dishion, T. J., & Tipsord, J. M. (2011). Peer contagion in child and adolescent social and emotional development. *Annual Review of Psychology, 62*, 189–214. https://doi.org/10.1146/annurev.psych.093008.100412

Doidge, J., Saoud, W., & Toplak, M. E. (2019). Executive functions and decision making in adolescents with ADHD. In S. P. Becker (Ed.), *ADHD in adolescents: Development, assessment, and treatment* (pp. 56–76). Guilford Publications.

Dymond, S., Bailey, R., Willner, P., & Parry, R. (2010). Symbol labelling improves advantageous decision making on the Iowa Gambling Task in people with intellectual disabilities. *Research in Developmental Disabilities, 31*(2), 536–544. https://doi.org/10.1016/j.ridd.2009.12.003

Egger, S., Nicolay, P., Huber, C., & Müller, C. (2021). Increased openness to external influences in adolescents with intellectual disability: Insights from an experimental study on social judgments. *Research in Developmental Disabilities, 113*, 1–12. https://doi.org/10.1016/j.ridd.2021.103918

Einfeld, S. L., Tonge, B. J., Gray, K. M., & Brereton, A. V. (2002). *Manual for the developmental behaviour checklist* (2nd ed.). University of New South Wales and Monash University.

Farmer, T. W., Dawes, M., Hamm, J. V., Lee, D., Mehtaji, M., Hoffman, A. S., & Brooks, D. S. (2018). Classroom social dynamics management: Why the invisible hand of the teacher matters for special education. *Remedial and Special Education, 39*(3), 177–192. https://doi.org/10.1177/0741932517718359

Farmer, T. W., McAuliffe Lines, M., & Hamm, J. V. (2011). Revealing the invisible hand: The role of teachers in children's peer experiences. *Journal of Applied Developmental Psychology, 32*(5), 247–256. https://doi.org/10.1016/j.appdev.2011.04.006

Fisher, Z., Bailey, R., & Willner, P. (2012). Practical aspects of a visual aid to decision making. *Journal of Intellectual Disability Research, 56*(6), 588–599. https://doi.org/10.1111/j.1365-2788.2011.01498.x

Forbes, E. E., & Dahl, R. E. (2010). Pubertal development and behavior: Hormonal activation of social and motivational tendencies. *Brain and Cognition, 72*(1), 66–72. https://doi.org/10.1016/j.bandc.2009.10.007

Foulkes, L., & Blakemore, S. J. (2016). Is there heightened sensitivity to social reward in adolescence? *Current Opinion in Neurobiology, 40*, 81–85. https://doi.org/10.1016/j.conb.2016.06.016

Gardner, M., & Steinberg, L. (2005). Peer influence on risk taking, risk preference, and risky decision making in adolescence and adulthood: An experimental study. *Developmental Psychology, 41*(4), 625–635. https://doi.org/10.1037/0012-1649.41.4.625

Gardner, T. W., Dishion, T. J., & Connell, A. M. (2008). Adolescent self-regulation as resilience: Resistance to antisocial behavior within the deviant peer context. *Journal of Abnormal Child Psychology, 36*(2), 273–284. https://doi.org/10.1007/s10802-007-9176-6

Greenspan, S. (2006). Functional concepts in mental retardation: Finding the natural essence of an artificial category. *Exceptionality, 14*(4), 205–224. https://doi.org/10.1207/s15327035ex1404_3

Greenspan, S. (2017). Borderline intellectual functioning: An update. *Current Opinion in Psychiatry, 30*(2), 113–122. https://doi.org/10.1097/YCO.0000000000000317

Guralnick, M. J. (2017). Early intervention for children with intellectual disabilities: An update. *Journal of Applied Research in Intellectual Disabilities, 30*(2), 211–229. https://doi.org/10.1111/jar.12233

Hamadi, L., & Fletcher, H. K. (2019). Are people with an intellectual disability at increased risk of attachment difficulties? A critical review. *Journal of Intellectual Disabilities, 1969*, 174462951986477. https://doi.org/10.1177/1744629519864772

Harrison, P. L., & Oakland, T. (2015). *ABAS-3: Adaptive behavior assessment system* (3rd ed.). Western Psychological Services.

Hatfield, E., Rapson, R. L., & Le, Y. L. (2009). Emotional contagion and empathy. In J. Decety & W. J. Ickes (Eds.), *The social neuroscience of empathy* (pp. 19–30). The MIT Press.

Hickson, L., & Khemka, I. (2013). Problem solving and decision making. In M. L. Wehmeyer (Ed.), *The Oxford handbook of positive psychology and disability* (pp. 198–225). Oxford University Press. https://doi.org/10.1093/oxfordhb/9780195398786.013.013.0015

Hobson, C. W., Scott, S., & Rubia, K. (2011). Investigation of cool and hot executive function in ODD/CD independently of ADHD. *Journal of Child Psychology and Psychiatry and Allied Disciplines, 52*(10), 1035–1043. https://doi.org/10.1111/j.1469-7610.2011.02454.x

Hopko, D. R., Lejuez, C. W., Daughters, S. B., Aklin, W. M., Osborne, A., Simmons, B. L., & Strong, D. R. (2006). Construct validity of the Balloon Analogue Risk Task (BART): Relationship with MDMA use by inner-city drug users in residential treatment. *Journal of Psychopathology and Behavioral Assessment, 28*(2), 95–101. https://doi.org/10.1007/s10862-006-7487-5

Hunt, M. K., Hopko, D. R., Bare, R., Lejuez, C. W., & Robinson, E. V. (2005). Construct validity of the Balloon Analog Risk Task (BART): Associations with psychopathy and impulsivity. *Assessment, 12*(4), 416–428. https://doi.org/10.1177/1073191105278740

Jacobs, E., & Nader-Grosbois, N. (2020). Affective and cognitive theory of mind in children with intellectual disabilities: How to train them to foster social adjustment and emotion regulation? *Journal of Education and Training Studies, 8*(4), 80. https://doi.org/10.11114/jets.v8i4.4757

Jahromi, L. B., Gulsrud, A., & Kasari, C. (2008). Emotional competence in children with down syndrome: Negativity and regulation. *American Journal on Mental Retardation, 113*(1), 12–32. https://doi.org/10.1352/0895-8017(2008)113[32:ECICWD]2.0.CO;2

Jenkinson, J., & Nelms, R. (1994). Patterns of decisionmaking behaviour by people with intellectual disability: An exploratory study. *Australia and New Zealand Journal of Developmental Disabilities, 19*(2), 99–109. https://doi.org/10.1080/07263869400035141

Kelly, J. R., & Barsade, S. G. (2001). Mood and emotions in small groups and work teams. *Organizational Behavior and Human Decision Processes, 86*(1), 99–130. https://doi.org/10.1006/obhd.2001.2974

Khemka, I., & Hickson, L. (2006). The role of motivation in the decision making of adolescents with mental retardation. *International Review of Research in Mental Retardation, 31*, 73–115. https://doi.org/10.1016/S0074-7750(05)31003-2

Khemka, I., Hickson, L., Casella, M., Accetturi, N., & Rooney, M. E. (2009). Impact of coercive tactics on the decision making of adolescents with intellectual disabilities. *Journal of Intellectual Disability Research, 53*(4), 353–362. https://doi.org/10.1111/j.1365-2788.2009.01152.x

Khemka, I., Hickson, L., & Mallory, S. B. (2016). Evaluation of a decision making curriculum for teaching adolescents with disabilities to resist negative peer pressure. *Journal of Autism and Developmental Disorders, 46*(7), 2372–2384. https://doi.org/10.1007/s10803-016-2770-0

Kilford, E. J., Garrett, E., & Blakemore, S. J. (2016). The development of social cognition in adolescence: An integrated perspective. *Neuroscience and Biobehavioral Reviews, 70*, 106–120. https://doi.org/10.1016/j.neubiorev.2016.08.016

Larson, R. W., Moneta, G., Richards, M. H., Holmbeck, G., & Duckett, E. (1996). Changes in adolescents' daily interactions with their families from ages 10 to 18: Disengagement and transformation. *Developmental Psychology, 32*(4), 744–754. https://doi.org/10.1037/0012-1649.32.4.744

Leshem, R., De Fano, A., & Ben-Soussan, T. D. (2020). The implications of motor and cognitive inhibition for hot and cool executive functions: The case of Quadrato motor training. *Frontiers in Psychology, 11*, 1–7. https://doi.org/10.3389/fpsyg.2020.00940

Logan, K. R., Jacobs, H. A., Gast, D. L., Murray, A. S., Daino, K., & Skala, C. (1998). The impact of typical peers on the perceived happiness of students with profound multiple disabilities. *Journal of the Association for Persons with Severe Handicaps, 23*(4), 309–318. http://journals.sagepub.com/doi/10.2511/rpsd.23.4.309

Luna, B., Padmanabhan, A., & O'Hearn, K. (2010). What has fMRI told us about the development of cognitive control through adolescence? *Brain and Cognition, 72*(1), 101–113. https://doi.org/10.1016/j.bandc.2009.08.005

McIntyre, L. L., Blacher, J., & Baker, B. L. (2006). The transition to school: Adaptation in young children with and without intellectual disability. *Journal of Intellectual Disability Research, 50*(5), 349–361. https://doi.org/10.1111/j.1365-2788.2006.00783.x

Megens, K. C. I. M., & Weerman, F. M. (2012). The social transmission of delinquency. *Journal of Research in Crime and Delinquency, 49*(3), 420–443. https://doi.org/10.1177/0022427811408432

Mercer, S. H., McMillen, J. S., & DeRosier, M. E. (2009). Predicting change in children's aggression and victimization using classroom-level descriptive norms of aggression and prosocial behavior. *Journal of School Psychology, 47*(4), 267–289. https://doi.org/10.1016/j.jsp.2009.04.001

Monahan, K. C., Steinberg, L., & Cauffman, E. (2009). Affiliation with antisocial peers, susceptibility to peer influence, and antisocial behavior during the transition to adulthood. *Developmental Psychology, 45*(6), 1520–1530. https://doi.org/10.1037/a0017417

Müller, C. M., Hofmann, V., & Arm, S. (2017). Susceptibility to classmates' influence on delinquency during early adolescence. *Journal of Early Adolescence, 37*(9), 1221–1253. https://doi.org/10.1177/0272431616653475

Müller, C. M., Hofmann, V., Begert, T., & Cillessen, A. H. N. (2018). Peer influence on disruptive classroom behavior depends on teachers' instructional practice. *Journal of Applied Developmental Psychology, 56*, 99–108. https://doi.org/10.1016/j.appdev.2018.04.001

Müller, C., Cillessen, T., Egger, S. & Hofmann, V. (2021). Peer influence on problem behaviors among students with intellectual disabilities. *Research in Developmental Disabilities, 114*, 1–12. https://doi.org/10.1016/j.ridd.2021.103994

Müller, C. M., Hofmann, V., Fleischli, J., & Studer, F. (2016). Effects of classroom composition on the development of antisocial behavior in lower secondary school. *Journal of Research on Adolescence, 26*(2), 345–359. https://doi.org/10.1111/jora.12195

Müller, C. & Minger, M. (2013). Which children and adolescents are most susceptible to peer influence? A systematic reviewregarding antisocial behavior. *Empirische Sonderpädagogik,*

2, 107–129. https://www.pedocs.de/volltexte/2014/8912/pdf/ESP_2013_2_Mueller_Minger_Welche_Kinder_und_Jugendliche.pdf

Müller, C., Hofmann, V. & Arm, S. (2020). Peer influence on the development of internalizing behaviors at school – Classroom and gender effects. *Empirische Sonderpädagogik*, 12, 91–111.

Müller, C., Hofmann, V., & Arm, S. (2021). Peer influence on the development of internalizing behaviors at school – Classroom and gender effects. *Empirische Sonderpädagogik.*

Nazli, & Chavan, B. (2017). Menstrual pattern among females with intellectual disability: A cross sectional study. *Journal of Disability Management and Rehabilitation, 2*(1), 8–12. http://jdmronline.org/index.php/jdmr/article/view/16

Nijs, S., & Maes, B. (2014). Social peer interactions in persons with profound intellectual and multiple disabilities: A literature review. *Education and Training in Autism and Developmental Disabilities, 49*(1), 153–165. https://www.jstor.org/stable/23880662

Papousek, I., Freudenthaler, H. H., & Schulter, G. (2008). The interplay of perceiving and regulating emotions in becoming infected with positive and negative moods. *Personality and Individual Differences, 45*(6), 463–467. https://doi.org/10.1016/j.paid.2008.05.021

Peake, S. J., Dishion, T. J., Stormshak, E. A., Moore, W. E., & Pfeifer, J. H. (2013). Risk-taking and social exclusion in adolescence: Neural mechanisms underlying peer influences on decision making. *NeuroImage, 82*, 23–34. https://doi.org/10.1016/j.neuroimage.2013.05.061

Penolazzi, B., Gremigni, P., & Russo, P. M. (2012). Impulsivity and reward sensitivity differentially influence affective and deliberative risky decision making. *Personality and Individual Differences, 53*(5), 655–659. https://doi.org/10.1016/j.paid.2012.05.018

Poon, K. (2018). Hot and cool executive functions in adolescence: Development and contributions to important developmental outcomes. *Frontiers in Psychology, 8*, 18–61. https://doi.org/10.3389/fpsyg.2017.02311

Prencipe, A., Kesek, A., Cohen, J., Lamm, C., Lewis, M. D., & Zelazo, P. D. (2011). Development of hot and cool executive function during the transition to adolescence. *Journal of Experimental Child Psychology, 108*(3), 621–637. https://doi.org/10.1016/j.jecp.2010.09.008

Prinstein, M. J., Boergers, J., & Spirito, A. (2001). Adolescents' and their friends' health-risk behavior: Factors that alter or add to peer influence. *Journal of Pediatric Psychology, 26*(5), 287–298. https://doi.org/10.1093/jpepsy/26.5.287

Rolison, J. J., Hanoch, Y., & Wood, S. (2012). Risky decision making in younger and older adults: The role of learning. *Psychology and Aging, 27*(1), 129–140. https://doi.org/10.1037/a0024689

Scherler, N., & Müller, C. M. (2018). *Peer Influence in adolescents with severe multiple disabilities – Measuring heart rate changes* [Poster Presentation]. https://doi.org/10.13140/RG.2.2.33141.22244

Schuengel, C., de Schipper, J. C., Sterkenburg, P. S., & Kef, S. (2013). Attachment, intellectual disabilities and mental health: Research, assessment and intervention. *Journal of Applied Research in Intellectual Disabilities, 26*(1), 34–46. https://doi.org/10.1111/jar.12010

Schuiringa, H., van Nieuwenhuijzen, M., Orobio de Castro, B., & Matthys, W. (2017). Executive functions and processing speed in children with mild to borderline intellectual disabilities and externalizing behavior problems. *Child Neuropsychology, 23*(4), 442–462. https://doi.org/10.1080/09297049.2015.1135421

Shin, H., & Ryan, A. M. (2017). Friend influence on early adolescent disruptive behavior in the classroom: Teacher emotional support matters. *Developmental Psychology, 53*(1), 114–125. https://doi.org/10.1037/dev0000250

Stanovich, K. (2011). *Rationality and the reflective mind.* Oxford University Press.

Steinberg, L. (2004). Risk taking in adolescence: What changes, and why? *Annals of the New York Academy of Sciences: Vol. 1021. Adolescent brain development: Vulnerabilities and opportunities.* New York Academy of Sciences.

Steinberg, L. (2008). A social neuroscience perspective on adolescent risk-taking. *Developmental Review, 28*(1), 78–106. https://doi.org/10.1016/j.dr.2007.08.002

Steinberg, L., & Monahan, K. C. (2007). Age differences in resistance to peer influence. *Developmental Psychology, 43*(6), 1531–1543. https://doi.org/10.1037/0012-1649.43.6.1531

Stroud, L. R., Foster, E., Papandonatos, G. D., Handwerger, K., Granger, D. A., Kivlighan, K. T., & Niaura, R. (2009). Stress response and the adolescent transition: Performance versus peer rejection stressors. *Development and Psychopathology, 21*(1), 47–68. https://doi.org/10.1017/S0954579409000042

Switzky, H. N. (2006). The importance of cognitive-motivational variables in understanding mental retardation in the 21st century. A personal view from the early twenty-first century. *International Review of Research in Mental Retardation, 31*, 1–29.

Tanaka, M., Malakoff, M. E., Bennett-Gates, D., & Zigler, E. (2001). Development of an outerdirected style of problem solving in individuals with and without mental retardation. *Journal of Applied Developmental Psychology, 22*(2), 191–198. https://doi.org/10.1016/S0193-3973(01)00076-4

Van der Molen, M. J., Van Luit, J. E. H., Van der Molen, M. W., Klugkist, I., & Jongmans, M. J. (2010). Effectiveness of a computerised working memory training in adolescents with mild to borderline intellectual disabilities. *Journal of Intellectual Disability Research, 54*(5), 433–447. https://doi.org/10.1111/j.1365-2788.2010.01285.x

Van der Put, C. E., Asscher, J. J., Stams, G. J. J. M., & Moonen, X. M. H. (2014). Differences between juvenile offenders with and without intellectual disabilities in the importance of static and dynamic risk factors for recidivism. *Journal of Intellectual Disability Research, 58*(11), 992–1003. https://doi.org/10.1111/jir.12078

Van Duijvenvoorde, A. C. K., Achterberg, M., Braams, B. R., Peters, S., & Crone, E. A. (2016). Testing a dual-systems model of adolescent brain development using resting-state connectivity analyses. *NeuroImage, 124*, 409–420. https://doi.org/10.1016/j.neuroimage.2015.04.069

Van Gelder, J. L. (2013). Beyond rational choice: The hot/cool perspective of criminal decision making. *Psychology, Crime and Law, 19*(9), 745–763. https://doi.org/10.1080/1068316X.2012.660153

Van Nieuwenhuijzen, M., Orobio De Castro, B., Van Aken, M. A. G., & Matthys, W. (2009). Impulse control and aggressive response generation as predictors of aggressive behaviour in children with mild intellectual disabilities and borderline intelligence. *Journal of Intellectual Disability Research, 53*(3), 233–242. https://doi.org/10.1111/j.1365-2788.2008.01112.x

Van Rest, M. M., Matthys, W., Van Nieuwenhuijzen, M., De Moor, M. H. M., Vriens, A., & Schuengel, C. (2019). Social information processing skills link executive functions to aggression in adolescents with mild to borderline intellectual disability. *Child Neuropsychology, 25*(5), 573–598. https://doi.org/10.1080/09297049.2018.1495186

Van Rest, M. M., Van Nieuwenhuijzen, M., Kupersmidt, J. B., Vriens, A., Schuengel, C., & Matthys, W. (2020). Accidental and ambiguous situations reveal specific social information processing biases and deficits in adolescents with low intellectual level and clinical levels of externalizing behavior. *Journal of Abnormal Child Psychology*. https://doi.org/10.1007/s10802-020-00676-x

Veenstra, R., Dijkstram, J. K., & Kreager, D. A. (2019). Pathways, networks, and norms. A sociological perspective on peer research. In W. M. Bukowski, B. Laursen, & K. H. Rubin (Eds.), *Handbook of peer interactions, relationships, and groups* (2nd ed., p. 748). Guilford Publications.

Wagemaker, E., Huizenga, H. M., Dekkers, T. J., Collot d'Escury-Koenigs, A. L., Salemink, E., & Bexkens, A. (2020). When do those "risk-taking adolescents" take risks? The combined effects of risk encouragement by peers, mild-to-borderline intellectual disability and sex. *Journal of Abnormal Child Psychology, 48*(4), 573–587. https://doi.org/10.1007/s10802-020-00617-8

Wallsten, T. S., Pleskac, T. J., & Lejuez, C. W. (2005). Modeling behavior in a clinically diagnostic sequential risk-taking task. *Psychological Review, 112*(4), 862–880. https://doi.org/10.1037/0033-295X.112.4.862

Warr, M. (2002). *Companions in crime: The social aspects of criminal conduct*. Cambridge University Press.

Wright, B. R., Caspi, A., Moffitt, T. E., & Silva, P. A. (2001). The effects of social ties on crime vary by criminal propensity: A life-course model of interdependence. *Criminology, 39*(2), 321–348. https://doi.org/10.1111/j.1745-9125.2001.tb00925.x

Willner, P., Bailey, R., Parry, R., & Dymond, S. (2010). Performance in temporal discounting tasks by people with intellectual disabilities reveals difficulties in decision making and impulse control. *American Journal on Intellectual and Developmental Disabilities, 115*(2), 157–171. https://doi.org/10.1352/1944-7558-115.2.157

Zelazo, P. D., & Carlson, S. M. (2012). Hot and cool executive function in childhood and adolescence: Development and plasticity. *Child Development Perspectives, 6*(4), 354–360. https://doi.org/10.1111/j.1750-8606.2012.00246.x

Zic, A., & Igri, L. (2001). Self-assessment of relationships with peers in children with intellectual disability. *Journal of Intellectual Disability Research, 45*(3), 202–211. https://doi.org/10.1046/j.1365-2788.2001.00311.x

Chapter 10
Decision Making and Culturally Diverse Individuals with IDD and Their Families: A Call for Research

Sandra B. Vanegas, Sandy Magaña, Weiwen Zeng, and Carlos Pavon

This chapter explores decision making among individuals with intellectual and developmental disabilities (IDD) from culturally diverse backgrounds. We provide an overview of decision making among individuals with IDD, outline how cultural diversity might influence decision making among individuals with IDD and their families, and present a scoping review of the literature on decision making among culturally diverse individuals with IDD and their families. In the scoping review, we map the existing literature using a broad lens of decision making, summarize the research findings, and identify gaps in the literature to date on culturally diverse communities (Arksey & O'Malley, 2005; Tricco et al., 2018). We then discuss where the field stands on these issues and provide recommendations for research and practice.

What Is Decision Making?

Making decisions is an important skill that an individual carries out across the lifespan and that can, directly and indirectly, affect their quality of life. Decision making, broadly defined, involves committing to a judgment or action in the service of a specific goal (Wehmeyer et al., 2010). Others have distinguished decision making from problem-solving in that decision making addresses issues with unclear outcomes or goals and no definitive path, whereas problem-solving may have a clearly

S. B. Vanegas (✉)
Texas State University, San Marcos, TX, USA
e-mail: svanegas@txstate.edu

S. Magaña · W. Zeng · C. Pavon
The University of Texas at Austin, Austin, TX, USA
e-mail: smagana@austin.utexas.edu; weiwenzeng@utexas.edu; cjpavon@utexas.edu

I. Khemka, L. Hickson (eds.), *Decision Making by Individuals with Intellectual and Developmental Disabilities*, Positive Psychology and Disability Series,
https://doi.org/10.1007/978-3-030-74675-9_10

defined path and outcome (Hickson & Khemka, 1999, 2013; Short & Evans, 1990). We briefly review theoretical constructs on decision making in IDD as a foundation for understanding decision making in culturally diverse individuals with IDD.

Theoretical Constructs Relevant to the Study of Decision Making and IDD

Decision making can be viewed as falling within the broader construct of self-determination. Within the Developmental Disabilities Assistance and Bill of Rights Act of 2000, decision making is considered a primary component of self-determination activities, specifically, "the ability and opportunity to communicate and make personal decisions" and "the ability and opportunity to communicate choices and exercise control over the type and intensity of services, supports, and other assistance the individuals receive" (Developmental Disabilities Assistance and Bill of Rights Act of 2000, 2000). A person who engages in self-determination acts purposefully to achieve a specific outcome or objective (Wehmeyer, 1996). We consider self-determination within our review of decision making skills as studies often incorporate decision making skills within the broader definition of self-determination.

Research on self-determination has been extensive among youth with IDD, including the development of specific assessment tools to measure self-determination (Shogren et al., 2008; Shogren et al., 2020a; Wehmeyer, 1995) and curriculums to enhance self-determination (Hagiwara et al., 2020; Raley et al., 2018). Overall, the research suggests that the type of disability, cognitive ability, and age serve as significant contributors to self-determination among individuals with IDD (Garrels & Granlund, 2018; Lee et al., 2012; Mumbardó-Adam et al., 2017; Shogren et al., 2013b). Contextual factors, such as socioeconomic status, family, and culture, serve as important factors in developing and manifesting self-determination (Cavendish, 2017; Taylor et al., 2019). Thus, there is a need for additional research to investigate how self-determination, or more specifically, decision making, may be subject to cultural influences that have yet to be fully captured by the literature to date.

Decision-Making Processes

Decision making has often been described as being comprised of four basic steps: (1) framing or defining the problem or issue to resolve, (2) generating alternatives or viable options, (3) appraising consequences as they relate to the alternatives or options, and (4) choosing the best course of action (Hickson & Khemka, 2013). This process outlines potential areas to support individuals with IDD. For example, identifying or defining the problem can be a challenge for some individuals with IDD (Short & Evans, 1990); thus, interventions or supports can be developed to enhance this ability. Consequences may be weighed differently across diverse cultural groups, depending on the priorities of the decision-maker. The decision-making process can be viewed as led by the individual, led by the family, or shared across multiple parties (i.e., shared decision making; Charles et al., 1997). Shared decision making involves at least two individuals who share information and their perspectives to arrive at a mutual agreement. In the scope of IDD, shared decision making may include the individual with IDD, the family of the individual with IDD, service providers, or other parties with a stake or role in the life of the individual with IDD (Adams & Levy, 2017). This captures the nuances of decision making among culturally diverse individuals with IDD, as some may need accommodations or may involve additional people in this process.

Why Is Cultural Diversity Important in Understanding Decision Making?

As described previously, decision making is an important skill for individuals with IDD due to the association with long-term outcomes, including quality of life, employment, and health (Shogren et al., 2013a; Shogren & Shaw, 2016). Although the research suggests that there may be many interactional influences between decision making, cognitive ability, disability status, and socioeconomic status, cultural diversity has not been adequately addressed. As defined in this chapter, cultural diversity includes racial/ethnic minorities in majority-white countries, individuals/families in low- and middle-income countries (LMIC), and individuals/families from underrepresented regions (e.g., Asia, Latin America) in research on individuals with IDD. Considering the impact that decision making may have on individuals with IDD's life and well-being, we must understand the role of cultural diversity. In the USA, the overall population is highly diverse, with 39.6% reported as non-white, racial/ethnic minority and 13.5% reported as foreign-born. Over 21.5% of households speak a non-English language at home (US Census Bureau, n.d.). With the vast diversity in the USA and rapid globalization, it is vital to examine how cultural diversity interacts with decision making among individuals with IDD.

What Is the Role of Culture in Decision Making?

To date, few papers have described decision making in diverse populations of individuals with IDD (Mumbardó-Adam et al., 2017; Shogren, 2011; Shogren & Wehmeyer, 2017). Shogren (2011) conducted a thematic review on self-determination across cultures, specifically within special education settings, finding only ten articles (research, theoretical, review) published before 2010. These articles recognized that self-determination as a construct should be inclusive of values and norms that extend beyond the values and norms of individualistic societies (e.g., the

USA). Building on this initial review, Shogren and Wehmeyer (2017) described self-determination as a universal construct when considered as self-determined causal action; however, culture shapes the what, where, when, and how of self-determination for individuals. Mumbardó-Adam et al. (2017) conducted a meta-analytic review to examine personal and contextual factors related to self-determination measures and an intellectual disabilities (ID) diagnosis in research published between 2002 and 2015. Overall, they reviewed 16 studies, finding that ethnicity was a significant factor in the relation between self-determination and ID diagnosis. However, due to the limited inclusion of race/ethnicity as a variable reported by the studies reviewed, the authors combined all racial/ethnic minority groups into a non-white category for their analyses. The consensus from these papers and our review of the literature finds that the empirical research on race/ethnicity involving individuals with IDD to date has been limited. There is a significant need to recognize, acknowledge, and include these factors in the study of self-determination and decision making.

To broaden our understanding of the role of cultural diversity in decision making in individuals with IDD and their families, we explore cultural differences within the Ecological Systems Theory originally proposed by Bronfenbrenner (Bronfenbrenner, 1977, 1986). Bronfenbrenner's Ecological Systems Theory has been applied to decision making in diverse individuals with IDD (Shogren, 2013; Small et al., 2013). The Ecological Systems Theory posits that human development can be viewed through multiple systems that influence and are influenced by the individual. According to Bronfenbrenner's Ecological Systems model, the microsystem, mesosystem, macrosystem, and chronosystem can shape the contextual lens through which we examine decision making among culturally diverse individuals.

Microsystem

The microsystem involves the direct contacts a person has with siblings, family members, and teachers, among others. Cultural differences may afford or limit opportunities for individuals with IDD to develop and practice decision-making skills. Analyses of the National Longitudinal Transition Study 2 (NLTS-2) showed that parent expectations for their youth with disabilities in graduating from high school, pursuing post-secondary education, and obtaining a paid job were significant predictors of the youth achieving these outcomes (Doren et al., 2012). The authors also reported that parents from diverse racial/ethnic minority backgrounds had lower expectations than white parents. However, the NLTS-2 may not fully capture culturally diverse parents' expectations for their children with disabilities as they reflect autonomy and independent living, outcomes that may not be considered high priorities for diverse parents.

Some studies suggest that adults with IDD who live in their family homes have greater opportunities for self-determination and report a greater quality of life than adults with IDD who live in group homes (Duvdevany et al., 2002; Kostikj-Ivanovikj

& Chichevska-Jovanova, 2016). Interviews with parents of racial/ethnic minority individuals with IDD find that many parents fear the uncertainty about the quality of care that the individual with IDD may encounter outside of the family home (Magaña & Smith, 2006). This may be attributed to *familismo* or the strong affiliation and importance placed on the family in the Latino culture (Sabogal et al., 1987). For individuals with IDD from Latino cultures and other cultures who place similar values on family, this cultural value and its related practices may impart protective health factors, including greater family support for the individual with IDD in making decisions (Perez & Cruess, 2014).

Another microsystem that may indirectly influence decision making among culturally diverse children and youth with IDD is the school setting, including the classroom and teachers. Racial/ethnic minority students with IDD tend to be disproportionately represented in more restrictive special education classrooms than in general education classrooms (Carter et al., 2012). Studies have also found that students with ID in special education classrooms score lower on decision-making and self-determination skills than students without disabilities (Wehmeyer, 1993). Thus, culturally diverse students may be at a disadvantage in obtaining adequate supports and opportunities for decision making due to their classroom placements. Several studies have also reported that teacher interactions with and perceptions of culturally diverse children and children with disabilities may affect how much autonomy and self-determination the child/youth may demonstrate in the classroom (Linton, 2015; Shogren et al., 2007). Parker et al. (2020) interviewed African American youth, with and without a disability, on their experience with self-determination across home and school settings. Youth reported that they felt low levels of support from their teachers to exert self-determination skills, such as self-advocacy and decision making. They utilized decision making and self-advocacy to counteract the negative expectations (e.g., misbehavior) from their teachers. Cavendish (2017) also examined the influence of school and family supports on youth goals and self-determination of diverse high-school students with disabilities (i.e., learning disability, ADHD, and emotional-behavioral disorder, intellectual disability, visual impairment, hearing impairment). The author reported that both family and school support predicted self-determination scores as measured by the Self-Determination Scale (Wehmeyer & Kelchner, 1995). These results, when considered with Parker et al. (2020), highlight the importance of the family and the school setting as independent and interactional factors to increase the opportunities for decision making in youth with IDD.

Mesosystem

The mesosystem involves interactions with all the systems at the microsystem level, such as the relationship between parents and teachers. Children and youth with IDD often receive services and supports in the school setting through Individualized Education Programs (IEPs) and 504 plans, which provide specific accommodations to support learning. Parents are vital participants in determining services and supports. However, qualitative and quantitative studies have found that interactions supporting decision making between parents and teachers (or school administrators) are often challenging for all parties involved (Park et al., 2001; Rossetti et al., 2018). Overall, the cultural, religious, and family background of individuals with a disability can factor in decision-making opportunities.

Macrosystem

Beyond the mesosystem, the macrosystem captures the social, political, and cultural factors that influence individuals. Specific policies can provide or limit opportunities available, such as higher education and supported employment. Within the macrosystem, key constructs include individualistic and collective cultural patterns. In general, many collective and high-context cultures place a greater focus on interdependence and relationships within the family and community (Leake & Boone, 2007; Zheng et al., 2015). The focus of decision making extends beyond the individual; that is, the final decision may be made by more than one person or by a senior member of the family or community.

Additionally, the priorities considered for the decision at hand may shift from the individual to the family or community. This process conflicts with self-determination theory, where the individual is in charge of making decisions about the self independently (Ryan & Deci, 2004), although some models consider support in decision making (Shogren et al., 2015). This view (independent decision making) aligns with individualistic cultures where there is a greater emphasis on the individual as a separate entity from others and a greater promotion of independence or autonomy (Schwartz et al., 2010). Thus, evaluations of decision making among culturally diverse populations may be inaccurate or inconclusive as many assessment tools, and even outcomes, are evaluated against the positive outcomes defined by individualistic cultures. For example, many policies regarding IDD services in adulthood are based on the premise that individuals with IDD reside outside of their family home. This expectation may be incompatible with cultural expectations for the individual with IDD, regardless of opportunity and capacity for autonomy and decision making (Geenen et al., 2005; Kim & Morningstar, 2005; Powers et al., 2009). These perceived ideal outcomes shape how we evaluate and measure decision making and outcomes of individuals with IDD. The standard outcome measures focus on the individual and not on the individual and family. However, it is important to note that the concepts of independence and collectivism are not mutually exclusive categories, but rather two spectrums that may overlap and shift across time and contexts (Greenfield & Cocking, 2014).

Chronosystem

The chronosystem captures the essence of time and its impact on development. The chronosystem is critical as it can describe variations in opportunities and capacities to make decisions by culturally diverse individuals with IDD. For example, schools may play a prominent role for youth with IDD; however, as they mature, this may be replaced with an employer. Shifts in immigration policies may influence the opportunities available for decision making by the individual with IDD and their families. Undocumented immigrants in the USA may experience different fears in pursuing services or support in their communities, limiting those opportunities for self-determination, autonomy, and decision making. Societal/professional definitions of IDD may also influence decision making across time. Two major classification systems, the International Classification of Diseases (ICD; World Health Organization, 2018) and the Diagnostic and Statistical Manual for Mental Disorders (DSM; American Psychiatric Association, 2013), have undergone significant changes over the past 40 years in how disorders are described and evaluated within healthcare and education systems (McKenzie et al., 2016). These changes may affect how individuals, families, and professionals perceive individuals with IDD, including the capacity for self-advocacy and decision making among individuals with IDD. Research on culturally diverse individuals with IDD must examine cultural diversity and contextual factors (e.g., religion, socioeconomics, immigration, and disability) as dynamic and related influences on decision making.

Current Study

Although extensive research on decision making within the context of self-determination among individuals with IDD has been conducted (Chou et al., 2017; Taylor et al., 2019), little research has focused explicitly on decision making among individuals with IDD from culturally diverse populations or included large enough samples to evaluate the role of cultural diversity on decision making (Frankland et al., 2004; Mumbardó-Adam et al., 2017). Culture plays a vital role in how decision making is viewed by individuals with IDD and their families (Wehmeyer et al., 2011). The self-determination framework provides a foundation for understanding decision making and related constructs, yet this framework may not fully capture the role that racial/ethnic identity and culture may have on the opportunities and capacity for decision making among culturally diverse individuals with IDD.

In this chapter, we build on Bronfenbrenner's model and propose the Socio-Ecological Model of Decision making in Culturally Diverse Individuals with IDD (see Fig. 10.1). Bronfenbrenner's model recognizes the influence of multiple systems (i.e., microsystem, mesosystem, macrosystem) on the individual with IDD and

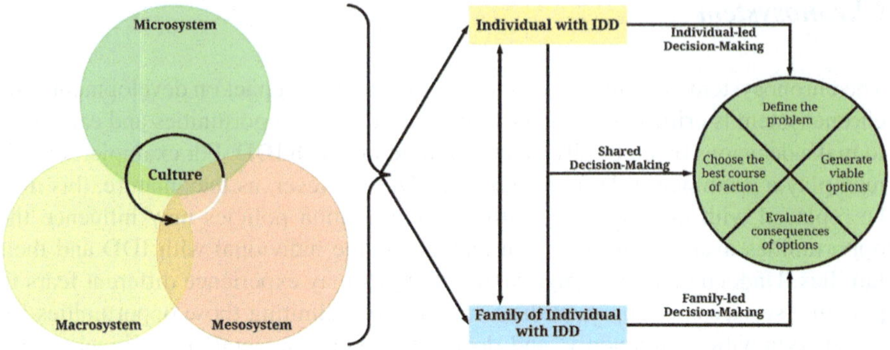

Fig. 10.1 Socio-ecological model of decision making in culturally diverse individuals with IDD

their family and the bidirectional effects between the individual with IDD and their family, recognizing that this model may not fully capture the distinct pathways that decision making may occur for individuals with IDD. The Socio-Ecological Model of Decision making in Culturally Diverse Individuals with IDD allows for multiple paths to decision making, including shared decision making, and recognizes that decision making involves numerous steps by which an individual may enact their decision-making skills. This model is important for the field to ensure access to culturally informed supports and services by individuals with IDD. This model acknowledges that the process of decision making is dynamic, involves multiple steps, and allows for independent decision making, individual-led decision making, family-led decision making, or shared decision making.

We focused on transition-age youth and adults with IDD as this developmental stage may afford increased opportunities for decision making by the individual with IDD. We also defined cultural diversity as inclusive of racial/ethnic minorities in majority-white countries, individuals/families in LMIC, and regions that are underrepresented in research on individuals with IDD, such as Asia and Latin America.

Objectives

In this chapter, we reviewed the current literature on decision making among individuals with IDD from racial/ethnic minority backgrounds in the United States and diverse cultures worldwide. Through a scoping review of the literature, we examined the process of decision making among individuals with IDD and their families across multiple settings and contexts. We use this review to address the following research questions: (1) What can we learn from the existing literature about decision making for individuals with IDD and their families across racial/ethnic groups? (2) Does the literature on decision making provide sufficient knowledge about racial/ethnic minority groups?

Method

Scoping Review Protocol

In a scoping review, the identification of the literature needs to achieve in-depth and broad results (Arksey & O'Malley, 2005). We adopted Arksey and O'Malley's (2005) five stages framework:

1. Identifying the research questions. The authors met to decide research questions that are being addressed by this chapter. Multiple relevant keywords were brainstormed.
2. Identifying relevant studies. A list of search terms was finalized, and searches on electronic databases were performed with the consultation of a librarian who has extensive knowledge in systematic review methodology. Table 10.1 documents the final search terms.
3. Study selection. A set of carefully defined inclusion and exclusion criteria was applied to search results to identify eligible studies. To identify potential missing research, an ancestral search was also performed by hand-searching key researchers in this field on Google Scholar and Scopus (Wilt & Morningstar, 2018).

Table 10.1 Finalized search terms

Line	Key construct	Search terms
1	Decision making, self-determination, autonomy	"Decision making" OR decision making OR "self direction" OR self-directed OR "self-determination" OR "self determination" OR "autonomy" OR "person centered planning" OR "person-centered planning" OR choice behavior* OR heuristics OR volition
2	IDD	((intellectual* OR mental* OR developmental* OR cognitive*) AND (disab* OR retard* OR handicap* OR delay* OR deficiency)) OR autis* OR "cerebral palsy" OR "down syndrome" OR "down's syndrome"
3	Racial/ethnic minorities in the USA or from diverse cultures/countries	Minorit* OR ethnicity OR "ethnic group*" OR race OR racism OR "racial group*" OR "racially diverse" OR "people of colo?r" OR "person* of colo?r" OR Hispanic OR Latin* OR Spanish OR Black OR "African American" OR Asian OR "Asian Pacific islander" OR "American Indian*" OR "native american*" OR tribal OR "Alaska* Native" OR foreigner OR "foreign-born" OR immigrant OR refugee OR aliens OR migrants OR international* OR "non-western" OR other countr* OR developing countr* OR third-world OR "low income countr*" OR "underdeveloped countr*"

4. Charting the data. The authors discussed and agreed upon a coding scheme, which was used to help map out relevant themes from eligible articles.
5. Collating, summarizing, and reporting the results. The authors paired up, and each pair independently conducted full article review to extract and report data. Further details of the protocol are presented in subsequent sections.

Inclusion and Exclusion Criteria

Four inclusion criteria were applied to select eligible studies. First, we only included peer-reviewed articles that focused on decision making of transition-age (> = 14 years old) youth or adults with IDD and their families across or about racial/ethnic minorities in majority-white countries or individuals/families in LMIC or regions underrepresented in research, such as Asia and Latin America. Second, articles were limited to those published between January 2000 and March 2020 (when the final search was performed) to reflect a more up-to-date overview of research evidence in the field. Third, we included studies that reported empirical results on predominantly racial/ethnic minority samples with quantitative, qualitative, or mixed-methods research designs. Given our topic's nascent nature, we also included studies if they offered analyses specific to racial/ethnic minority participants in the USA or other low- and middle-income countries. Lastly, we only included studies that were published in English.

Based on the same set of considerations, the following kinds of articles were excluded: (1) articles that did not report concrete data, examples including single case reports, theoretical articles, government briefs, and organizational reports; (2) gray literature such as theses, dissertations, and conference abstracts and presentations; (3) other systematic/scoping/literature reviews; and (4) articles that were not written in English.

Search and Screening Procedures

We systematically conducted our searches based on procedures outlined by Arksey and O'Malley (2005). In phase one, a search was performed on EBSCOhost using six electronic databases, including PubMed, PsycInfo, CINAHL, Health Source: Nursing/Academic Edition, Psychology and Behavioral Sciences Collection, and SocINDEX. In phase two, a complementary search using the same search terms and limiters (peer-reviewed and publication years) was performed on Web of Science. The searches in phase one and two yielded a total number of 2707 search results. The search results were then imported into Rayyan, a free and user-friendly web application that allows researchers to collaborate on systematic review projects (Ouzzani et al., 2016).

After removing all duplicates, 2059 unique records were entered into the screening phase. One of the authors performed the title screening to assess if the studies addressed decision making, IDD, or racial/ethnic minorities. As a result, 1833 records were screened out, and 226 records were entered into abstract screening. The same author read the abstracts and was able to exclude 174 records. All the authors then reassessed the remaining 52 abstracts and further narrowed them down to 31 studies. Full texts were downloaded and carefully reviewed for these studies. The authors independently rated the studies based on inclusion and exclusion criteria. Any disagreements were resolved by consensus when deciding on a final list of included studies. Finally, full-text reviews yielded eight studies, complemented by seven additional studies identified through hand-search that met all the inclusion criteria. Figure 10.2 presents the screening procedures for the included studies.

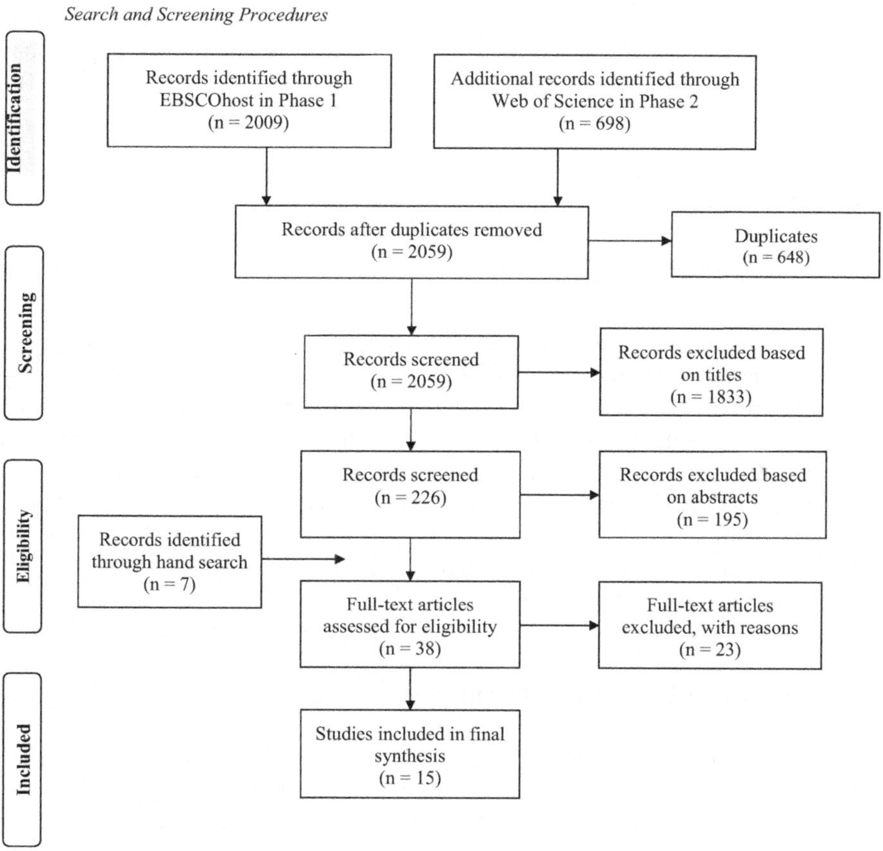

Fig. 10.2 Search and screening procedures

Data Extraction and Reporting Procedures

After assembling the final list of included studies, all the authors met and decided on the data extraction and reporting procedures. The PRISMA-P 2015 checklist (Moher et al., 2009, 2015) was used to extract data and report results. Specifically, the authors designed a coding sheet using an Excel file and defined the five domains of decision making for which data would be extracted. These five elements include housing and independent living, daily living skills and activities, safety, employment and postsecondary education, and healthcare and IDD-related services. Each of the authors independently performed data extraction using the coding sheet. Data from the 15 included studies were synthesized and categorized using the five elements. Then, each of the authors wrote a summary of the data source, sample, methodology, and major findings for each study. To confirm findings were consistent across all authors, the coding sheets and summaries were combined and compared. Discrepancies were resolved by consensus. Lastly, the synthesized results were compiled into comprehensive overview of current research and to pinpoint implications for future research, practice, and policy reforms (Levac et al., 2010).

Results

Research Question 1

In our first research question, we explored what we can learn from the existing literature about decision making for individuals with IDD and their families across racial and ethnic groups. We expanded the idea of decision making to concepts such as self-direction, self-determination, and autonomy. We organized our findings by contexts in which decision making would be important for youth and adults with IDD, including housing and independent living, daily living skills, safety, employment or education, and healthcare and IDD-specific services. Table 10.2 shows the sample sizes and study methodology. Below we primarily describe findings related to race/ethnicity and/or culture.

Decision Making About Housing and Independent Living

This section highlights those studies that focused on decision making related to living independently and residential status. In a qualitative study examining self-determination, Lin (2006) found that youth in Taiwan aged 14–16 years old had relatively low self-determination related to independence, goal setting, and planning, all essential aspects of decision making. The author reported that youth were heavily influenced by their parents, family members, and teachers, attributing this to Taiwanese culture (e.g., such as Confucianism, family values, and patriarchy).

Table 10.2 Characteristics of articles included in scoping review

Authors	Year	Topics addressed	Sample size	Description of participants	Type of disability	Study design
Chou, Schalock, Tzou, Lin, Chang, Lee & Chang	2007	Quality of life; one domain of the scale was self-determination	233	Individuals with mild intellectual disability in Taiwan, age 16 and older	ID	Quantitative analysis of survey
Griffin, Taylor, Urbano, & Hodapp	2014	Student involvement in transition planning; participation in IEP about transition planning decisions	320	High school students: White 66.5% African American 16.8% Latino 10.8% Asian/Pacific Islander 3.8%	ASD	Quantitative analysis of National Longitudinal Transition Study-2 (NLTS2)
Khemka, Hickson, & Reynolds	2005	Abuse prevention and empowerment	36	Women with ID: White 33% African American 50% Latino 16.7%	ID	Randomized control intervention study
Larkin, Unwin, Iyer, Tsimopoulou, Zahid, Kroese, & Rose	2018	Independence, identity, social care of adults with ID from minority backgrounds in the UK	29	Adults with IDD: Black Caribbean 30% Pakistani 25% Other East Indian 45%	Mild/ moderate LD	Qualitative
Lin	2005	Self-determination related to Maslow's hierarchy	8	Adolescents from Taiwan with ID ages 14–16	ID	Qualitative
Nasser, Sachs, & Sa'ar	2017	Decision making by parents in Palestinian towns about residential placement	18	Palestinian parents of youth with ID placed in residential facilities, ages 11 to 30	ID	Qualitative
Rueda, Monzo, Shapiro, Gomez, & Blacher	2005	Latina mothers' views of transition, cultural models	16	Latina mothers of youth with ID ages 14 to 31	DS ID ASD	Qualitative

(continued)

Table 10.2 (continued)

Authors	Year	Topics addressed	Sample size	Description of participants	Type of disability	Study design
Shogren	2012	Latina mothers' perceptions of self-determination	7	Latina mothers of transition-aged youth ages 14 to 21	ASD ID Multiple disabilities	Qualitative
Shogren, Kennedy, Dowsett, Villareal & Little	2014	Self-determination of youth from diverse racial/ethnic groups	Not reported for this analysis	Adolescents with disabilities ages 16 to 18: White, African Americans, and Latinos (Ns not reported for each group)	Various disabilities including ID, ASD, sensory impairments, emotional, health and cognitive impairments	Quantitative analysis of NLTS2
Shogren, Rifenbark, Wehmeyer, Dean, Kileen, & Karsevar	2020	Supported decision making inventory	152	Adults with disabilities: White 10% African American 80% Latino 9%	Various disabilities including ID, ASD, sensory, and mental health	Quantitative analysis of survey
Shogren & Shaw	2017a	Self-determination and early adult outcomes	Not reported for this analysis	Adults with disabilities: White, African American, Latino (Ns not reported for each group)	High incidence, cognitive and intellectual disabilities	Quantitative analysis of NLTS2
Shogren, Shaw, Raley, & Wehmeyer	2018a	Self-determination and personal characteristics	3044	Youth ages 13–22: White 26% African American 30% Latino 31% Other 12%	No disabilities, LD, and ID	Quantitative analysis of survey

(continued)

Table 10.2 (continued)

Authors	Year	Topics addressed	Sample size	Description of participants	Type of disability	Study design
Shogren, Wehmeyer, Little, Forber-Pratt, Palmer, & Seo	2017b	Validity and reliability of self-determination inventory, student reported	311	Adolescents ages 12 to 22: White 74.3% African American 6.4% American Indian/Pacific islander 2.6% Asian 2.9% Other/missing 13.9%	No disabilities, LD, and ID	Quantitative analysis of survey
Shogren, Wehmeyer, Shaw, Grigal, Hart, Smith, & Khamsi	2018b	Factors associated with self-determination	251	Students in higher education: White 70% African Americans 12% Latinos 10%	IDD	Quantitative analysis of data from universities hosting TPSID programs
Simonsen & Neubert	2013	Employment, self-determination, self-management	338	Transitioning youth: White 48.8 African American 44.4 Latino 2.7% Asian 2.7%	IDD	Quantitative analysis of survey

Note: ASD Autism spectrum disorder, *DS* Down syndrome, *ID* Intellectual disability; *IDD* Intellectual and developmental disabilities, *LD* Learning disabilities, *Ns* Sample size

A qualitative study in the UK (Larkin et al., 2018) examined racial/ethnic minorities with a learning disability (i.e., considered equivalent to ID in the USA; Agrawal et al., 2019) and reported themes related to the concept of independence, which as an outcome implies living independently either from one's family or from a residential setting. The participants in the study were Caribbean Blacks and from South Asian countries such as Pakistan and India. Those who lived with their families seemed content and stable in this arrangement and did not desire to live independently. Participants who lived in residential settings reported being encouraged to live independently, but some were threatened by this as they understood they would lose needed supports. In these cases, participants felt that they had not been allowed to make their own decision about living independently.

A study conducted in Israel with Palestinian parents of children with ID focused on parental decision making about their children's placement in residential settings (Nasser et al., 2017). This study used a grounded theory approach and interviewed families of children with ID between the ages of 11 and 30 who lived in residential facilities in the Galilee region. The authors reported that Palestinians have higher rates of ID but lower rates of institutionalization than the Israeli population and found a core theme that parents felt conflicted about their decision to place their child. These parents perceived placing their child as abandonment, but a necessary evil, given the limited resources and support available to them. Authors raised the issue of contextual and cultural issues such as low income and limited support and a core value of familism, highlighting the importance of collectivism versus individualism.

Similarly, a focus group study in the USA found that Latina mothers of young adults with IDD discussed the issue of individualism and living independently. They indicated that in their culture, living independently once their child was an adult was not the norm until marriage, even for typically developing young adults (Rueda et al., 2005). Most of the decision making was led by the mothers, who considered their adult child's perspective with IDD. They reported frustration over not being taken seriously by professionals as decision-makers.

A different study focused on perceptions of self-determination among Latina mothers of transition-age youth with disabilities using a phenomenological qualitative design reported similar findings (Shogren, 2012). These mothers believed that self-determination only applied to the idea of their son or daughter living independently, which conflicted with their cultural expectation that their children do not move out at age 18. They reported that they placed great value in teaching their child to indicate preferences, make decisions, and advocate for themselves, all behaviors of self-determination. The mothers reported that these behaviors were primarily taught at home and not at school.

A study in which 80% of the sample were African Americans (the remaining 20% were white or Latino) with primarily ID focused on the refinement of an instrument to measure supported decision making among individuals with disabilities (Shogren, Rifenbark, et al., 2020). Only 39% lived at home with their parent or guardian. The remaining individuals lived in group homes (31%) or independently with or without support (27%). These authors did not compare racial or ethnic groups or discuss any cultural or contextual issues that may be specific to African Americans with disabilities. They found that those who lived independently scored higher on the measure than those who lived in group homes. There were no differences between those living at home and other living arrangements; however, those who had legal guardians scored lower than those who did not.

Overall, there were several common themes related to independent living across studies. Studies that included populations that were Taiwanese, South Indian, Caribbean Black, Palestinian, or Latino reported the importance of parents and family in the decision-making process for the person with ID related to living independently or in other settings (Larkin et al., 2018; Lin, 2006; Nasser et al., 2017; Rueda et al., 2005; Shogren, 2012). Some of these studies referred to cultural aspects such as familism and collectivism as important values for these families. Another common theme is that living at home with the family is not inconsistent with decision making. The study of minority groups in the UK found that those who lived at home with their families were satisfied. The study of predominantly African American adults with ID found no differences in levels of decision making between those living at home and those in other living arrangements (Larkin et al., 2018; Shogren et al., 2020b).

Decision Making About Daily Living Skills and Activities

Some of the studies cited in the previous section also reported the importance of decision making about daily living skills and activities. In the qualitative study of minority persons with ID in the UK (Larkin et al., 2018), participants discussed decision making in their daily activities such as taking a walk to the store or going to the movies. While the service system encouraged *independence*, participants complained when they were not allowed to make these kinds of decisions independently. In the qualitative study on self-determination among youth in Taiwan, investigators used Maslow's Hierarchy of Needs to categorize results. The category "Self-determinations of self-achievement essentials" referred to concepts of decision making such as independence/autonomy, goal setting, and planning/implementation (Lin, 2006). This author reported that the students made decisions about daily living routines and goal setting; however, parents made decisions on what the authors referred to as major important matters. In the two studies of Latina mothers of transition-aged youth with disabilities, mothers emphasized teaching their youth life skills as a form of independence (Rueda et al., 2005; Shogren, 2012). Both studies noted that the schools did not significantly teach these skills, and instead, they were taught in the family home. In a study that analyzed national data, authors compared racial and ethnic groups of youth with disabilities using a self-determination measure (Shogren & Shaw, 2017). They categorized the youth into three groups, high incidence, cognitive, and intellectual disabilities. Groups were defined based on similarities identified in self-determination and disability characteristics and followed terms used by the IDEA (Individuals with Disabilities Education Act, 2004; Shogren et al., 2014). The high-incidence disabilities group included specific learning disabilities, emotional disturbance, speech-language impairments, and other health impairments. The cognitive disabilities group included autism, multiple disabilities, and deaf-blindness. The intellectual disabilities group was comprised of intellectual disabilities alone as this disability was distinctive from other disabilities in self-determination and disability characteristics. They found that African American youth from the cognitive group had lower financial independence and advocacy levels than the white youth. In the cognitive and intellectual disability groups, Latinos had lower social relationship outcomes than whites.

Studies in this section emphasize the importance of decision making about daily activities and that these are viewed as a form of independence by both individuals with disabilities and their parents. The findings regarding African Americans and financial independence and Latinos and social relationships need further exploration to contextualize them.

Decision Making About Safety

Safety was often discussed in the literature as parents being concerned that their son or daughter might not make safe decisions or worried about their safety. The two studies that interviewed Latina mothers of youth with disabilities emphasized the need to protect their son or daughter from the dangers of the outside world or teach their child skills related to safety (Rueda et al., 2005; Shogren, 2012). The qualitative study of youth with ID in Taiwan found that they lacked awareness about safety issues and the ability to take precautions (Lin, 2006). The research that focused on youth with ID in Palestine reported that the safety of the youth with IDD and their siblings was an important factor in parental decision making about the placement of the youth with IDD into a home (Nasser et al., 2017). A randomized controlled trial aimed at increasing decision-making skills in women with ID, focusing on safety issues, such as abuse prevention, in a diverse group (50% African American, 33% white, and 17% Latino) of participants (Khemka et al., 2005). These investigators found that women with ID in the intervention group reported greater knowledge and produced more effective decision-making responses than the control group. This latter study extended beyond understanding opportunities for decision making by supporting the development of independent decision-making skills.

In sum, these studies demonstrate the concern parents have for safety that youth from diverse backgrounds may lack awareness and knowledge about safety issues and that interventions can increase knowledge and skills in these areas among youth with IDD from diverse backgrounds.

Decision Making About Education and Employment

This section includes articles that reported findings of decision making in secondary or post-secondary educational settings or related to employment. A study that analyzed data from the National Longitudinal Transition Study-2 (NLTS2) focused on youth with autism spectrum disorder (ASD) in high school and their participation in transition planning IEP meetings. The authors found that white students reported greater attendance and participation at IEP meetings than African Americans (Griffin et al., 2014). The authors highlighted the need to focus on the greater involvement of African American students in transition intervention research. A study of adults with disabilities aimed to refine an instrument of supported decision making and compared outcomes on the instrument across various factors, including employment (Shogren et al., 2020b). A large majority of the sample was African American (80%), and the remaining participants were Latino or white. Investigators found that participants who were engaged in competitive employment scored higher on decision-making experiences than those who were unemployed. Authors suggest that competitive employment may offer more decision-making opportunities to individuals with disabilities. A study that examined data from universities that hosted Transition and Postsecondary Programs for Students with Intellectual Disabilities (TPSID) programs found that Latino students with ID enrolled in higher education received lower scores on the self-regulation subscale of the Self-Determination Scale than white students with ID (Shogren et al., 2018b). The self-regulation subscale involved problem-solving questions that may be similar to decision making. A study examining community employment outcomes of youth with IDD who transitioned out of school interviewed employees of state-funded rehabilitation agencies about their clientele (Simonsen & Neubert, 2013). This study examined factors that were related to working in the community versus being in a shelter. The authors found that Latino adults with IDD were more likely to be working in the community versus in a sheltered workshop or unemployed than white adults with IDD. While the decision making of individuals with IDD was not examined in this study, living with their family and the family expressing a preference for community work was associated with working in the community.

To summarize this section, we found only three articles that focused on or mentioned decision making or self-determination by underrepresented groups related to education or employment and one article that examined employment outcomes, but not decision making. One article analyzed race and documented differences between African Americans and whites, two articles documented differences between Latinos and whites, and the other included a majority African American sample but did not examine issues specific to this population. These studies only give us a glimpse of issues related to race/ethnicity and education or employment that might be explored in future research.

Decision Making About Healthcare and Disability-Related Services

This section includes articles that focused on people with IDD or their families, making decisions about healthcare or disability-related services. While some studies involved youth and adults with disabilities and their families engaged in disability-related service systems (Larkin et al., 2018; Nasser et al., 2017; Rueda et al., 2005; Shogren, 2012), none of the studies investigated decision making of the youth or adult in whether to utilize these services. One of the studies involved parent decision making about placement in a disability-related residential setting (Nasser et al., 2017). Furthermore, none of the studies we reviewed focused on youth or adults with IDD or their families making decisions about healthcare. Clearly, research is needed on youth with IDD from diverse backgrounds with respect to decision making about healthcare and services.

Self-Determination

A few articles were not about decision making in any of the above categories but were about self-determination more broadly. As described earlier, decision making can be a component of self-determination. A study in Taiwan administered a quality-of-life scale to youth and adults 16 years of age and older, and this measure included items on self-determination, rights, and social inclusion, which have some relationship to decision making (Chou et al., 2007). Respondents indicated the importance of each item and whether they were in use. Respondents rated these items the lowest in importance and use, compared to other items. The use of self-determination was the lowest item out of all other items on this measure. A study that focused on self-determination among youth with either high incidence or cognitive disabilities in the NLTS2 found that African American students in the cognitive group reported greater autonomy and self-realization than white and Latino students in this group. They also reported higher psychological empowerment than their white counterparts (Shogren et al., 2014). Another study that examined personal characteristics of the Self-Determination Inventory among adolescents with and without disabilities found that among females with ID, African American and Latino females scored lower on the Self-Determination Scale than white females (Shogren et al., 2018a). Another finding related to race and ethnicity was that older adolescents with disabilities scored higher than younger ones on self-determination, but this was not true for African American adolescents. A different study by the same author reported the validity and reliability of the Self-Determination Inventory among youth ages 12–22 in secondary education settings (Shogren et al., 2017). These authors compared results by race and ethnicity but did not find any significant differences.

These studies examined either the validity or predictors of standardized measures, including self-determination and quality of life. The study in Taiwan was focused on quality of life with minimal information about decision making or self-determination. The other studies in this section examined self-determination and included race/ethnicity in their examination but lacked explanations for their findings regarding race/ethnicity. This may be due to the definitions being used with regard to self-determination as they may not reflect how self-determination is defined and valued by diverse populations.

Research Question 2

In our second research question, we ask whether the literature on decision making provides sufficient knowledge about racial/ethnic minority groups. We found only 15 studies that met our criteria of including some element related to decision making (self-determination, autonomy, independence) and that either included race/ethnicity in their analysis, had a substantial proportion of minorities in their sample, or were about a particular racial or ethnic group (including those from countries underrepresented in research). Table 10.2 describes each of the 15 studies and lists authors, year, topics addressed, sample size, description of participants, type of disability, and study design. Most of these studies were not specifically about decision making but included related elements. Nine of them focused on self-determination, and many of these examined predictors of a self-determination measure and included race and ethnicity among many other characteristics. Most studies were based in the US (n = 11). Five of these included racial and ethnic minority groups in their samples but were majority white, while four studies had diverse samples in which white respondents were not the majority. Two studies included participants from one ethnic group. We found four studies that were conducted in other countries; two were from Taiwan, one was from the UK and focused on minorities there, and one focused on Palestinians, a minority population in Israel. Although some studies addressed the consideration of preferences of the individual with IDD (e.g., Rueda et al., 2005), no studies specifically investigated shared decision making by culturally diverse individuals with IDD. Overall, our search indicates that (1) while there is an emerging body of research on self-determination with respect to race and ethnicity, this area needs to be fleshed out to understand better the contextual issues and potential explanations for findings; and (2) there is insufficient research about youth and adults with IDD from racially diverse populations or LMIC, with respect to decision making, the extent to which they make life decisions in each of the areas discussed, and their perceptions about their role in making important life decisions.

Discussion

From the scoping review of the current literature, we see that, across several domains, we uncovered elements of decision making that are important to racial/ethnically diverse individuals with IDD. Cultural differences and their respective values play a crucial role in determining what is important when it comes to decision making and what is considered appropriate for individuals with IDD. This includes the desired outcomes of the individual with IDD, along with families' preferences. The findings presented here shed light on the experience of decision making among culturally diverse individuals with IDD and their families. We review differences and similarities in our findings with the general literature and describe an adapted model of decision making that considers socio-ecological and cultural influences.

One of the more prominent differences observed between our results and the general literature on decision making in IDD was that of cultural differences in the defining characteristics of decision making and more broadly self-determination. Several studies cited in our review identified that the idea of adult children with IDD living away from the family home did not align with the cultural expectations for adult children generally. That is, families did not anticipate or expect their adult children to depart the home after completing schooling or reaching 18 years of age, regardless of diagnosis (Lin, 2006; Nasser et al., 2017; Rueda et al., 2005; Shogren, 2012). This aligns with other studies of cultural differences between collectivist and individualistic value systems (Leake & Boone, 2007; Zheng et al., 2015). Moreover, these cultural differences in expectations and definitions of what it means to be independent call into question whether transition planning adequately addresses the needs of culturally diverse students with disabilities. Reports from parents, our review, and other sources have consistently documented inequities in access to information related to transition planning for culturally diverse youth with disabilities (Sheehey, 2006). Furthermore, families have also reported that decision making and self-determination skills were taught in the home as schools were not adequately addressing these skills (Rueda et al., 2005; Shogren, 2012). Surveys of transition planning teams across five states reported low use of research-based practices for culturally and linguistically diverse students and families, with some school districts reporting implementation of the same practices for all students regardless of diverse status (Gothberg et al., 2019). Thus, the training and support available for transition-age youth with IDD often neglects the inclusion of families as a key ingredient for self-determination and decision making for culturally diverse youth. This absence disregards the value and significance of shared decision making and family-led decision making for youth with additional support needs, as demonstrated in our model (see Fig. 10.1). Although several papers have addressed this topic and included recommendations on steps to improve the transition-planning process for youth and families, our review suggests more work is needed.

This also raises concerns about assessment measures used to track transition outcomes that may not adequately capture the outcomes that are valued by individuals with IDD from diverse backgrounds. Although two studies included in our review focused on cultural diversity, the authors did not specifically consider cultural factors in the development of these items (Shogren et al., 2017, 2020b). This has significant implications for individuals with IDD who may depend on these metrics to qualify for services. Furthermore, when instruments are not culturally informed or normed across diverse populations, our knowledge and understanding of the constructs and the research that follows will inevitably be limited in its applicability and generalizability beyond the culture for which it was developed.

Despite these differences in decision making, there are some findings reported that were consistent with the general literature on decision making in IDD. Concerns about the safety of individuals with IDD have been cited by numerous studies (Lin, 2006; Rueda et al., 2005; Shogren, 2012), with many families identifying limited cognitive capacity and opportunities to learn and practice safety as the main reasons for concerns. Of the studies that focused on the family perspective on decision making, families reported a desire to honor their child's interests and needs, recognizing the need for their child with IDD to learn independent skills and develop practical work or employment skills (Rueda et al., 2005; Shogren, 2012). This finding aligns with other studies that indicate families are interested in having their adult children with IDD develop decision-making skills.

As presented earlier, we propose the Socio-Ecological Model of Decision making in Culturally Diverse Individuals with IDD as a way of understanding the complex cultural influences on decision making in culturally diverse individuals with IDD (see Fig. 10.1). The model begins by identifying the cultural factors within the various systems originally described as part of Bronfenbrenner's Ecological Systems Theory (i.e., microsystem, mesosystem, macrosystem). Inclusion of these systems provides a structure to frame the multiple cultural factors that influence individuals with IDD and their families, from opportunities for decision making in the home at the microsystem level to cultural values aligned with collectivism and interdependence at the macrosystem level. These factors interact with each other and continue throughout the individual's life. The model then describes the bidirectional influences between the individual with IDD and their family, recognizing that each may contribute to the other's opportunity and ability to make decisions. Based on this component, decision making may take one or more paths: individual-led decision making, family-led decision making, or shared decision making where both parties participate in the process. We conclude the model in recognizing that making decisions can involve multiple steps and that the individual with IDD and their families may play distinct roles across each step. This model incorporates our current understanding of decision making in culturally diverse individuals with IDD and may be updated as our knowledge and understanding increases with additional research and practice.

Limitations

There were several limitations to this scoping review chapter. First, a methodological concern is that the rigor of research design was not appraised (Arksey & O'Malley, 2005). Given the emergent nature of research in this topic, this review aimed to provide a map of current empirical evidence (Wilt & Morningstar, 2018). Second, we may have missed out on important historical contexts that affected the lifelong decision-making process for racial/ethnic minority individuals with IDD and their families by limiting our scope to peer-reviewed journal articles published 2000–2020. Finally, we recognize that the exclusion of non-English language studies likely led to an underrepresentation of research conducted in languages other than English. As the literature continues to burgeon in this field, we will be able to see a more holistic picture of research evidence in other cultures as well as countries across the world.

Future Directions

Implications for Research

We found the body of research on the decision making of youth and adults with IDD from diverse backgrounds to be lacking, suggesting the need for substantially more research going forward. Most of the studies we reviewed did not assess the extent to which these individuals were making decisions in important life areas. Some of the studies suggested that parents were involved in making key decisions, but only one study explored decision making among parents in depth. Studies that referred to decision making about housing and independent living had the most substantive information; however, most reflected the parent's perspective. Research is needed in which individuals with IDD from diverse backgrounds are interviewed about their living arrangements and preferences and what independent living means to them. Evaluating these preferences can inform providers and family members on how to best support their decision making to work toward these outcomes. This research must incorporate diverse cultural perspectives, such as interdependence, shared decision making, and the inclusion of family members, rather than focusing on the strict definitions of "independence" often seen in research conducted in the USA. In the area of safety, parents from diverse backgrounds expressed concern about keeping their son or daughter with IDD safe in the studies reviewed. This concern may hinder independent living and decision making of the youth or adult with IDD. As such, research on culturally tailored interventions that provide information about safety and that teach safety skills, needs to be developed for individuals with IDD and their parents. Studies are needed on decision making about education and employment among youth and adults with IDD from diverse backgrounds. How are diverse youth with IDD given opportunities to engage in their own educational processes, such as transition planning, and what are the interventions that can enhance this engagement? How do diverse families and adults with IDD make decisions about employment options and what employment opportunities are available to them? An emerging body of research is studying how individuals with disabilities engage in decision making about their own healthcare. We found virtually no research on this topic among racial and ethnic minorities with IDD. This is an area of research that is open for future study.

Implications for Practice

Research on decision making in individuals with IDD has consistently shown that an individual's ability to make decisions is related to positive life outcomes. However, as demonstrated in our review, the research on individuals with IDD from diverse backgrounds remains limited. In practice, it is important that individuals who work with and who support individuals with IDD and their families receive training and professional development on culturally informed practices. This may involve an initial review of cultural and linguistic groups being served within the school, organization, or agency to understand the scope of cultural competencies needed. Cultural brokers may help facilitate the connections between professionals and diverse individuals and families. Cultural brokers are individuals from the community who are trusted and respected; who share a common knowledge of values, beliefs, and cultural practices; and who have experience navigating between systems of care (Georgetown University National Center for Cultural Competence, n.d.). Other culturally sensitive practices may involve partnering with other organizations that the individual with IDD and their families interact with, ensuring that all information and materials are available and provided in the family's native language, and practicing flexibility when possible to engage culturally diverse families (Croke & Thompson, 2011; Ratto et al., 2017; Zamora et al., 2016).

As interventions and programs are being developed and disseminated to support self-determination and decision making in individuals with IDD (Burke et al., 2020; Luckner et al., 2020), this work must address the needs of culturally diverse individuals and their families. Several interventions and support programs have been developed or adapted specifically for diverse youth with disabilities and their families to enhance self-advocacy, family empowerment, and problem-solving skills (Hasnain et al., 2003; Kuhn et al., 2019). Programs that incorporate comprehensive family-centered approaches across educational, vocational, and healthcare settings will be an important next step in supporting individuals with IDD and their families.

Conclusions

In this chapter, we explored decision making, and more broadly self-determination, among racially/ethnically, culturally diverse individuals with IDD and their families. We reviewed the literature on decision making and provided a model through which to evaluate and consider cultural factors that might influence the decision-making process for culturally diverse individuals with IDD and their families. We conducted a scoping review to capture the current state of the research, finding that very few studies have been published that directly evaluate decision making in culturally diverse populations with IDD. Even when studies included diverse samples, comparisons of decision-making skills and their related outcomes were often not the focus. Given the emerging work in this area, there is substantial opportunity for growth in research and practice.

References

Adams, R. C., & Levy, S. E. (2017). Shared decision making and children with disabilities: Pathways to consensus. *Pediatrics, 139*(6), 1–9. https://doi.org/10.1542/peds.2017-0956

Agrawal, J., Barrio, B. L., Kressler, B., Hsiao, Y.-J., & Shankland, R. K. (2019). International policies, identification, and Services for Students with learning disabilities: An exploration across 10 countries. *Learning Disabilities: A Contemporary Journal, 17*(1), 95–113.

American Psychiatric Association. (2013). *Diagnostic and statistical manual of mental disorders* (5th ed.). American Psychiatric Association.

Arksey, H., & O'Malley, L. (2005). Scoping studies: Towards a methodological framework. *International Journal of Social Research Methodology, 8*(1), 19–32. https://doi.org/10.1080/1364557032000119616

Bronfenbrenner, U. (1977). Toward an experimental ecology of human development. *American Psychologist, 32*, 513–531.

Bronfenbrenner, U. (1986). Ecology of the family as a context for human development. Research perspectives. *Developmental Psychology, 22*(6), 723–742. https://doi.org/10.1037/0012-1649.22.6.723

Burke, K. M., Raley, S. K., Shogren, K. A., Hagiwara, M., Mumbardó-Adam, C., Uyanik, H., & Behrens, S. (2020). A meta-analysis of interventions to promote self-determination for students with disabilities. *Remedial and Special Education, 41*(3), 176–188. https://doi.org/10.1177/0741932518802274

Carter, E. W., Austin, D., & Trainor, A. A. (2012). Predictors of postschool employment outcomes for young adults with severe disabilities. *Journal of Disability Policy Studies, 23*(1), 50–63. https://doi.org/10.1177/1044207311414680

Cavendish, W. (2017). The role of gender, race/ethnicity, and disability status on the relationship between student perceptions of school and family support and self-determination. *Career Development and Transition for Exceptional Individuals, 40*(2), 113–122. https://doi.org/10.1177/2165143416629359

Census Bureau, U. S. (n.d.). *Quick facts*. Retrieved June 1, 2020, from https://www.census.gov/quickfacts/fact/table/US/RHI825218

Charles, C., Gafni, A., & Whelan, T. (1997). Shared decision making in the medical encounter: What does it mean? (Or it takes at least two to tango). *Social Science and Medicine, 44*(5), 681–692. https://doi.org/10.1016/S0277-9536(96)00221-3

Chou, Y. C., Schalock, R. L., Tzou, P. Y., Lin, L. C., Chang, A. L., Lee, W. P., & Chang, S. C. (2007). Quality of life of adults with intellectual disabilities who live with families in Taiwan. *Journal of Intellectual Disability Research, 51*(11), 875–883. https://doi.org/10.1111/j.1365-2788.2007.00958.x

Chou, Y.-C., Palmer, S. B., Wehmeyer, M. L., & Skorupski, W. P. (2017). Comparison of self-determination of students with disabilities: Multivariate and discriminant function analyses. *Journal of Intellectual Disability Research, 61*(2), 144–154. https://doi.org/10.1111/jir.12297

Croke, E. E., & Thompson, A. B. (2011). Person centered planning in a transition program for Bronx youth with disabilities. *Children and Youth Services Review, 33*(6), 810–819. https://doi.org/10.1016/j.childyouth.2010.11.025

Developmental Disabilities Assistance and Bill of Rights Act of 2000, Pub. L. No. 114 (2000). https://www.congress.gov/bill/106th-congress/senate-bill/1809

Doren, B., Gau, J. M., & Lindstrom, L. E. (2012). The relationship between parent expectations and postschool outcomes of adolescents with disabilities. *Exceptional Children, 79*(1), 7–23. https://doi.org/10.1177/001440291207900101

Duvdevany, I., Ben-Zur, H., & Ambar, A. (2002). Self-determination and mental retardation: Is there an association with living arrangement and lifestyle satisfaction? *Mental Retardation, 40*(5), 379. https://doi.org/10.1352/0047-6765(2002)0402.0.CO;2

Frankland, H. C., Turnbull, A. P., Wehmeyer, M. L., & Blackmountain, L. (2004). An exploration of the self-determination construct and disability as it relates to the dine (Navajo) culture. *Education and Training in Developmental Disabilities, 39*(3), 191–205.

Garrels, V., & Granlund, M. (2018). Measuring self-determination in Norwegian students: Adaptation and validation of the AIR self-determination scale. *European Journal of Special Needs Education, 33*(4), 466–480. https://doi.org/10.1080/08856257.2017.1342420

Geenen, S., Powers, L. E., & Lopez-Vasquez, A. (2005). Barriers against and strategies for promoting the involvement of culturally diverse parents in school-based transition planning. *Journal for Vocational Special Needs Education, 27*(3), 4–14.

Georgetown University National Center for Cultural Competence. (n.d.). *What Is the role of cultural brokers in health care delivery?* Bridging the Cultural Divide in Health Care Settings: The Essential Role of Cultural Broker Programs. Retrieved November 1, 2019, from https://nccc.georgetown.edu/culturalbroker/2_role/index.html

Gothberg, J. E., Greene, G., & Kohler, P. D. (2019). District implementation of research-based practices for transition planning with culturally and linguistically diverse youth with disabilities and their families. *Career Development and Transition for Exceptional Individuals, 42*(2), 77–86. https://doi.org/10.1177/2165143418762794

Greenfield, P. M., & Cocking, R. R. (Eds.). (2014). *Cross-cultural roots of minority child development* (1st ed.). Psychology Press.

Griffin, M. M., Taylor, J. L., Urbano, R. C., & Hodapp, R. M. (2014). Involvement in transition planning meetings among high school students with autism spectrum disorders. *Journal of Special Education, 47*(4), 256–264. https://doi.org/10.1177/0022466913475668

Hagiwara, M., Shogren, K. A., Lane, K. L., Raley, S. K., & Smith, S. A. (2020). Development of the self-determined learning model of instruction coaching model: Implications for research and practice. *Education and Training in Autism and Developmental Disabilities, 55*(1), 17–27.

Hasnain, R., Sotnik, P., & Ghiloni, C. (2003). Person-centered planning: A gateway to improving vocational rehabilitation services for culturally diverse individuals with disabilities. *Journal of Rehabilitation, 69*(3), 10.

Hickson, L., & Khemka, I. (1999). Decision making and mental retardation. *International Review of Research in Mental Retardation, 22*(C), 227–265. https://doi.org/10.1016/S0074-7750(08)60135-4

Hickson, L., & Khemka, I. (2013). Problem solving and decision making. In *The Oxford handbook of positive psychology and disability* (pp. 198–225). Oxford University Press.

Individuals with Disabilities Education Act, Pub. L. No. 20 U.S.C. § 1400 (2004).

Khemka, I., Hickson, L., & Reynolds, G. (2005). Evaluation of a decision-making curriculum designed to empower women with mental retardation to resist abuse. *American Journal on Mental Retardation, 110*(3), 193–204. https://doi.org/10.1352/0895-8017(2005)110<193:EOADCD>2.0.CO;2

Kim, K.-H., & Morningstar, M. E. (2005). Transition planning involving culturally and linguistically diverse families. *Career Development for Exceptional Individuals, 28*(2), 92–103. https://doi.org/10.1177/08857288050280020601

Kostikj-Ivanovikj, V., & Chichevska-Jovanova, N. (2016). Relation between quality of life, choice making, and future expectations in adults with intellectual disability. *Journal of Special Education & Rehabilitation, 17*(3/4), 29–45. http://10.0.74.113/jser.2016.8

Kuhn, J. L., Vanegas, S. B., Salgado, R., Borjas, S. K., Magaña, S., & Smith DaWalt, L. (2019). The cultural adaptation of a transition program for Latino families of youth with autism Spectrum disorder. *Family Process*. https://doi.org/10.1111/famp.12439

Larkin, M., Unwin, G., Iyer, M., Tsimopoulou, I., Zahid, S., Malik, K., Stenfert Kroese, B., & Rose, J. L. (2018). Cultural affordance, social relationships, and narratives of independence: Understanding the meaning of social care for adults with intellectual disabilities from minority ethnic groups in the UK. *International Journal of Developmental Disabilities, 64*(3), 195–203. https://doi.org/10.1080/20473869.2018.1469807

Leake, D., & Boone, R. (2007). Multicultural perspectives on self-determination from youth, parent, and teacher focus groups. *Career Development for Exceptional Individuals, 30*(2), 104–115. https://doi.org/10.1177/08857288070300020101

Lee, Y., Wehmeyer, M. L., Palmer, S. B., Williams-Diehm, K., Davies, D. K., & Stock, S. E. (2012). Examining individual and instruction-related predictors of the self-determination of students with disabilities: Multiple regression analyses. *Remedial and Special Education, 33*(3), 150–161. https://doi.org/10.1177/0741932510392053

Levac, D., Colquhoun, H., & O'Brien, K. K. (2010). Scoping studies: Advancing the methodology. *Implementation Science, 5*(1). https://doi.org/10.1186/1748-5908-5-69

Lin, H.-C. (2006). The self-determination for youth with intellectual disabilities in Taiwan. *The International Journal of Diversity in Organizations, Communities, and Nations: Annual Review, 5*(1), 1–16. https://doi.org/10.18848/1447-9532/CGP/v05i01/38857

Linton, K. (2015). Differential ratings of specific behaviors of African Americans children in special education. *Child and Adolescent Social Work Journal, 32*(3), 229–235. https://doi.org/10.1007/s10560-014-0363-3

Luckner, J. L., Banerjee, R., Movahedazarhouligh, S., & Millen, K. (2020). A systematic review of replicative self-determination intervention studies. *Journal of Special Education, 54*(1), 29–39. https://doi.org/10.1177/0022466919850188

Magaña, S., & Smith, M. J. (2006). Psychological distress and Well-being of Latina and non-Latina white mothers of youth and adults with an autism spectrum disorder: Cultural attitudes towards coresidence status. *American Journal of Orthopsychiatry, 76*(3), 346–357. https://doi.org/10.1037/0002-9432.76.3.346

McKenzie, K., Milton, M., Smith, G., & Ouellette-Kuntz, H. (2016). Systematic review of the prevalence and incidence of intellectual disabilities: Current trends and issues. *Current Developmental Disorders Reports, 3*(2), 104–115. https://doi.org/10.1007/s40474-016-0085-7

Moher, D., Liberati, A., Tetzlaff, J., & Altman, D. G. (2009). Preferred reporting items for systematic reviews and meta-analyses: The PRISMA statement. *BMJ: British Medical Journal Publishing Group, 339*, 7716. https://doi.org/10.1136/bmj.b2535

Moher, D., Shamseer, L., Clarke, M., Ghersi, D., Liberati, A., Petticrew, M., Shekelle, P., Stewart, L. A., Group, P.-P., Altman, D. G., Booth, A., Chan, A.-W., Chang, S., Clifford, T., Dickersin, K., Egger, M., Gøtzsche, P. C., Grimshaw, J. M., Groves, T., … Whitlock, E. (2015). Preferred reporting items for systematic review and meta-analysis protocols (PRISMA-P) 2015 statement. *Systematic Reviews, 4*(1). https://doi.org/10.1186/2046-4053-4-1

Mumbardó-Adam, C., Guárdia-Olmos, J., Adam-Alcocer, A. L., Carbó-Carreté, M., Balcells-Balcells, A., Giné, C., & Shogren, K. A. (2017). Self-determination, intellectual disability, and context: A meta-analytic study. *Intellectual and Developmental Disabilities, 55*(5), 303–314. https://doi.org/10.1352/1934-9556-55.5.303

Nasser, K., Sachs, D., & Sa'ar, A. (2017). A necessary evil: Residential placement of people with intellectual disability among the Palestinian minority in Israel. *Research in Developmental Disabilities, 60*, 115–124. https://doi.org/10.1016/j.ridd.2016.11.018

Ouzzani, M., Hammady, H., Fedorowicz, Z., & Elmagarmid, A. (2016). Rayyan–a web and mobile app for systematic reviews. (Report). *Systematic Reviews, 5*(1). https://doi.org/10.1186/s13643-016-0384-4

Park, J., Turnbull, A. P., & Park, H.-S. (2001). Quality of partnerships in service provision for Korean American parents of children with disabilities: A qualitative inquiry. *Research and Practice for Persons with Severe Disabilities, 26*(3), 158–170. https://doi.org/10.2511/rpsd.26.3.158

Parker, J. S., Amabile, A., Oliver, E., Garnes, J., & Sarathy, A. (2020, February). Facilitators and barriers to African American high school students' self-determination skill expression. In *Psychology in the schools* (pp. 1–20). https://doi.org/10.1002/pits.22368.

Perez, G. K., & Cruess, D. (2014). The impact of familism on physical and mental health among Hispanics in the United States. *Health Psychology Review, 8*(1), 95–127. https://doi.org/10.1080/17437199.2011.569936

Powers, K., Geenen, S., & Powers, L. E. (2009). Similarities and differences in the transition expectations of youth and parents. *Career Development for Exceptional Individuals, 32*(3), 132–144. https://doi.org/10.1177/0885728809346302

Raley, S. K., Shogren, K. A., & McDonald, A. (2018). Whole-class implementation of the self-determined learning model of instruction in inclusive high school mathematics classes. *Inclusion, 6*, 164–174.

Ratto, A. B., Anthony, B. J., Pugliese, C., Mendez, R., Safer-Lichtenstein, J., Dudley, K. M., Kahn, N. F., Kenworthy, L., Biel, M., Martucci, J. L., & Anthony, L. G. (2017). Lessons learned: Engaging culturally diverse families in neurodevelopmental disorders intervention research. *Autism, 21*(5), 622–634. https://doi.org/10.1177/1362361316650394

Rossetti, Z., Redash, A., Sauer, J. S., Bui, O., Wen, Y., & Regensburger, D. (2018). Access, accountability, and advocacy: Culturally and linguistically diverse families' participation in IEP meetings. *Exceptionality*, 1–16. https://doi.org/10.1080/09362835.2018.1480948

Rueda, R., Monzo, L., Shapiro, J., Gomez, J., & Blacher, J. (2005). Cultural models of transition: Latina mothers of young adults with developmental disabilities. *Exceptional Children, 71*(4), 401–414. https://doi.org/10.1177/001440290507100402

Ryan, R. M., & Deci, E. L. (2004). Autonomy is no illusion: Self-determination theory and the empirical study of authenticity, awareness, and will. In *Handbook of experimental existential psychology* (pp. 449–479). Guilford Press.

Sabogal, F., Marín, G., Otero-Sabogal, R., Marín, B. V., & Perez-Stable, E. J. (1987). Hispanic Familism and acculturation: What changes and what Doesn't? *Hispanic Journal of Behavioral Sciences, 9*(4), 397–412. https://doi.org/10.1177/07399863870094003

Schwartz, S. J., Weisskirch, R. S., Hurley, E. A., Zamboanga, B. L., Park, I. J. K., Kim, S. Y., Umaña-Taylor, A., Castillo, L. G., Brown, E., & Greene, A. D. (2010). Communalism, familism, and filial piety: Are they birds of a collectivist feather? *Cultural Diversity & Ethnic Minority Psychology, 16*(4), 548–560. https://doi.org/10.1037/a0021370

Sheehey, P. (2006). Parent involvement in educational decision making: A Hawaiian perspective. *Rural Special Education Quarterly, 25*(4), 3–15.

Shogren, K. A. (2011). Culture and self-determination: A synthesis of the literature and directions for future research and practice. *Career Development for Exceptional Individuals, 34*(2), 115–127. https://doi.org/10.1177/0885728811398271

Shogren, K. A. (2012). Hispanic mothers' perceptions of self-determination. *Research and Practice for Persons with Severe Disabilities, 37*(3), 170–184.

Shogren, K. A. (2013). A social-ecological analysis of the self-determination literature. *Intellectual and Developmental Disabilities, 51*(6), 496–511. https://doi.org/10.1352/1934-9556-51.6.496

Shogren, K. A., & Shaw, L. A. (2016). The role of autonomy, self-realization, and psychological empowerment in predicting outcomes for youth with disabilities. *Remedial and Special Education, 37*(1), 55–62. https://doi.org/10.1177/0741932515585003

Shogren, K. A., & Shaw, L. A. (2017). The impact of personal factors on self-determination and early adulthood outcome constructs in youth with disabilities. *Journal of Disability Policy Studies, 27*(4), 223–233. https://doi.org/10.1177/1044207316667732

Shogren, K. A., & Wehmeyer, M. L. (2017). Culture and self-determination. In *Development of self-determination through the life-course* (pp. 159–168). https://doi.org/10.1007/978-94-024-1042-6_12

Shogren, K. A., Wehmeyer, M. L., Palmer, S. B., Soukup, J. H., Little, T. D., Garner, N., & Lawrence, M. (2007). Examining individual and ecological predictors of the self-determination of students with disabilities. *Exceptional Children, 73*(4), 488–509.

Shogren, K. A., Wehmeyer, M. L., Palmer, S. B., Soukup, J. H., Little, T. D., Garner, N., & Lawrence, M. (2008). Understanding the construct of self-determination: Examining the relationship between the Arc's Self-Determination Scale and the American Institutes for Research Self-Determination Scale. *Assessment for Effective Intervention, 33*(2), 94–107. https://doi.org/10.1177/1534508407311395

Shogren, K. A., Forber-Pratt, A. J., Nittrouer, C., & Aragon, S. R. (2013a). The emergence of a human services cooperative to support families and young adults with disabilities: Implications for disability services and supports. *Research and Practice for Persons with Severe Disabilities, 38*(4), 259–273. https://doi.org/10.1177/154079691303800405

Shogren, K. A., Wehmeyer, M. L., Palmer, S. B., & Paek, Y. (2013b). Exploring personal and school environment characteristics that predict self-determination. *Exceptionality, 21*(3), 147–157. https://doi.org/10.1080/09362835.2013.802231

Shogren, K. A., Kennedy, W., Dowsett, C., Garnier Villarreal, M., & Little, T. D. (2014). Exploring essential characteristics of self-determination for diverse students using data from NLTS2. *Career Development and Transition for Exceptional Individuals, 37*(3), 168–176.

Shogren, K. A., Wehmeyer, M. L., Palmer, S. B., Forber-Pratt, A. J., Little, T. J., & Lopez, S. (2015). Causal agency theory: Reconceptualizing a functional model of self-determination. *Education and Training in Autism and Developmental Disabilities, 50*(3), 251–263.

Shogren, K. A., Wehmeyer, M. L., Little, T. D., Forber-Pratt, A. J., Palmer, S. B., & Seo, H. (2017). Preliminary validity and reliability of scores on the self-determination inventory: Student report version. *Career Development and Transition for Exceptional Individuals, 40*(2), 92–103. https://doi.org/10.1177/2165143415594335

Shogren, K. A., Shaw, L. A., Raley, S. K., & Wehmeyer, M. L. (2018a). The impact of personal characteristics on scores on the self-determination inventory: Student report in adolescents with and without disabilities. *Psychology in the Schools, 55*(9), 1013–1026. https://doi.org/10.1002/pits.22174

Shogren, K. A., Wehmeyer, M. L., Shaw, L. A., Khamsi, S., Grigal, M., Hart, D., & Smith, F. A. (2018b). Predictors of self-determination in postsecondary education for students with intellectual and developmental disabilities. *Education and Training in Autism and Developmental Disabilities, 53*(2), 146–159.

Shogren, K. A., Little, T. D., Grandfield, E., Raley, S., Wehmeyer, M. L., Lang, K. M., & Shaw, L. A. (2020a). The self-determination inventory–student report: Confirming the factor structure of a new measure. *Assessment for Effective Intervention, 45*(2), 110–120. https://doi.org/10.1177/1534508418788168

Shogren, K. A., Rifenbark, G. G., Wehmeyer, M. L., Dean, E. E., Killeen, M. B., & Karsevar, J. (2020b). Refining the supported decision making inventory. *Journal of Policy and Practice in Intellectual Disabilities*. https://doi.org/10.1111/jppi.12335

Short, E. J., & Evans, S. W. (1990). Individual differences in cognitive and social problem-solving skills as a function of intelligence. *International Review of Research in Mental Retardation, 16*(C), 89–123. https://doi.org/10.1016/S0074-7750(08)60093-2

Simonsen, M. L., & Neubert, D. A. (2013). Transitioning youth with intellectual and other developmental disabilities: Predicting community employment outcomes. *Career Development and Transition for Exceptional Individuals, 36*(3), 188–198. https://doi.org/10.1177/2165143412469399

Small, N., Raghavan, R., & Pawson, N. (2013). An ecological approach to seeking and utilising the views of young people with intellectual disabilities in transition planning. *Journal of Intellectual Disabilities, 17*(4), 283–300. https://doi.org/10.1177/1744629513500779

Taylor, W. D., Cobigo, V., & Ouellette-Kuntz, H. (2019). A family systems perspective on supporting self-determination in young adults with intellectual and developmental disabilities. *Journal of Applied Research in Intellectual Disabilities, 32*(5), 1116–1128. https://doi.org/10.1111/jar.12601

Tricco, A. C., Lillie, E., Zarin, W., O'Brien, K. K., Colquhoun, H., Levac, D., Moher, D., Peters, M. D. J., Horsley, T., Weeks, L., Hempel, S., Akl, E. A., Chang, C., McGowan, J., Stewart, L., Hartling, L., Aldcroft, A., Wilson, M. G., Garritty, C., … Straus, S. E. (2018). PRISMA extension for scoping reviews (PRISMA-ScR): Checklist and explanation. *Annals of Internal Medicine, 169*(7), 467–473. https://doi.org/10.7326/M18-0850

Wehmeyer, M. L. (1993). Perceptual and psychological factors in career decision making of adolescents with and without cognitive disabilities. *Career Development for Exceptional Individuals, 16*(2), 135–146. https://doi.org/10.1177/088572889301600203

Wehmeyer, M. L. (1995). The use of assistive technology by people with mental retardation and barriers to this outcome: A pilot study. *Technology and Disability, 4*(3–4), 195–204. https://doi.org/10.3233/TAD-1995-43-403

Wehmeyer, M. L., & Kelchner, K. (1995). The Arc's Self-Determination Scale. The Arc of the United States.

Wehmeyer, M. L. (1996). Self-determination as an educational outcome: Why is it important to children, youth and adults with disabilities? In D. J. Sands & M. L. Wehmeyer (Eds.), *Self-determination across the life span: Independence and choice for people with disabilities* (pp. 17–36). Brookes Publishing.

Wehmeyer, M. L., Shogren, K. A., Zager, D., Smith, T. E. C., & Simpson, R. (2010). Research-based principles and practices for educating students with autism: Self-determination and social interactions. *Education and Training in Autism and Developmental Disabilities, 45*(4), 475–486.

Wehmeyer, M. L., Abery, B. H., Zhang, D., Ward, K., Willis, D., Hossain, W. A., Balcazar, F., Ball, A., Bacon, A., Calkins, C., Heller, T., Goode, T., Dias, R., Jesien, G. S., McVeigh, T., Nygren, M. A., Palmer, S. B., & Walker, H. M. (2011). Personal self-determination and moderating variables that impact efforts to promote self-determination. *Exceptionality, 19*(1), 19–30. https://doi.org/10.1080/09362835.2011.537225

Wilt, C. L., & Morningstar, M. E. (2018). Parent engagement in the transition from school to adult life through culturally sustaining practices: A scoping review. *Intellectual and Developmental Disabilities, 56*(5), 307–320. https://doi.org/10.1352/1934-9556-56.5.307

World Health Organization. (2018). *International classification of diseases for mortality and morbidity statistics*. (11th Revis). World Health Organization.

Zamora, I., Williams, M. E., Higareda, M., Wheeler, B. Y., & Levitt, P. (2016). Brief report: Recruitment and retention of minority children for autism research. *Journal of Autism and Developmental Disorders, 46*(2), 698–703. https://doi.org/10.1007/s10803-015-2603-6

Zheng, Y., Maude, S. P., Brotherson, M. J., Summers, J. A., Palmer, S. B., & Erwin, E. J. (2015). Foundations for self-determination perceived and promoted by families of young children with disabilities in China. *Education and Training in Autism and Developmental Disabilities, 50*(1), 109–122.

Part III
Profiles of Decision Making

Chapter 11
Understanding Decision Making and Intellectual and Developmental Disabilities

Ishita Khemka

Introduction

Decision making plays a central role in most aspects of life. However, it is apparent that decision making is an area of difficulty for many individuals with intellectual and developmental disabilities (IDD). The chapters in this volume address these shortcomings in a variety of ways – by identifying risk factors, by advocating and providing supports for expanded decision-making opportunities, and by examining the variables that contribute to decision-making effectiveness. Based on the literature, we (Khemka and Hickson) know that individual differences in decision making might relate to differences in capacities to carry out decision tasks or to differences in decision styles – the characteristic ways in which individuals with IDD approach decision making. Our own overriding commitment to the study of decision making over the past 25 years has been in the interest of deepening our understanding of decision making by individuals with IDD as a basis for developing interventions to strengthen their decision-making capabilities.

In particular, the primary focus of our research has been upon studying the decision making of adolescents and adults with IDD in interpersonal situations which involve a risk of coercion or abuse (see Chapter 18 in this volume). Interpersonal decision making is distinct from decision making in other domains (e.g., vocational or lifestyle choices) in that the decision-making process in interpersonal situations is essentially a non-iterative process with minimal scope for correction within the course of decision making. Decision making in an interpersonal situation can be influenced by a gamut of social relationship factors (e.g., peer dynamics, relational aggression, etc.), all of which can pose challenges for individuals with IDD,

I. Khemka (✉)
St. John's University, Queens, NY, USA
e-mail: khemkai@stjohns.edu

I. Khemka, L. Hickson (eds.), *Decision Making by Individuals with Intellectual and Developmental Disabilities*, Positive Psychology and Disability Series,
https://doi.org/10.1007/978-3-030-74675-9_11

especially during adolescence. However, in this chapter, we review relevant research on decision making and related factors for adolescents and adults with IDD across a wide range of decision tasks. We believe that findings from studies using controlled laboratory-based and probability reasoning tasks can inform our understanding of the decision-making processes and challenges faced by individuals with IDD in a range of interpersonal and real-life situations. With the primary purpose of examining the underlying component processes involved in different types of decisions and identifying individual factors that may influence how a person with IDD will approach a particular decision-making situation, we have, in this chapter, focused our attention primarily on research studies that shape our current views and understanding of decision making in individuals with IDD and those that point toward future research.

The Pathways Model of Decision Processing

Since we are most interested in understanding why different people may approach the same decision task in different ways, we have proposed an updated version of our *Pathways Model of Decision Processing* (referred to as *Pathways Model)* (presented in Chapter 6 in this volume) which offers a framework for examining individual differences in decision making across the broad range of individuals with IDD. In the current chapter, we examine the decision-making capabilities and information processing styles of individuals with IDD. Through consideration of available research findings, we evaluate mechanisms that might underlie the decision-making patterns and difficulties of adolescents and adults with various types of IDD. We then assess the applicability of different pathways of decision processing that can be considered when developing training approaches to strengthen their decision making.

In the present chapter, we start by examining the components at the base of the *Pathways Model* which include (1) situational/environmental factors, (2) neurodevelopmental/personal factors, and (3) the three basic processes of cognition, motivation, and emotion. Later in the chapter, we shift our focus to the upper part of the *Pathways Model* to consider how individuals with different types of IDD might be expected, considering the extant research, to navigate the pathways of decision processing specified in the model. Figure 11.1 highlights only the base components of the *Pathways Model* to allow us to examine the possible roles of these components for adolescents and adults with various forms of IDD. The upper part of the model containing the pathways is shaded out.

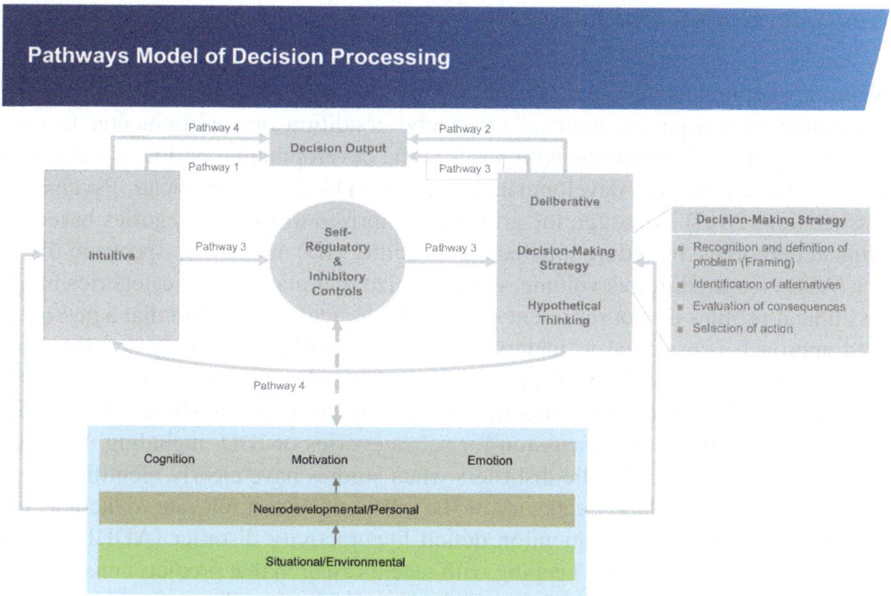

Fig. 11.1 Base components of the *Pathways Model of Decision Processing*

Impact of Situational/Environmental Factors

A critical consideration in decision making is the complexity of the situation or task itself. This is represented in the bottom layer of the base section of the model as shown in Fig. 11.1. Most of the decisions we make in everyday life take place in the context of our environmental surroundings, taking into consideration a multitude of factors. Consideration of the impact of each of these variables on the decision making of individuals with IDD is clearly warranted. Situational factors like time pressure can pose additional challenges. The situational/environmental context impacts decision processing in ways that give rise to different types of decisions.

Impact of Neurodevelopmental/Personal Factors

We include an emphasis on neurodevelopmental and personal factors in the *Pathways Model* (see middle bar of the base components of the *Pathways Model* in Fig. 11.1) to recognize the origins of the deficits that might produce impairments in areas of personal, social, academic, or occupational functioning for individuals with IDD and likely impact their decision-making performance.

Neurodevelopmental Factors

Neurodevelopmental disorders (see the second level of the base section of Fig. 11.1), associated with a known medical or genetic condition or environmental factors, include a group of conditions with onset in the developmental period or those that can be traced to the developmental period (DSM-5) (American Psychiatric Association, 2013). Although the group is comprised of many categories based on clinical outcomes, including intellectual disability (ID) and autism spectrum disorder (ASD) featured in this volume, we recognize that many of the categories based on clinical profiles do not represent singular biological entities and that a personalized approach is essential to understanding the underlying etiology and levels of functioning (Mitchell, 2015). Within this broader umbrella of neurodevelopmental disorders, we look at the IDD classification as one group due to shared characteristics. We differentiate our discussion for subcategories of IDD, including ASD and ID, as well as specific genetic disorders when studies have clearly identified these subgroups within the larger category of IDD. We consider it relevant to include and examine characteristics of attention deficit hyperactivity disorder (ADHD) along with those of ID and ASD in keeping with recent research that predicts considerable overlap in the characteristics of these disorders due to shared genetic and environmental risk factors and that suggests an etiological and neurodevelopmental continuum of functioning (see Morris-Rosendahl & Crocq, 2020). From this neurodevelopmental perspective, we present examples from studies to illustrate how individual neurodevelopmental and personal characteristics can yield distinct patterns of performance and inform our study of decision-making processing.

Autism Spectrum Disorder

Individuals with ASD tend to represent a diverse set of phenotype characteristics relating to neurodevelopmental conditions which manifest in social communication and behavioral difficulties (Tager-Flusberg, 2007). Most decision-making studies with individuals with ASD have involved participants with IQs that fall in the average or above average range (referred to as high functioning autism) (e.g., Bernard-Opitz et al., 2001; Hillier et al., 2007). While overall intellectual functioning in individuals with ASD can fall in the average to superior range, a sizeable number of individuals with ASD (approximately 50–70%) are also diagnosed with an ID (Matson & Shoemaker, 2009). Since ASD frequently co-occurs in individuals with ID (ranging from 28–40%) (see Bryson et al., 2008), it is important to consider how this comorbidity may impact the interpersonal decision-making experiences of those with ID and ASD. The extremely variable cognitive and behavioral functioning influences how the core *social* diagnostic symptoms get manifested for this group. Therefore, assessing the range of functioning within the ASD group is important in order to understand whether certain decision-making patterns are related to ASD-specific symptoms or interrupted by cognitive deficiencies related to co-occurring ID or other developmental disorder, such as ADHD. Research is also focused on studying whether certain characteristics specific to ASD, such as a tendency to focus more on details and less on the bigger picture on perceptual tasks (requiring low-level cognitive processing), might extend to a reduced sensitivity to contextual information on decision-making tasks (requiring higher-level cognitive processing) (see Farmer et al., 2017).

Intellectual Disability and Singular Disorders

Over the past two decades, research on the implications of specific genotypes and phenotypes of singular disorders associated with ID has evolved. However, studies that have focused on the impact of these specific disorders on decision-making performance are limited. A look at key defining characteristics of the disorders within ID can help to explain and predict specific decision-making behaviors of individuals in some of these subgroups. In general, however, having an ID, regardless of the cause, presents challenges for higher-order thinking and cognitive processing during decision making.

Williams syndrome (WS) is an example of a genetically determined singular disorder associated with ID. Individuals with WS show intellectual functioning in the borderline to moderate ID range with largely socio-communicative language difficulties (e.g., low conceptual/relational language and pragmatic skills). In comparison to other forms of IDD, individuals with WS exhibit a phenotype profile of marked social-emotional and behavioral predispositions toward an overfriendly social personality. Marked with many fears and anxieties relating to social interactions, they tend to face substantial problems in social adjustment and in forming and sustaining friendships (Dykens, 2003). A more detailed discussion of decision making by individuals with singular disorders and ID is provided in the chapter by *Goscicki* et al. in this volume.

Attention Deficit Hyperactivity Disorder (ADHD)

ADHD, characterized by a persistent pattern of inattention and/or hyperactivity and impulsivity that interferes with functioning and/or development, is a frequently co-occurring neurodevelopmental condition in individuals with IDD or ASD. With inattention and hyperactivity/impulsivity as its main symptom domains, ADHD has also been related to a broad range of poor health outcomes, such as substance abuse, gambling, sexual risk-taking, etc. underscored by decision-making deficits (Nigg, 2013). These real-life decision-making deficits have received much attention because of the potential for negative impact, both at an individual and societal level. Decision-making research specific to individuals with ADHD is discussed in detail in the chapter by *Lyon* et al. in this volume.

Personal Factors

In examining the relationship between IDD and decision making, researchers have identified several personal characteristics that appear to play a key role in the independent decision-making capabilities of individuals with IDD (see the second level of the base section of the *Pathways Model* in Fig. 11.1). These include gender, age, level of social participation, communication, and language functioning.

Gender

In the Hickson et al. (1998) study, we compared the decision-making performance of women and men and found that women with and without ID were generally more likely than men to recommend that the protagonist takes self-protective action in situations of abuse. Women were also more likely than men to state possible negative consequences of failing to take self-protective action. It was suggested that women may have been more sensitive to the risks inherent in the situations, possibly enhancing the emotional significance of the task and increasing motivation to take self-protective action. Very few studies have looked at the impact of gender differences in decision making. We believe that this is an important variable, which we have begun to explore (see Khemka & Hickson, 2017a) and include it here in the hope of stimulating further research.

Age

Researchers have noted a profound impact of age on decision-making performance. In particular, the changes that happen at adolescence are dramatic. Age effects on decision making can be understood in terms of how certain important underlying factors evolve or mature with age, thereby changing their influences on decision making over time. According to Poon (2018), the bell-shaped development of *hot* executive function suggests a period of heightened risk-taking propensity in middle adolescence.

Social Participation

Interpersonal decision making of individuals with IDD, especially during the adolescent years, is likely influenced by a host of factors (e.g., participation in social activities, friendships) in the social domain that may contribute to overall knowledge and understanding of social interactions. Research has shown that the friendships of youth with ASD or ID are characterized differently than those of typically developing (TD) adolescents. Individuals with ASD have been found to have significantly fewer reciprocal relationships than their TD peers, due to the impairments in social communication and restricted interests (Howlin et al., 2004; Orsmond et al., 2004). Similar to youth with ASD, adolescents with ID tend to experience reduced participation in social activities and have fewer and less sophisticated friendships (McVilly et al., 2006; Solish et al., 2010). With limited conflict resolution skills and social competence, the social decision-making experience of these groups tends to be impacted (Larkin et al., 2011). These social deficits also exacerbate feelings of loneliness as well as internalizing problems like depression and anxiety for adolescents with IDD and render them more susceptible to bullying and victimization. Although the development of social skills in WS appears to be delayed compared to same-aged peer norms (see Mervis et al., 2001), comparisons with individuals with other singular disorders associated with ID indicate that the social competence of those with WS might be stronger than in those with Prader-Willi syndrome but poorer when compared to those with Down syndrome (Rosner et al., 2004). Additional empirical research to investigate the relationship of social deficits and decision-making difficulties in interpersonal situations involving friendships, peer relationships, and victimization among individuals with IDD, and by type of IDD, is warranted.

Communication and Language Functioning

Communication difficulties in individuals with ID tend to be wide ranging and can impact diverse areas of receptive, expression, and pragmatic language functioning. They are influenced by complex factors, both individual and environmental. Communication limitations can exclude individuals with IDD from meaningful decision-making experiences and participation in their community. A focus on communicative competence allows for assessment of skills needed to express desires, ideas, choices, etc. and prevents individuals from being excluded in decision making (Brady et al., 2016; Garcia et al., 2020). Smith et al. (2020) measured communication skills of 601 adults (40+ years) with ID using data from a national cross-sectional study in Ireland and found that 57.9% of participants experienced some communication difficulties and more than half found communicating with professionals and non-familiar partners difficult. Several key factors were significantly associated with communication difficulties: level of ID, low social participation, challenging behaviors, and a diagnosis of Down syndrome.

Roles of the Three Basic Processes

It has been long recognized that decision making involves the interplay of cognitive, motivational, and emotional processes (highlighted in the top layer of the base section of the *Pathways Model* shown in Fig. 11.1) and that while each of the three processes performs unique functions in decision making, these processes do not operate in isolation but rather interact with each other in complex ways to produce decision outcomes (Kuhl, 1986). Recently, increasingly specific information has been emerging about the neural substrates of these processes because of intensified interest in decision making in fields like neuropsychology and neurodevelopment. In our research on interpersonal decision making in individuals with IDD, we have examined the inputs of cognitive, motivational, and emotional processes in different types of interpersonal decisions (see Hickson & Khemka, 2014; Khemka & Hickson, 2017b). We understand that these processes play a central role in defining the various pathways to decision making that individuals can adopt. Acknowledging that these three basic processes are essential to any discussion of decision making allows us to focus on studies aimed at illuminating their relative contributions in decision-making situations involving adolescents and adults with IDD. It should be noted, however, that the number of relevant studies varies widely for various types of IDD.

Cognitive Considerations in Decision Making

Much of the research on cognitive considerations in decision making and IDD has focused on individuals, mostly adults, with ASD. These studies are summarized first followed by a discussion of studies involving individuals with ID and related disorders.

Autism Spectrum Disorder

Individuals with ASD report difficulties in making routine decisions (e.g., Grandin, 2000; Johnson et al., 2006). There is also evidence to suggest that individuals with ASD can make more rational decisions compared to individuals with typical development. Therefore, identifying the conditions under which individuals with ASD tend to make rational decisions and exploring the cognitive mechanisms that underlie their decision making are critical to gaining a full picture of their decision-making strengths and weaknesses. A few consistently observed shortcomings relate to when decisions are to be made quickly or when they involve a change of routine (e.g., Luke et al., 2012). For example, Levin et al. (2015) compared a group of high-functioning, college-aged individuals with ASD with a same-aged group without ASD on a battery of decision tasks. They found that the groups differed significantly in decision style. Individuals in the ASD group were less inclined to use, and less successful at using, intuitive impressions to make decisions than were individuals in the non-ASD group. The groups did not differ in their ability to use rational/deliberative thinking in making decisions.

Framing

Framing effects, initially introduced by Tversky and Kahneman (1981), reflect susceptibility to a cognitive bias that leads people to shift their choices between two equivalent options, depending upon how those options are presented. Susceptibility to the effects of the distracting options leads the decision-maker to select the option that feels more desirable than others. There appears to be a reduced influence of framing effects on decision making by individuals with ASD (Mosner et al., 2017; Shah et al., 2016; Wu et al., 2018). In an often-cited study by De Martino et al. (2008), young adults with ASD (mean IQ = 112.1) showed significantly less sensitivity to a contextual framing effect (financial tasks presented in either a gain or a loss frame) with a more logical pattern of decision making, in comparison to age- and IQ-matched young adults with typical development. This suggests that individuals with ASD make choices that are more consistent and conventionally rational than those in the general population and rely less on how choices make them feel.

Decision Information Search and Processing Style

On probabilistic reasoning tasks, observations of individuals with typical development (TD) suggest that decision probabilities are routinely updated after learning the outcome of initial draws. In general, individuals with ASD pay less attention to new information from contextual stimuli during decision making, relative to neurotypical individuals. A recent study by Fujino et al. (2020) adds to the previous evidence of reduced sensitivity to context stimuli observed in individuals with ASD, for example, the impact of sunk costs on decision making (relatively less impact in individuals with ASD).

Overall, research suggests an underlying neural and cognitive mechanism that differentiates the decision-making process of individuals with ASD from that of neurotypical people. Few studies have examined decision making in persons with ASD using the Iowa Gambling Task (IGT), a computerized card game that simulates real-life decision making, under conditions of ambiguity or unpredictability, with the goal of earning money by deciding among decks of cards (Mussey et al., 2015; Zhang et al., 2015).

In general, findings have suggested that individuals with ASD tend to be risk-averse (see Gosling & Moutier, 2018) in their decision making, demonstrating a lack of future planning. Use of distinctly different cognitive learning styles by individuals with ASD (e.g., Johnson et al., 2006) and a slower rate of learning, (e.g., Farmer et al., 2017) suggest that the differences observed in individuals with ASD, relative to TD controls, might be more in the way they approach and process decisions, rather than in their ability to arrive at successful decision outcomes.

Although not using emotions to inform one's decision making might be advantageous in situations where being logical (deciding between two mathematically equivalent options) is desirable, this indifference to emotional cueing might become problematic in other settings where consideration of the emotions of oneself and others might be vital to the decision situation. Difficulty in decision making in social environments where some degree of unpredictability is inevitable was indicated in a study by Robic et al. (2015). Utilizing a laboratory task, decision-making performance of a group of adults with ASD was compared to that of adults without ASD in a social versus nonsocial environment and with a predictable versus unpredictable task context. The two groups did not differ in their decision-making performance in the predictable condition in the nonsocial context. However, the participants with ASD displayed greater decision-making difficulty, relative to the group without ASD, with an unpredictable condition and when a social cue was used to simulate a social context. These results suggest that impaired social cue processing and unpredictability associated with social environments can cause decision making difficulty for people with ASD.

In a recent study (Zeif & Yechiam, 2020), involving a meta-analysis of 14 studies of the IGT decision performance of 433 participants with high-functioning ASD and 500 controls, no differences in decision-making outcomes between the participants with and without ASD were observed. Greater use of an explorative strategy in initial trials was noticed for individuals with ASD. The researchers highlighted a need for more research on the use and purpose of explorative strategies and careful evaluation by type of decision tasks for a full understanding of the performance of individuals with ASD. It is also noted that most IGT and probability-based decision studies involve participants who have high-functioning autism and that the patterns of decision making might be different across individuals with ASD with lower intellectual functioning.

Overall, the decision-making approach most consistently observed in individuals with ASD has been one that is individual-centered with limited consideration of the perspectives of other people or being able to readily incorporate and integrate new information and influences. Findings that show reduced selection of options that predict better long-term outcomes, in favor of short-term gains on the IGT task, carry implications for how well individuals with ASD might function with decision making in interpersonal social situations especially in adolescent contexts, such as those involving drug use, gambling, drinking, etc., where evaluation of longer-term consequences, as opposed to short-term gains, might be pivotal to the selection of one's choices (Kluwe-Schiavon et al., 2020).

Executive Function

A key set of specific mental abilities, known as executive functions (EF), related to activity in the ventromedial areas of the brain, are necessary for an individual to adapt to new, non-routine situations (Rabbitt, 1997) for which there are no ready-made solutions (Damasio, 1995). Connections between EF and decision making have been observed in adolescents and adults with IDD and related disorders. Disturbances in ventromedial activity can result in inadequate bioregulatory processes that work to inform the cognitive system about the prospective consequences of an action (Bechara et al., 2005). Therefore, we can expect EF to have a significant role in decision situations that require adapting to new situations or rapid shifting of attention across multiple solutions in unpredictable environments. Given these lines of evidence, decision-making impairments associated with EF might be expected in individuals with IDD, but empirical data are limited.

Kloosterman et al. (2014) investigated whether EF would be a significant predictor of peer victimization for three groups of adolescents, students with ASD, students with other special education needs, and TD students. They found EF to be a significant predictor of victimization for all three groups, with higher EF being associated with less victimization. Since EF involves the processes used in achieving goals, including skills such as planning, organizing, and emotional control, it is likely to influence levels of victimization via key components of decision making involved in these situations.

Intellectual Disability and Singular Disorders

Given that ID is characterized by significant limitations in the intellectual functioning of the individual, the disability is broadly associated with differences in thought processes, reasoning, memory, and comprehension, all of which can have an impact on an individual's decision making. The adaptive behavior deficits intrinsic to ID are reflected in difficulties with social problem solving, credulity (i.e., lack of wariness), and extreme gullibility (Greenspan et al., 2001) and are likely to overlap with the limitations imposed by cognitive dysfunctions during decision making.

Many of the early studies on decision making focused on the cognitive shortcomings of individuals with ID from the perspective of stage-based models (see Hickson & Khemka, 1999, 2014). It was observed that individuals (mostly adults) with ID experienced difficulty with each stage in the decision-making process showing limited comprehension of decision situations, generating few alternative solutions and possible consequences, and not selecting an effective course of action (e.g., Castles & Glass, 1986; Tymchuk et al., 1990; Wehmeyer & Kelchner, 1994). Adults with ID were noticed to be more focused on one-dimensional strategies, (i.e., focusing on one aspect of a decision while ignoring other potentially important aspects) indicating difficulties balancing competing dimensions in selecting a decision action (Jenkinson & Nelms, 1994; Willner et al., 2010a). It was observed that the hypothetical thinking required for considering the possible consequences associated with various courses of action presented high demands for cognitive functioning. The hypothetical thinking needed to play out the possible consequences of alternative courses of action is an important step in the decision-making process with high demands on verbal fluency, memory, abstract thinking, and overall information-processing capacity, all areas of potential difficulty for individuals with ID (Hickson & Khemka, 2013, 2014; Stanovich, 2011).

Ruble et al. (2008) found that individuals with ASD with lower cognitive functioning (Verbal IQ 57-101) had difficulty with generating alternatives during decision making. The researchers posited that this difficulty was perhaps related to their tendency toward repetitive (or rigid) behavior and issues with flexibility. The authors reported that more than half of the study participants improved in their ability to generate multiple solutions to problems after participation in a cognitive/behavioral social skills group, implying that the ability for generation of alternatives for individuals with ASD can be improved through practice with flexible solution generation.

The evaluation of consequences involves predicting immediate and long-term consequences of a particular action. Research on the ability of people with ID to perform temporal discounting tasks (e.g., Willner et al., 2010b) suggests difficulties with forecasting delayed rewards and a greater tendency for impulsive decision making in favor of short-term gains. Although temporal discounting has been studied largely in the context of monetary/rewards-based decision-making tasks for individuals with ID, the patterns of interference of the temporal dimension in decision making are likely to prevail across different domains of decision making, including situations involving risky interpersonal decisions where long-term gains (or losses) may need to be weighed against immediate gains. For instance, individuals with ID may engage in risky decision making by choosing behaviors that are high in immediate subjective desirability or excitement, but that may carry the potential for injury or loss.

Bexkens et al. (2016) examined use of *cool* (cognitive-based) decision-making strategies using the paper-and-pencil Gambling Machine Task (GMT) in adolescents (12–18 years) with mild borderline ID (IQ 50-85) (MBID). Adolescents with MBID showed lower decision accuracy and use of suboptimal decision strategies relative to a TD group and a comparison group including adolescents with behavior disorders (BD). When compared to a group of adolescents with both MBID and BD, the performance of adolescents with MBID was almost equivalent for utilization of decision-making strategies on the GMT task that did not have the complexity to take account of all decision elements for optimal decision making.

Attention Deficit Hyperactivity Disorder

Adolescents with ADHD show a propensity toward risk-taking and reckless behavior, suggesting difficulties with decision making and impulsivity. Decision making difficulties, relating mostly to the generation of alternative solutions to support decision making, have been attributed to deficits in cognitive processing or EF (e.g., Stanovich & West, 2008; Toplak et al., 2011). Dekkers et al. (2020) investigated the decision making of adolescents with ADHD ($n = 81$) and TD adolescents ($n = 100$) using laboratory gambling tasks and found decision-making deficits in adolescents with ADHD to be a result of suboptimal patterns of decision responding, including the use of less complex strategies, rather than from general risk-seeking tendencies. In comparison to TD adolescents, those with ADHD were less likely to adopt strategies that integrated all relevant task characteristics required for optimal decision making and, in addition, they were less sensitive to integrating information from direct feedback on performance to influence and improve their decisions.

Motivational Considerations in Decision Making

Motivational differences that may affect decision making include goal priorities and the relatively weak personal agency beliefs that have been described as *learned helplessness* in individuals with IDD (e.g., Jenkinson, 1999). Although there is convincing evidence that internal perceptions of control play a critical role in how individuals with IDD respond to decision making situations, especially interpersonal situations, more research is needed to explore the temporal ordering of the relationship between self-variables and decision-making (i.e., whether less successful decision outcomes result in lower self-perceptions or lower self-perceptions increase chances for ineffective decision making).

Positive decision outcomes were found to be significantly related to participants' internal control perceptions in a pilot study involving transition-aged adolescents (ages 18–21 years) with moderate to severe IDD from a self-contained special education school in a large metropolitan city (Khemka, 2016). Study participants were assessed for their ability to suggest independent, prevention-focused decision actions (e.g., verbally speak up right away, sign off the Internet/walk away, stop being friends with the perpetrator) in response to six situations conveying threats of cyberbullying as presented as hypothetical vignettes. Effective decision responding was fairly limited with the participants proposing an independent and time effective decision response only 51% of the time, with the remaining decision responses spanning other decision categories (e.g., seeking help, *Call for help*; being avoidant, *Just ignore it*; or complacent, *Go along with it*). Participants also completed the internality scale (measuring level of internal locus of control) on the Levenson IPC scales (Levenson, 1973). Although the sample was relatively small and findings need further investigation, a significant correlation ($r = 0.52$, $p < 0.01$) was found between internality scores and effective decision making that involved resisting the cyberbullying. Findings corroborate growing evidence supporting the importance of personal agency beliefs in shaping independent decision-making behaviors. Research studies involving individuals with specific types of IDD are presented below.

The impact of motivational factors such as the role of external locus of control and outer-directedness has been well documented in individuals with ID. Outer-directedness refers to a behavioral tendency to rely more on external rather than internal cues during problem solving. Outer-directedness in individuals with ID exists across all life stage but has been observed to be at its greatest in adults (Bybee & Zigler, 1998; Tanaka et al., 2001). Shogren et al. (2010) compared locus of control orientations of students with ID to their peers with learning disabilities and those with no disabilities and found that at 8 years, those with ID tend to be more externally oriented than their peers with learning disabilities and no disabilities. Further, while their peers showed ability to develop more adaptive perceptions of control over time, students with ID did not significantly change (increase or decrease) in their perceptions of their ability to exert control over their lives from 8 to 20 years of age. Higher external locus of motivation (Switzky, 2006) among adolescents with ID has also been linked to more openness to peer influences as well as greater risk-taking in decision making under peer pressure (Greenspan, 2017). These longer-term pervasive deficits in adaptive perceptions of control are associated with an other-dependent style in problem solving and decision making, particularly in interpersonal situations involving others.

Individuals with WS are known to have a distinct social personality that is reflected in them being overly friendly, prone to indiscriminately approaching strangers, and highly empathetic in their social interactions. These attributes offer strong motivational input in social situations with decisions being driven by an eagerness for social interaction. Consequently, a high drive for social interactions predisposes individuals with WS to erratic relationships and social vulnerability (Jawaid et al., 2012). Together, these findings suggest that decision making in social situations, including risk-taking, might be impacted by these strong social motivations.

Some people with ASD, with strong special interests, appear to show relatively low flexibility in their decisions to pursue their restricted interests, as was observed in the Ruble et al. (2008) study described earlier. The highly salient and consistent behavioral feature of *rigidity* in individuals with ASD is associated with atypical behavioral motivations that underpin aspects of their restrictive and repetitive interests and behaviors. This is likely to influence their goal selection during decision-making and their level of effort (and self-efficacy) toward optimal goal-directed decision making behavior. The rigidity might also interfere with the ability to make effective intuitive/automatic decisions, lengthening typical reaction time or approaching situations from a restricted repertoire of responses.

Emotional Considerations in Decision Making

There is evidence that emotional patterns observed in individuals with IDD (e.g., Wishart et al., 2007) also play a key role in decision making, although research in this area is nascent. Decision-making effectiveness in individuals with IDD may be limited by difficulties in accurately identifying the emotions and intentions of others (i.e., emotional context) as well as a lack of the emotional self-regulation needed to formulate a reasoned response in the face of a threatening situation.

Affect and Heuristic Processing

As discussed earlier, individuals with ASD who are more rational in their decision making will resist framing effects by consistently choosing the sure option, no matter how the decision is framed. In ASD, a reduction of framing effects might explain both enhanced analytic and impoverished social abilities, the latter reflecting an inability to deploy an *affect heuristic* in complex and uncertain social contexts. This pattern appears consistent with the core neurobiological deficits seen in ASD and is in line with theories that posit that individuals with ASD have enhanced analytical-logical processing compared to their TD peers (e.g., Baron-Cohen, 2009). According to De Martino et al. (2008), an inability to incorporate a broad range of contextual cues into the decision process carries a clear disadvantage when making decisions in social environments in which multiple subtle contextual cues carry critical information. In handling decision-making situations involving social uncertainty, competent decision-makers tend to rely on affect heuristics. Psychophysiological evidence collected in the De Martino et al. (2008) study (skin conductance response measuring emotional sensitivity) corroborated evidence from previous studies by De Martino et al. (2006) and Kahneman and Frederick (2007). It pointed toward a potential core neurobiological deficit in ASD that interfered with the ability to integrate emotional context easily into the decision-making process. Deliberative and less intuitive thinking among persons with autism serves them well when it comes to decisions that require deep thinking but interferes with decisions that most people can make spontaneously and with little effort. ASD-related social deficits affect some decisions but not others.

These deficits can assume considerable importance during social interactions where information about others is often ambiguous and the need to absorb emotional contextual information into the decision process outweighs the need for inferential reasoning processes. For optimal decision making in social environments, the ability to incorporate a broad range of contextual cues into the decision process operates as an affect heuristic that allows the decision-maker to evaluate multiple sources of critical and subtle information and to respond rapidly without having to engage in a demanding analytic process (Stanovich & West, 2002). On this basis, a failure by individuals with ASD to deploy an *affect heuristic* in complex and uncertain social contexts might limit their intuitive reasoning and thereby their social competence. There has been growing attention to the study of emotion, including biosomatic markers related to anxiety or other personality conditions, and etiological influences.

Emotion Regulation and Anxiety

Emotions have a considerable influence on the way decisions are made. Social decision making is often difficult, anxiety provoking, and exhausting for individuals with ASD (Ahlstrom & Wentz, 2014; Hull et al., 2017). Emotion regulation and EF are plausible influences on the decision making of individuals with ASD given their prototypical delays in social functioning. However, the extent and nature of such contributions are currently unclear from the extant literature.

In a study by Woodcock et al. (2020) examining the social decision-making profiles of adolescents with ASD using standardized game theory decision tasks, the impact of deficits in emotion regulation was observed between adolescents with and without ASD. Findings showed that the main difference emerged in the way the two groups regulated their emotions while deciding on accepting unfair outcomes, with adolescents with ASD experiencing more negative emotion following *acceptance* of unfair outcomes relative to typical adolescents. The groups did not differ in their emotional experience following *rejection* of unfair outcomes. However, adolescents with ASD failed to downregulate their negative emotion in response to unfair outcomes before making their decisions to accept in the same way as neurotypical counterparts. This is consistent with prior research that shows lower adaptability in regulating emotions in individuals with ASD (e.g., Zantinge et al., 2017). The study thus shows that atypical processes may underpin even apparently typical decisions in adolescents with ASD. It also suggests that individuals with ASD might learn to make socially appropriate decisions, but they might not be employing adaptive mechanisms that psychologically help to control emotional reactions in the same way as their TD peers.

The commonly observed strict adherence to rules and inflexibility in the presence of new and conflicting information during decision making appears to be related to the core deficits in ASD (American Psychiatric Association, 2013). Research suggests that persons with ASD tend to take longer to make decisions and are more likely to report avoidance of decision making (Brosnan et al., 2014; Luke et al., 2012). Luke et al. (2012) compared the decision making of adults with (mean verbal IQ = 116.4) and without ASD (mean verbal IQ = 114.2) and found distinct differences in the experiences of the two groups. In the absence of cognitive differences between the two groups, the differences in decision-making likely reflect the impact of characteristics linked to the autism phenotype. In this study, adults with ASD reported greater reliance on an avoidant decision making style and significant difficulties with decisions that had to be made quickly, decisions that involved a change of routine, and decisions that involved talking to others, in comparison to their non-disabled counterparts. Interference with decision-making performance was also associated with higher levels of anxiety and depression. On a measure of decision-making style, Khemka et al. (2013) found adolescents with ASD to be more likely than adolescents with other forms of IDD to report that when they had a big decision to make, they worried about the decision and they tried to remember past decisions. A significant negative correlation was found between worrying about decisions and making effective decision-making responses that involved resisting coercion in simulated interpersonal situations.

The ability to accurately interpret others' emotions allows us to monitor others' intentions and to predict others' behaviors. These underlying emotion and applied perspective taking mechanisms are essential to adaptive decision-making performance, but they are acknowledged to be problematic in individuals with ASD. Peters and Thompson (2018) emphasize that applied perspective-taking skills affect social skills development in individuals with ASD and must be targeted for direct teaching for improving socially important behavior. It is therefore reasonable to assume that these underlying psychobiological inputs might alter the way individuals with ASD interpret subjective information when they make decisions in interpersonal situations. Severe impairments in emotional processing may be related to a coexisting condition of alexithymia often seen in ASD. It has been shown to negatively affect ability to integrate understanding of social conventions or moral intuitions during decision making in social situations (Hill et al., 2004; Komeda et al., 2016; see

Komeda chapter in this volume. Dempsey et al. (2020) after a recent review of studies examining moral decision making in children and adults with autism concluded that an alternative framework for assessing moral decision making should be explored to account for the functioning of individuals with ASD, instead of using norms based on rational social moral conventions that assume a deficit-based perspective for those with ASD.

Individuals with WS exhibit key personality traits that can negatively impact their social judgment and decision-making functioning in interpersonal situations. Significant socio-communicative difficulties that affect their ability to interact well with others have been observed, with some indication that there might be overlap with an ASD profile, in areas of communicative and pragmatic language (Klein-Tasman et al., 2011). Individuals with WS experience higher levels of anxiety, emotional difficulties, and limited adaptive behaviors that often interfere with their ability to establish meaningful peer relationships, despite their highly sociable, affiliative nature which forms a distinct part of their personality profile (Dykens, 2003; Järvinen et al., 2013; Leyfer et al., 2006). When compared to individuals with ID, individuals with WS were four times more likely to experience anxiety (Royston et al., 2017). Riby et al. (2014) found that a predisposition to anxiety was related to dysfunctionality in social abilities as measured on the Social Responsiveness Scale for a group of individuals with WS (age 6–36 years). There is also evidence of a high prevalence of attention problems (see Leyfer et al., 2006) and difficulties with sociability and empathy (Doyle et al. 2004; Klein-Tasman & Mervis, 2003) among individuals with WS that can all potentially influence perspective taking and social cognition for understanding a social situation for decision making.

Implications for Interpersonal Decision Making

Although most of the studies described in the previous sections have involved laboratory decision-making tasks requiring the integration of contextual information with perceptual stimuli, these studies carry important implications for understanding decision difficulties in interpersonal situations. Decision making in interpersonal situations typically requires a degree of deliberative thinking to identify not only one's own goal preferences (e.g., staying safe in a potentially abusive situation) but also the motives of the other(s) in the situation (e.g., recognizing use of a lure as a tactic for coercion) and to calibrate possible alternative decision actions accordingly. It also involves the ability to think flexibly, often under time constraints, and to generate alternative options and evaluate them based on the demands of a specific situation.

In our research that has focused on the interpersonal decision-making effectiveness of adolescents and adults with IDD (see Hickson & Khemka, 2014; Khemka & Hickson, 2017b), we have typically presented vignettes depicting hypothetical situations in which a protagonist is faced with a decision involving the possibility of coercive peer pressure or abuse. Overall, we have found that the decision-making responses of adults with ID are significantly less likely to be effective than those of their non-disabled counterparts (see Khemka & Hickson, 2017b). On average, the level of decision effectiveness (i.e., responses that appeared to be in the best interest of the protagonist) by individuals with ID has been at about 50–60% effective. In comparison, when adults without disabilities have been included, their performance tends to be at a significantly higher level of above 90% effective (Hickson & Khemka, 1999).

Taken as a whole, the series of studies suggests that the decision making of adolescents and adults with ID is constrained by limitations associated with comprehension, reasoning ability, and overall processing capacity. For example, cognitive limitations may include not only reduced processing capacity and memory but also limitations in the comprehension required for fully understanding the nature of a situation requiring a decision (e.g., friendly or threatening) (e.g., Khemka & Hickson, 2017b). When people with ID fail to apply a systematic decision-making process and rely on a limited number of solutions drawn from their experience, their decisions tend to involve some sort of intuitive processing, but not necessarily one that leads to effective decisions. In addition, if the past experiences of individuals with ID with decision making are not optimal, their reliance on past negative outcomes does not necessarily create conditions for better adaptive functioning and improved decision making for the future (see Hickson & Khemka, 1999).

In an exploratory study, we examined the interpersonal decision-making performance of a group with an ASD diagnosis ($n = 18$) and a group with IDD and no ASD diagnosis ($n = 31$) (Khemka et al., 2013). Correlational analyses were performed between number of effective decisions (prevention-focused decisions in response to a set of 12 hypothetical situations of peer pressure) and several cognitive variables (IQ, language functioning, and risk perception). The relationship between effective decision making and self-actualized goal orientation, a motivational variable, was also assessed. Significant positive correlations were found between effective decision-making scores and language functioning scores ($r = 0.54$, $p < 0.01$) and between effective decision-making scores and correct risk perception scores ($r = 0.80$, $p < 0.01$) for the combined sample, as well as for the two subgroups, separately. In addition, for adolescents in the ASD group only, a significant relationship was obtained between effective decision making and IQ ($r = 0.47$, $p < 0.05$) and effective decision making and self-actualized goal orientation ($r = 0.52$, $p < 0.05$). These findings corroborate the robust role of cognitive abilities, such as risk perception, in the decision making of adolescents with IDD. Since IQ and self-actualized goal orientation were significantly associated with effective decision-making scores only for adolescents with ASD in this study, we suggest that the decision making of adolescents with ASD may be particularly subject to various cognitive and motivational influences relative to the broader IDD group. This holds importance for identifying ways in which we can support adolescents with ASD by increasing their awareness of adaptive goal priorities.

In our studies, we have also found individuals with IDD to show an overreliance on other-dependent decisions (e.g., Khemka et al., 2009) that consist of seeking support or help from someone as opposed to independent decisions aimed at actively confronting and addressing a decision problem situation. This type of decision making might be related to their feelings of learned helplessness and/or outer-directedness and therefore a tendency to rely more on others' ideas than their own.

Applications of Pathways Model of Decision Processing

In our *Pathways Model of Decision Processing* (introduced in Chapter 6 of this volume), the four alternative pathways for decision processing are highlighted in the upper part of the model in Fig. 11.2 with a different colored line for each pathway. In outlining the possible pathways of decision processing that individuals with IDD could follow, we provide for a range of ways in which an individual with IDD can adapt to respond optimally to a decision situation. In Fig. 11.2, the base section of the model, which has been the focus of this chapter up until now, is left in shadow so that we can focus on the pathways.

Pathway 1 (blue line) is the intuitive pathway that is applied with little effort for most everyday decisions. *Pathway 2* (green line) is the deliberative pathway that is typically applied for serious decisions with long-term consequences. *Pathway 3* (yellow line) involves a shift from intuitive processing to deliberative processing when self-regulatory mechanisms (shown in the light blue circle) detect a need for a more careful, deliberative process. Finally, *Pathway 4* (magenta line) begins with deliberative processing which can change over time, with repeated practice in a

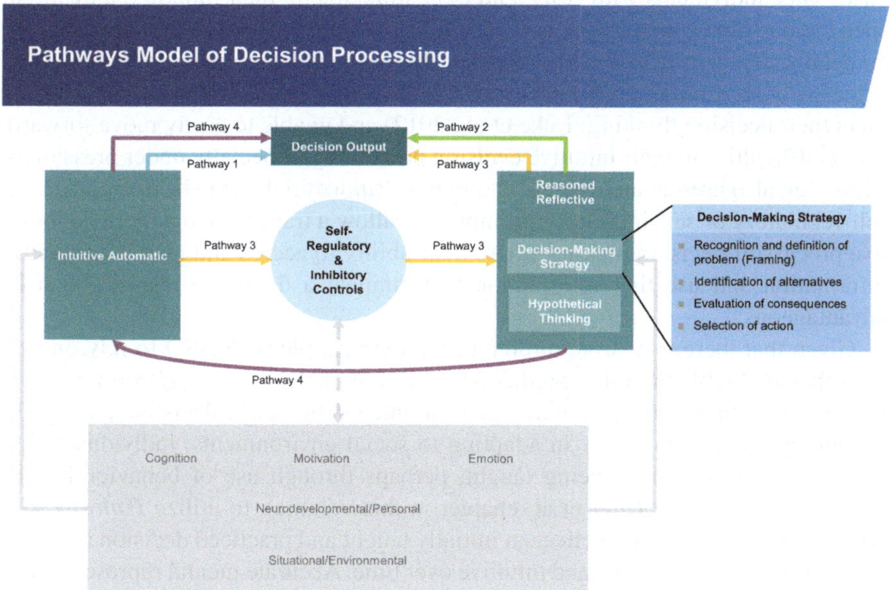

Fig. 11.2 Decision pathways from the *Pathways Model of Decision Processing*

specific domain so that, eventually, decisions in that domain can be made quickly using an intuitive process to retrieve the familiar decision action from memory. Drawing upon the research reviewed in this chapter, examples of how these pathways might be utilized by individuals with various types of IDD are presented below. While there is no one best way to tackle any decision situation, the model allows for exploring pathways of decision processing, and potential difficulties in reaching an effective decision outcome, that can help explain gaps and limitations in an individual's repertoire of decision-making skills.

Autism Spectrum Disorder

In decision situations requiring decisions under time pressure, or unexpected or rapidly changing contexts as in social interactions, individuals with ASD might find it difficult to make quick-paced, intuitive (*Pathway* 1) decisions due to their need to have control over changing events. They might also feel anxious or confused if rushed to make an on-the-spot decision because of their preference for considering all details and a desire to maintain consistency with previous experiences and formally learned rules (Allman et al., 2005; Baron-Cohen, 2006; Brosnan et al., 2016). Since individuals with ASD tend to interpret events in very concrete terms and remain detail oriented, they might find *Pathway 1*'s intuitive processing hard to rely upon, especially when they have to actively hold and process all the information they need to access in a given time, as in time-pressured or urgent decision making. Also, since individuals with ASD tend to be inflexible in their thinking, they might spend undue time comparing a given decision situation with past situations for similar details, etc. getting lost in the process. This can result in them searching memory extensively for similar experiences to use as a model and getting stuck (i.e., locked up in their decision thinking; Luke et al., 2012) and unable to easily move forward. These difficulties in their intuitive decision processing, especially under pressure or stress, can also present an obstacle in utilizing *Pathway 3*, but can be overcome with self-regulatory or self-inhibitory prompts that allow a transition to a more deliberative process of decision making where their ability to access and take in long-term information and use that information to formulate a definitive response can be advantageous.

Given that there is a strong preference among people with ASD to rely on systems that are highly lawful or predictable, they are likely to adopt a *Pathway* 2 style of processing that involves a more detail-oriented deliberative decision making. To encourage more spontaneity in adapting to social environments, individuals with ASD might benefit from being taught, perhaps through use of behavioral skills training methods (see *Uher* et al. chapter in this volume), to utilize *Pathway 4*, if possible, where through practice, an initially taught and practiced decision response becomes more spontaneous and intuitive over time. Accurate mental representations of social situations, indicating a varied set of uncertainties in the social domain, could be taught. Through repeated consideration these learned representations may serve as formal rules or social conventions to rely upon when navigating fluctuating social environments. To learn to deal with the unexpected is addressing the very core deficit in ASD linked to incomplete understanding of social situations and inflexibility to adjusting with change (Gomot & Wicker, 2012).

Intellectual Disabilities

The selection of a pathway depends on the type and complexity of the decision problem. Decisions made intuitively or those derived from gist interpretations based on past experiences draw upon memory functions. This can be particularly challenging for individuals with ID who might encounter not only memory difficulties but also may be prone to making faulty interpretations based on their past experiences with decision making which might have been problematic. With a high likelihood of having acquired rigid patterns of learned helplessness and other-dependent or avoidant decision making or low confidence in one's ability to make decisions due to limited successes in past decision making, it is plausible that many individuals with ID will approach decisions along *Pathways 1* and *4*, but their decisions may not be effective due to past learned responses that were faulty or incomplete. Further, individuals with ID might find decision processing on *Pathway 3* difficult due to their overall poor inhibitory and self-regulatory skills, where, in the absence of systematic training, a shift from original intuitive processing to more deliberative decision making might not proceed in a predictable manner or get initiated on time with a decision situation. More complex decisions requiring a thorough search of available alternatives and careful evaluation of consequences due to their importance, for instance, decisions in health care or finances, might require a more deliberative approach as in *Pathway 2*. Due to the cognitive demands of this type of reasoned, analytic processing, individuals with ID might require supports to fully execute this deliberative pathway of decision processing. Empirical studies have shown that systematic decision-making strategy instruction can improve preparation for approaching decision making in a more planful way (see Khemka, 2000; Khemka et al., 2005, 2016).

For individuals with ID who have severe cognitive or communication difficulties, or show extreme deficits in attention or emotional regulation, the model can help elucidate which specific skills or stages within a pathway should be targeted for supports or training. Such individual accommodations will prepare individuals with ID to exercise personal agency in their decision making when full independent participation is not feasible. By simplifying the decision process and removing certain demands for cognitive or emotional processing, individuals with severe limitations due to ID might acquire simplified decision rules that, when learned and rehearsed, can develop into an adaptive intuitive response. In the context of highly pressured situations or those involving risk to personal safety, such decision rules might be particularly beneficial. The use of behavioral skills training methods might be particularly effective in such contexts.

A Singular Disorder: Williams Syndrome

Individuals with Williams syndrome (WS) tend to have well-developed language skills and demonstrate a hyper-social personality trait that makes them overfriendly. Given underlying cognitive functioning deficits, individuals with WS are likely to face general limitations in cognitive aspects of decision making, although a main source of difficulty would likely relate to impairments with emotion regulation and elevated levels of anxiety (Dykens, 2003; Leyfer et al., 2006). Given their dominant social phenotype and greater likelihood of acting on impulse and being overly spontaneous (related to a desire to be friendly), they are likely to be swayed by exaggerated emotional impulses during decision processing, particularly in interpersonal situations involving both known and unknown people (i.e., strangers). This naturally inclines them to lean toward an intuitive style of decision processing, such as in *Pathway 1,* but with anticipated difficulties due to their emotional personality and tendency for increased social engagement (see Jarvinen et al. 2013). With heightened approach tendencies toward strangers as some of the core features of the WS social phenotype, intuitive decisions might be overly biased toward satisfying prosocial goals. Moreover, Jarvinen et al. (2013) state that, "the social behavior of individuals with WS is however often inappropriate and is accompanied by marked deficits in social skills, such as difficulties in social adjustment, social judgment, with an inflexible, repetitive, and pragmatically insensitive social repertoire" (p 3). In regard to this unique social-behavioral profile combined with difficulties with response inhibition (Menghini et al., 2010), rapid, automatic decision processing might be problematic for individuals with WS as many aspects of the decision would need to come together all at once. *Pathway 3*-type processing where intuitive mechanisms are deliberately slowed down for a more reasoned consideration of a decision might be more adaptive in the long run and could serve as a useful training goal for individuals with WS.

Attention Deficit Hyperactivity Disorder

Decision-making difficulties in adults with ADHD appear to be particularly evident in situations requiring a high level of cognitive control, which is likely to lessen efficiency with deliberative processing of information as indicated in *Pathway 2*. Pervasive impairments in EF often found in adults with ADHD could make following *Pathway 2* a challenge for them. Monitoring of initial affective input and shifting toward a more reasoned, thought-out mode of information processing would require effective regulatory mechanisms, such as working memory or response inhibition, which have been found to be impaired for individuals with ADHD (Mäntylä et al., 2012). It is therefore predicted that people with ADHD will most automatically draw upon decision processing in ways that are consistent with *Pathway 1* mechanisms, relying more on their affect or intuition. Further, decision-making in this group may be mediated by problems in reward processing and delay aversion, resulting in a higher motivation to be impulsive or risk-taking in favor of immediate, short-term gains as opposed to larger long-term gains. This suggests possible difficulties with affective/experiential processing in *Pathway 1*, with greater chances for dysregulated and impulsive decision making. Over time, with training and supports, the integration of both affective/experiential and deliberative/analytic processing systems can be better coordinated, possibly resulting in greater decision-making proficiency (Kahneman, 2003; Peters et al., 2007).

Conclusion

The literature reviewed in this chapter extends our awareness and understanding of decision-making abilities and profiles of persons with IDD and the types of constraints that might limit an individual's decision-making capabilities. In recent years, a growing body of research has expanded understanding of decision-making difficulties experienced by individuals with IDD, to noncognitive factors that can profoundly influence decision outcomes. The early sections of the chapter draw upon the base components of the *Pathways Model* to organize and analyze the sources of these constraints, by looking at key roles of cognition, motivation, and emotion in decision making, along with a myriad of internal (personal) and external (situational/environmental) factors through a neurodevelopmental lens. Although much of the research reviewed was based on impersonal and laboratory decision tasks, the implications for interpersonal decision making are apparent and far-reaching. In later sections of the chapter, examples are provided of how the pathways might be applied by individuals with different types of IDD, including ASD, ID, WS, and ADHD. In addition, it lays out a framework for training considerations to help individuals with IDD to acquire a range of decision strategies for greater flexibility and effectiveness in their decision making in real-life situations. The chapter offers direction in identifying specific areas of weakness that would be important to address in intervention efforts. Finally, the chapter reveals gaps in research in studying decision making in real-life situations that are often ambiguous and unpredictable and can require integration of several influences at one time, making decision making for individuals with IDD more complex and demanding.

References

Ahlström, B. H., & Wentz, E. (2014). Difficulties in everyday life: Young persons with attention-deficit/hyperactivity disorder and autism spectrum disorders perspectives. A chat-log analysis. *International Journal of Qualitative Studies on Health and Well-Being, 9*, 23376. https://doi.org/10.3402/qhw.v9.23376

Allman, J. M., Watson, K. K., Tetreault, N. A., & Hakeem, A. Y. (2005). Intuition and autism: A possible role for Von Economo neurons. *Trends in Cognitive Sciences, 9*(8), 367–373. https://doi.org/10.1016/j.tics.2005.06.008

American Psychiatric Association (Ed.). (2013). *Diagnostic and statistical manual of mental disorders* (5th ed.). American Psychiatric Association.

Baron-Cohen, S. (2006). The hyper-systemizing, assortative mating theory of autism. *Progress in Neuro-Psychopharmacology and Biological Psychiatry, 30*(5), 865–872.

Baron-Cohen, S. (2009). Autism: The empathizing-systemizing (E-S) theory. *Annals of the New York Academy of Sciences, 1156*, 68–80. https://doi.org/10.1111/j.1749-6632.2009.04467.x

Bechara, A., Damasio, H., & Damasio, A. (2000). Emotion, decision making and the orbitofrontal cortex. *Cerebral Cortex, 10*(3), 295–307. https://doi.org/10.1093/cercor/10.3.295

Bechara, A., Damasio, H., Tranel, D., & Damasio, A. R. (2005). The Iowa gambling task and the somatic marker hypothesis: Some questions and answers. *Trends in Cognitive Sciences, 9*(4), 159–162.; discussion 162-164. https://doi.org/10.1016/j.tics.2005.02.002

Bernard-Opitz, V., Sriram, N., & Nakhoda-Sapuan, S. (2001). Enhancing social problem solving in children with autism and normal children through computer-assisted instruction. *Journal of Autism and Developmental Disorders, 31*(4), 377–384. https://doi.org/10.1023/a:1010660502130

Bexkens, A., Jansen, B. R., Van der Molen, M. W., & Huizenga, H. M. (2016). Cool decision making in adolescents with behavior disorder and/or mild-to-borderline intellectual disability. *Journal of Abnormal Child Psychology, 44*(2), 357–367. https://doi.org/10.1007/s10802-015-9996-8

Brady, N. C., Bruce, S., Goldman, A., Erickson, K., Mineo, B., Ogletree, B. T., Paul, D., et al. (2016). Communication services and supports for individuals with severe disabilities: Guidance for assessment and intervention. *American Journal of Intellectual Developmental Disability, 121*, 121–138.

Brosnan, M., Chapman, E., & Ashwin, C. (2014). Adolescents with autism spectrum disorder show a circumspect reasoning bias rather than 'jumping-to-conclusions'. *Journal of Autism and Developmental Disorders, 44*(3), 513–520. https://doi.org/10.1007/s10803-013-1897-5

Brosnan, M., Lewton, M., & Ashwin, C. (2016). Reasoning on the autism spectrum: A dual process theory account. *Journal of Autism and Developmental Disorders, 46*(6), 2115–2125. https://doi.org/10.1007/s10803-016-2742-4

Bryson, S., Bradley, E., Thompson, A., & Wainwright, A. (2008). Prevalence of autism among adolescents with intellectual disabilities. *The Canadian Journal of Psychiatry, 53*(7), 449–459.

Bybee, J., & Zigler, E. (1998). Outerdirectedness in individuals with and without mental retardation: A review. In J. Burack, R. Hodapp, & E. Zigler (Eds.), *Handbook of mental retardation and development* (pp. 434–461). Cambridge University Press.

Castles, E. E., & Glass, C. R. (1986). Training in social and interpersonal problem-solving skills for mildly and moderately mentally retarded adults. *American Journal on Mental Deficiency, 91*, 35–42.

Damasio, A. R. (1995). *On some functions of the human prefrontal cortex*. New York Academy of Sciences.

De Martino, B., Kumaran, D., Seymour, B., & Dolan, R. J. (2006). Frames, biases, and rational decision making in the human brain. *Science, 313*, 684–687.

De Martino, B., Harrison, N. A., Knafo, S., Bird, G., & Dolan, R. J. (2008). Explaining enhanced logical consistency during decision making in autism. *The Journal of Neuroscience, 28*(42), 10746–10750. https://doi.org/10.1523/JNEUROSCI.2895-08.2008

Dekkers, T. J., Popma, A., Sonuga-Barke, E. J., et al. (2020). Risk taking by adolescents with attention-deficit/hyperactivity disorder (ADHD): A behavioral and psychophysiological investigation of peer influence. *Journal of Abnormal Child Psychology, 48*, 1129–1141. https://doi.org/10.1007/s10802-020-00666-z

Dempsey, E., Moore, C., Johnson, S., Stewart, S., & Smith, I. (2020). Morality in autism spectrum disorder: A systematic review. *Development and Psychopathology, 32*(3), 1069–1085. https://doi.org/10.1017/S0954579419001160

Doyle, T. F., Bellugi, U., Korenberg, J. R., & Graham, J. (2004). "Everybody in the world is my friend". Hypersociability in young children with Williams syndrome. *American Journal of Medical Genetics, 124A*, 263–273.

Dykens, E. M. (2003). Anxiety, fears, and phobias in persons with Williams syndrome. *Developmental Neuropsychology, 23*, 291–316.

Farmer, G. D., Baron-Cohen, S., & Skylark, W. J. (2017). People with autism Spectrum conditions make more consistent decisions. *Psychological Science, 28*(8), 1067–1076.

Fujino, J., Tei, S., Itahashi, T., Aoki, Y. Y., Ohta, H., Kubota, M., Hashimoto, R. I., Nakamura, M., Kato, N., & Takahashi, H. (2020). Impact of past experiences on decision making in autism spectrum disorder. *European Archives of Psychiatry and Clinical Neuroscience, 270*(8), 1063–1071. https://doi.org/10.1007/s00406-019-01071-4

García, J. C., Díez, E., Wojcik, D. Z., & Santamaría, M. (2020). Communication support needs in adults with intellectual disabilities and its relation to quality of life. *International Journal of Environmental Research and Public Health, 17*(20), 7370. MDPI AG. https://doi.org/10.3390/ijerph17207370

Gomot, M., & Wicker, B. (2012). A challenging, unpredictable world for people with autism spectrum disorder. *International Journal of Psychophysiology, 83*(2), 240–247. https://doi.org/10.1016/j.ijpsycho.2011.09.017

Gosling, C. J., & Moutier, S. (2018). Brief Report: Risk-Aversion and Rationality in Autism Spectrum Disorders. *Journal of autism and developmental disorders, 48*(10), 3623–3628. https://doi.org/10.1007/s10803-018-3616-8.

Grandin, T. (2000). My mind is a web browser: How people with autism think. *Cerebrum, 2*, 14–22.

Greenspan, S. (2017). Borderline intellectual functioning: An update. *Current Opinion in Psychiatry, 32*(2), 113–122. https://doi.org/10.1097/yco.0000000000000317

Greenspan, S., Loughlin, G., & Black, R. S. (2001). Credulity and gullibility in people with developmental disorders: A framework for future research. In L. M. Glidden (Ed.), *International review of research in mental retardation* (Vol. 24, pp. 101–135). Academic Press.

Hickson, L., & Khemka, I. (1999). Decision making and mental retardation. In L. M. Glidden (Ed.), *International review of research in mental retardation* (Vol. 22, pp. 227–265). Academic Press.

Hickson, L., & Khemka, I. (2013). Problem solving and decision making. In M. Wehmeyer (Ed.), *The Oxford handbook of positive psychology and disability*. Oxford University Press.

Hickson, L., & Khemka, I. (2014). The psychology of decision making. In R. Hodapp (Ed.), *International review of research in developmental disabilities* (Vol. 47, pp. 185–229). Elsevier.

Hickson, L., Golden, H., Khemka, I., Urv, T., & Yamusah, S. (1998). A closer look at interpersonal decision making in adults with and without mental retardation. *American Journal on Mental Retardation, 103*(3), 209–224.

Hill, E., Berthoz, S., & Frith, U. (2004). Brief report: Cognitive processing of own emotions in individuals with autistic spectrum disorder and in their relatives. *Journal of Autism and Developmental Disorders, 34*, 229–235.

Hillier, A., Fish, T., Cloppert, P., & Beversdorf, D. Q. (2007). Outcomes of a social and vocational skills support Group for Adolescents and Young Adults on the autism Spectrum. *Focus on Autism and other Developmental Disabilities, 22*(2), 107–115. https://doi.org/10.1177/10883576070220020201

Howlin, P., Goode, S., Hutton, J., & Rutter, M. (2004). Adult outcome for children with autism. *Journal of Child Psychology and Psychiatry, 45*(2), 212–229. https://doi.org/10.1111/j.1469-7610.2004.00215.x

Hull, L., Petrides, K. V., Allison, C., Smith, P., Baron-Cohen, S., Lai, M. C., & Mandy, W. (2017). "Putting on my best Normal": Social camouflaging in adults with autism Spectrum conditions.

Journal of Autism and Developmental Disorders, 47(8), 2519–2534. https://doi.org/10.1007/s10803-017-3166-5

Järvinen, A., Korenberg, J. R., & Bellugi, U. (2013). The social phenotype of Williams syndrome. *Current Opinion in Neurobiology, 23*, 414–422. https://doi.org/10.1016/j.conb.2012.12.006

Jawaid, A., Riby, D. M., Owens, J., White, S. W., Tarar, T., & Schulz, P. E. (2012). 'Too withdrawn' or 'too friendly': Considering social vulnerability in two neuro-developmental disorders. *Journal of Intellectual Disability Research, 56*(4), 335–350. https://doi.org/10.1111/j.1365-2788.2011.01452.x

Jenkinson, J. C. (1999). Factors affecting decision-making by young adults with intellectual disabilities. *American Journal on Mental Retardation, 104*, 320–329.

Jenkinson, J. C., & Nelms, R. (1994). Patterns of decision making behavior by people with intellectual disability: An exploratory study. *Australia and New Zealand Journal of Developmental Disabilities, 19*, 99–109.

Johnson, S. A., Yechiam, E., Murphy, R. R., Queller, S., & Stout, J. C. (2006). Motivational processes and autonomic responsivity in Asperger's disorder: Evidence from the Iowa Gambling Task. *Journal of the International Neuropsychological Society, 12*, 668–676.

Kahneman, D. (2003). A perspective on judgment and choice: Mapping bounded rationality. *American Psychologist, 58*, 697–720. https://doi.org/10.1037/0003-066X.58.9.697

Kahneman, D., & Frederick, S. (2007). Frames and brains: Elicitation and control of response tendencies. *Trends in Cognitive Sciences, 11*, 45–46.

Khemka, I. (2000). Increasing independent decision-making skills of women with mental retardation in simulated interpersonal situations of abuse. *American Journal on Mental Retardation, 105*, 387–401.

Khemka, I. (2016). Examining decision making and locus of control patterns of transition-aged students with intellectual and developmental disabilities in response to simulated situations of cyberbullying and abuse: *Two pilot studies*. Unpublished manuscript.

Khemka, I., & Hickson, L. (2017a). Empowering women with intellectual disabilities to resist abuse in interpersonal relationships: Promising interventions and practices. In A. J. Johnson, J. R. Nelson, & E. M. Lund (Eds.), *Religion, disability, and interpersonal violence* (pp. 67–86). Springer.

Khemka, I., & Hickson, L. (2017b). The role of positive psychology in interpersonal decision making. In K. A. Shogren, M. Wehmeyer, & N. N. Singh (Eds.), *Handbook of positive psychology in intellectual and developmental disabilities: Translating research into practice* (pp. 145–171). Springer.

Khemka, I., Hickson, L., & Reynolds, G. (2005). Evaluation of a decision-making curriculum designed to empower women with mental retardation to resist abuse. *American Journal on Mental Retardation, 110*(3), 193–204.

Khemka, I., Hickson, L., Casella, M., Accetturi, N., & Rooney, M. E. (2009). Impact of coercive tactics on the decision making of adolescents with intellectual disabilities. *Journal of Intellectual Disability Research, 53*(4), 353–362.

Khemka, I., Hickson, L., Mallory, S., & Zealand, R. (2013, March). Social vulnerability of adolescents with autism spectrum disorders and adolescents with other developmental disabilities in situations of negative peer pressure. In *Symposium paper presented at the 46th annual gatlinburg conference on research and theory in intellectual and developmental disabilities*, San Antonio, TX.

Khemka, I., Hickson, L., & Mallory, S. B. (2016). Evaluation of a decision-making curriculum for teaching adolescents with disabilities to resist negative peer pressure. *Journal of Autism and Developmental Disorders, 46*(7), 2372–2384.

Klein-Tasman, B. P., & Mervis, C. B. (2003). Distinctive personality characteristics of children with Williams syndrome. *Developmental Neuropsychology, 23*, 271–292.

Klein-Tasman, B. P., Li-Barber, K. T., & Magargee, E. T. (2011). Honing in on the social phenotype in Williams syndrome using multiple measures and multiple raters. *Journal of Autism and Developmental Disorders, 41*, 341–351.

Kloosterman, P. H., Kelley, E. A., Parker, J. D. A., & Craig, W. M. (2014). Executive functioning as a predictor of peer victimization in adolescents with and without an Autism Spectrum Disorder. *Research in Autism Spectrum Disorders, 8*, 244–254.

Kluwe-Schiavon, B., Kexel, A., Manenti, G., Cole, D. M., Baumgartner, M. R., Grassi-Oliveira, R., Tobler, P. N., & Quednow, B. B. (2020). Sensitivity to gains during risky decision making differentiates chronic cocaine users from stimulant-naïve controls. *Behavioural Brain Research, 379*, 112386. https://doi.org/10.1016/j.bbr.2019.112386

Komeda, H., Osanai, H., Yanaoka, K., Okamoto, Y., Fujioka, T., Arai, S., Inohara, K., Koyasu, M., Kusumi, T., Takiguchi, S., Kawatani, M., Kumazaki, H., Hiratani, M., Tomoda, A., & Kosak, H. (2016). Decision-making processes based on social conventional rules in early adolescents with and without autism spectrum disorders. *Scientific Reports*. https://doi.org/10.1038/srep37875

Kuhl, J. (1986). Motivation and information processing: A new look at decision making, dynamic change, and action control. In M. Sorrentino & E. T. Higgins (Eds.), *Handbook of motivation and cognition: Foundations of social behavior*. Guilford.

Larkin, P., Jahoda, A., Mac Mahon, K., & Pert, C. (2011). Interpersonal sources of conflict in young people with and without mild to moderate intellectual disabilities at transition from adolescence to adulthood. *Journal of Applied Research in Intellectual Disabilities, 25*(1), 29–38. https://doi.org/10.1111/j.1468-3148.2011.00652.x

Levenson, H. (1973). Multidimensional locus of control in psychiatric patients. *Journal of Consulting and Clinical Psychology, 41*, 397–404.

Levin, I. P., Gaeth, G. J., Foley-Nicpon, M., Yegorova, V., Cederberg, C., & Yan, H. (2015). Extending decision-making competence to special populations: A pilot study of persons on the autism spectrum. *Frontiers in Psychology, 6*, 539. https://doi.org/10.3389/fpsyg.2015.00539

Leyfer, O. T., Woodruff-Borden, J., Klein-Tasman, B. P., Fricke, J. S., & Mervis, C. B. (2006). Prevalance of psychiatric disorders in 4 to 16 year olds with William syndrome. *American Journal of Medical Genetics, Part B, Seminars in Medical Genetics, 141 B*, 615–622.

Luke, L., Clare, I. C., Ring, H., Redley, M., & Watson, P. (2012). Decision making difficulties experienced by adults with autism spectrum conditions. *Autism, 16*(6), 612–621.

Mäntylä, T., Still, J., Gullberg, S., & Del Missier, F. (2012). Decision making in adults with ADHD. *Journal of Attention Disorders, 16*(2), 164–173. https://doi.org/10.1177/1087054709360494

Matson, J. L., & Shoemaker, M. (2009). Intellectual disability and its relationship to autism spectrum disorders. *Research in Developmental Disabilities, 30*(6), 1107–1114. https://doi.org/10.1016/j.ridd.2009.06.003

Matson, J. L., & Williams, L. W. (2013). Differential diagnosis and comorbidity: Distinguishing autism from other mental health issues. *Neuropsychiatry, 3*(2), 233.

McVilly, K. R., Stancliffe, R. J., Parmenter, T. R., & Burton-Smith, R. M. (2006). I get by with a little help from my friends': Adults with intellectual disability discuss loneliness. *Journal of Applied Research in Intellectual Disabilities, 19*(2), 191–203. https://doi.org/10.1111/j.1468-3148.2005.00261.x

Menghini, D., Addona, F., Costanzo, F., & Vicari, S. (2010). Executive functions in individuals with Williams syndrome. *Journal of Intellectual Disability Research, 54*(5), 418–432. https://doi.org/10.1111/j.1365-2788.2010.01287.x

Mervis, C. B., Klein-Tasman, B. P., & Mastin, M. E. (2001). Adaptive behavior of 4- through 8-year-old children with Williams syndrome. *American Journal of Mental Retardation, 106*(1), 82–93. https://doi.org/10.1352/0895-8017(2001)106<0082:ABOTYO>2.0.CO;2

Mitchell, K. J. (2015). The genetic architecture of neurodevelopmental disorders. In K. J. Mitchell (Ed.), *The genetics of neurodevelopmental disorders*. Wiley. http://biorxiv.org/content/early/2014/09/19/009449

Morris-Rosendahl, D. J., & Crocq, M. A. (2020). Neurodevelopmental disorders-the history and future of a diagnostic concept. *Dialogues in Clinical Neuroscience, 22*(1), 65–72. https://doi.org/10.31887/DCNS.2020.22.1/macrocq

Mosner, M. G., Kinard, J. L., McWeeny, S., Shah, J. S., Markiewitz, N. D., Damiano-Goodwin, C. R., Burchinal, M. R., Rutherford, H., Greene, R. K., Treadway, M. T., & Dichter, G. S. (2017). Vicarious effort-based decision making in autism Spectrum disorders. *Journal of Autism and Developmental Disorders, 47*(10), 2992–3006. https://doi.org/10.1007/s10803-017-3220-3

Mussey, J. L., Travers, B. G., Klinger, L. G., & Klinger, M. R. (2015). Decision making skills in ASD: Performance on the Iowa gambling task. *Autism Research, 8*(1), 105–114. https://doi.org/10.1002/aur.1429

Nigg, J. T. (2013). Attention-deficit/hyperactivity disorder and adverse health outcomes. *Clinical psychology review, 33*(2), 215–228. https://doi.org/10.1016/j.cpr.2012.11.005

Orsmond, G. I., Krauss, M. W., & Seltzer, M. M. (2004). Peer relationships and social recreational activities among adolescents and adults with autism. *Journal of Autism and Developmental Disorders, 34*(3), 245–256. https://doi.org/10.1023/B:JADD.0000029547.96610.df

Peters, L. C., & Thompson, R. H. (2018). How teaching perspective taking to individuals with autism Spectrum disorders affects social skills: Findings from research and suggestions for practitioners. *Behavior Analysis in Practice, 11*(4), 467–478. https://doi.org/10.1007/s40617-018-0207-2

Peters, E., Hess, T. M., Västfjäll, D., & Auman, C. (2007). Adult age differences in dual information processes: Implications for the role of affective and deliberative processes in older adults' decision making. *Perspectives on Psychological Science, 2*(1), 1–23. https://doi.org/10.1111/j.1745-6916.2007.00025.x

Poon, K. (2018). Hot and cool executive functions in adolescence: Development and contributions to important developmental outcomes. *Frontiers in Psychology, 8*, 2311.

Rabbitt, P. (1997). Methodologies and models in the study of executive function. In *Methodology of frontal and executive functions* (pp. 1–38). Psychology Press.

Riby, D. M., Hanley, M., Kirk, H., et al. (2014). The interplay between anxiety and social functioning in Williams syndrome. *Journal of Autism and Developmental Disorders, 44*, 1220–1229. https://doi.org/10.1007/s10803-013-1984-7

Robic, S., Sonie, S., Fonlupt, P., Henaff, M.-A., Touil, N., Coricelli, G., … Schmitz, C. (2015). Decision making in a changing world: A study in autism spectrum disorders. *Journal of Autism and Developmental Disorders, 45*(6), 1603–1613. https://doi.org/10.1007/s10803-014-2311-7

Rosner, B. A., Hodapp, R. M., Fidler, D. J., Sagun, J. N., & Dykens, E. M. (2004). Social competence in persons with Prader-Willi, Williams and Down's syndromes. *Journal of Applied Research in Intellectual Disabilities, 17*(3), 209–217. https://doi.org/10.1111/j.1468-3148.2004.00200.x

Royston, R., Howlin, P., Waite, J., & Oliver, C. (2017). Anxiety disorders in Williams syndrome contrasted with intellectual disability and the general population: A systematic review and meta-analysis. *Journal of Autism and Developmental Disorders, 47*, 3765–3777. https://doi.org/10.1007/s10803-016-2909-z

Ruble, L. A., Willis, H., & Crabtree, V. (2008). Social skills group therapy for autism spectrum disorders. *Journal of Clinical Case Studies, 7*, 287–300.

Shah, P., Catmur, C., & Bird, G. (2016). Emotional decision making in autism spectrum disorder: The roles of interoception and alexithymia. *Molecular Autism, 7*, 43. https://doi.org/10.1186/s13229-016-0104-x

Shogren, K. A., Bovaird, J. A., Palmer, S. B., & Wehmeyer, M. L. (2010). Locus of control orientations in students with intellectual disability, learning disabilities, and no disabilities: A latent growth curve analysis. *Research and Practice for Persons with Severe Disabilities, 35*(3–4), 80–92. https://doi.org/10.2511/rpsd.35.3-4.80

Smith, M., Manduchi, B., Burke, É., Carroll, R., McCallion, P., & McCarron, M. (2020). Communication difficulties in adults with intellectual disability: Results from a national cross-sectional study. *Research in Developmental Disabilities, 97*, 103557. https://doi.org/10.1016/j.ridd.2019.103557

Solish, A., Perry, A., & Minnes, P. (2010). Participation of children with and without disabilities in social, recreational and leisure activities. *Journal of Applied Research in Intellectual Disabilities, 23*(3), 226–236. https://doi.org/10.1111/j.1468-3148.2009.00525.x

Stanovich, K. E. (2011). *Rationality and the reflective mind.* Oxford University Press.

Stanovich, K. E., & West, R. F. (2002). Individual differences in reasoning: Implications for the rationality debate? In T. Gilovich, D. W. Griffin, & D. Kahneman (Eds.), *Heuristics and biases: The psychology of intuitive judgment* (pp. 421–440). Cambridge University Press.

Stanovich, K. E., & West, R. F. (2008). On the relative independence of thinking biases and cognitive ability. *Journal of Personality and Social Psychology, 94*(4), 672–695. https://doi.org/10.1037/0022-3514.94.4.672

Switzky, H. N. (2006). The importance of cognitive-motivational variables in understanding mental retardation in the 21st century. In H. N. Switzky & S. Greenspan (Eds.), *What is mental retardation: Ideas for an evolving disability in the 21st century* (pp. 267–282). American Association on Mental Retardation.

Tager-Flusberg, H. (2007). Evaluating the theory-of-mind hypothesis of autism. *Current Directions in Psychological Science, 16*(6), 311–315. https://doi.org/10.1111/j.1467-8721.2007.00527.x

Tanaka, M., Malakoff, M., Bennett-Gates, D., & Zigler, E. (2001). Development of an outerdirected style of problem solving in individuals with and without mental retardation. *Journal of Applied Developmental Psychology, 22*, 191–198.

Toplak, M. E., West, R. F., & Stanovich, K. E. (2011). The cognitive reflection test as a predictor of performance on heuristics-and-biases tasks. *Memory & Cognition, 39*(7), 1275–1289. https://doi.org/10.3758/s13421-011-0104-1

Tversky, A., & Kahneman, D. (1981). The framing of decisions and the psychology of choice. *Science (New York, N.Y.), 211*(4481), 453–458. https://doi.org/10.1126/science.7455683

Tymchuk, A. J., Yokota, A., & Rahbar, B. (1990). Decision making abilities of mothers with mental retardation. *Research in Developmental Disabilities, 11*(1), 97–109. https://doi.org/10.1016/0891-4222(90)90007-U

Wehmeyer, M. L., & Kelchner, K. (1994). Interpersonal cognitive problem-solving skills of individuals with mental retardation. *Education and Training in Mental Retardation and Developmental Disabilities, 29*, 265–278.

Willner, P., Bailey, R., Parry, R., & Dymond, S. (2010a). Evaluation of executive functioning in people with intellectual disabilities. *Journal of Intellectual Disability Research, 54*(4), 366–379. https://doi.org/10.1111/j.1365-2788.2010.01249.x

Willner, P., Bailey, R., Parry, R., & Dymond, S. (2010b). Performance in temporal discounting tasks by people with intellectual disabilities reveals difficulties in decision making and impulse control. *American Journal of Intellectual Developmental Disabilities, 115*(2), 157–171.

Wishart, J. G., Cebula, K. R., Willis, D. S., & Pitcairn, T. K. (2007). Understanding of facial expressions of emotion by children with intellectual disabilities of differing aetiology. *Journal of Intellectual Disability Research, 51*(7), 551–563.

Woodcock, K. A., Cheung, C., González Marx, D., & Mandy, W. (2020). Social Decision making in Autistic Adolescents: The Role of Theory of Mind, Executive Functioning and Emotion Regulation. *Journal of autism and developmental disorders, 50*(7), 2501–2512. https://doi.org/10.1007/s10803-019-03975-5.

Wu, H. C., White, S., Rees, G., & Burgess, P. W. (2018). Executive function in high-functioning autism: Decision making consistency as a characteristic gambling behaviour. *Cortex; A Journal Devoted to the Study of the Nervous System and Behavior, 107*, 21–36. https://doi.org/10.1016/j.cortex.2018.01.013

Zantinge, G., van Rijn, S., Stockmann, L., & Swaab, H. (2017). Psychophysiological responses to emotions of others in young children with autism spectrum disorders: Correlates of social functioning. *Autism Research, 10*(9), 1499–1509. https://doi.org/10.1002/aur.1794

Zeif, D., & Yechiam, E. (2020). Autism is not associated with poor or enhanced performance on the Iowa gambling task: A meta-analysis. *Neuroscience & Biobehavioral Reviews, 113*, 440–447.

Zhang, L., Tang, J., Dong, Y., Ji, Y., Tao, R., Liang, Z., Chen, J., Wu, Y., & Wang, K. (2015). Similarities and differences in decision-making impairments between autism Spectrum disorder and schizophrenia. *Frontiers in Behavioral Neuroscience, 9*, 259. https://doi.org/10.3389/fnbeh.2015.00259

Chapter 12
Common Sense in Persons with Intellectual Disabilities: The Challenge of Deciding to Say *No* to Social Manipulation

Stephen Greenspan

Introduction

In the animal ethology literature, there is a method called "motivation analysis" (Lorenz, 1979), in which a particular action (e.g., a dog or bird responding in a friendly or aggressive manner to another animal from its own species) is explained as the result of the strength of two or more congruent or incongruent motivational factors operating together. Such a multifactor method improves behavior prediction in two ways: (a) few behaviors can be explained by a single motive, and (b) a slight shift in the balance of competing motives (e.g., fear and hostility) explains how an organism such as a dog can switch instantly from one behavior (e.g., non-aggression) to its opposite. In the human psychology literature, such an approach to explaining or predicting a specific action or non-action is uncommon, as there is a tendency to rely on a single independent variable which however powerful (an example being IQ) rarely predicts even a third of the variance on any given dependent variable, even when a strong relationship might be expected.

Three reasons for one's inability to use IQ to predict adequately the behavior of someone with intellectual disabilities (ID) are the following: (a) there are other aspects of cognition (such as executive functions and social intelligence) that are imperfectly aligned with IQ and which likely have equal or greater importance in predicting non-academic behavior, such as most forms of decision making (Sternberg, 2019), (b) situational factors (such as time pressure) and personality factors (such as friendliness) independent of IQ play an important role and are generally not entered into the equation (Cole, 2003), and (c) individual self-regulatory

S. Greenspan (✉)
University of Connecticut, Mansfield, CT, USA
e-mail: stephen.greenspan@gmail.com

I. Khemka, L. Hickson (eds.), *Decision Making by Individuals with Intellectual and Developmental Disabilities*, Positive Psychology and Disability Series,
https://doi.org/10.1007/978-3-030-74675-9_12

factors, such as emotion (e.g., greed, fear, anger) and biological state (e.g., alertness, exhaustion) that vary within the ID population (and within an individual at a particular point in time), likely mediate the relationship between person and situation for any given person (Baumeister et al., 2007).

In this chapter, I take a *whole-person* (Zigler, 1973) approach to address multiple aspects of personality and other competence variables that play a role in the decision-making behavior of people with ID. The view of ID here is not as a unitary class of people who all have the same external personality trait, but rather as a diverse population with considerable variability on a range of personality domains and in the quality of decisions about non-academic challenges. Despite the fact that people with ID are more at risk for making poor everyday decisions, there are some people with ID who generally refrain from making poor decisions, and even when someone with ID makes a poor decision, there likely are other occasions when the same person has not done so. It is my intention in this chapter to provide an explanation for why I believe this to be the case.

The Meaning and Centrality of Common Sense

The key underlying factor that is addressed in poor decision making (which I term foolishness) is *common sense*, which can be defined as awareness of obvious risk (Greenspan, 2009a, 2019). Within the relatively small literature on common sense and foolishness (and the much larger literature on wisdom), one will find almost no discussion of risk awareness as a factor, let alone a central one. In contrast, common sense in my view can be placed on a continuum, with unawareness of obvious risk (termed foolishness) at the low end, awareness of non-obvious risk (termed wisdom) at the high end, and common sense (awareness of obvious but not of non-obvious risk) in the middle. This developmental continuum is depicted in Fig. 12.1.

People who frequently demonstrate foolishness can, thus, be described as lacking common sense, while wisdom can be described as a more advanced form of common sense. This formulation is different from that of Robert Sternberg (2002), who has written the most on these topics, as he uses all three terms—common

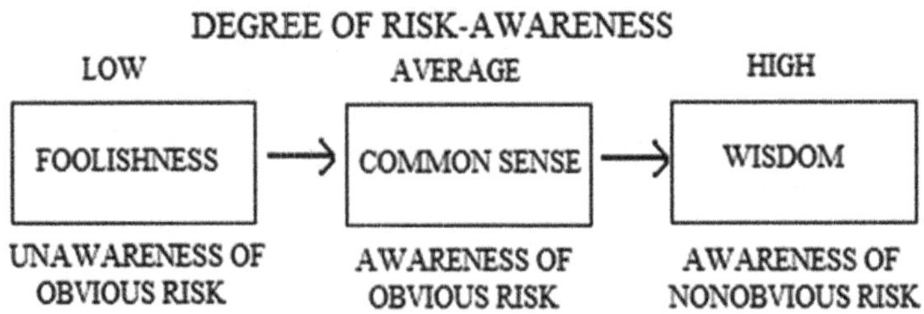

Fig. 12.1 Placing common sense on a developmental continuum. (Source: Greenspan 2019)

sense, foolishness, and (un)wisdom—as roughly identical and does not pay special (or really any) attention to risk or risk awareness.

Another way of thinking about these constructs (see Fig. 12.2) is in terms of a hypothesized normal curve distribution as follows: a small portion (let us say 10%) of the population can be expected to demonstrate wisdom (awareness of both obvious and non-obvious risk), the vast majority (let us say 80%) can be expected to demonstrate common sense (awareness of obvious risk but not of non-obvious risk), while another small portion (again, let us say 10%) can be expected to demonstrate foolishness (unawareness of obvious risk). As described below, however, an appreciation of context, biological state and task complexity also sheds light on the possibility that a person with (or without) ID could demonstrate common sense on one occasion but an absence of common sense on a different occasion.

An aspect of common sense that has been emphasized since it was first written about by the ancient Greek philosopher Aristotle is that it is an intuitive and incidentally learned form of knowledge, rather than specialized expertise that can only be acquired through formal instruction. That is why this basic form of practical intelligence is so widely possessed in the general population and why its absence in an individual can be concerning enough to warrant protective arrangements such as conservatorships and group homes.

From a public health standpoint, absence of common sense is a serious problem, as unawareness of obvious risk can get one into a wide range of difficulties, including going to jail, being swindled or molested, getting into accidents, being assaulted or killed, becoming seriously ill, etc. The literature on decision making in economics (less so in disability) has mainly focused on the processes involved in optimal decisions (Szpiro, 2020) about relatively trivial matters (e.g., maximizing profits), with risk being conceptualized in minor ways (e.g., making less money). Common sense, viewed as awareness of obvious risk, has more survival value, such as avoiding the worst (e.g., death) kinds of outcomes. The study of common sense thus is more likely to lead one to studying poor rather than optimal decisions and to focus on those (such as people with ID) who are more likely to demonstrate a dangerous absence of common sense.

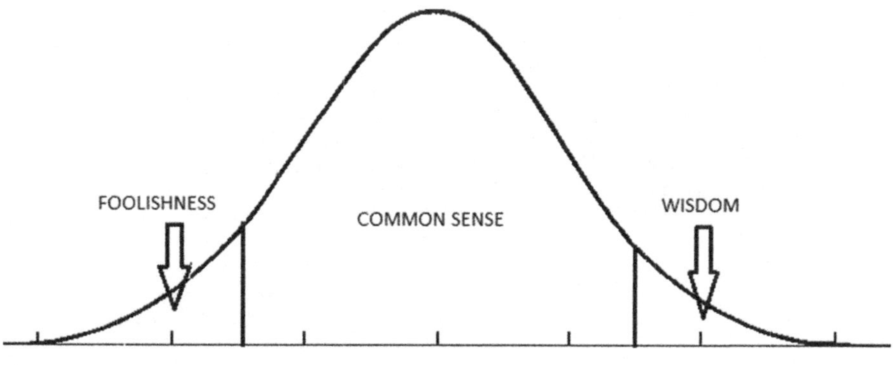

Fig. 12.2 Hypothesized normal distribution of risk awareness. (Source: Greenspan 2019)

While people who qualify as having ID (i.e., who have a full-scale IQ below 70–75 and significant limitations in adaptive behavior) are more likely to fall in that category, a focus on risk obliviousness as a critical factor for people in need of formal services and supports leads one away from a narrow focus on officially labeled ID and toward a broader focus on developmental disability (DD), or what has been termed *ID equivalence* (Greenspan et al., 2015). DD is a more inclusive class of brain-impaired people than is ID and incorporates many brain-based conditions that could be characterized as common sense deficit disorders. That is because absence of risk awareness (a central aspect of adaptive behavior when it is viewed as everyday judgment) is generally more likely to identify support-needing people with brain-based disorders than is either below-70 IQ or deficiencies in routine activities of daily living (e.g., cooking, bed-making, toileting) that are so heavily represented in the most widely used adaptive behavior instruments (Saulnier & Klaiman, 2018).

Two broad types of risk face all of us, and these are depicted in Fig. 12.3. They fall roughly into practical/physical risk (such as getting run over or starving) and social risk (such as getting fired or incarcerated). Social risk can be broken down further into *induced* and *non-induced*, with induced risk occurring when one or more other people use pressure or deception in order to get the target of the manipulation to do something undesired or not in their best interest. Foolish action occurring in response to such inducement can be described as *gullible,* the subject of a book by this author (Greenspan, 2009a, 2009b) and something that has begun to be recognized as especially problematic for people with ID, as reflected in the fact that the Vineland Adaptive Behavior Scale (VABS) went from zero gullibility items in its first edition to one in its second edition to four in the VABS-3 (Sparrow et al., 2016).

People with ID, even the subcategory of mild ID, typically are in need of protective arrangements (conservatorships, supported living, etc.) for one primary reason: their increased vulnerability to being harmed by social or physical danger (Finlayson

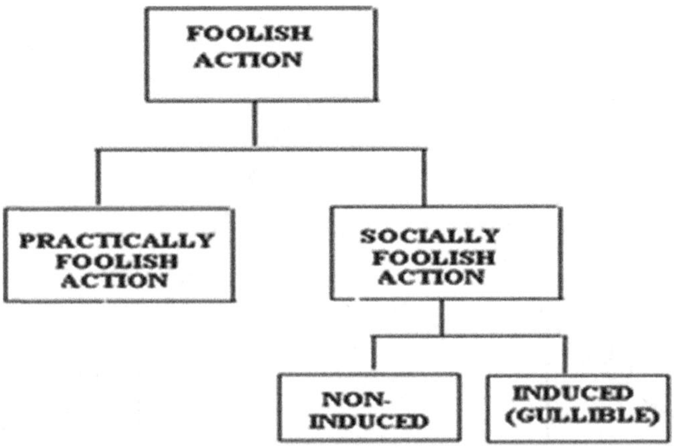

Fig. 12.3 Types of foolish action. (Source: Greenspan 2019)

et al., 2010). The ability to recognize and avoid danger, especially when it is fairly obvious, is, as mentioned earlier, what I am calling *common sense*. A relative lack of common sense thus is central to almost all people with ID, and understanding of this vulnerability is a central concern to caregivers and is a major reason for protective arrangements.

Factors Contributing to Intelligent or Unintelligent Decisions

Cognition obviously plays a major role in explaining decisions that can be described as intelligent or unintelligent, but it is not the whole story. A four-factor explanatory model, based loosely on the motivational theory of Martin Ford (1992), was developed by this author (Greenspan, 2009b, 2019) as a framework for explaining *foolish action*, i.e., behavior which fails to avoid or deal effectively with social or practical risk. Here it is used specifically as a framework for explaining decision making.

A linear action model is depicted in Fig. 12.4. In this model, foolish action is viewed as an outcome which occurs as the result of the intersection of four broad factors, which come together at a moment in time and which in some combination cause a person to act foolishly or non-foolishly (i.e., with common sense). The four factors each could be subdivided further, so it should be understood that these are broad-based and include many possible sub-factors. One of the four factors (situation) is external to the actor, while the other three factors (cognition, affect/state, and personality) are internal to the person. The weight to be attributed to each factor varies according to each specific act and each actor, so for a given instance of behavior some factors may be very strong while the remaining factors may provide a weak, nonexistent, or countervailing influence.

In most instances, fewer than four factors may be involved in motivating a specific foolish act; there may even be some forces that are very weak or actually working in the opposite direction. In such a case, a foolish outcome is less predictable than it would be where all four factors are present, strong, and all pointing in the same direction. Of course, we generally do not know all of the factors or their strengths operating on an individual at a given moment in time and thus cannot effectively quantify these factors. At this point we can mainly use the model descriptively, to do post hoc analyses (*behavioral autopsies*) of foolish or non-foolish acts.

Fig. 12.4 Explanatory model of foolish action. (Source: Greenspan 2019)

Situation

The first prong, "situation," is the problem posed to the actor, which they have to deal with, either successfully or unsuccessfully. Situations vary in terms of their complexity and the extent to which they pull or push the actor toward or away from acting foolishly. As unawareness of obvious risk is central to my theory of foolish behavior, the situation must contain some degree of risk, although the degree of obviousness can of course vary from one situation to another. Aspects of a situation that create or ameliorate stress, including the presence or absence of others who encourage (eggers on) or reduce (support persons) incompetent behavior, are important factors in explaining foolish action, especially in those (such as children or adults with ID) who rely heavily on others for guidance.

Cognition

This is the collection of general and specific knowledge, thinking, and language skills that an individual brings to the risky situation and which enables them to fully recognize or evade the risks facing them. Although cognition traditionally has been identified as another word for intelligence, it has long been understood (Anastasi, 1983) that full-scale IQ scores mainly tap academic potential, while there are other largely independent content aspects of intelligence, such as practical intelligence and social intelligence. Social intelligence is especially important in the behavior of people with ID, as foolish behavior typically involves social functioning and it is a core problem for people with ID and related disorders (Thiébaut et al., 2016). Although not generally termed a measure of intelligence, neuropsychological tests of *executive functions* (planning, reasoning, problem-solving cause-effect thinking, etc.) are being seen as in some ways excellent measures of intelligence, with particular relevance to everyday adaptive behavior (Zelazo & Müller, 2002).

Affect/State

This refers to aspects of one's biological self-regulatory system. Affect involves emotions or feelings (such as fear or anger), while state refers to one's level of equilibrium or disequilibrium (such as exhaustion, intoxication, or sexual arousal). When self-regulatory processes are non-optimal, vulnerability to foolish action is increased (Heatherton & Wagner, 2011; Martin & Delgado, 2011), although affect and state work in somewhat different ways: high affect pushing one toward acting foolishly, while *disequilibrated* state sub-factors may pull one to behave foolishly by reducing one's ability to resist.

Personality

This refers to behavioral tendencies or needs that are relatively stable and which differentiate one person from another. It is typically described in terms of traits, such as the *Big Five* (Digman, 1990) while I find a needs approach, as in the Reiss 16-Needs Profile (Reiss, 2000), to be more useful in explaining behavior. There are many personality variables that could affect one's ability to make safe or unsafe decisions. Here are some, from the Reiss profile, that are particularly relevant to decision-makers who have a developmental disorder: (a) status (need to feel significant): an important motivator for people with ID, as they typically feel very insignificant; (b) social contact (need for friends): important for people with ID as they often are very socially isolated; (c) honor (need to follow traditional rules): important for people with ID as they are generally able to survive by concretely following clear-cut rules; (d) romance (need for a love relationship): can be a major motivator for people who may have difficulty finding an intimate partner; (e) order (need for sameness): can be a motivator for people with ID who may be more likely to be bothered by change or uncertainty; (f) acceptance (need to feel appreciated): important for people who are often rejected by non-disabled age-mates; and (g) tranquility (need to feel secure): important for people who are so dependent on others.

Two Decision-Making Case Studies

To illustrate the utility of the model, and the four factors in explaining foolish or non-foolish acts, I have chosen to analyze two case examples (lightly fictionalized accounts of real incidents), both involving a person with ID confronted with a situation that contains some risk. The first of these examples depicts a decision that had an unfortunate outcome, while the other example depicts a decision that had a better outcome. My purpose in doing these analyses is partly to explain the elements contributing to poor decisions by people with ID but more importantly to (a) show how and why decisions made by people with ID are not always poor ones and also (b) draw lessons regarding what could be done to help people with ID to make better and safer decisions. The four-factor model is thus used to explain the behaviors of the target individuals, but will also serve as the organizing framework in a later section focusing on practical suggestions for improving the decisions of people with ID.

Case 1: A Risky Dating Decision

This case involves *Annie*, a woman in her early 20s with mild ID who lives alone in an apartment provided by a public agency. Annie mainly gets informal drop-in support from her mother, who lives in the same city. The agency funding the apartment had cut back substantially on their monitoring of Annie, in part because she was doing well, but also because the young woman resented what she considered their snooping and told them she did not want their services. In the same apartment building there was living a 30-year-old man named *James* who took an interest in Annie and started asking her out. James did not have a developmental disorder, but he had been arrested a few times. Annie had never dated and very excitedly told her mother about James. The mom had serious reservations, but she decided to stay out of it, as she did not want to squelch her daughter's happiness. That changed later when a serious problem emerged.

During the few weeks before Annie's mother called the police, consensual sexual activity between James and Annie (who had been fitted at her mother's urging with a birth control device a year earlier) took place. But the main problem that emerged was financial. Annie had a bank card tied to an account with a balance of around $2000 and James managed to drain all of it, both by making her pay for meals for him as well as others and also by soliciting daily cash payments, which he told her were a requirement for him to be her boyfriend. The police decided against arresting James, as Annie was of legal age and there was insufficient evidence of physical coercion. But the mother moved Annie back into her home, ordered her to end the relationship with James, and discouraged discussion of any further out-of-home placements.

When using the four-factor model to deconstruct a foolish or non-foolish act, it is essential to address a specific micro-decision. There are several such micro-decisions by Annie that could be addressed within the context of this story: agreeing to go on a date with James, agreeing to have sex with him, agreeing to let him use her bank card for meals and other purchases, and agreeing to pay him to be her boyfriend. With regard to the first two (dating and sex), I do not necessarily consider those acts foolish, as adults with ID have a right to date as well as have sex and showing poor judgment in picking a romantic partner is hardly limited to people with ID. Furthermore, one would need information about the specific tactics used by James to get Annie in bed with him, and such information is lacking.

The main foolishness here, at least for purposes of my analysis, involves Annie's agreeing to allow James to use her bank card for purchases and to pay for his companionship, in the process depleting all her savings. While more micro knowledge here would also be helpful (e.g., about the exact words James used to convey the boyfriend subsidy idea), the practice is so outside the norm (except perhaps for gigolos or mistresses kept by very rich people) that an analysis of the action is impossible without knowing all the details.

At least two situational sub-factors contributed to this unfortunate outcome. The first one is that Annie was living in an apartment by herself and with essentially no official supports. This made it possible for her to be confronted by a socially risky situation that she lacked the skills to navigate on her own. It is a little unusual for people with ID to not have a housemate, but the regional director of this agency was operating on the overly concrete idea that sharing a home would violate the *normalization* principle (in fact, for most adults in the United States, living with another non-related person is very normal, and living alone makes many people both with and without disabilities lonely and anxious). This placement, along with the pulling of formal supports, also reflected an overestimate of Annie's adaptive functioning level, stemming from the fact that the instrument the agency used mainly tapped daily living skills with very incomplete assessment of her social competence (this has been a consistent theme of my scholarly writing for over four decades). In fact, Annie's lack of common sense (recognition of obvious risk) is much more likely to manifest in social than in everyday practical situations. The other situational sub-factor operating here is that Annie had the misfortune to be courted by a man with a very malevolent agenda, namely, to exploit her sexually and financially. While not a sophisticated person, James recognized Annie's cognitive and emotional vulnerabilities and knew how to manipulate her to get his own needs met with no concern for her interests or well-being. If Annie had hooked up with someone with benevolent qualities, this story could have had a much happier ending.

The three within-person factors in the model also entered into the mix. Cognition played an important role, as Annie lacked an understanding that paying someone to be one's boyfriend is very unusual and a major red flag indicating that one is being exploited. She also did not understand, apparently, the risk that her generosity toward James was posing to her financial solvency. Affect/state was a major contributor here, particularly affect. Specifically, Annie desperately wanted to have a romantic partner and was afraid that if she said no to paying James, that the relationship would end. Personality entered in here, as Annie has a tendency to take her cues from others (Switzky, 2001), and saying no to James would have been in conflict with that need. In sum, all three of the within-person factors were pulling Annie to a foolish decision and the situational factors (pressure from James, insufficient protection from caregivers) also contributed to pushing her toward a very unfortunate outcome.

Case 2: Close Call with a Criminal Scheme

This case involved *Peter*, a man of 19, also with mild ID, who was still enrolled in a special education program at his high school. On weekends and weekday afternoons, Peter would spend time in an outdoor mall where young people in the town would congregate. One day Peter ran into a former classmate named Andy at the mall. Andy asked Peter to help him teach a lesson to another young man named Stan who had disrespected Andy's sister. The idea was that Peter would be the lookout outside Stan's apartment, while Andy went in and slapped him around a little. Peter agreed with this plan after Andy told him that if he did not help he would no longer be his friend. Peter was not enthused about this as he was not an aggressive person and he had no reason to dislike Stan, a person he barely knew. But he believed the story about Stan mistreating Andy's sister and he did not want to lose Andy's friendship. As they were walking up the sidewalk to the proposed victim's home, Andy pulled a metal pipe out of his backpack and Peter realized that Andy planned to do more than slap Stan around. Peter told Andy he changed his mind and would not help. Andy threatened to harm Peter if he did not help, but Peter ran away and called his father who immediately notified the police. As a result Peter avoided being involved in an assault that could very well have resulted in Stan's death.

This scenario (a person with ID being sucked into participating in a severe assault against someone they had no beef with) is not uncommon. My knowledge comes from two or three similar criminal cases (Greenspan, 2016) I have worked on as an expert witness, where the assault resulted in death and the person with ID was facing the most serious of criminal penalties. In fact, Peter was very fortunate that he did not suffer a similar fate. Here is my analysis, using the four-factor model, regarding why this story took the fortunate turn that it did.

One reason why Peter made the very wise decision to abstain from the assault is that the situational factors were not as compelling as they might have been. More typically, initiators do a better job of keeping their true motives (to kill rather than just beat up a victim) secret until after the assault has begun, and the same is also true about the skill and ferocity with which they threaten the person who they lure into the scheme. So it is very possible that Peter (who actually started to participate before he got cold feet) would have become fully ensnared in the planned crime if the situational pressures had been more compelling. Still, there are within-person factors that had a lot to do with explaining Peter's sensible decision to walk away from this very risky situation.

Although Peter was a person with ID, he was at the high end of the ID continuum, which helped him to figure out Andy's true intentions (severe assault and possible killing of Stan) and place the scenario in the category of activities he should not be involved in. Personality also contributed in two ways: (a) Peter is a kind person and hurting someone fundamentally violated his values, and (b) Peter is somewhat inner-directed and thus able on occasion to assert his autonomy in the face of countervailing situational pressures. Finally, as with the first case, affect/state played a major role, but here it worked in a non-foolish direction, by moving Peter away from rather than toward making a poor decision. Specifically, any anxiety that Peter may have felt about disappointing Andy was greatly overshadowed by the fear he felt at the prospect of participating in a serious crime.

The Importance of Deciding to Say *No*

Much of the emphasis in the decision-making literature is on the importance of generating optimal solutions to problems (Kahneman & Tversky, 2000), but a theme of my own research program has been on the importance of recognizing and avoiding the worst possible solutions (Greenspan, 2011). In terms of common sense, this involves saying *no* to courses of action (often suggested by others) that could significantly impact, if not ruin, one's life. While these kinds of decisions sometimes involve money—the focus of decision-making research in economics (see Ingersoll, 1987)—they almost always involve other people, specifically people who have a malign agenda, whether trying to get ahold of your money (as in case 1 above), to involve you in potential misconduct (as in case 2), or to use you for their own sexual or other purposes (as in both cases).

Everyone who reads this chapter, as well as all people who live with any degree of freedom in a community setting (including people with ID who more and more have such freedom), will at some point in their lives likely be required to call upon their common sense to say *no* to an unwanted and socially dangerous demand. But given the typically hidden nature of the true motives of malign demand-makers, the exercise of such common sense requires some cognitive skills (recognizing motives and dangers) that people with cognitive impairments are less likely than the general population to possess. Manipulation situations also make affective and character demands that can be challenging, especially to those who have less opportunity to assert their autonomy or who have fewer friends and thus are more susceptible to entreaties from someone pretending to be a friend.

It is important to consider what, if anything, can be done to decrease the likelihood that someone with ID will be tempted to make such a dangerous decision and to protect them from the full set of negative consequences when such an action is taken. Such decisions may not occur very often, but it should be kept in mind that it takes only one flunked major *decision test* to move one's life trajectory from success to failure or even tragedy.

Is Common Sense Teachable to People with ID?

Training programs aimed at improving decision-making skills for adolescents and adults with ID have mainly utilized a social problem-solving methodology (see Hickson & Khemka, 2014) in which hypothetical situations are used both to assess and teach disengagement tactics. Khemka (2000) and Hickson and Khemka (2016) have had success in helping individuals with ID to utilize a stepwise decision-making strategy to improve self-protective decision making in hypothetical situations, but having that improved skill generalize into saying *no* to real-life manipulators is more difficult to measure or achieve.

The Generalization Problem

As an example of the difficulty in translating advice into behavior, look at the ease with which typically developing children can be talked by a stranger into getting into their car with them, even shortly after being warned by their parents never to do such a thing (Hadden, 2015). In one notable case, when a child's parent asked why he got in the car of a stranger (a research confederate playing the role), the child responded "but he was looking for his dog." The two reasons typically given for why abduction prevention advice doesn't work are: (a) the word *stranger* connotes *bad person* while successful abductors make an effort to come across as warm and friendly, and (b) emotion trumps cognition, as the repeated success of any cover story involving puppies demonstrates.

This problem of inoculating potential victims against social victimization is not limited to normally developing children, but can be found in neurotypical adults as well. As example, shortly after completing the first (and still only) comprehensive book on human gullibility, I learned I had been swindled by master con artist Bernard Madoff (Greenspan, 2009a). This caused a Canadian financial columnist to write "the first Greenspan, Alan, will be remembered as the economist who didn't see it [financial collapse] coming while the other Greenspan, Stephen, will be remembered as the gullibility expert who forgot to read his own book" (Fulford, 2008, p. 52). In my analysis of why I was hoodwinked, I used the four-factor model to show how situation (a very skillful scam, plus the endorsement by so many gulls), cognition (my ignorance of finance), personality (a trusting and compliant nature), and affect (greed, plus the excitement of finding an attractive investment strategy) all came together to derail my usual skepticism.

My reason for bringing this up is to indicate that (a) gullibility is an ever-present possibility for all humans, (b) no abuse-inoculation curriculum can anticipate all of the scenarios that manipulators will come up with to ensnare their victims, and (c) individual differences in personality, cognition, and affect, in combination with compelling situational factors, can put anyone, especially people with ID, in danger of being taken in by a manipulator, regardless of their performance on decision-making analogue exercises. That said, there undoubtedly is some benefit in helping adolescents and adults with ID to recognize socially dangerous situations, to learn decisional tactics for dealing with such situations, and to buttress their willingness to stand up to malefactors. A more detailed account of issues of gullibility and how they impacts decision making and training in ID is presented in Greenspan et al. (2001).

Social Decision-Making Training and the Four-Factor Model

The theory underlying the decision-making training model used by Hickson and Khemka (2014) bears many similarities to the four-factor model of common sense, especially the three within-person factors in that model. The main difference is in terminology, with the word *motivation* used by them to refer to what I term *personality*, while the other two terms *cognition* and *emotion* (affect) are virtually identical. However, in line with the theory of Martin Ford (1992) from whom my model was derived, I use the word motivation to refer to all four factors working together and not just to one of those factors. The fourth element—*situations*—while addressed in the scenarios used as training materials, is not framed as a curricular focus. In a final section, in something of a nod to the paternalistic origins of this field, I argue that situational interventions should not be abandoned in the effort to keep people with ID safe from manipulation and exploitation.

Decision-making curricula utilizing hypothetical situations understandably are mainly cognitive, in that they seek to improve the ability of participants to (a) recognize the manipulative aspects of depicted scenarios and (b) generate and evaluate tactics that can be used to deal with such scenarios. These curricula aim to neutralize the emotional pull of manipulative situations through the use of group discussion and stop-and-think methods, in which participants are helped to identify the affective obstacles to saying *no* and switch from an impulsive to a deliberative problem-solving mode. Finally, the curricula address the role of personality by using assertion training methods to strengthen the willpower of subjects and help them to understand that saying *no* is their moral and legal right.

Conclusion: Risk Is Not Always Dignified

Robert Perske (1972) coined the expression *dignity of risk* when writing about what Bradley and Knoll (1995) termed the *community revolution in disability services*. This refers to the fact that the welcome move away from highly protective segregated programs such as large congregate facilities has a downside, and that is that individuals with ID have become much freer to make decisions that could put them in harm's way. A prime example, which occupied much of Perske's (1972) and my (Greenspan, 2016) time over the past 25 years, is the problem of wrongful conviction for murder based on too-easily obtained false confessions (Leo, 1996). In fact, my rediscovery of the centrality of gullibility in ID (Greenspan et al., 2001) grew out of my involvement in a case of coerced false confession to murder where the brain-impaired defendant, Richard Lapointe (finally freed after 26 years of wrongful imprisonment), was described by a family member as "the most gullible person on the planet."

In an interrogation, gullible compliance is made likely through the use of multiple interrogators creating a coercive situation based on the *Reid technique (*a highly effective accusatory interrogation method relying often on false evidence to convince a subject that resistance is futile), in combination with intraindividual factors, specifically cognition (failure to understand that one does not have to speak to police and that interrogations are exercises in legally sanctioned lying), personality (a high degree of credulity and interpersonal trust), and affect/state (terror induction combined with sleep deprivation).

No amount of training is likely to prove effective in preventing most people with (or many without) ID from facing a real threat of being hoodwinked in such a highly coercive situation. So for the most serious kinds of situations, such as a police interrogation (where the consequence of one's gullibility could be life in prison), the only reliable intervention is situational. Specifically, it should be a rule (as it is in the United Kingdom) that cognitively impaired individuals can be interrogated only in the presence of a lawyer or other trained advocate who will ensure that the person is aware of their rights and understands what is happening. Another safeguard, which is often followed but not typically mandated, is to videotape all such conversations, in order that a judge or jury can decide if incriminating statements are knowing, voluntary, and free of coercion.

A very large percentage of adults with mild ID living in the community do not have conservators or guardians, although representative payees for handling of social security benefits are commonplace. The methods used to evaluate whether someone needs a guardian are similar to adaptive behavior assessments, and as with such assessments, they are very heavy on practical skills (such as balancing a checkbook) but very light on social skills (such as knowing when to resist writing a check). Yet common sense, like its opposite (risk-oblivious action), is primarily an exercise of social judgment, and social judgment is critical for making potentially life-altering decisions. Thus, a very important situational protection against making poor decisions is to make certain that every person with ID who is socially vulnerable is properly assessed and provided with a guardian or its equivalent if needed (Millar, 2013).

The topic of people with ID exercising common sense or its opposite in the making of important decisions is one that opens itself to many research directions. One topic that seems particularly important is exploring what Reyna and Zayas (2014) termed the *neuroscience of risky decision making*. Specifically, much remains to be learned about the way in which the brain impairments of people with ID and related disorders restrict them when they make potentially dangerous decisions. One hoped-for outcome of such research would be to shed light on how people with ID could be helped to best exercise the common sense that they already possess.

References

Anastasi, A. (1983). What do intelligence tests measure? In S. B. Anderson & J. S. Hemlick (Eds.), *On educational testing: intelligence, performance standards, test anxiety, and latent traits* (pp. 5–28). Jossey-Bass.

Baumeister, R. F., Vohs, K. D., DeWall, C. N., & Zhang, L. (2007). How emotion shapes behavior: Feedback, anticipation, and reflection, rather than direct causation. *Personality and Social Psychology Review, 11*, 167–203.

Bradley, V., & Knoll, J. (1995). Shifting paradigms in services to people with disabilities. In O. C. Karan & S. Greenspan (Eds.), *Community rehabilitation services for people with disabilities*. Butterworth Heinemann.

Cole, M. (2003, June 22). *Vygotsky and context*. Paper presented at the International Society for Theoretical Psychology, Istanbul. http://lchc.ucsd.edu/People/MCole/lsvcontext.html

Digman, J. M. (1990). Personality structure: Emergence of the five-factor model. *Annual Review of Psychology, 41*, 417–440.

Finlayson, J., Morrison, J., Jackson, A., Mantry, D., & Cooper, S.-A. (2010). Injuries, falls and accidents among adults with intellectual disabilities. Prospective cohort study. *Intellectual Disability Research, 54*(11). https://doi.org/10.1111/j.1365-2788.2010.01319

Ford, M. E. (1992). *Motivating humans: Goals, emotions, and personal agency beliefs*. Sage.

Fulford, R. (2008, December 27). *Ponzi would have been proud*. The National Post. http://www.robertfulford.com/2008-12-27-ponzi.html

Greenspan, S. (2009a, January 3). Why we keep falling for financial scams. *Wall Street Journal*. www.wsj.com/articles/SB123093987596650197

Greenspan, S. (2009b). Foolish action in adults with intellectual disabilities: The forgotten problem of risk-unawareness. In L. M. Glidden (Ed.), *International review of research in mental retardation*. Elsevier.

Greenspan, S. (2011). Execution of Joe Arridy: Comments of a forensic expert. *Intellectual & Developmental Disabilities, 49*(3), 197–202.

Greenspan, S. (2016). Discriminatory killing and execution of people with ID. In J. R. Lutzker, K. Guastaferro, & M. Graham (Eds.), *Maltreatment of people with intellectual and developmental disabilities*. AAIDD.

Greenspan, S. (2019). *Anatomy of foolishness: The overlooked problem of risk-unawareness*. Hamilton Books/ Rowman & Littlefield.

Greenspan, S., Loughlin, G., & Black, R. (2001). Credulity and gullibility in persons with mental retardation. In L. M. Glidden (Ed.), *International review of research in mental retardation* (Vol. 24, pp. 101–135). Academic.

Greenspan, S., Brown, N. N., & Edwards, W. (2015). FASD and the concept of "intellectual disability equivalence". In M. Nelson & M. Trussler (Eds.), *Law and ethics in fetal alcohol Spectrum disorder*. Springer.

Hadden, J. (2015, May 4). Social experiment sheds light on how easy child abductions can be. *Global News*. www.globalnews.ca/news/1977284

Heatherton, T. F., & Wagner, D. D. (2011). Cognitive neuroscience of self-regulation failure. *Trends in Cognitive Science, 15*(3), 132–139. https://doi.org/10.1016/j.tics.2010.12.005

Hickson, L., & Khemka, I. (2014). The psychology of decision making. In R. Hodapp (Ed.), *International review of research in developmental disabilities* (Vol. 47, pp. 185–229). Elsevier.

Hickson, L., & Khemka, I. (2016). Prevention of maltreatment of adults with intellectual and developmental disabilities. In J. R. Lutzker, K. Guastaferro, ; M. L. Benka-Coker (Eds.), *Maltreatment of people with intellectual and developmental disabilities* (pp. 233–261). American Association on Intellectual and Developmental Disabilities.

Ingersoll, J. (1987). *Theory of financial decision making*. Rowman & Littlefield.

Kahneman, D., & Tversky, A. (2000). *Choices, values and frames*. Cambridge University Press.

Khemka, I. (2000). Increasing independent decision-making skills of women with mental retardation in simulated interpersonal situations of abuse. *American Journal on Mental Retardation, 105*(5), 387–401.

Leo, R. A. (1996). Miranda's revenge: Police interrogation as a confidence game. *Law and Society Review, 30*, 259–260.

Lorenz, K. (1979). *On aggression*. Methuen.

Martin, L. N., & Delgado, M. R. (2011). The influence of emotion regulation on decision making under risk. *Journal of Cognitive Neuroscience, 23*(9), 2569–2581.

Millar, D. S. (2013). Guardianship alternatives: Their use affirms self-determination of individuals with intellectual disabilities. *Education and Training in Autism and Developmental Disabilities, 48*(3), 291–305. www.jstor.org/stable/23880988

Perske, R. (1972). The dignity of risk and the mentally retarded. *Mental Retardation, 10*(1), 24–27.

Reiss, S. (2000). *Who am I? The 16 basic desires that motivate our actions and define our personalities*. Tarcher/Putnum.

Reyna, V. F., & Zayas, V. (2014). *The neuroscience of risky decision making*. American Psychological Association.

Saulnier, C. A., & Klaiman, C. (2018). *Essentials of adaptive behavior assessment of neururodevelopmental disorders*. Wiley.

Sparrow, S. S., Cicchetti, D. V., & Saulnier, C. A. (2016). *vineland adaptive behavior scales* (3rd ed.). Pearson Assessment.

Stanghellini, G. (2000). Vulnerability to schizophrenia and lack of common sense. *Schizophrenia Bulletin, 26*(4), 775–787.

Sternberg, R. J. (Ed.). (2002). *Why smart people can be so stupid*. Yale University Press.

Sternberg, R. J. (2019). A theory of adaptive intelligence and its relation to general intelligence. *Journal of Intelligence, 7*(4), 23. https://doi.org/10.3390/jintelligence7040023

Switzky, H. N. (Ed.). (2001). *Personality and motivational differences in persons with mental retardation*. Routledge.
Szpiro, G. G. (2020). *Risk, choice, and uncertainty: Three centuries of economic decision making*. Columbia University Press.
Thiébaut, F. I., White, S. J., Walsh, A., Klargaard, S. K., Wu, H.-C., Rees, G., & Burgess, P. W. (2016). Does faux pas detection in adult autism reflect differences in social cognition or decision-making abilities? *Journal of Autism and Developmental Disorders, 46*, 103–112. https://doi.org/10.1007/s10803-015-2551-1
Zelazo, P. D., & Müller, U. (2002). Executive function in typical and atypical development. In U. Goswami (Ed.), *Handbook of childhood cognitive development* (pp. 445–469). Blackwell.
Zigler, E. (1973). The retarded child as a whole person. In D. K. Routh (Ed.), *The experimental psychology of mental retardation*. Aldine.

Chapter 13
Judgment and Decision-Making Paradigms in Adolescents and Adults with ADHD and Associations with Cognitive Abilities

Rachael E. Lyon, Elizabeth A. Wanstall, and Maggie E. Toplak

Attention-Deficit/Hyperactivity Disorder (ADHD) is a heritable neurodevelopmental disorder characterized by levels of inattention, impulsivity, and hyperactivity that cause clinically significant impairment across settings (American Psychiatric Association, 2013). Numerous adverse health outcomes are related to ADHD (Nigg, 2013) as well as impairments related to life domains including occupational functioning and management of money (Barkley, 2015). Individuals with ADHD tend to be more likely to engage in risky driving, substance abuse, gambling, and sexual risk-taking [see Pollak et al. (2019) for a review]. Many of the behaviors related to these potential negative outcomes implicate judgment and decision-making (JDM) skills. Clinical characterizations of ADHD implicate difficulties in decision making, which may underlie why many of these individuals report difficulties with achieving goals (Barkley, 2006; 2015). The idea of "smart people doing foolish things" has progressed from a folk psychological concept to an area of study grounded in empirical research in decision making (Stanovich, 2009b; Sternberg, 2002), with particular relevance to clinical populations such as those with ADHD. There has been an emerging distinction between cognitive abilities (such as intelligence and executive functions) and JDM skills in the cognitive science literature (Stanovich, 1999; Stanovich, 2009b), which provides a useful framework for understanding judgment and decision making in ADHD.

ADHD is one of the most common and costly neurodevelopmental disorders (Doshi et al., 2012; Polanczyk et al., 2007). It is a disorder with marked heterogeneity across multiple levels of analysis (Castellanos & Tannock, 2002) as well as heterogeneity across several cognitive and neuropsychological functions (Coghill et al., 2018; Lambek et al., 2010; Nigg et al., 2005; Sonuga-Barke, 2002; 2005;

R. E. Lyon · E. A. Wanstall · M. E. Toplak (✉)
Department of Psychology, LaMarsh Centre for Child and Youth Research, York University, Toronto, ON, Canada
e-mail: mtoplak@yorku.ca

I. Khemka, L. Hickson (eds.), *Decision Making by Individuals with Intellectual and Developmental Disabilities*, Positive Psychology and Disability Series,
https://doi.org/10.1007/978-3-030-74675-9_13

Sonuga-Barke et al., 2008; Willcutt et al., 2005). In these studies, it has been reported that only a small proportion of those categorized as having ADHD demonstrated a *deficit* on any one neuropsychological task. The empirical study of JDM in ADHD provides new perspectives and paradigms to measure the heterogeneity of behaviors in individuals with ADHD, especially to operationalize behaviors that implicate poor choices and decisions. These conceptualizations are in keeping with the notion that real-life decision making involves choices that are based on expected but uncertain rewards and penalties and that optimal choices are based on well-considered strategies. As such, it seems plausible to assume that impaired decision making may reflect some of the difficulties often clinically described in ADHD (Barkley, 2015), such as impulse control problems (Morgan et al., 2006). In particular, temporal discounting (de Castro Paiva et al., 2019; Doidge et al., 2021; Jackson & MacKillop, 2016; Patros et al., 2016) and risky decision making (Dekkers et al., 2016) have been the most well-studied paradigms in children, adolescents, and adults with ADHD. Temporal discounting involves choosing between immediate and delayed rewards, and risky decision making involves choices between uncertain rewards and risks.

Given the complex heterogeneous neuropsychological profiles of individuals with ADHD, it is important to consider the contribution of various cognitive abilities to JDM difficulties identified in this population. Several studies have demonstrated that JDM tasks correlate positively with cognitive abilities, such as intelligence and executive functioning, in both clinical and non-clinical samples (Shamosh & Gray, 2008; Stanovich, 1999; Stanovich, 2009b; Stanovich & West, 2008; Toplak et al., 2011). These positive associations are likely due to the fact that many cognitive processing requirements (e.g., nonverbal reasoning, inhibition, interference control, working memory) are needed to generate alternative solutions to support decision-making performance. However, the size of these associations varies depending on the JDM skill under study (Stanovich & West, 2008), suggesting that JDM skills are conceptually and measurably separate from other cognitive abilities (Stanovich, 2009b). In fact, correlations between JDM skills and certain cognitive abilities are generally small to modest for temporal discounting and risky decision-making tasks based on a meta-analysis and empirical review of these literatures that include both clinical and non-clinical samples (Shamosh & Gray, 2008; Toplak et al., 2010). Therefore, it is important to understand the associations between JDM skills and cognitive abilities in individuals with ADHD to characterize these domains of competence and to further understand the cognitive heterogeneity of individuals with ADHD.

ADHD is not considered an intellectual or developmental disability (IDD), like autism spectrum disorder (ASD). However, ADHD, like ASD, is classified as a neurodevelopmental disorder. Neurodevelopmental disorders are first present in childhood and often persist into adulthood. ADHD is important to consider in the context of IDD given the symptom overlap and high rates of co-occurrence of ADHD and ASD across the life span (e.g., Hartman et al., 2016; Panagiotidi et al., 2019; Zablotsky et al., 2020). Although previous studies have found that ASD is separate and distinct from ADHD, the core symptoms of attentional difficulties, impulsivity, and hyperactivity are often part of the autism phenotype (Mayes et al., 2012). After the publication of *DSM-5*, the diagnoses of ASD and ADHD are no longer mutually exclusive, and many individuals are given both diagnoses when warranted. Furthermore, ASD and ADHD share many similar impairments in developmental and cognitive domains (e.g., executive functions, pragmatic language difficulties; Leitner, 2014). Evidence from neuropsychological and neuroimaging studies also suggests common structural brain abnormalities that are shared in those with ASD and ADHD (Dougherty et al., 2016; Jacobs et al., 2021; Lukito et al., 2020). ASD and ADHD are both highly heritable neurodevelopmental disorders, and genetic factors account for 70% to 80% of the phenotypic variance of each disorder (Faraone et al., 2006; Freitag et al., 2010). As such, we aim to provide a review of the research on the association between JDM skills and cognitive abilities in individuals with ADHD in this chapter, with consideration of whether these JDM constructs have also been implicated in understanding IDDs, such as ASD.

JDM Paradigms in ADHD

Considering the complexity of clinical difficulties related to decision making that have been reported in children, adolescents, and adults with ADHD (Pollak et al., 2019), some relevant JDM paradigms from the cognitive science literature have been studied in this population. In particular, temporal discounting (Doidge et al., 2021; Jackson & MacKillop, 2016; Patros et al., 2016) and risky decision making (Dekkers et al., 2016) have been the most well-studied paradigms in adolescents and adults with ADHD.

Temporal Discounting-Related Paradigms

One of the JDM paradigms that are commonly studied in individuals with ADHD is temporal discounting. In temporal discounting tasks, individuals are asked to make several choices between smaller immediate rewards and larger delayed rewards, where length of delay and size of reward are systematically varied in order to map an individual's preferences. If individuals tend to attribute less value to larger delayed rewards in the future when presented with these choices, they are considered to *temporally discount* (Shamosh & Gray, 2008). For example, participants may be offered the choice between receiving $10 today and $15 in a month's time. Given this choice, many people may choose the option of $10 today. Though the larger $15 in a month option may be in our best long-term interest, people often "discount" this choice because temporally it is too far off. Ainslie (2001) describes how an exponential function is more optimal than a hyperbolic function with respect to patterns of choices that take into account the value of a reward over time. That is, it is expected that over time individuals will attribute less value for a fixed reward amount, but many individuals tend to discount too steeply, which explains why hyperbolic functions tend to better fit individual's preference patterns than exponential functions (Ainslie, 2001; Green et al., 1997; Johnson & Bickel, 2002; Myerson et al., 2003; Rachlin et al., 1991; Robles & Vargas, 2007). The most common dependent measures to assess degree of temporal discounting are the *k*-value (Mazur, 1987), area under the curve (Myerson et al., 2003), and indifference point (Critchfield & Kollins, 2001). In the field of JDM, prudent temporal discounting has been defined as an indicator of rational thinking, based on scoring participant's choices on items where the rate of return warrants selecting the larger delayed reward (Basile & Toplak, 2015; Stanovich, 2009b; Stanovich et al., 2016).

Similar variants of the temporal discounting paradigm have also been studied, including delay of gratification and delay aversion paradigms. The core difference is that delay of gratification and delay aversion paradigms require participants to actually wait out the complete delay time of the trial in order to get their reward and continue onto the next trial. The difference between "choosing only" (i.e., temporal discounting) and "choosing and waiting" (i.e., delay of gratification and delay aversion) makes these tasks somewhat different experimentally and conceptually (Stanovich, 2011). Specifically, temporal discounting paradigms index prudent discounting of the future, but delay of gratification and delay aversion paradigms have the additional requirement of utilizing self-control skills to resist the immediate reward and/or tolerating the delay. In addition to the difference between choosing and choosing and waiting, the task instructions and task set-up of the delay of gratification task also make the benefits of waiting more salient and concrete to the examinee (Toplak et al., 2016). While the salient and concreteness of instructions for the delay of gratification task may have been designed to make this task as concrete and understandable as possible to 3-5-year-olds who were first administered this task (Mischel & Ebbesen, 1970), the task instructions that explain the waiting component of this task may provide explicit cues to the participant that the delayed reward is worthwhile (Toplak et al., 2016). Across all of these paradigms, the selection of the larger delayed rewards is typically scored as more optimal (Basile & Toplak, 2015).

In the ADHD literature, the delay aversion hypothesis has been used to characterize the consistent observation that children with ADHD tend to prefer the smaller immediate reward over the larger delayed reward (Sonuga-Barke, 1994, 2002; Sonuga-Barke et al., 1992). The Choice-Delay Task is a common task (C-DT; Sonuga-Barke et al., 1992). In the classic version of this task, participants are told that they will play a game where they must earn points, which will then be exchanged for nickels at the end. They must choose between two shapes, labeled as "1 point" and "2 points," which are then followed by a pre-reward delay of two or 30 seconds, respectively. They are given 20 trials with no time limit. Within the delay aversion conceptualization, a similar distinction has been made between choice impulsivity (temporal discounting) and alternative indices of delay aversion (delay of gratification). In particular, Paloyelis et al. (2009) compared choices on delay paradigms in a large sample of children using a post-reward delay condition and a no post-reward delay condition. Membership in the group with extreme inattention ratings was significantly associated with choices in both conditions, but the lack of differences across these conditions may have been attributable to the fact that these conditions were a within-subject manipulation instead of a between-subject manipulation. Delay discounting or temporal discounting tasks have also been examined in adults with ADHD (Hurst et al., 2011). Although the exact format can vary, all of these tasks involve having participants make choices between two monetary amounts at different delay periods: (1) an amount that is immediately available that changes with each trial and (2) a fixed, unchanging amount after a delay.

When comparing individuals with ADHD to those without ADHD, it has generally been found that they make more choices for immediate rewards on temporal discounting, delay aversion, and delay of gratification tasks based on empirical reviews and meta-analyses of this literature (Doidge et al., 2021; Jackson & MacKillop, 2016; Patros et al., 2016). These differences have primarily been explained by the delay aversion hypothesis in ADHD, which predicts their response to delay-related situations based on alterations in signaling of delayed rewards and an acquired motivational attitude to escape or avoid when possible (Sonuga-Barke, 1994; Sonuga-Barke et al., 2008). In addition to behavioral differences on these tasks, more recent work in neuroimaging supports differential brain activation during these tasks in individuals with ADHD in regions such as the amygdala and the dorsolateral prefrontal cortex (Van Dessel et al., 2018). While many or most of these studies have been conducted in children with ADHD (see Patros et al., 2016 for a review), several studies have also been conducted with adolescents and adults with ADHD.

Adolescence

One meta-analytic review on temporal discounting-related paradigms reported a medium effect size for this age group, indicating that adolescents with ADHD tend to prefer immediate smaller rewards over larger delayed rewards (Patros et al., 2016). That said, some studies have not shown any differences in temporal discounting choices between adolescents with ADHD and non-ADHD controls (Chantiluke et al., 2014; Scheres et al., 2006; 2010). Some have suggested that these inconsistent findings may be due in part to variations in features of the tasks used, such as using variable rather than fixed delays and using hypothetical rather than real rewards (Chantiluke et al., 2014). When looking at age effects, Patros et al. (2016) found that the effect size for children and adolescents with ADHD's preference for small immediate rewards was nearly equivalent, suggesting that differences in temporal discounting preferences persist into adolescence.

Adulthood

In adulthood, studies examining temporal discounting for individuals with ADHD have produced mixed results. Mowinckel et al.'s (2015) meta-analysis reported no significant differences between adults with and without ADHD on temporal discounting tasks. However, several other studies have shown that symptoms of ADHD in adulthood are related to performance on temporal discounting tasks, such that higher symptoms of ADHD in adults are related to a preference for smaller immediate rewards (Hurst et al., 2011; Scheres et al., 2008; Taylor et al., 2017). Additionally, another meta-analytic review found that developmental level (i.e., whether participants were over or under 18 years of age) was not a significant moderator of performance on temporal discounting tasks, indicating that individuals with ADHD across the life span continued to have difficulty on these tasks (Jackson & MacKillop, 2016). As such, preference for immediate over delayed rewards in temporal discounting likely persists in a subset of individuals with ADHD across the life span. While these findings overall seem mixed, this may also reflect greater heterogeneity in performance patterns on temporal discounting tasks, especially in adults with ADHD.

Influencing factors
In addition to developmental period, there are other factors that have been suggested to play a role in temporal discounting abilities in individuals with ADHD. First, participant characteristics have been considered, such as sex differences and comorbidities. In a meta-analysis by Doidge et al. (2021), they found that females with ADHD preferred smaller immediate rewards on tasks of temporal discounting more often than males with ADHD. Other factors such as task type, age, and reward types did not significantly predict these sex differences. Some studies have also investigated the role of comorbid disorders, such as oppositional defiant disorder (ODD), but Antonini et al. (2015) found that ODD symptoms were not a significant predictor of temporal discounting performance. Second, task characteristics have also been examined as potential influencing factors for temporal discounting abilities in individuals with ADHD. For example, in a review by Utsumi and Miranda (2018), they found that temporal discounting tasks that used hypothetical rewards were more commonly used in children with ADHD than tasks with real rewards. However, they found that both these types of tasks displayed comparable findings for children with and without ADHD.

Risky Decision-Making Paradigms

Individuals with ADHD have been documented to engage in more risk-taking than individuals without ADHD, including elevated rates of substance use, risky sexual behaviors, criminal activities, gambling problems, and risky driving (Faregh & Derevensky, 2011; Flory et al., 2006; Jerome et al., 2006; Lee et al., 2011; Pratt et al., 2002). Given these risk-taking behaviors in ADHD samples, risky decision-making tasks provide a useful operationalization to characterize these tendencies.

Relying on the probabilistic nature of risk-taking behavior, some researchers have developed laboratory-based probabilistic decision-making tasks to operationalize risk-taking behaviors (Schonberg et al., 2011). A popular laboratory procedure used for studying risk-taking in ADHD involves gambling tasks, where subjects are asked to choose between no risk and risky alternatives. Participants are typically offered the choice between options that differ in magnitude and probability of gains and losses. From a decision theory perspective and expected utility framework, the expected value of a risky alternative comprises its subjective potential payoff weighted by its probability. A rational decision-maker should opt for the alternative with the highest expected value (Schonberg et al., 2011). Generally, making choices that result in a high probability of a large loss is indicative of risky decision making. Individual differences in risky decision making performance have been examined in typically developing adults (Steingroever et al., 2013) but also extensively in clinical samples (see Toplak et al., 2010 for a review), demonstrating that many clinical samples display more risky decision making than non-clinical participants.

Several different paradigms have been used to assess risky decision making in ADHD samples (Dekkers et al., 2016). The Iowa Gambling Task (IGT) is among the most commonly used laboratory gambling tasks (Bechara et al., 1994). In the IGT, participants are asked to collect as much money as possible by choosing cards from one of four decks (A, B, C, or D). The amount of money won or lost systematically varies between decks. For example, decks A and B represent disadvantageous choices that result in overall loss, while decks C and D represent advantageous choices that result in overall gain. The Cambridge Gambling Task (CGT) from the Cambridge Neuropsychological Test Automated Battery (CANTAB) is also used to assess risky decision making. The CGT uses two different conditions: ascending, where the subject must wait to make larger bets, and descending, where larger bets can be made quickly. On each trial, participants are presented with an array of ten boxes, colored red or blue, and the participant must guess which color conceals a token. On the Balloon Analogue Risk Task (BART), participants are told that they will have the opportunity to win money by inflating balloons on a computer screen without popping them. Participants may choose to stop pumping and collect the money earned from the trial at any time. Since each pump increases the risk of the balloon popping, resulting in loss of money, greater numbers of balloon pumps indicate greater levels of risk-taking. Across all of these tasks, participants are asked to make multiple choices that over the course of the task may result in an overall gain or loss. Dekkers et al. (2016) conducted a meta-regression analysis of 37 studies and reported that individuals with ADHD displayed riskier decision making than controls across decision-making paradigms. The effect size of these effects ranged from small to medium. While many of these studies have been conducted with children who have ADHD, some studies have also been conducted with adolescents and adults with ADHD.

Adolescence

Of the few studies that have been conducted in adolescent samples, the available research suggests that adolescents with ADHD tend to make riskier choices on risky decision-making tasks than adolescents without ADHD. Five studies have examined adolescent populations using the CGT, IGT, or the BART. In most of these studies, adolescents with ADHD were consistently reported to make riskier choices on the laboratory gambling tasks than adolescents without ADHD (Ernst et al., 2003; Hobson et al., 2011; Toplak et al., 2005), with the exception of one study (Weafer et al., 2011). On the CGT, adolescents with ADHD were found to risk smaller sums and chose the unfavorable condition more frequently than non-ADHD controls but had the same speed of decision and risk adjustment (Kroyzer et al., 2014). One interpretation offered by Kroyzer et al. (2014) is that this conservative behavior indicates that adolescents with ADHD tend less toward risk-taking in general; however, they acknowledge that this interpretation seems at odds with evidence indicating that ADHD is associated with risk-taking behavior. Alternatively, authors suggest that participants with ADHD might have been less confident in their initial choice and consequently decided to risk smaller bets. The ability to make long-term advantageous choices increased with age in typically developing adolescents (Crone & van der Molen, 2004); however, adolescents with ADHD tend to be more likely to make more risky choices on these tasks.

Adulthood
Mowinckel et al. (2015) conducted a meta-analysis of 59 studies examining decision making and attention in adults with ADHD. Only nine of these studies specifically assessed risky decision making. The results indicated decision-making differences between adults with ADHD and non-ADHD controls, reporting a small effect size for risky decision making (Hedge's $g = 0.226$). The small effect size for risky decision making is not surprising given the results of a prior meta-analysis (Groen et al., 2013), which indicated risky decision-making differences in children with ADHD but not in adults with ADHD.

Influencing factors
Participant and task characteristics have been identified as influencing factors for understanding risky decision-making performance. Participant characteristics have included comorbid conditions, age and sex differences, and task characteristics, such as feedback and reward contingencies.

In the study of risky decision making, the presence of conduct disorder (CD) and oppositional defiant disorder (ODD) have been examined as relevant comorbidities. Externalizing symptoms may increase risky choices in individuals with ADHD, such as ODD (Luman et al., 2010). Dekkers et al. (2016) demonstrated that disruptive behavior disorders (DBD) had a significant moderating influence on risky decision making, as individuals with ADHD and comorbid DBD tended to take more risks than individuals with ADHD alone. Hobson et al. (2011) demonstrated that although adolescents with ADHD and comorbid ODD/CD made riskier choices than adolescents without ADHD, ODD/CD symptoms were independently related to risky decision making, while ADHD symptoms were not.

Many studies have examined the effect of age on risky decision making in ADHD. Goren et al. (2013) conducted a literature review of 14 studies and reported that children and adolescents with ADHD made riskier choices on risky decision-making tasks than controls, but that no differences were obtained in adults. In contrast, the Dekkers et al. (2016) meta-regression of 37 studies reported that developmental level (less than 12 years of age, 12–18 years of age, or over 18 years of age) did not have a significant impact on findings, suggesting that individuals with ADHD tend to make riskier choices on laboratory tasks regardless of age.

In addition to these developmental considerations, there is evidence to suggest that reinforcement contingencies may operate differently in ADHD than controls (Luman et al., 2005). It was reported in a follow-up study that boys with ADHD were unaffected by frequency and magnitude of reward relative to typically developing controls on a stimulus-response learning task (Luman et al., 2009). Experimental gambling tasks suggest that children with ADHD pay more attention to rewards and less attention to punishments compared to controls (Masunami et al., 2009). The interface between the development of risky decision making and the possibility of altered reward mechanisms is an area for further research in ADHD.

Examining the effect of feedback on task performance, Pollak and Shoham (2015) found that adolescents with ADHD and controls performed similarly on the IGT when no feedback was provided, whereas under the feedback condition, subjects with ADHD chose the unfavorable outcomes more frequently and risked smaller sums than controls. The authors note that this seems to be at odds with the broader literature. They propose an alternative interpretation, similar to Krozyer et al. (2014) mentioned above, that due to choosing the unfavorable option, participants with ADHD may have realized that they were losing and consequently chose to decrease their bets (Pollak & Shoham, 2015). A recent study by Dekkers et al. (2020) included a gambling machine task and found that adolescent boys with ADHD were less likely to adopt strategies in which all characteristics relevant to make an optimal decision were integrated, both with and without feedback. Based on these findings, the authors suggested that ADHD-related decision-making deficits do not originate in increased risk seeking but in suboptimal decision making.

In addition to performance differences between adolescents and adults with ADHD and non-clinical controls on temporal discounting and risky decision-making paradigms, correlations between these tasks and other cognitive abilities inform our understanding of the separability of these psychological constructs, both conceptually and empirically.

Associations Between Cognitive Abilities and Judgment and Decision making in ADHD

Recent models in cognitive science have differentiated between decision making and cognitive abilities, such as intelligence and performance-based measures of executive function (Stanovich, 2009a, 2009b; Toplak et al., 2010). Neuropsychological assessments measure cognitive abilities, including intelligence and executive function performance. These cognitive ability measures capture computational resources that are assessed under highly structured conditions where the examiner sets the parameters for optimal performance (Stanovich, 2009a). Alternatively, judgment and decision-making skills are conceptually unique from cognitive abilities. Cognitive science theories of decision making and rational thinking use experimental methods to illustrate thinking errors where people fail to successfully track the world (such as knowledge calibration) or fail to achieve their goals (such as maximize expected utility; Stanovich, 2009a, 2009b, 2011).

Stanovich and West (2008) have described process considerations that can explain the degree of association between intelligence and performance on JDM tasks. Some measures of JDM show modest correlations with cognitive ability (in the range of .20 to .35), but many JDM tasks show no association with cognitive ability. JDM tasks that are more likely to show correlations with cognitive ability are ones where the processing requirements of inhibition (or interference control) and working memory are required for the generation of alternative responses. However, given that correlations tend to be modest, this suggests that there is room for dissociation between cognitive abilities and decision-making performance.

Shamosh and Gray (2008) performed a meta-analysis to examine the association between delay discounting and intelligence in the general population. Across the 24 studies identified in this meta-analysis, they found that individuals with higher intellectual abilities had significantly less of a tendency to prefer "shorter sooner" (SS) rewards to "larger longer" (LL) rewards, with a small to medium effect size. Several moderating variables were also examined. Studies that used payoffs that were subject to chance found a weaker relationship between delay discounting and intelligence compared to those that used all hypothetical or all real payoffs. Other moderators, including the delay discounting measure, the type of delay discounting paradigm, and the type of intelligence test, did not reveal any significant effects.

Toplak et al. (2010) examined 43 studies that reported associations between IGT performance and cognitive abilities in samples of participants with neurological disorders and psychiatric disorders, as well as non-clinical child and adult samples. Specifically, they included studies with inhibition, set-shifting, working memory, and intelligence tasks. Across all four domains, only 20.8% of correlational comparisons were statistically significant, and actual correlations were generally low, with median values in the four domains of $r = .18$, .15, .06, and .23. Similar to the findings with temporal delay discounting, the correlation between risky decision making and cognitive abilities is small across clinical and non-clinical samples.

Table 13.1 summarizes the studies that have assessed the associations between temporal discounting and risky decision making with neuropsychological measures of cognitive abilities (intelligence and executive function (EF) tasks) in adolescents and adults with ADHD. As limited studies to examine these associations have been conducted to date with adolescents and adults with ADHD, studies in ADHD child samples were also included.

Associations Between Cognitive Abilities and Temporal Discounting-Related Paradigms in ADHD

In total, five studies that have examined correlations between cognitive abilities and temporal discounting in individuals with ADHD (or individuals with ADHD ratings) were identified. Of these, four studies included children and adolescents with ADHD and one study included only adolescents. The findings from these studies are somewhat inconsistent, with some studies noting a significant association between tasks of intelligence and temporal discounting (Bitsakou et al., 2009; Marco et al., 2009; Toplak et al., 2016), while others have failed to detect significant associations (Antonini et al., 2015; Barkley et al., 2001; Solanto et al., 2001).

In addition to these studies, some studies have also entered intelligence as a covariate in their analyses, which indirectly helps to elucidate the relationship between intelligence and temporal discounting. For example, Paloyelis et al. (2009) found that when entering intelligence as a covariate, the relationship between higher inattention ratings and the preference for smaller, immediate rewards was attenuated. Similarly, Wilson et al. (2011) found that when controlling for intelligence, the trend for children with ADHD to discount more steeply than controls was no longer statistically significant. However, others have found that children with ADHD's preference for shorter, smaller rewards remained significant even when controlling for intelligence (Scheres et al., 2013; Sonuga-Barke et al., 2010; Yang et al., 2011). As such, the findings about the association between intelligence and temporal discounting in the context of ADHD remain mixed.

Studies that have assessed the association between EF tasks and temporal discounting generally have reported significant relationships between these constructs (Antonini et al., 2015; Karalunas & Huang-Pollock, 2011; Lambek et al., 2018; Thorell, 2007; Toplak et al., 2016; Van Dessel et al., 2018). In these studies, some of the EF abilities that were studied include verbal working memory, spatial working memory, inhibition, and complex EF. For example, Van Dessel et al. (2018) found that when looking across the entire sample (i.e., children with ADHD and control children) and also when the analysis was conducted in the ADHD-only group, participants' performance on the temporal discounting task was significantly related to inhibition. One pattern that emerged in the review of correlations between temporal discounting-related paradigms and cognitive abilities in Table 13.1 is that sometimes EF tasks correlated with temporal discounting choices, but not with intelligence scores (Antonini et al., 2015; Solanto et al., 2001) or that statistically controlling for intelligence attenuated the relationship between temporal discounting choices and EF task performance (Barkley et al., 2001).

These findings may suggest that the pattern of correlations with temporal discounting choices may differ for IQ and EF tasks, but it may also be attributable to the fact that some of these studies tend to use age-corrected scaled scores as the

Table 13.1 Correlations between temporal discounting and risky decision-making paradigms and cognitive abilities in adolescent and Adult ADHD samples

Authors	Participants	Measure and dependent variables (DV): decision-making tasks	Measure and dependent variables (DV): cognitive ability tasks examined in relation to decision-making tasks	Results summary
Temporal discounting and delay aversion paradigms				
Barkley, Edwards, Laneri, Fletcher, & Metevia (2001)	101 adolescents with ADHD and ODD (mean age 14.7 years; 12 females) and 39 controls (mean age 14.8 years; 6 females); IQ >80	Simplified Reward Discounting Task DV: Indifference point	*Vigilance and response inhibition*: Conners Continuous Performance Test DV: Total omissions; commissions; hit rate; standard error of hit rate; variability in standard error of hit rate *Verbal working memory*: Digit Span Reversed DV: Longest correct span length *Nonverbal working memory*: Simon Game DV: Longest correctly reproduced sequence *Verbal fluency*: Controlled Oral Word Association Test DV: Total number of words generated *Ideational fluency*: Object Usage Test DV: Total number of uses for objects *Nonverbal fluency*: Form Fluency Task DV: Total number of objects *Intelligence*: KBIT DV: Estimated full-scale IQ	- ADHD group discounted the delay reward significantly more than the control group for the $100 delay task, but not the $1000 delay task - Association between discounting task and EFs was not assessed - The main effect of IQ on performance on the discounting task was not significant and did not interact significantly with group membership

Authors	Participants	Measure and dependent variables (DV): decision-making tasks	Measure and dependent variables (DV): cognitive ability tasks examined in relation to decision-making tasks	Results summary
Bitsakou, Psychogeiou, Thompson, & Sonuga-Barke (2008)	77 children and adolescents with ADHD Combined subtype (6–17 years; 17–19% female), 65 siblings unaffected by ADHD, and 50 controls (6–17 years; 24–42% female)	Maudsley Index of Childhood Delay Aversion DV: % of choice for the 2-point delayed reward	*Intelligence*: Wechsler Intelligence Scales for Children (3rd edition) DV: Estimated IQ from Vocabulary and Block Design subtests	- ADHD group less likely to wait for larger later rewards compared to controls, across all ages - IQ moderated performance on the MIDA, with lower IQ associated with less preference for the delayed large reward
Scheres, Tontsch, & Lee Thoeny (2013)	25 children and adolescents with combined ADHD (mean age 11.4, 6 females), 20 children and adolescents with inattentive ADHD (mean age 12.9, 8 females), 37 controls (mean age 11.5, 16 females); IQ >75	Three temporal discounting tasks with varied reward magnitude and session length DV: Area under the curve	*Intelligence*: Not specified DV: Estimated IQ	- ADHD-combined group showed steep temporal reward discounting, even after entering IQ as a covariate

Table 13.1 (continued)

Authors	Participants	Measure and dependent variables (DV): decision-making tasks	Measure and dependent variables (DV): cognitive ability tasks examined in relation to decision-making tasks	Results summary
Sonuga-Barke, Bitsakou, & Thompson (2010)	71 children and adolescents with ADHD (10–14 years of age; 15% female), 65 unaffected siblings (9–14 years of age; 52% female), and 50 controls (10–13 years; 24% female); IQ >70	Maudsley Index of Childhood Delay Aversion DV: % of choice for the 2-point delayed reward	*Inhibition*: Stop-Signal Task, Go/No-Go Task, Modified Stroop Task DV: Stop-Signal Reaction Time, probability of correct inhibition *Working memory*: Forward and backward digit span (WISC-III) DV: Level at which child failed to correctly repeat numbers on two consecutive trials at one level of difficulty *Intelligence*: WISC-III (Vocabulary and Block Design) DV: Estimated full-scale IQ	- Children with ADHD preferred shorter smaller rewards than controls, which remained significant when controlling for IQ
Toplak, Hosseini, & Basile (2016)	249 children and adolescents (mean age 11.01, 121 females) with ADHD ratings	Temporal discounting task DV: *k*-value, area under the curve, mean indifference point, interest rate total score, response profile	*Intelligence:* WASI (vocabulary and matrices) DV: z-score composite measure *Working memory*: Sentence Span DV: Recall accuracy *Inhibition*: Stroop test DV: Interference score *Set-shifting*: Trail-Making Task DV: Total completion on Part B minus Part A	- Ratings of ADHD symptoms had small but significant correlations with performance on the temporal discounting task - Poorer performance on intelligence and executive function task measures were significantly associated with performance on the temporal discounting task
Risky decision-making paradigms				

Authors	Participants	Measure and dependent variables (DV): decision-making tasks	Measure and dependent variables (DV): cognitive ability tasks examined in relation to decision-making tasks	Results summary
Dekkers, Huizenga, Popma, Bexkens, Zadelaar, & Jansen (2020)	81 adolescents with ADHD (mean age = 15.0 years, all male) and 100 typically developing adolescents TD (mean age = 15.1 years, all male); IQ ≥ 80	Gambling Machine Task (three versions: risky vs. suboptimal; decision strategy; feedback) DV: Task 1, proportion of risky choices, proportion of optimal choices; task 2, decision-making strategies; task 3, number of optimal choices	*Intelligence*: WISC-III-NL (Block Design, Vocabulary); WAIS-IV for adolescents >16yrs (Vocabulary and matrix reasoning DV: IQ proxy	- Adolescents with ADHD made less optimal choices than TD adolescents on both gambling conditions with feedback on performance - Significant effects of intelligence and age, such that more intelligent and older participants made more optimal decisions
Duarte, Woods, Rooney, Atkinson, Grant, TMARC (2012)	25 participants with methamphetamine (MA) dependence and ADHD (mean age = 40.0 years); 23 participants with MA dependence but no symptoms of ADHD (mean age = 40.2 years); 22 controls (mean age = 43.0 years)	Computerized version of the Iowa Gambling Task – DV: Total net score (i.e., the total number of advantageous choices minus the total disadvantageous choices)	*Working memory*: Paced Auditory Serial-Addition Task DV: Range from 1 (above average, T-score < 55) to 9 (severe, T-score < 20)	- MA+ADHD+ group was significantly more likely to select cards from the disadvantageous decks - MA+ADHD+ participants who were working memory impaired had significantly lower scores than group members not WM impaired - Working memory deficits interact with ADHD to amplify risky decision making in MA-dependent individuals

Table 13.1 (continued)

Authors	Participants	Measure and dependent variables (DV): decision-making tasks	Measure and dependent variables (DV): cognitive ability tasks examined in relation to decision-making tasks	Results summary
Ernst, Grant, London, Contoreggi, Kimes, & Spurgeon (2003)	33 adolescents with externalizing behavior disorders (primarily ADHD; mean age = 12.6 years; 3 females) and 31 controls (mean age = 12.8 years; 14 females) *IQ ≥ 80	Iowa Gambling Task – DV: Global score (net score) and a score for each consecutive block of 20 cards	*Complex executive function*: Wisconsin Card Sorting Task (WSCT) DV: Number of completed sets, number of perseverative errors, number of non-perseverative errors, number of trials to complete first category, failure to maintain set	-Adolescents with behavior disorders scored lower on the IGT than controls at week 2 administration, but not on week 1 administration -WCST performance not associated with IGT performance
Matthies, Philipsen, & Svaldi (2012)	Study 1: 15 adults with ADHD (mean age = 38.1 years; 7 females) and 16 controls (mean age = 32.4 years; 8 females); IQ ≥ 85 Study 2: 14 adults with ADHD (mean age = 32.8 years; 6 females) and 13 controls (mean age = 38.0 years, 7 females)	Game of Dice Task DV: Safe minus unsafe choices Study 2, boredom induced prior to task	*Set-shifting:* Trail-Making Test, A and B DV: Seconds to complete form A and seconds to complete form B	-Adults with ADHD scored significantly lower than adult controls. Adults with ADHD chose the risky two dice more often and the safe four number option less often than controls - No significant correlation between Game of Dice Task and Trail-Making -No significant group differences were found due to an increase in risky decision making in CG following the boredom induction
Toplak, Jain, & Tannock (2005)	44 adolescents with ADHD (mean age = 15.6 years; 5 females) and 34 controls (mean age = 15.4 years; 20 females); IQ > 70	Iowa Gambling Task DV: Advantageous deck choices minus disadvantageous deck choices	*Intelligence:* DV: Estimated FSIQ *Working memory:* DV: Auditory working memory scaled score DV: Visual-spatial working memory scaled score	-Adolescents with ADHD chose more cards from the disadvantageous deck and fewer cards from the advantageous deck than controls -Performance on IGT unrelated to IQ or verbal and visual spatial working memory

dependent variable for intelligence tests and non-age-corrected scores as the dependent variable for EF tasks, given that age-corrected scores are so widely available for intelligence measures. For those studies conducted in developmental samples where there are rapid age-related changes in cognitive abilities, the use of age-corrected intelligence scores may explain any differences in these data patterns (Rizeq et al., 2017). That is, these correlations may have been attenuated of age-corrected intelligence scores that were used in child and youth samples. However, it should be noted that even when significant correlations were noted between cognitive abilities (i.e., EF and intellectual abilities) and temporal discounting in individuals with ADHD, they were generally small or moderate. Overall, correlations between temporal discounting and cognitive abilities have been inconsistent or small in ADHD samples, consistent with the empirical literature that has also reported similar findings in general population samples (Shamosh & Gray, 2008).

Associations Between Cognitive Abilities and Risky Decision-Making Paradigms in ADHD

Table 13.1 includes five studies that examined the associations between risky decision-making paradigms and cognitive abilities in adolescents with ADHD (Dekkers et al., 2020; Ernst et al., 2003; Toplak et al., 2010) and adults with ADHD (Duarte et al., 2012; Matthies et al., 2012). Although many studies of risky decision making have included measures of cognitive abilities (intelligence and executive function tasks), few of these studies have examined associations between cognitive abilities and risky decision making on laboratory gambling tasks in an ADHD population. Overall, performance on the risky decision-making tasks was not correlated with working memory, set-shifting, inhibition, planning, and verbal fluency in samples of ADHD and controls across development (Antonini et al., 2015; Drechsler et al., 2008, 2010; Ernst et al., 2003; Geurts et al., 2006; Matthies et al., 2012; Skogli et al., 2014; Sørensen et al., 2017; Toplak et al., 2005).

Inhibition has also been demonstrated to be unrelated to risky decision making in ADHD samples. Specifically, inhibition was not significantly correlated with the Game of Dice (Drechsler et al., 2010) or IGT task (Geurts et al., 2006) in children with ADHD. Similarly, the relationship between risky decision making and various measures of set-shifting, such as the Trail-Making Test (Matthies et al., 2012) and Card Sort Tests (Ernst et al., 2003), was also not statistically significant. Antonini et al. (2015) and Skogli et al. (2014) found that working memory performance in children with ADHD was not related to performance on a child-friendly version of the IGT. Similar results were found between working memory and the CGT in children with ADHD (Sørensen et al., 2017). Likewise, Toplak et al. (2005) reported that IGT performance in adolescents with ADHD was not correlated with verbal or visual working memory. In one study, working memory difficulties were found to interact with ADHD symptoms and risky decision-making performance in methamphetamine-dependent adults (Duarte et al., 2012). With respect to intelligence, while some studies have reported a lack of correlations with risky decision-making performance in ADHD samples (Drechsler et al., 2008; Sørensen et al., 2017; Toplak et al., 2005). Dekkers et al. (2020) found significant effects of intelligence and age in their study of adolescents with and without ADHD, such that more intelligent and older participants made more optimal decisions on a gambling task. Overall, risky decision-making performance has not displayed consistent significant correlations with cognitive abilities in ADHD samples, consistent with the empirical literature with non-clinical samples (Toplak et al., 2010).

Temporal Discounting and Risky Decision-Making Paradigms in Other Intellectual and Developmental Disabilities

Temporal Discounting

Relative to studies in ADHD samples, temporal discounting paradigms have been less well-studied in ASD and intellectual disability (ID) samples. It has been hypothesized that individuals with ASD may experience difficulties with temporal discounting due to their difficulties with perspective taking, as temporal discounting involves taking the perspective of our future self (Peters & Büchel, 2010). However, even the literature on future-oriented perspective taking in individuals with ASD has been mixed (Ciaramelli et al., 2018; Crane et al., 2013). Similarly, findings on temporal discounting abilities in individuals with ASD have also been mixed. While some studies have identified a preference for smaller immediate rewards in individuals with ASD (Carlisi et al., 2017; Murphy et al., 2017; Warnell et al., 2019), other studies have failed to display differences in temporal discounting in individuals with ASD (Demurie et al., 2012; Karalunas et al., 2018). Additionally, a few studies have directly compared individuals with ADHD and individuals with ASD on tasks of temporal discounting, also producing mixed results. For example, Chantiluke et al. (2014) found that only boys with ASD-only or ASD and comorbid ADHD discounted delayed rewards more steeply than boys with ADHD-only and controls. However, others have found that individuals with ASD do not display the same preference for smaller immediate rewards that individuals with ADHD do and actually perform more similarly to controls (Antrop et al., 2006; Demurie et al., 2012).

Temporal discounting paradigms have rarely been studied in individuals with (ID). In fact, many studies that investigate temporal discounting routinely exclude participants with a lower IQ (i.e., less than 70 or 80), which systematically excludes individuals with ID. Willner et al. (2010) did administer temporal discounting tasks to individuals with and without ID. They found that more than half of participants with ID performed inconsistently on the temporal discounting tasks. However, when participants with ID were provided explicit training on these tasks, they displayed more consistent responding. Additionally, performance on the temporal discounting tasks was significantly related to EF abilities, but not intelligence.

Risky Decision Making

Similar to temporal discounting paradigms, risky decision making has been less well -studied in individuals with ASD and ID. Similar paradigms have been used to assess risky decision making in individuals with ASD as in individuals with ADHD. In a recent meta-analysis by Zeif and Yechiam (2020), differences between high-functioning individuals with ASD and controls were examined using the IGT. A total of 14 studies were included. The results showed virtually no difference in IGT performance between groups, except for a slight disadvantage in the first block of trials for the ASD group (Zeif & Yechiam, 2020). Notably, these studies included only individuals without ID. Although little research has been done directly comparing individuals with ADHD and ASD, preliminary evidence suggests that those with ADHD performed worse compared to individuals with Asperger's syndrome on the IGT (Gonzalez-Gadea et al., 2013).

Gambling tasks have been minimally used to explore risky decision making in ID populations. In a study using a gambling machine task, adolescents with mild to borderline ID used suboptimal decision-making strategies as compared to controls (Bexkens et al., 2016). On the adapted two-choice IGT developed by Garon and Moore (2004), people with ID chose advantageously and showed high levels of subjective awareness about the relative goodness and badness of the decks (Dymond et al., 2010). The learning profile of controls increased throughout the task whereas the profile for individuals with ID remained constant. To date, no research has explicitly investigated group differences between individuals with ID and those with ADHD.

Summary, Conclusions, and Applied Clinical Practice Considerations

Empirically, there has been an increasing interest in the study of JDM paradigms in ADHD samples. In particular, temporal discounting and risky decision-making paradigms have been examined most frequently empirically, likely because they capture performance tendencies and clinical observations related to impulsivity and engagement in risky behaviors often observed in individuals with ADHD (Barkley, 2015). Notably, there has been study of other JDM paradigms in ADHD samples, such as broader measures of decision-making competence (Del Missier et al., 2012), estimations of competence or performance calibration [see Wanstall et al. (2019) for a review], and reward-based/social decision making (Humphreys et al., 2016; Ma et al., 2017), suggesting that other paradigms may also be relevant, but just have been less of a focus empirically to date.

In order to advance our understanding of what JDM paradigms can tell us about clinical populations such as ADHD, we undertook this review of studies that have examined associations between performance on temporal discounting or risky decision-making paradigms and cognitive abilities (intelligence and executive function tasks), primarily in adolescents and adults with ADHD. Consistent with research findings in reviews that have included non-clinical samples (Shamosh & Gray, 2008; Toplak et al., 2010), the correlations between temporal discounting or risky decision-making paradigms and cognitive abilities are small to modest in ADHD samples. This consistency in overall correlation patterns provides useful convergence across clinical and non-clinical samples. The small to modest correlations between JDM paradigms and cognitive abilities have been described in cognitive science models of rational thinking (Stanovich et al., 2008), reinforcing both the conceptual and empirical separability of JDM and cognitive abilities.

The convergence across non-clinical and ADHD samples in correlations between judgment and decision-making paradigms and cognitive abilities is important to build a cumulative literature to develop models that explain performance on these tasks. These relatively small effect sizes, however, still leave considerable theoretical and empirical work to understand potential dissociations between these different abilities or competencies. The discrepancy of "smart people doing foolish things" has been commonly recognized and described in typically developing individuals (Sternberg, 2002) and has been operationalized and empirically studied as the discrepancies between competencies in intellectual abilities and rational thinking (Stanovich, 2009b). While we often think of this as a paradox, such as when intelligent adolescents make poor choices by subjecting themselves to risk factors such as substance abuse, sexually transmitted diseases, and unsafe driving (Reyna & Farley, 2006), this growing literature is documenting how and why such discrepancies may not be so unexpected. This paradox in competencies has also been described in the popular press literature on ADHD (such as Brown, 2005; Kelly & Ramundo, 2006).

The conceptual separation of JDM from cognitive abilities in the cognitive science literature is consistent with the pattern of findings reviewed in this chapter in samples of individuals with ADHD. Frameworks that begin to organize and explain how cognitive failures on JDM tasks are separable from neuropsychological measures of cognitive ability will provide ways to better operationalize the heterogeneity of cognitive performance in ADHD. For example, conceptually separating our understanding of information processing limitations related to capacity and efficiency, often captured on measures of cognitive ability, from mechanisms that may signal impulsive or nonoptimal responding, such as on JDM tasks. For example, in ADHD, it is of interest whether performance differences on these JDM paradigms are attributable to the impulsive and hyperactive symptoms that may underlie risk-seeking or risky behaviors or whether they are attributable to cognitive ability difficulties that are often more associated with inattentive symptoms

(Chhabildas et al., 2001). Some evidence suggests that risk-seeking tendencies do not explain performance in risky decision-making performance in participants with ADHD (Dekkers et al., 2021) and that instead, performance differences may be attributable to adopting strategies to determine optimal performance (Dekkers et al., 2020). Mäntylä et al. (2012) have also suggested that the cognitive information processing requirements may explain performance differences in adults with ADHD and non-ADHD controls, as their participants with ADHD only displayed significant differences on the applying decision rules task from the Decision Making Competence battery. However, JDM measures provide an additional set of paradigms to measure and assess the heterogeneity of performance on several indicators of cognitive competence reported in ADHD samples.

There are several clinical implications and future directions of this work; here we focus on some perspectives to inform our understanding of individuals with ADHD. Understanding JDM in ADHD and its relationship to other cognitive abilities is an important endeavor with implications for how we conceptualize difficulties in ADHD and intervention efforts. The growing empirical work on JDM constructs in ADHD reflects the conceptual separability of these skills and abilities from intellectual abilities and executive function task performance, as has been done in the cognitive science research on these skills and abilities (Stanovich, 2009a, 2009b). It will be important to continue to empirically parse these different skills and abilities that have been described in models of ADHD (such as Barkley, 2006, 2015), in order to better understand the heterogeneity in presentation in individuals with ADHD (Doidge et al., 2019). JDM paradigms provide a further way to operationalize and measure these abilities in individuals with ADHD and to differentiate domains of functioning that may be affected in these individuals. This may permit the development of more individualized recommendations and treatments for individuals with ADHD. For example, Sonuga-Barke (2004) has suggested that reorganizing incentive structures may facilitate delay tolerance on delay aversion tasks.

Relatively less research has been done on temporal discounting and risky decision-making paradigms in ID and ASD samples, but this might reflect two different possibilities. First, the study of JDM paradigms, such as temporal discounting and risky decision making, may be relatively understudied perhaps partly due to the fact that individual differences in intelligence are strongly implicated in understanding performance on these paradigms, which will impact its relevance in individuals with ID and for individuals with ASD who also have ID [DSM-5, APA, 2013)]. Even in studies with ADHD samples, intelligence cutoffs are conventionally used and individuals with IQ less than 80 or 85 (sometimes less than 70) are routinely excluded from studies examining JDM paradigms in ADHD populations. Second, it may be that temporal discounting and risky decision-making paradigms may more reliably characterize individuals with ADHD than individuals with ID or ASD. There is evidence emerging that other JDM constructs, such as analogical reasoning, counterfactual and false-belief reasoning, and pragmatic reasoning, may be more relevant constructs for understanding the clinical presentation of ASD (Morsanyi & Byrne, 2019). These bodies of research may eventually provide novel assessment tools to help us better differentiate neurodevelopmental conditions such as ADHD and ASD in the context of clinical assessments. The use of JDM paradigms may be useful for differential diagnosis and/or determining comorbidity particularly for ADHD and ASD (APA, 2013), as overlap in cognitive characteristics has been documented for these conditions (Karalunas et al., 2018).

The growing study of JDM paradigms in ADHD provides another set of indicators of competence to examine the heterogeneity of performance on neuropsychological measures in samples of participants with ADHD. Temporal discounting and risky decision-making paradigms have been the most well-studied paradigms in ADHD; however, there are indeed other conceptualizations and paradigms of decision making that have been studied to a lesser extent in ADHD. Conversely, these JDM paradigms have been relatively less well studied in ASD and ID samples. Consistent with the cognitive science literature of JDM paradigms, small to modest associations have been reported between JDM paradigms and cognitive abilities, including intelligence and executive function task performance in ADHD samples. This pattern of findings reinforces the conceptual separation between JDM paradigms and cognitive abilities in the cognitive science literature. The measurement of JDM paradigms may offer a separable domain to assess and to understand the heterogeneity in cognitive performance that has been observed in individuals with ADHD.

References

Ainslie, G. (2001). *Breakdown of will.* Cambridge University Press.

Albert, D., & Steinberg, L. (2011). Judgment and decision making in adolescence. *Journal of Research on Adolescence, 21*(1), 211–224. https://doi.org/10.1111/j.1532-7795.2010.00724.x

American Psychiatric Association. (2013). *Diagnostic and Statistical Manual of Mental Disorders, 5th Ed. (DSM-V).* American Psychiatric Association. https://doi.org/10.1176/appi.books.9780890425596.744053

Antonini, T. N., Becker, S. P., Tamm, L., & Epstein, J. N. (2015). Hot and cool executive functions in children with attention-deficit/hyperactivity disorder and comorbid oppositional defiant disorder. *Journal of the International Neuropsychological Society, 21*(8), 584–595. https://doi.org/10.1017/S1355617715000752

Antrop, I., Stock, P., Verte, S., Wiersema, J. R., Baeyens, D., & Roeyers, H. (2006). ADHD and delay aversion: The influence of non-temporal stimulation on choice for delayed rewards. *Journal of Child Psychology and Psychiatry and Allied Disciplines, 47*(11), 1152–1158. https://doi.org/10.1111/j.1469-7610.2006.01619.x

Barkley, R. A. (2006). *Attention-deficit hyperactivity disorder: A handbook for diagnosis and treatment* (3rd ed.). Guilford Press.

Barkley, R. A. (2015). *Attention-deficit hyperactivity disorder: A handbook for diagnosis and treatment* (4th ed.). The Guilford Press.

Barkley, R. A., Edwards, G., Laneri, M., Fletcher, K., & Metevia, L. (2001). Executive functioning, temporal discounting, and sense of time in adolescents with attention deficit hyperactivity disorder (ADHD) and oppositional defiant disorder (ODD). *Journal of Abnormal Child Psychology, 29*(6), 541–556. https://doi.org/10.1023/a:1012233310098

Basile, A. G., & Toplak, M. E. (2015). Four converging measures of temporal discounting and their relationships with intelligence, executive functions, thinking dispositions, and behavioral outcomes. *Frontiers in Psychology, 6*, 728. https://doi.org/10.3389/fpsyg.2015.00728

Bechara, A., Damasio, A. R., Damasio, H., & Anderson, S. W. (1994). Insensitivity to future consequences following damage to human prefrontal cortex. *Cognition, 50*(1-3), 7–15. https://doi.org/10.1016/0010-0277(94)90018-3

Bexkens, A., Jansen, B. R., Van der Molen, M. W., & Huizenga, H. M. (2016). Cool decision making in adolescents with behavior disorder and/or mild-to-borderline intellectual disability. *Journal of Abnormal Child Psychology, 44*(2), 357–367. https://doi.org/10.1007/s10802-015-9996-8

Bitsakou, P., Psychogiou, L., Thompson, M., & Sonuga-Barke, E. J. (2008). Inhibitory deficits in attention-deficit/hyperactivity disorder are independent of basic processing efficiency and IQ. *Journal of Neural Transmission (Vienna), 115*(2), 261–268. https://doi.org/10.1007/s00702-007-0828-z

Bitsakou, P., Psychogiou, L., Thompson, M., & Sonuga-Barke, E. J. (2009). Delay aversion in attention deficit/hyperactivity disorder: An empirical investigation of the broader phenotype. *Neuropsychologia, 47*(2), 446–456. https://doi.org/10.1016/j.neuropsychologia.2008.09.015

Brown, T. E. (2005). *Attention Deficit Disorder: The unfocused mind in children and adults*. Yale University Press.

Carlisi, C. O., Norman, L., Murphy, C. M., Christakou, A., Chantiluke, K., Giampietro, V., Simmons, A., Brammer, M., Murphy, D. G., Mataix-Cols, D., & Rubia, K. (2017). Comparison of neural substrates of temporal discounting between youth with autism spectrum disorder and with obsessive-compulsive disorder. *Psychological Medicine, 47*(14), 2513–2527. https://doi.org/10.1017/s0033291717001088

Castellanos, F. X., & Tannock, R. (2002). Neuroscience of attention-deficit/hyperactivity disorder: the search for endophenotypes. *Nature Reviews: Neuroscience, 3*(8), 617–628. https://doi.org/10.1038/nrn896

Chantiluke, K., Christakou, A., Murphy, C. M., Giampietro, V., Daly, E. M., Ecker, C., Brammer, M., Murphy, D. G., Consortium, M. A., & Rubia, K. (2014). Disorder-specific functional abnormalities during temporal discounting in youth with attention deficit hyperactivity disorder (ADHD), autism and comorbid ADHD and autism. *Psychiatry Research, 223*(2), 113–120. https://doi.org/10.1016/j.pscychresns.2014.04.006

Chhabildas, N., Pennington, B. F., & Willcutt, E. G. (2001). A comparison of the neuropsychological profiles of the DSM-IV subtypes of ADHD. *Journal of Abnormal Child Psychology, 29*(6), 529–540. https://doi.org/10.1023/a:1012281226028

Ciaramelli, E., Spoglianti, S., Bertossi, E., Generali, N., Telarucci, F., Tancredi, R., Muratori, F., & Igliozzi, R. (2018). Construction of past and future events in children and adolescents with ASD: Role of self-relatedness and relevance to decision making. *Journal of Autism and Developmental Disorders, 48*(9), 2995–3009. https://doi.org/10.1007/s10803-018-3577-y

Coghill, D., Toplak, M., Rhodes, S., & Adamo, N. (2018). Cognitive functioning in ADHD: Inhibition, memory, temporal discounting, decision making, timing and reaction time variability. In T. Banaschewski, D. Coghill, & A. Zuddas (Eds.), *Oxford textbook of Attention Deficit Hyperactivity Disorder*. Oxford University Press.

Crane, L., Lind, S. E., & Bowler, D. M. (2013). Remembering the past and imagining the future in autism spectrum disorder. *Memory, 21*(2), 157–166. https://doi.org/10.1080/09658211.2012.712976

Critchfield, T. S. & Kollins, S. H. (2001). Temporal discounting: Basic research and the analysis of socially important behavior. *Journal of applied behavior analysis, 34*(1), 101–122. https://doi.org/10.1901/jaba.2001.34-101

Crone, E. A., & van der Molen, M. W. (2004). Developmental changes in real life decision making: Performance on a gambling task previously shown to depend on the ventromedial prefrontal cortex. *Developmental Neuropsychology, 25*(3), 251–279. https://doi.org/10.1207/s15326942dn2503_2

de Castro Paiva, G. C., de Souza Costa, D., Malloy-Diniz, L. F., Marques de Miranda, D. & Jardim de Paula, J. (2019). Temporal reward discounting in children with attention deficit/hyperactivity disorder (ADHD), and children with autism spectrum disorder (ASD): A systematic review. *Developmental Neuropsychology, 44*(6), 468–480. https://doi.org/10.1080/87565641.2019.1667996

Dekkers, T. J., Popma, A., Agelink van Rentergem, J. A., Bexkens, A., & Huizenga, H. M. (2016). Risky decision making in attention-deficit/hyperactivity disorder: A meta-regression analysis. *Clinical Psychology Review, 45*, 1–16. https://doi.org/10.1016/j.cpr.2016.03.001

Dekkers, T. J., Agelink van Rentergem, J. A., Huizenga, H. M., Raber, H., Shoham, R., Popma, A., & Pollak, Y. (2021). Decision making deficits in ADHD are not related to risk seeking but to suboptimal decision making: Meta-analytical and novel experimental evidence. *Journal of Attention Disorders, 25*(4), 486–501. https://doi.org/10.1177/1087054718815572

Dekkers, T. J., Huizenga, H. M., Popma, A., Bexkens, A., Zadelaar, J. N., & Jansen, B. R. J. (2020). Decision making deficits in adolescent boys with and without attention-deficit/hyperactivity disorder (ADHD): An experimental assessment of associated mechanisms. *Journal of Abnormal Child Psychology, 48*(4), 495–510. https://doi.org/10.1007/s10802-019-00613-7

Del Missier, F., Mäntylä, T., & De Bruin, W. B. (2012). Decision making competence, executive functioning, and general cognitive abilities. *Journal of Behavioral Decision making, 25*(4), 331–351.

Demurie, E., Roeyers, H., Baeyens, D., & Sonuga-Barke, E. (2012). Temporal discounting of monetary rewards in children and adolescents with ADHD and autism spectrum disorders. *Developmental Science, 15*(6), 791–800. https://doi.org/10.1111/j.1467-7687.2012.01178.x

Doidge, J. L., Flora, D. B., & Toplak, M. E. (2021). A meta-analytic review of sex differences on delay of gratification and temporal discounting tasks in ADHD and typically developing samples. *Journal of Attention Disorders, 25*(4), 540–561. https://doi.org/10.1177/1087054718815588

Doidge, J., Saoud, W., & Toplak, M. E. (2019). Executive functions and decision making in adolescents with ADHD. In S. P. Becker (Ed.), *ADHD in Adolescents: Development, Assessment, and Treatment* (pp. 56–76). Guliford Press.

Doshi, J. A., Hodgkins, P., Kahle, J., Sikirica, V., Cangelosi, M. J., Setyawan, J., Erder, M. H., & Neumann, P. J. (2012). Economic impact of childhood and adult attention-deficit/hyperactivity disorder in the United States. *Journal of the American Academy of Child and Adolescent Psychiatry, 51*(10), 990–1002. e1002. https://doi.org/10.1016/j.jaac.2012.07.008

Dougherty, C. C., Evans, D. W., Myers, S. M., Moore, G. J., & Michael, A. M. (2016). A comparison of structural brain imaging findings in autism spectrum disorder and attention-deficit hyperactivity disorder. *Neuropsychology Review, 26*(1), 25–43. https://doi.org/10.1007/s11065-015-9300-2

Drechsler, R., Rizzo, P., & Steinhausen, H. C. (2008). Decision making on an explicit risk-taking task in preadolescents with attention-deficit/hyperactivity disorder. *J Neural Transm (Vienna), 115*(2), 201–209. https://doi.org/10.1007/s00702-007-0814-5

Drechsler, R., Rizzo, P., & Steinhausen, H. C. (2010). Decision making with uncertain reinforcement in children with attention deficit/hyperactivity disorder (ADHD). *Child Neuropsychology, 16*(2), 145–161. https://doi.org/10.1080/09297040903190774

Duarte, N. A., Woods, S. P., Rooney, A., Atkinson, J. H., Grant, I., & Translational Methamphetamine, A. R. C. G. (2012). Working memory deficits affect risky decision making in methamphetamine users with attention-deficit/hyperactivity disorder. *Journal of Psychiatric Research, 46*(4), 492–499. https://doi.org/10.1016/j.jpsychires.2012.01.006

Dymond, S., Bailey, R., Willner, P., & Parry, R. (2010). Symbol labelling improves advantageous decision making on the Iowa Gambling Task in people with intellectual disabilities. *Research in Developmental Disabilities, 31*(2), 536–544. https://doi.org/10.1016/j.ridd.2009.12.003

Ernst, M., Grant, S. J., London, E. D., Contoreggi, C. S., Kimes, A. S., & Spurgeon, L. (2003). Decision making in adolescents with behavior disorders and adults with substance abuse. *American Journal of Psychiatry, 160*(1), 33–40. https://doi.org/10.1176/appi.ajp.160.1.33

Faraone, S. V., Biederman, J., Spencer, T., Mick, E., Murray, K., Petty, C., Adamson, J. J., & Monuteaux, M. C. (2006). Diagnosing adult attention deficit hyperactivity disorder: Are late onset and subthreshold diagnoses valid? *American Journal of Psychiatry, 163*(10), 1720–1729.

Faregh, N., & Derevensky, J. (2011). Gambling behavior among adolescents with attention deficit/hyperactivity disorder. *Journal of Gambling Studies, 27*(2), 243–256. https://doi.org/10.1007/s10899-010-9211-3

Flory, K., Molina, B. S., Pelham, W. E., Jr., Gnagy, E., & Smith, B. (2006). Childhood ADHD predicts risky sexual behavior in young adulthood. *Journal of Clinical Child and Adolescent Psychology, 35*(4), 571–577. https://doi.org/10.1207/s15374424jccp3504_8

Freitag, C. M., Staal, W., Klauck, S. M., Duketis, E., & Waltes, R. (2010). Genetics of autistic disorders: Review and clinical implications. *European Child and Adolescent Psychiatry, 19*(3), 169–178.

Garon, N., & Moore, C. (2004). Complex decision making in early childhood. Brain *Cognition, 55*(1), 158–170. https://doi.org/10.1016/s0278-2626(03)00272-0

Geurts, H. M., van der Oord, S., & Crone, E. A. (2006). Hot and cool aspects of cognitive control in children with ADHD: Decision making and inhibition. *Journal of Abnormal Child Psychology, 34*(6), 813–824. https://doi.org/10.1007/s10802-006-9059-2

Gonzalez-Gadea, M. L., Baez, S., Torralva, T., Castellanos, F. X., Rattazzi, A., Bein, V., Rogg, K., Manes, F., & Ibanez, A. (2013). Cognitive variability in adults with ADHD and AS: Disentangling the roles of executive functions and social cognition. *Research in Developmental Disabilities, 34*(2), 817–830. https://doi.org/10.1016/j.ridd.2012.11.009

Green, L., Fry, A. F., & Myerson, J. (1994). Discounting of delayed rewards: A life-span comparison. *Psychological Science, 5*(1), 33–36. https://doi.org/10.1111/j.1467-9280.1994.tb00610.x

Green, L., Myerson, J., & McFadden, E. (1997). Rate of temporal decreases with amount of reward. *Memory & Cognition, 25*(5), 715–723.

Groen, Y., Gaastra, G. F., Lewis-Evans, B., & Tucha, O. (2013). Risky behavior in gambling tasks in individuals with ADHD – A systematic literature review. *PloS One, 8*(9), e74909. https://doi.org/10.1371/journal.pone.0074909

Hartman, C. A., Geurts, H. M., Franke, B., Buitelaar, J. K., & Rommelse, N. N. J. (2016). Changing ASD-ADHD symptom co-occurrence across the lifespan with adolescence as crucial time window: Illustrating the need to go beyond childhood. *Neuroscience & Biobehavioral Reviews, 71*, 529–541. https://doi.org/10.1016/j.neubiorev.2016.09.003

Hobson, C. W., Scott, S. & Rubia, K. (2011), Investigation of cool and hot executive function in ODD/CD independently of ADHD. *Journal of Child Psychology and Psychiatry and Allied Disciplines*, 52(10), 1035–1043. https://doi.org/10.1111/j.1469-7610.2011.02454.x

Humphreys, K. L., Galán, C. A., Tottenham, N., & Lee, S. S. (2016). Impaired social decision making mediates the association between ADHD and social problems. *Journal of Abnormal Child Psychology, 44*(5), 1023–1032.

Hurst, R. M., Kepley, H. O., McCalla, M. K., & Livermore, M. K. (2011). Internal consistency and discriminant validity of a delay-discounting task with an adult self-reported ADHD sample. *Joural of Attention Disorders, 15*(5), 412–422. https://doi.org/10.1177/1087054710365993

Jackson, J. N., & MacKillop, J. (2016). Attention-deficit/hyperactivity disorder and monetary delay discounting: A meta-analysis of case-control studies. *Biological Psychiatry: Cognitive Neuroscience and Neuroimaging, 1*(4), 316–325. https://doi.org/10.1016/j.bpsc.2016.01.007

Jacobs, G. R., Voineskos, A. N., Hawco, C., Stefanik, L., Forde, N. J., Dickie, E. W., Lai, M. C., Szatmari, P., Schachar, R., Crosbie, J., Arnold, P. D., Goldenberg, A., Erdman, L., & Ameis, S. H. (2021). Integration of brain and behavior measures for identification of data-driven groups cutting across children with ASD, ADHD, or OCD. *Neuropsychopharmacology, 46*, 643-653. https://doi.org/10.1038/s41386-020-00902-6

Jerome, L., Segal, A., & Habinski, L. (2006). What we know about ADHD and driving risk: A literature review, meta-Analysis and critique. *Journal of the Canadian Academy of Child and Adolescent Psychiatry. Journal de l'Académie Canadienne de Psychiatrie de L'enfant et de l'Adolescent, 15*(3), 105–125.

Johnson, M. W., & Bickel, W. K. (2002). Within-subject comparison of real and hypothetical money rewards in delay discounting. *Journal of the Experimental Analysis of Behavior, 77*, 129–146.

Karalunas, S. L., & Huang-Pollock, C. L. (2011). Examining relationships between executive functioning and delay aversion in attention deficit hyperactivity disorder. *Journal of Clinical Child and Adolescent Psychology, 40*(6), 837–847. https://doi.org/10.1080/15374416.2011.614578

Karalunas, S. L., Hawkey, E., Gustafsson, H., Miller, M., Langhorst, M., Cordova, M., Fair, D., & Nigg, J. T. (2018). Overlapping and distinct cognitive impairments in attention-deficit/hyperactivity and autism spectrum disorder without intellectual disability. *Journal of Abnormal Child Psychology, 46*(8), 1705–1716.

Kelly, K., & Ramundo, P. (2006). *You mean I'm not lazy, stupid or crazy?!: The classic self-help book for adults with Attention Deficit Disorder*. Scribner.

Kroyzer, N., Gross-Tsur, V., & Pollak, Y. (2014). Risk taking in adolescents with attention deficit hyperactivity disorder on a probabilistic choice task. *Journal of Nervous and Mental Disease, 202*(3), 247–252. https://doi.org/10.1097/NMD.0000000000000100

Lambek, R., Tannock, R., Dalsgaard, S., Trillingsgaard, A., Damm, D., & Thomsen, P. H. (2010). Validating neuropsychological subtypes of ADHD: How do children with and without an executive function deficit differ? *Journal of Child Psychology and Psychiatry, 51*(8), 895–904.

Lambek, R., Sonuga-Barke, E., Tannock, R., Sørensen, A. V., Damm, D., & Thomsen, P. H. (2018). Are there distinct cognitive and motivational sub-groups of children with ADHD? *Psychological Medicine, 48*(10), 1722–1730. https://doi.org/10.1017/s0033291717003245

Lee, S. S., Humphreys, K. L., Flory, K., Liu, R., & Glass, K. (2011). Prospective association of childhood attention-deficit/hyperactivity disorder (ADHD) and substance use and abuse/dependence: A meta-analytic review. *Clinical Psychology Review, 31*(3), 328–341. https://doi.org/10.1016/j.cpr.2011.01.006

Leitner, Y. (2014). The co-occurrence of autism and attention deficit hyperactivity disorder in children - What do we know? *Frontiers in Human Neuroscience, 8*, 268. https://doi.org/10.3389/fnhum.2014.00268

Lukito, S., Norman, L., Carlisi, C., Radua, J., Hart, H., Simonoff, E., & Rubia, K. (2020). Comparative meta-analyses of brain structural and functional abnormalities during cognitive control in attention-deficit/hyperactivity disorder and autism spectrum disorder. *Psychological Medicine, 50*(6), 894–919. https://doi.org/10.1017/s0033291720000574

Luman, M., Oosterlaan, J., & Sergeant, J. A. (2005). The impact of reinforcement contingencies on AD/HD: A review and theoretical appraisal. *Clinical Psychology Review, 25*(2), 183–213. https://doi.org/10.1016/j.cpr.2004.11.001

Luman, M., Van Meel, C. S., Oosterlaan, J., Sergeant, J. A., & Geurts, H. M. (2009). Does reward frequency or magnitude drive reinforcement-learning in attention-deficit/hyperactivity disorder? *Psychiatry Research, 168*(3), 222–229. https://doi.org/10.1016/j.psychres.2008.08.012

Luman, M., Sergeant, J. A., Knol, D. L., & Oosterlaan, J. (2010). Impaired decision making in oppositional defiant disorder related to altered psychophysiological responses to reinforcement. *Biological Psychiatry, 68*(4), 337–344. https://doi.org/10.1016/j.biopsych.2009.12.037

Ma, I., Lambregts-Rommelse, N. N. J., Buitelaar, J. K., Cillessen, A. H. N., & Scheres, A. P. J. (2017). Decision making in social contexts in youth with ADHD. *European Child and Adolescent Psychiatry, 26*(3), 335–344. https://doi.org/10.1007/s00787-016-0895-5

Mantyla, T., Still, J., Gullberg, S., & Del Missier, F. (2012). Decision making in adults with ADHD. *Journal of Attention Disorders, 16*(2), 164–173. https://doi.org/10.1177/1087054709360494

Marco, R., Miranda, A., Schlotz, W., Melia, A., Mulligan, A., Müller, U., Andreou, P., Butler, L., Christiansen, H., Gabriels, I., Medad, S., Albrecht, B., Uebel, H., Asherson, P., Banaschewski, T., Gill, M., Kuntsi, J., Mulas, F., Oades, R., Roeyers, H., Steinhausen, H. C., Rothenberger, A., Faraone, S. V., & Sonuga-Barke, E. J. (2009). Delay and reward choice in ADHD: an experimental test of the role of delay aversion. *Neuropsychology, 23*(3), 367–380. https://doi.org/10.1037/a0014914

Masunami, T., Okazaki, S., & Maekawa, H. (2009). Decision making patterns and sensitivity to reward and punishment in children with attention-deficit hyperactivity disorder. *International Journal of Psychophysiology, 72*(3), 283–288. https://doi.org/10.1016/j.ijpsycho.2009.01.007

Matthies, S., Philipsen, A., & Svaldi, J. (2012). Risky decision making in adults with ADHD. *Journal of Behavior Therapy and Experimental Psychiatry, 43*(3), 938–946. https://doi.org/10.1016/j.jbtep.2012.02.002

Mayes, S. D., Calhoun, S. L., Mayes, R. D., & Molitoris, S. (2012). Autism and ADHD: Overlapping and discriminating symptoms. *Research in Autism Spectrum Disorders, 6*(1), 277–285. https://doi.org/10.1016/j.rasd.2011.05.009

Mazur, J. E. (1987). An adjusting procedure for studying delayed reinforcement. In M. L. Commons, J. E. Mazur, J. A. Nevin, & H. Rachlin (Eds.), *Quantitative analyses of behavior: Vol. 5. The effect of delay and intervening events on reinforcement value* (pp. 56–73). Erlbaum.

Mischel, W., & Ebbesen, E. B. (1970). Attention in delay of gratification. *Journal of Personality and Social Psychology, 16*(2), 329.

Morgan, M. J., Impallomeni, L. C., Pirona, A., & Rogers, R. D. (2006). Elevated impulsivity and impaired decision making in abstinent Ecstasy (MDMA) users compared to polydrug and drug-naive controls. *Neuropsychopharmacology, 31*(7), 1562–1573. https://doi.org/10.1038/sj.npp.1300953

Morsanyi, K., & Byrne, R. M. (2019). *Thinking, Reasoning, and Decision making in Autism.* Routledge.

Mowinckel, A. M., Pedersen, M. L., Eilertsen, E., & Biele, G. (2015). A meta-analysis of decision making and attention in adults with ADHD. *Journal of Attention Disorders, 19*(5), 355–367. https://doi.org/10.1177/1087054714558872

Murphy, C. M., Christakou, A., Giampietro, V., Brammer, M., Daly, E. M., Ecker, C., Johnston, P., Spain, D., Robertson, D. M., Murphy, D. G., & Rubia, K. (2017). Abnormal functional activation and maturation of ventromedial prefrontal cortex and cerebellum during temporal discounting in autism spectrum disorder. *Human Brain Mapping, 38*(11), 5343–5355. https://doi.org/10.1002/hbm.23718

Myerson, J., Green, L., Hanson, S., Holt, D. D., & Estle, S. J. (2003). Discounting delayed and probabilistic rewards: Process and traits. *Journal of Economic Psychology, 24*, 619–635.

Nigg, J. T. (2013). Attention-deficit/hyperactivity disorder and adverse health outcomes. *Clinical Psychology Review, 33*(2), 215–228. https://doi.org/10.1016/j.cpr.2012.11.005

Nigg, J. T., Stavro, G., Ettenhofer, M., Hambrick, D. Z., Miller, T., & Henderson, J. M. (2005). Executive functions and ADHD in adults: Evidence for selective effects on ADHD symptom domains. *Journal of Abnormal Psychology, 114*(4), 706–717. https://doi.org/10.1037/0021-843X.114.3.706

Paloyelis, Y., Asherson, P., & Kuntsi, J. (2009). Are ADHD symptoms associated with delay aversion or choice impulsivity? A general population study. *Journal of the American Academy of Child and Adolescent Psychiatry, 48*(8), 837–846. https://doi.org/10.1097/CHI.0b013e3181ab8c97

Panagiotidi, M., Overton, P. G., & Stafford, T. (2019). Co-occurrence of ASD and ADHD traits in an adult population. *Journal of Attention Disorders, 23*(12), 1407–1415. https://doi.org/10.1177/1087054717720720

Patros, C. H., Alderson, R. M., Kasper, L. J., Tarle, S. J., Lea, S. E., & Hudec, K. L. (2016). Choice-impulsivity in children and adolescents with attention-deficit/hyperactivity disorder (ADHD): A meta-analytic review. *Clinical Psychology Review, 43*, 162–174. https://doi.org/10.1016/j.cpr.2015.11.001

Peters, J., & Büchel, C. (2010). Episodic future thinking reduces reward delay discounting through an enhancement of prefrontal-mediotemporal interactions. *Neuron, 66*(1), 138–148. https://doi.org/10.1016/j.neuron.2010.03.026

Polanczyk, G., de Lima, M. S., Horta, B. L., Biederman, J., & Rohde, L. A. (2007). The worldwide prevalence of ADHD: a systematic review and metaregression analysis. *American Journal of Psychiatry, 164*(6), 942–948. https://doi.org/10.1176/ajp.2007.164.6.942

Pollak, Y., & Shoham, R. (2015). Feedback may harm: Role of feedback in probabilistic decision making of adolescents with ADHD. *Journal of Abnormal Child Psychology, 43*(7), 1233–1242. https://doi.org/10.1007/s10802-015-0016-9

Pollak, Y., Dekkers, T. J., Shoham, R., & Huizenga, H. M. (2019). Risk-taking behavior in attention deficit/hyperactivity disorder (ADHD): A review of potential underlying mechanisms and of interventions. *Current Psychiatry Report, 21*(5), 33. https://doi.org/10.1007/s11920-019-1019-y

Pratt, T. C., Cullen, F. T., Blevins, K. R., Daigle, L., & Unnever, J. D. (2002). The relationship of attention deficit hyperactivity disorder to crime and delinquency: A meta-analysis. *International Journal of Police Science & Management, 4*(4), 344–360. https://doi.org/10.1350/ijps.4.4.344.10873

Rachlin, H., Raineri, A., & Cross, D. (1991). Subjective probability of delay. *Journal of The Experimental Analysis of Behavior, 55*, 233–244.

Reyna, V. F., & Farley, F. (2006). Risk and rationality in adolescent decision making: Implications for theory, practice, and public policy. *Psychological Science in the Public Interest, 7*(1), 1–44. https://doi.org/10.1111/j.1529-1006.2006.00026.x

Reynolds, B., & Schiffbauer, R. (2005). Delay of gratification and delay discounting: A unifying feedback model of delay-related impulsive behavior. *The Psychological Record, 55*(3), 439–460.

Rizeq, J., Flora, D. B., & Toplak, M. E. (2017). Changing relations among cognitive abilities across development: Implications for measurement and research. *Clinical Neuropsychologist, 31*(8), 1353–1374. https://doi.org/10.1080/13854046.2017.1317034

Robles, E., & Vargas, P. A. (2007). Functional parameters of delay discounting assessment tasks: Order of presentation. *Behavioral Processes, 75*, 237–241.

Scheres, A., de Water, E., & Mies, G. W. (2013). The neural correlates of temporal reward discounting. *WIREs Cognitive Science, 4*(5), 523–545. https://doi.org/10.1002/wcs.1246

Scheres, A., Dijkstra, M., Ainslie, E., Balkan, J., Reynolds, B., Sonuga-Barke, E., & Castellanos, F. X. (2006). Temporal and probabilistic discounting of rewards in children and adolescents: Effects of age and ADHD symptoms. *Neuropsychologia, 44*(11), 2092–2103. https://doi.org/10.1016/j.neuropsychologia.2005.10.012

Scheres, A., Lee, A., & Sumiya, M. (2008). Temporal reward discounting and ADHD: task and symptom specific effects. *Journal of Neural Transmission, 115*(2), 221–226. https://doi.org/10.1007/s00702-007-0813-6

Scheres, A., Tontsch, C., Thoeny, A. L., & Kaczkurkin, A. (2010). Temporal reward discounting in attention-deficit/hyperactivity disorder: The contribution of symptom domains, reward magnitude, and session length. *Biological Psychiatry, 67*(7), 641–648. https://doi.org/10.1016/j.biopsych.2009.10.033

Scheres, A., Tontsch, C., & Thoeny, A. L. (2013). Steep temporal reward discounting in ADHD-combined type: Acting upon feelings. *Psychiatry Research, 209*(2), 207–213. https://doi.org/10.1016/j.psychres.2012.12.007

Schonberg, T., Fox, C. R., & Poldrack, R. A. (2011). Mind the gap: Bridging economic and naturalistic risk-taking with cognitive neuroscience. *Trends in Cognitive Sciences, 15*(1), 11–19. https://doi.org/10.1016/j.tics.2010.10.002

Shamosh, N. A., & Gray, J. R. (2008). Delay discounting and intelligence: A meta-analysis. *Intelligence, 36*(4), 289–305.

Skogli, E. W., Egeland, J., Andersen, P. N., Hovik, K. T., & Øie, M. (2014). Few differences in hot and cold executive functions in children and adolescents with combined and inattentive subtypes of ADHD. *Child Neuropsychology, 20*(2), 162–181. https://doi.org/10.1080/09297049.2012.753998

Solanto, M. V., Abikoff, H., Sonuga-Barke, E., Schachar, R., Logan, G. D., Wigal, T., Hechtman, L., Hinshaw, S., & Turkel, E. (2001). The ecological validity of delay aversion and response inhibition as measures of impulsivity in AD/HD: A supplement to the NIMH multimodal treatment study of AD/HD. *Journal of Abnormal Child Psychology, 29*(3), 215–228. https://doi.org/10.1023/a:1010329714819

Sonuga-Barke, E. J. (1994). On dysfunction and function in psychological theories of childhood disorder. *Journal of Child Psychology and Psychiatry and Allied Disciplines, 35*(5), 801–815. https://doi.org/10.1111/j.1469-7610.1994.tb02296.x

Sonuga-Barke, E. J. (2002). Psychological heterogeneity in AD/HD–a dual pathway model of behaviour and cognition. *Behavioural Brain Research, 130*(1-2), 29–36. https://doi.org/10.1016/s0166-4328(01)00432-6

Sonuga-Barke, E. J. (2004). On the reorganization of incentive structure to promote delay tolerance: A therapeutic possibility for AD/HD? *Neural Plasticity, 11*(1-2), 23–28. https://doi.org/10.1155/np.2004.23

Sonuga-Barke, E. J. (2005). Causal models of attention-deficit/hyperactivity disorder: From common simple deficits to multiple developmental pathways. *Biological Psychiatry, 57*(11), 1231–1238. https://doi.org/10.1016/j.biopsych.2004.09.008

Sonuga-Barke, E. J., Taylor, E., Sembi, S., & Smith, J. (1992). Hyperactivity and delay aversion: I. The effect of delay on choice. *Journal of Child Psychology and Psychiatry and Allied Disciplines, 33*(2), 387–398. https://doi.org/10.1111/j.1469-7610.1992.tb00874.x

Sonuga-Barke, E. J., Sergeant, J. A., Nigg, J., & Willcutt, E. (2008). Executive dysfunction and delay aversion in attention deficit hyperactivity disorder: Nosologic and diagnostic implications. *Child and Adolescent Psychiatric Clinics of North America, 17*(2), 367–384., ix. https://doi.org/10.1016/j.chc.2007.11.008

Sonuga-Barke, E., Bitsakou, P., & Thompson, M. (2010). Beyond the dual pathway model: Evidence for the dissociation of timing, inhibitory, and delay-related impairments in attention-deficit/hyperactivity disorder. *Journal of the American Academy of Child and Adolescent Psychiatry, 49*(4), 345–355. https://doi.org/10.1016/j.jaac.2009.12.018

Sørensen, L., Sonuga-Barke, E., Eichele, H., van Wageningen, H., Wollschlaeger, D., & Plessen, K. J. (2017). Suboptimal decision making by children with ADHD in the face of risk: Poor risk adjustment and delay aversion rather than general proneness to taking risks. *Neuropsychology, 31*(2), 119–128. https://doi.org/10.1037/neu0000297

Stanovich, K. E. (1999). *Who is rational?: Studies of individual differences in reasoning*. Psychology Press.

Stanovich, K. E. (2009a). Distinguishing the reflective, algorithmic, and autonomous minds: Is it time for a tri-process theory? In J. Evans & K. Frankish (Eds.), *In two minds: Dual processes and beyond* (pp. 55–88). Oxford University Press.

Stanovich, K. E. (2009b). *What intelligence tests miss: The psychology of rational thought*. Yale University Press.

Stanovich, K. (2011). *Rationality and the reflective mind*. Oxford University Press.

Stanovich, K. E., & West, R. F. (2008). On the relative independence of thinking biases and cognitive ability. *Journal of Personality and Social Psychology, 94*(4), 672–695. https://doi.org/10.1037/0022-3514.94.4.672

Stanovich, K. E., Toplak, M. E., & West, R. F. (2008). The development of rational thought: A taxonomy of heuristics and biases. *Advances in Child Development and Behavior, 36*, 251–285. https://doi.org/10.1016/s0065-2407(08)00006-2

Stanovich, K. E., West, R. F., & Toplak, M. E. (2012). Judgment and decision making in adolescence: Separating intelligence from rationality. In *The adolescent brain: Learning, reasoning, and decision making* (pp. 337–378). American Psychological Association. https://doi.org/10.1037/13493-012

Stanovich, K. E., West, R. F., & Toplak, M. E. (2016). *The Rationality Quotient: Toward a test of rational thinking*. MIT press.

Steinberg, L. (2010). A dual systems model of adolescent risk-taking. *Developmental Psychobiology, 52*(3), 216–224. https://doi.org/10.1002/dev.20445

Steinberg, L., Graham, S., O'Brien, L., Woolard, J., Cauffman, E., & Banich, M. (2009). Age differences in future orientation and delay discounting. *Child Development, 80*(1), 28–44. https://doi.org/10.1111/j.1467-8624.2008.01244.x

Steingroever, H., Wetzels, R., Horstmann, A., Neumann, J., & Wagenmakers, E. J. (2013). Performance of healthy participants on the Iowa Gambling Task. *Psychological Assessment, 25*(1), 180. https://doi.org/10.1037/a0029929

Sternberg, R. J. (2002). *Why smart people can be so stupid*. Yale University Press.

Taylor, M. R., Boden, J. M., Rucklidge, J. J., & Farmer, R. R. (2017). The function of reward sensitivity and temporal discounting in the relationship between risk and ADHD in adults. *New Zealand Journal of Psychology, 46*(1), 36–46.

Thorell, L. B. (2007). Do delay aversion and executive function deficits make distinct contributions to the functional impact of ADHD symptoms? A study of early academic skill deficits. *Journal of Child Psychology and Psychiatry and Allied Disciplines, 48*(11), 1061–1070. https://doi.org/10.1111/j.1469-7610.2007.01777.x

Toplak, M. E., Jain, U., & Tannock, R. (2005). Executive and motivational processes in adolescents with attention-deficit-hyperactivity disorder (ADHD). *Behavioral and Brain Functions, 1*(1), 8. https://doi.org/10.1186/1744-9081-1-8

Toplak, M. E., Sorge, G. B., Benoit, A., West, R. F., & Stanovich, K. E. (2010). Decision making and cognitive abilities: A review of associations between Iowa Gambling Task performance, executive functions, and intelligence. *Clinical Psychology Review, 30*(5), 562–581. https://doi.org/10.1016/j.cpr.2010.04.002

Toplak, M. E., West, R. F., & Stanovich, K. E. (2011). The Cognitive Reflection Test as a predictor of performance on heuristics-and-biases tasks. *Memory & Cognition, 39*(7), 1275–1289. https://doi.org/10.3758/s13421-011-0104-1

Toplak, M. E., Hosseini, A., & Basile, A. G. (2016). Temporal discounting and associations with cognitive abilities and ADHD-related difficulties in a developmental sample. In M. E. Toplak & J. A. Weller (Eds.), *Individual differences in judgment and decision making: A developmental perspective* (pp. 85–106). Psychology Press.

Utsumi, D. A., & Miranda, M. C. (2018). Temporal discounting and attention-deficit/hyperactivity disorder in childhood: Reasons for devising different tasks. *Trends Psychiatry Psychotherapy, 40*(3), 248–252. https://doi.org/10.1590/2237-6089-2017-0094

Van Dessel, J., Sonuga-Barke, E., Mies, G., Lemiere, J., Van der Oord, S., Morsink, S., & Danckaerts, M. (2018). Delay aversion in attention deficit/hyperactivity disorder is mediated by amygdala and prefrontal cortex hyper-activation. *Journal of Child Psychology and Psychiatry and Allied Disciplines, 59*(8), 888–899. https://doi.org/10.1111/jcpp.12868

Wanstall, E. A., Doidge, J., Weiss, J., & Toplak, M. E. (2019). Estimations of competence in neurodevelopmental conditions: A review. *Psihologijske Teme, 28*(1), 193–232. https://doi.org/10.31820/pt.28.1.10

Warnell, K. R., Maniscalco, S., Baker, S., Yi, R., & Redcay, E. (2019). Social and delay discounting in autism spectrum disorder. *Autism Research, 12*(6), 870–877. https://doi.org/10.1002/aur.2085

Weafer, J., Milich, R., & Fillmore, M. T. (2011). Behavioral components of impulsivity predict alcohol consumption in adults with ADHD and healthy controls. *Drug Alcohol Dependence, 113*(2–3), 139–146. https://doi.org/10.1016/j.drugalcdep.2010.07.027

Willcutt, E. G., Doyle, A. E., Nigg, J. T., Faraone, S. V., & Pennington, B. F. (2005). Validity of the executive function theory of attention-deficit/hyperactivity disorder: A meta-analytic review. *Biological Psychiatry, 57*(11), 1336–1346. https://doi.org/10.1016/j.biopsych.2005.02.006

Willner, P., Bailey, R., Parry, R., & Dymond, S. (2010). Performance in temporal discounting tasks by people with intellectual disabilities reveals difficulties in decision making and impulse control. *American Journal on Intellectual and Developmental Disabilities, 115*(2), 157–171. https://doi.org/10.1352/1944-7558-115.2.157

Wilson, V. B., Mitchell, S. H., Musser, E. D., Schmitt, C. F., & Nigg, J. T. (2011). Delay discounting of reward in ADHD: Application in young children. *Journal of Child Psychology and Psychiatry and Allied Disciplines, 52*(3), 256–264. https://doi.org/10.1111/j.1469-7610.2010.02347.x

Yang, B. R., Chan, R. C., Gracia, N., Cao, X. Y., Zou, X. B., Jing, J., Mai, J. N., Li, J., & Shum, D. (2011). Cool and hot executive functions in medication-naive attention deficit hyperactivity disorder children. *Psychological Medicine, 41*(12), 2593–2602. https://doi.org/10.1017/S0033291711000869

Zablotsky, B., Bramlett, M. D., & Blumberg, S. J. (2020). The co-occurrence of autism spectrum disorder in children with ADHD. *Journal of Attention Disorders, 24*(1), 94–103. https://doi.org/10.1177/1087054717713638

Zeif, D., & Yechiam, E. (2020). Autism is not associated with poor or enhanced performance on the Iowa Gambling Task: A meta-analysis. *Neuroscience & Biobehavioral Review, 113*, 440–447. https://doi.org/10.1016/j.neubiorev.2020.04.016

Chapter 14
Social Functioning and Decision Making: From Group to Individual Differences Across the Autism Spectrum

Irwin P. Levin, Gary J. Gaeth, Aron M. Levin, and Shiyun Chen

Introduction

Autism spectrum disorder (ASD) is one of the most frequently occurring intellectual and developmental disabilities (IDD) and has a profound impact on the lives of both the person with the IDD and their providers of support including family, friends, and professionals. Various researchers have reported that persons with ASD have difficulties making life decisions and are less focused in their decision making. However, both the severity and the mix of characteristics for those on the autism spectrum are broad. Many people within the general population possess some degree of traits associated with autism. One goal of this chapter is to show that in addition to group differences, individual differences in the level of social functioning ability associated with ASD are predictive of differences in decision making. At the end of this chapter, we suggest how this approach might be applied with other categories of IDD.

I. P. Levin · G. J. Gaeth (✉) · S. Chen
University of Iowa, Iowa City, IA, USA
e-mail: irwin-levin@uiowa.edu; gary-gaeth@uiowa.edu; Shiyun-chen@uiowa.edu

A. M. Levin
Department of Marketing University of Northern, Kentucky Highland Heights, KY, USA

I. Khemka, L. Hickson (eds.), *Decision Making by Individuals with Intellectual and Developmental Disabilities*, Positive Psychology and Disability Series,
https://doi.org/10.1007/978-3-030-74675-9_14

Background

Autism has been described as a multifaceted spectrum of neurodevelopmental conditions displayed through varying levels of social communication and behavioral difficulties (American Psychiatric Association, 2013). Persons with ASD have extremely variable cognitive and behavioral functioning abilities that cannot be explained by a single construct (Hill, 2004; Tager-Flusberg, 2007; Van de Cruys, et al., 2014). A number of previous studies have shown that persons on the autism spectrum, even those who are high functioning, have difficulty with decision making in general, take longer to make decisions, and are more apt to avoid decisions altogether (Brosnan, et al., 2014; Luke et al., 2012). Luke et al. (2012) in particular concluded that persons with autism find decision making exhausting, with special difficulties when decisions involve a change in routine, when they must be completed quickly, or when they involve talking to others.

On the other hand, there has been growing recognition of the unique abilities and potential contributions of some persons on the autism spectrum. In fact, high-tech companies have recently sought out those on the autism spectrum as employees because of the unique skills that many of them possess related to attention to detail (Lindson, 2019).

Persons with ASD are especially known to have deficits in the area of social functioning or "theory of mind" (e.g., Baron-Cohen, et al., 1985; Baron-Cohen, et al., 1994) which affect their interpersonal decisions and evaluations. Theory of mind involves the ability to understand the mental states of others where deficits result in difficulty to predict and understand the mental states of others and the behaviors arising from these states. In terms of the constructs developed in this chapter, deficits in this area have a negative impact on social functioning and decisions related to interpersonal relationships.

Deficits in theory of mind are not the only distinguishing feature of autism. As we shall demonstrate, decision-making difficulties as well as strengths associated with autism are domain specific depending on the processes tapped by different tasks and measures.

Dual-process theories (e.g., Kahneman, 2003) assume two distinct processes, a rapid and automatic intuitive process and a more effortful and deliberative process. Brosnan et al. (2016; see also Ashwin & Brosnan, 2020) applied dual-process theory to better understand reasoning processes among persons with autism. A popular instrument for assessing different thinking styles is the Rational-Experiential Inventory (REI) (Epstein et al., 1996) which includes one set of items designed to assess one's engagement in deliberative reasoning (the Rational component) and one set of items designed to assess one's engagement in intuitive reasoning (the Experiential component). Brosnan et al. (2016) proposed a model in which autism is associated with more deliberative reasoning and less reliance on intuition.

A study with results consistent with this model employed the Iowa Gambling Task (IGT; Bechara et al., 1994). In this task, participants sample from four decks of cards where each draw of a card leads to a specific gain or loss. Even though outcomes vary across successive draws for each deck, two of the decks are "good" because in the long run the player will be rewarded and two are "bad" because in the long run the player will be punished. A key performance measure is how many draws it takes before the player consistently chooses a good deck. Johnson et al. (2006) showed that adolescents and young adults with Asperger syndrome were more likely than controls to vary their choices from draw to draw, and this over-deliberation led them to be slower to arrive at an optimal choice. However, a recent meta-analysis by Zeif and Yechiam (2020) concluded that although their strategy may differ, decision performance on the IGT as measured by net winnings is similar for persons with ASD and controls.

The greater reliance on analytic and deliberate processing may underlie the difficulty in making efficient and quick decisions which may be particularly important in the social world. However, these same mechanisms may serve well in other contexts. High-functioning persons on the autism spectrum do well on tasks that require thoughtful decisions. Frith and Frith (2003) suggested that those who are higher functioning may have learned to rely more on language, decision rules, and other nonsocial cognitive functioning skills rather than on social skills in solving problems.

Current Focus

Following previous research by others on the decision making of individuals on the autism spectrum, research by our team has focused on decision making by high-functioning adults on the autism spectrum. This population was selected because of increased awareness of its presence, its likely mixture of strengths and weaknesses, and because we believe that this is a group that has attained growing responsibility for making the decisions that affect not only their personal quality of life but also that of those around them.

In this chapter the following aspects of decision making are addressed: (1) assessing perceived difficulty in making routine everyday decisions such as when to go to bed, when to wake up, what clothes to wear, and what food to eat; (2) exploring perceived difficulty in forming personal relationships with fellow students, workers, and housemates; (3) assessing the consequences of these decisions in areas such as consumer behavior; (4) measuring aspects of decision-making competence such as the ability to incorporate numerical information into their decisions; and (5) using a wide range of tasks and measures in order to differentiate between different processes that underlie decision making in different domains. We will describe the results of prior research and extend them to current research.

During the progression of this research stream, we have come to view autism not as the subject for only dichotomous comparisons between individuals with autism and individuals in a control group but as a source of important data for determining individual differences in decision-making strengths and weaknesses. In particular, the quality of social functioning is identified as a key source of such differences where inclusion of persons with autism into the selection of research participants serves to increase the range of individual differences in this important domain as a unique vehicle for studying the role of social functioning on decision making in the general population.

The first part of this chapter focuses on group differences between high-functioning adults on the autism spectrum and controls. Based on these findings focus is then shifted to individual differences in the quality of social functioning. As briefly summarized below, there has been a rich history of research on individual differences in decision making.

History of the Study of Individual Differences in Human Judgment and Decision Making

The history of research in human judgment and decision making is marked by a progression from demonstration of interesting phenomena to a more in-depth analysis, often oriented toward a judgment error (or bias). Examples include base rate neglect, hindsight bias, and framing effects, where effort is made to provide a better understanding of the processes underlying these effects. This process-level analysis focuses on searching for task-related variables that moderate the effects (e.g., risk domain and amount and complexity of information) as well as individual characteristics of the decision-maker that serve as mediators (such as age, gender, and personality). It is natural to focus on judgment weaknesses, such as heuristics and biases, but it is important when looking at individual differences to also consider strengths that can improve decision making.

Focus on individual differences in our lab goes back over 20 years (Levin, 1999) and has been a recurring theme in our research over the years (e.g., Levin et al., 2014). A sampling of other labs with this focus includes Payne, Bettman, and Johnson's (1993) study of adaptive decision making, Stanovich and West's (1998) study of rational thought, and Baron and Ritov's (2004) work on decision biases. See also edited volume by Toplak and Weller (2016) with many examples.

Perhaps the most useful tool for tracking these developments has been the establishment of the Decision Making Individual Differences Inventory (DMIDI) (Appelt et al., 2011). This inventory includes a collection of a wide variety of examples of individual difference measures falling under categories such as decision style, cognitive ability, personality, attitude toward risk, motivation, impulsiveness, and time orientation.

Following the initial publication of the inventory, members of the Society for Judgment and Decision Making have been encouraged to add to the list, and these updates can be assessed through the website, http://www.sjdm.org/dimidi/.

In the present case we focus on our newly developed Quality of Social Functioning Index (QSFI) as a key continuous measure of individual differences underlying both variations along the autism spectrum and variations within the general population.

Introduction to Individual Differences in Social Functioning

Our initial study (Levin et al., 2015) was small in scope and conducted in the laboratory (N = 15 college students diagnosed with ASD) which allowed us to conduct face-to-face interviews. With what was to become the basis of later research, open-ended questions were asked about participants' social experiences: *the quality of personal relations at home, school, work, and with friendship networks*. Transcripts were scored on quality of social relations by two independent coders who were in close agreement with each other. On a scale of 1–5 for each question, the two coders were no more than one point apart on any question. The final score for each participant was the average of the coders for the total score across all questions. These "socially based" responses correlated with other measures of behaviors associated with autism. In later studies, the applicability of this scale was extended by developing a formal numeric index that measures the same concept. As we will illustrate later, an index of these items was designed as a measure that we called QSFI.

The larger follow-up study (Gaeth et al., 2016) used MTurk (Buhrmester, Kwang, & Gosling, 2011) survey techniques and included within-group analyses of the relation between social functioning, thinking style, and scores on an autism screener. For participants in the ASD group, social relationships were poorest for those showing the lowest levels of engagement in experiential/intuitive decision-making style. Those scoring at the most extreme levels on the autism screener were more apt to be deliberative thinkers and less apt to be intuitive/impulsive thinkers. These results are consistent with Ashwin and Brosnan's (2020) dual-process theory of autism.

In subsequent studies these analyses were expanded to examine the role of social functioning across a combined sample of persons with and without a formal autism diagnosis. These analyses provide new insight into how the spectrum of autistic traits extends to those not formally diagnosed. We start, however, with a description of studies comparing persons with and without autism.

Group Differences in Decision Making Between Persons with and without Autism

This section summarizes results from our prior research comparing a group of persons with autism and a control group. As will be described in more detail later, our basic recruitment procedure is to use MTurk online panels to recruit both a group of respondents who report diagnosis of autism or Asperger syndrome and an unrestricted group and then to verify classification based on clinically developed screener surveys such as Baron-Cohen et al.'s (2001) Autism Spectrum Quotient (AQ).

These results have been reported in papers by Levin et al. (2015), Gaeth et al. (2016), Levin et al. (2019), and Levin et al. (2020a, b) which include multiple tasks and measures designed to generate a profile of differences and similarities between persons in each group. Gaeth et al. (2016) included a sample consisting of 72 persons with ASD and 67 control participants. Mean age of participants in the ASD group tended to be somewhat lower than that of controls (30.0 vs. 37.3 years). In each group, females made up approximately 40% of the sample. Education level was somewhat higher, but not significantly, for persons in the ASD group; 43% of persons in the ASD group completed a 4-year college degree or higher compared to 38% for the controls. This is a primary reason why we classify persons in our ASD groups as high functioning. Nevertheless, participants with ASD were less apt to have full-time jobs compared to those in the control group (50% vs. 65%). When asked about current living arrangements, participants with ASD compared to those in the control group were less apt to be living with a spouse or significant other (28% vs. 40%) and more apt to be living with parents (35% vs. 19%).

Difficulty with Everyday Decisions

In the study by Gaeth et al. (2016), persons in the group with autism reported significantly more difficulty than those in the control group in making everyday routine decisions that include the following: *when to go to bed, when to wake up, what clothes to wear, what to eat, when to shower, what and when to take medications, when to pay bills, and making and keeping medical appointments.* (See Table 14.1.) From the point of view of the dual-process theory, these results reflect overreliance on deliberative processes.

A follow-up survey with the same samples measuring what we call "Bad Decision Outcomes" showed that the group with ASD reported higher incidence of bad outcomes in the following categories: *Rented movie but unwatched, bought clothes never worn, quit a job after a month, at least 2 weeks late on a rent payment, had a check bounce, took out a loan that was not paid back, borrowed money from parent, and used emergency credit* (see Table 14.2 from Gaeth et al. 2016). The bad outcomes

Table 14.1 Group comparisons of difficulty in everyday decisions: means (95% confidence interval)

Difficulty with	ASD (N = 72)	Control (N = 68)	Effect size	Significance of difference
Deciding when to go to bed	6.26 (±0.67)	8.60 (±0.55)	0.90	$p < 0.0001$
Deciding when to wake up	6.40 (±0.69)	7.89 (±0.55)	0.57	$p < 0.0100$
Deciding what clothes to wear	6.17 (±0.68)	8.03 (±0.55)	0.72	$p < 0.0001$
Deciding what to eat	4.83 (±0.54)	6.43 (±0.56)	0.69	$p < 0.0001$
Deciding when to shower	6.61 (±0.64)	8.93 (±0.50)	0.96	$p < 0.0001$
Deciding what and when to take medications	7.08 (±0.64)	8.78 (±0.49)	0.70	$p < 0.0001$
Deciding when to pay bills	5.94 (±0.65)	7.25 (±0.65)	0.48	$p < 0.0100$
Making and keeping medical appointments	5.90 (±0.63)	6.99 (±0.68)	0.40	$P = 0.0200$

Note: Gaeth et al. (2016; Table 14.3). Scale was 1 = extremely difficult to 10 = not difficult

Table 14.2 Group comparisons of social functioning: means (95% confidence interval)

Measure	ASD (SE) (N = 72)	Control (SE) (N = 68)	Effect size	Significance of difference
Satisfied with schooling	5.26 (±0.60)	6.65 (±0.48)	0.60	$p < 0.001$
Get along with teachers	5.73 (±0.62)	7.75 (±0.33)	0.95	$p < 0.0001$
Get along with fellow students	4.18 (±0.57)	7.21 (±0.52)	1.32	$p < 0.0001$
Satisfied with employment	6.31 (±0.62)	6.65 (±0.68)	0.14	*NS*
Get along with supervisors	6.46 (±0.55)	7.77 (±0.59)	0.52	$p < 0.01$
Get along with fellow workers	5.75 (±0.53)	7.95 (±0.49)	1.14	$p < 0.0001$
Satisfied with living arrangement	6.81 (±0.61)	7.86 (±0.61)	0.41	$p = 0.02$
Get along with people you live with	6.80 (±0.70)	8.33 (±0.62)	0.56	$p < 0.01$
Satisfied with friendship network	6.18 (±0.67)	6.57 (±0.69)	0.14	*NS*

Note: Gaeth et al. (2016; Table 14.2). Scale was 0–10

reported in Gaeth et al. (2016) follow from the difficulties reported in Table 14.1 and could have notable negative consequences on how persons function in society.

Social Functioning Disability

Perhaps our most dramatic results to date were those where persons with autism reported greater levels of difficulty in aspects of social functioning including *getting along with teachers, getting along with fellow students, getting along with supervisors, getting along with fellow workers, and getting along with people you live with.* As seen in Table 14.2, persons in the group with ASD reported greater difficulty for each item, and most of the differences were statistically significant in comparison to the control group. Later replications supported the reliability of each of these differences.

Social deficits associated with autism also emerged in two other tasks that required decision making. Using a variant of temporal construal (Trope & Liberman, 2003) where participants choose between a more moderate immediate outcome and a more extreme outcome at a designated future date, items concerning interpersonal contacts were added. For these new items, participants were asked to rate on a scale of 0–10 how likely they were to defer meeting a new neighbor and to postpone a job interview. Persons with ASD (N = 72) were more likely than controls (N = 68) to choose to postpone a job interview (mean = 3.22 vs. 2.17) and choose to postpone meeting a new neighbor (mean = 5.29 vs. 4.33). The job interview effect, in particular, could have dire consequences on employment status which is a recognized problem for persons with autism and was recognized earlier in this chapter as reflected in lower levels of full-time employment even for high-functioning persons on the spectrum.

Using the Domain-Specific Risk-Taking (DOSPERT) inventory (Weber et al., 2002), the reported tendency to take risks was separated into different domains. Across the domains of financial risks, recreational risks, health/safety risks, ethical risks, and social risks, persons with ASD (N = 72) reported significantly less risk-taking (on a scale from −3 extremely unlikely to +3 extremely likely for each item) than controls (N = 68) only in the social domain (e.g., *Disagreeing with an authority figure on a major issue, Moving to a city far from your extended family*). By contrast, they reported greater risk-taking in the ethical domain (e.g., *Passing off somebody else's work as your own, Not returning a wallet you found with $20*). DeGroot (2020) recently extended these findings by reporting a positive correlation between measures of autism and risk perception in the social domain but not in other domains. The avoidance of social risks, while not always a bad strategy, at its extreme could interfere with the ability to get ahead in the world. For example, one's likelihood of succeeding in a competitive job could well depend on their willingness to go outside the box in a competitive environment.

It is these results that ultimately led us to focus on social functioning as a key component of individual differences associated with autism. However, not all of our results supported differences between groups. For example, when rating the difficulty of making important life decisions such as choosing jobs, college majors, and living arrangements, persons in the ASD and control groups did not differ. We attribute this to the use of a deliberative decision style.

Two decision-making tasks involving risky choice analysis were conducted in our laboratory. One was a classic task that previewed the winning of a Nobel Prize by one of its creators, Daniel Kahneman. The other was a task developed more recently in our laboratory. As a preview, neither task detected deficits related to autism.

Framing Effects: The Unusual Disease Problem

The Unusual Disease Problem (originally referred to as the "Asian Disease Problem"; Tversky & Kahneman, 1981) is included for two main reasons. First, it was the forerunner of all subsequent studies that fall under the heading of Risky Choice Framing Effects (Levin, et al., 1998) and still stands as the gold standard for this area of research. Second, it highlights emotional factors in risky decision making by contrasting conditions with identical objective information but different emotionally charged labels, "lives saved" versus "lives lost." Previous research with gambling tasks suggests that the loss frame typically leads to more risk-taking than the gain frame, but that persons with autism show a reduced framing effect when gambles are alternatively described in terms of money lost versus money gained (DeMartino et al., 2008). This has been interpreted as a reduced focus on emotional information in the decision process and has been linked to a greater reliance on deliberative and rational decision style (Ashwin & Brosnan, 2020). However, it remains to be seen whether this extends to the extreme emotional labels of "lives saved" versus "lives lost."

High-functioning adults with ASD (N = 165) were presented with the following cover story (Levin et al., 2019): "The United States is preparing for the outbreak of an unusual disease that is expected to kill 600 people if untreated. Two treatment programs have been developed to combat this disease and the exact scientific estimates of the consequences of these programs are listed below."

In the positive frame condition, the two programs are described as follows:

A. *200 lives will be saved.*
B. *There is 1/3 probability that all 600 will be saved and 2/3 probability that none will be saved.*

In the negative frame condition, the two programs are described as follows:

C. *400 lives will be lost.*
D. *There is 1/3 probability that no lives will be lost and 2/3 probability that all 600 lives will be lost.*

Note that in addition to the fact that A and C are equivalent and B and D are equivalent, within each pair the expected values of the risky and riskless options are equated.

In the original version a binary response scale was employed, Select A or Select B. In later versions (Levin et al., 2002), a rating scale was employed, Much prefer A to Much prefer B. The latter was employed here with a range of 0–10. As in the Levin et al. (2002) study, a within-subject version of the task was used where each person received both the positive and negative versions of the problem separated by presenting one at the beginning of a long survey and the other at the end. The order of presentation was randomized across participants.

Analysis of Variance (ANOVA) included order of presentation of gain and loss framing conditions as a factor. Neither the main effect of order nor the interaction between order and frame was significant. These results were taken as support for combining data across orders. A difference score between framing conditions was then computed for each participant. Mean responses were 4.98 for losses and 3.95 for gains where higher numbers represent greater preferences for the risky options. The overall difference score was significantly different from zero ($t(164) = 4.43$, $p < 0.0001$), indicating that the typical finding of greater preference for risk in the loss domain than in the gain domain was replicated here.

The use of a within-subject design allowed us to examine the framing effect at the individual participant level. Eighty percent showed greater risk preference in the loss frame than in the gain frame. Mean AQ score did not differ between those who did and did not show the typical framing effect ($t(163) = 0.69$, $p = 0.493$). The main message here is that high-functioning adults with autism were not significantly less responsive to emotional outcome frames than were prior respondents.

This result seems to contradict evidence that persons on the spectrum are relatively insensitive to contextual frame. It may in fact be a sign that high-functioning adults on the spectrum have learned to overcome their inherent lack of attention to emotional cues, especially when these pertain to life or death.

Risky Decision Making and the Cups Task

As described by Parker and Fischhoff (2005), decision-making competence refers to the ability to adhere to a cluster of decision-making principles represented by different tasks. These include numeracy (a set of questions requiring the appropriate use of numerical information), applying decision rules (the ability to combine multiple sources of information to arrive at the best choice), resistance to sunk costs, resistance to framing effects, and under-/overconfidence. Of particular interest, persons with autism scored at least as high as the control group on numeracy and applying decision rules, each of which is a key to sound decision making (Gaeth et al., 2016).

As a new way of examining how numerical information is used in risky decision making, the Cups task (Levin & Hart, 2003; Weller et al., 2012, 2015) was developed to examine risky choice across both gain and loss domains where ability to differentiate between "good" and "bad" risks is measured by the extent to which a respondent makes a risky or riskless choice depending on which one has the higher expected value.

Each trial required a choice between a risky and a riskless option. On gain trials the choice was between a sure gain of one coin and varying probabilities of winning multiple coins, represented by varying numbers of cups one of which was the winning cup. Over trials, the number of cups and the amount to be won by selecting the winning cup were varied. On loss trials the choice was between a sure loss of one coin and varying probabilities of losing multiple coins, represented by varying numbers of cups one of which was the losing cup. Over trials the number of cups and the amount of possible loss by selecting the losing cup were varied.

Included are some trials in which the risky option (e.g., one chance out of three of winning five coins on a gains trial; one chance out of three of losing two coins on a loss trial) is more favorable in the long run than the riskless option (winning one coin for sure on gain trials, or losing one coin for sure on loss trials). We call these risk advantageous or RA trials. On other trials the risky option (e.g., one chance out of five of winning three coins or one out of two chances of losing three coins) is less favorable than the riskless choice (a sure gain of one coin on gain trials, a sure loss of one coin on loss trials). We call these risk disadvantageous or RD trials. Across multiple examples of both types of trials, as well as trials with equal expected value for risky and riskless options (risk neutral or RN trials, e.g., one out of three chances of winning three coins compared to winning one for sure), our key dependent measure is the difference in number of risks taken for RA and RD trials.

As summarized in Fig. 14.1 from Levin et al. (2019), we found no difference between persons with autism and controls (N = 80 in each group) on our key measure. Across gain and loss trials, out of a possible range of 0–6, mean response on RA trials was 4.80 for persons with autism versus 4.84 for controls, and mean

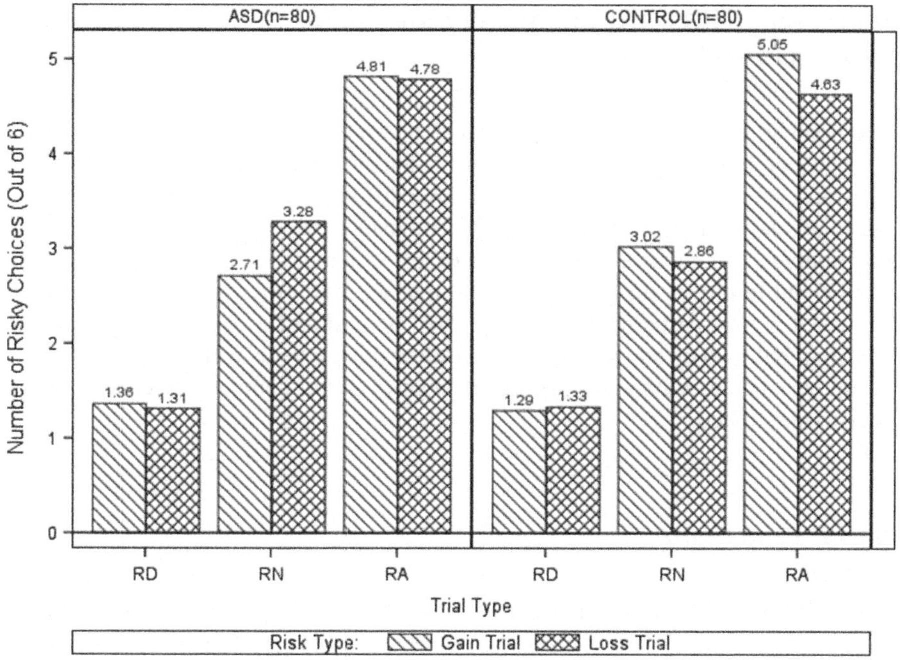

Fig. 14.1 Results from risk-taking for cups experiment (*Note.* From Levin et al. 2019)

response on RD trials was 1.34 for persons with autism and 1.31 for controls, resulting in mean (RA – RD) of 3.46 for individuals with autism and 3.53 for controls. The difference in means for RA – RD between groups did not approach statistical significance, $t < 1$. In other words, for this measure of decision-making competence, high-functioning persons with ASD were indistinguishable from persons in the control group.

The rather dramatic differences between results for different measures of decision making can be explained by the dual-process theory of autism (Ashwin & Brosnan, 2020) and the theory of mind account of deficits. The tendency to be more deliberative and less intuitive among persons with autism serves them well when it comes to decisions that require deep thinking but interferes with decisions that most people can make spontaneously and with little effort. Furthermore, social deficits affect some decisions but not others. In the case of the cups task, deliberative thinking would facilitate performance whereas social deficits would play no role.

Next, we describe our most recent research that focuses on individual differences in the quality of social functioning that predict decision-making processes in domains that affect quality of life.

Individual Differences in Social Functioning and Decision Making

The first part of this chapter emphasized dichotomous differences in decision making between high-functioning adults on the autism spectrum and controls without autism. A small pilot study showed that large differences occurred in the formation of social relationships. This motivated us to consider individual differences in social functioning as an important determinant of decision making across the general population and to pursue this in larger, more focused studies. In the ongoing research described below, participants were recruited who identify as being on the autism spectrum but with an added purpose. Including this group as well as an unrestricted group provides us with a unique wider range of social competence than has been previously studied.

Social Functioning

Earlier results show how persons with ASD differ from persons without ASD in terms of their responses to a series of questions assessing quality of social functioning such as self-reports of how well they got along with schoolmates, teachers, fellow workers, bosses, and housemates. It was concluded that persons with ASD reported dramatically lower levels of social functioning than persons in the control group and that this can be extended to individual differences.

In a recent study (Levin et al., 2020b), MTurk was used to recruit both a group with autism and a control group and we used the Autism Quotient (AQ; Baron-Cohen et al., 2001) Index to verify the classification (see Gaeth et al., 2016). The Quality of Social Functioning Index (QSFI) was developed by computing the mean for each respondent on the following nine items where each item was scored 0 to 10 on degree of agreement: *satisfied with school experience, get along with teachers, get along with fellow students, satisfied with employment environment, get along with supervisor, get along with fellow workers, satisfied with living arrangement, get along with people you live with,* and *satisfied with current friendship network.* The scale was designed so that the higher the score the higher the level of social competence.

This index serves two purposes. First, it verifies that the inclusion of a sample recruited from individuals with autism in our pool increased the range of social functioning beyond what would have been found with an open sample. Second, it provides us with the new index of individual differences that we call the QSFI. Based on current data, this index possesses the desirable property of high internal consistency (coefficient alpha = 0.86) where each item correlates at least 0.50 with the total for all nine items. Furthermore, while this index correlates highly with scores on autism screeners ($r = 0.47$ for the Iowa Screener (Gaeth et al., 2016) and $r = 0.42$ for AQ), it represents a purer measure of social functioning than the multidimensional screeners. In our initial large-scale study, we apply it to social media use.

Figure 14.2 plots the frequency distribution of index values for persons in each group separately and for the combined group. Summary statistics are $M = 6.02$ and $SD = 1.75$ for the 77 persons in the group with ASD, $M = 7.11$ and $SD = 1.50$ for the 135 persons in the control group, and $M = 6.71$ and $SD = 1.67$ for all 212 persons when the groups are combined. The figure shows that while there is a wide range of index values within each group, there are notable differences between groups at each end of the range. Approximately one-third of persons in the group with ASD scored at or below 5.5 on the index compared to about one-eighth for participants in the control group. Almost one-half of the participants in the control group scored 7.5 or higher compared to only about 1 in 5 of the participants in the group with ASD. The result of combining groups generates a distribution with greater range and normality than either of the separate groups.

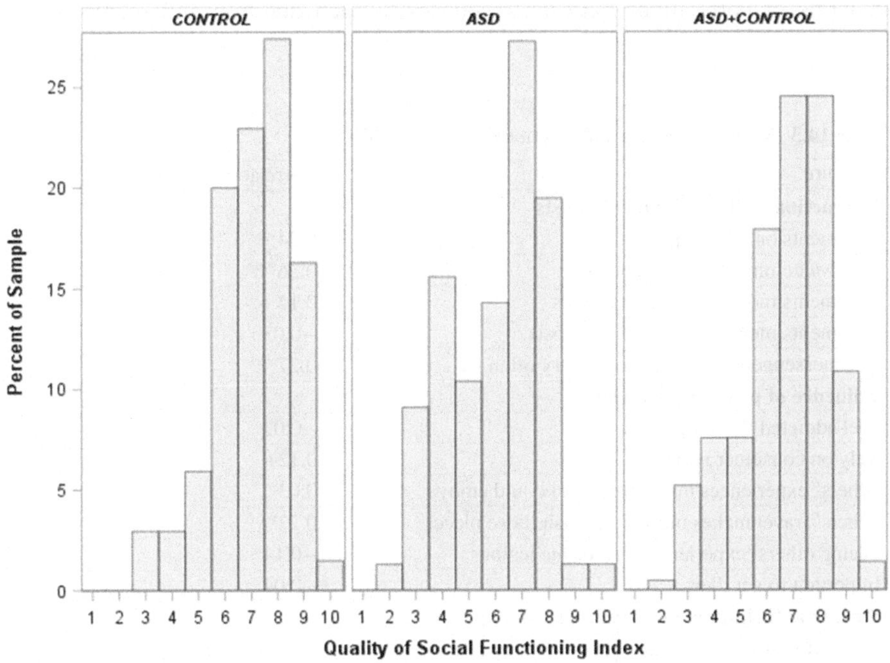

Fig. 14.2 Comparison of the distribution of the QSFI for participants recruited as ASD and control (*Note.* Levin et al. 2020b)

Impact of Social Functioning on Social Media Decision Making

One way in which social functioning ability plays out in the decisions made in daily life is through use of social media. The increasing role of the social media in today's society includes the formation and maintenance of personal relations as well as the transmission of information that could be valuable for everyday decisions such as consumer purchases. Social media use is itself a form of social communications, and social functioning can thus play a crucial role in the extent of social media use and the utility of the information conveyed. Persons with disabilities in social functioning could use social media as a substitute for personal interactions and thus benefit from it, or their disability could extend to misuse of social media. We thus consider the relation between the quality of social functioning and decisions to use social media to be a ripe area for study.

In the study by Levin et al. (2020b), 212 respondents were recruited from MTurk. Of these, 77 were recruited with the requirement they believed they had autism (males = 53, M_{age} = 32,4) and 135 were recruited with no restrictions as controls (males = 74 M_{age} = 33.9). The participants were administered a series of questions concerning social media use and reported reactions. Table 14.3 summarizes results in terms of correlations between responses to each question and scores on the

Table 14.3 Social media use and reactions related to QSFI

Measure	Correlation with QSFI
Interactions with social media posts	
Comments on others' posts	0.24***
Receive comments on own posts	0.26***
Comments more often on own posts	0.12 +
Comments more often on others' posts	−0.03
Use messengers on social media very often	0.27***
Influence of posts on behavior	
Feel addicted to smart phone	−0.02
Rely on consumer reviews	0.12+
Others' experiences make me wish I could enjoy	0.03
Others' travel makes me want to visit those places	0.22***
Seeing others' experiences makes me jealous	−0.14*
Important to get likes	−0.04
Social media impact on consumer behavior	
Connect to brands	0.29***
Friends impact buying	0.15*
Tell friends about good experiences	0.16*
Tell friends about bad experiences	0.11+
Use online deals	0.24***
Scale: 1 = strongly disagree to 7 = strongly agree	

+ $p < 0.10$, * $p < 0.05$, ** $p < 0.01$, *** $p < 0.001$

QSFI. Positive correlations signify that those who score higher on the index provide more affirmative responses.

Table 14.3 is divided roughly into items depicting interactions with social media posts, the influence of these posts, and the specific influence on consumer decision making. Here is a summary of the results: 1) Three of the five items dealing with interactions with social media yielded significant correlations. Persons who scored higher on our QSFI were more frequent users of social media overall and were more likely to both choose to comment on others' posts and to receive comments on their own posts. 2) Of the six items pertaining to the influence of social media, two yielded significant correlations. Those scoring higher on our index were more apt to want to visit places highlighted on social media. However, those scoring lower on the index were more apt to feel jealous of others' experiences depicted on social media. 3) Connections to consumer behavior were stronger for those scoring higher on our index on all five items in this category, with the strongest relationships occurring for connecting to brands and use of online deals, each of which strongly affects decision making in the consumer domain.

Overall, Table 14.3 provides strong evidence of individual differences in our QSFI as a predictor of social media use and its influence on consumer behavior, with lesser impact for those scoring lower on our index. This is of particular significance because we know that those with deficits in the social domain are prone to avoid personal contacts (Baron-Cohen et al., 1985) and conceivably could rely on social media as a substitute for personal interactions. However, decisions concerning the use of social media are themselves a form of social functioning and could thus be impaired among those who scored low on our index. Our data support the latter interpretation. Furthermore, they suggest that the social deficits of persons with ASD affect their decision making with regard to social media use as well as risk-taking in the social domain and postponement of social engagements.

These findings have relevance to quality of life. While relying on social media as a key source of information can have its drawbacks, overlooking reliable recommendations and sources of information or reluctance to share personal opinions can interfere with sound decision making, particularly in the domain of consumer choice.

Summary and Conclusions

This chapter provides evidence that high-functioning adults with ASD possess strengths that can help them make sound decisions in some domains but that they might also have weaknesses, particularly of a social nature, that can lead to bad decisions. Performance on the unusual disease task and the cups task, each with a heavy cognitive component, was equivalent to that of people without autism. Deliberative thinking was likely a key strength. However, their deficits in social functioning can be a huge hurdle to making decisions that impact quality of life. We extend this analysis to the role of social functioning across the general population, where our findings of correlations between our QSFI and decision making offer a potential explanation at both the group and individual level. Among the deficits attributed to persons with low social functioning ability and their potential consequences are the following:

1. Reluctance to make risky decisions in the social domain and decisions to postpone social interactions can prevent people from achieving satisfactory employment and advancing their professional goals.
2. Lower levels of communication through social media can inhibit potentially useful sources of information for decisions concerning which products and services are worthy of consideration and possible purchase.
3. Reliance on careful deliberation rather than intuition can be useful in many situations but, as we demonstrated, can inhibit everyday decision making and can have adverse effects on how one is perceived by others.

What can be done to reduce these problems? Because our findings are based on self-reports, we know that there is a high degree of self-awareness. Persons on the autism spectrum are often well aware of their social challenges and seek training to improve interpersonal communication such as making more eye contact, engaging in give-and-take conversations, and doing practice job interviews. We suggest that training opportunities for persons with autism be extended to reach out to all those who recognize that they have difficulty in making decisions that involve social contact. And, of course, increasing awareness of the possible negative consequences of their social difficulties can enable improvement. Understanding the consequences of one's limitations may be a starting point for many persons with autism toward greater motivation for self-improvement and openness to intervention for effective decision making and enhanced outcomes.

We also suggest that for both those diagnosed with autism and those not formally diagnosed but who have special difficulties in the social domain, training through the use of simulations can be employed to reinforce the ability to make frugal decisions within a fixed length of time and with consultation with others.

Finally, we hope that our approach of extending investigations from formal all-or-none classifications to continuous measures of individual differences traits will be applied to other forms of disability beyond autism.

Related disabilities include fragile X syndrome and ADHD. Fragile X syndrome, while involving different brain functions than ASD, shares the traits of social anxiety and social interaction deficits. Persons with fragile X syndrome could be important additions to studies that focus on the role of quality of social functioning in a variety of everyday decisions and behaviors.

Many of us who would not meet the formal criteria for ADHD suffer from some level of attention deficit that could interfere with our ability to make sound decisions in everyday life. An index of the ability to attend to relevant stimuli in the environment would be a useful tool for assessing the effectiveness of different educational programs aimed at increasing attention. We conclude this chapter by stating that the inclusion of samples of high-functioning persons on the autism spectrum allowed us to better understand both the strengths and weaknesses of this increasing segment of society and how these insights might extend to the general population. Studies such as these could increase our ability to understand this special population and the processes underlying both group and individual differences in decision making which can interfere with the quality of life.

Acknowledgments We thank the Marketing Science Institute (F6786 - 4-1948) and the Tippie College of Business, University of Iowa, for their support of this project and Benjamin Nedderman, Eleanor Burke, and Kerry Strum for their assistance in data collection. Special thanks to Itay Sisso, the Federmann Center for the Study of Rationality, the Hebrew University of Jerusalem, for his invaluable assistance in developing rules for identifying and eliminating "bot" respondents.

References

American Psychiatric Association. (2013). *Diagnostic and statistical manual of mental disorders* (5th ed.). Author.

Appelt, K. C., Milch, K. F., Handgraff, M. J. J., & Weber, E. U. (2011). The decision making individual differences inventory and guidelines for the study of individual differences in judgment and decision making research. *Judgment and Decision making, 6*, 252–262.

Ashwin, C., & Brosnan, M. (2020). The dual process theory of autism. In K. Morsanyi & R. M. J. Byrne (Eds.), *Thinking and reasoning in autism*. Psychology Press/Routledge series on Current Thinking and Reasoning.

Baron, J., & Ritov, I. (2004). Omission bias, individual differences, and normality. *Organizational Behavior and Human Decision Processes, 94*, 74–85.

Baron-Cohen, S., Leslie, A., & Frith, U. (1985). Does the autistic child have a "theory of mind"? *Cognition, 21*, 36–47. https://doi.org/10.1016/0010-0277(85)90022-8

Baron-Cohen, S., Tager-Flusberg, H., & Cohen, D. (1994). *Understanding other minds: Perspectives from autism*. Oxford University Press.

Baron-Cohen, S., Wheelwright, S., Skinner, R., Martin, J., & Clubely, E. (2001). The autism Spectrum quotient (AQ): Evidence from Asperger syndrome/high-functioning autism, male and female, scientists and mathematicians. *Journal of Autism Developmental Disorder, 31*, 5–17. https://doi.org/10.1023/A:1005653411471

Bechara, A. R., Damasio, H., & Damasio, S. W. (1994). Anderson insensitivity to future consequences following damage to prefrontal cortex. *Cognition, 50*, 7–15.

Brosnan, M., Chapman, E., & Ashwin, C. (2014). Adolescents with autism spectrum disorder show a circumspect reasoning bias rather than 'jumping-to-conclusions'. *Journal of Autism and Developmental Disabilities, 44*, 513–520.

Brosnan, M., Lewton, M., & Ashwin, C. (2016). Reasoning on the autism spectrum: A dual process theory account. *Journal of Autism Developmental Disorder, 46*, 2115–2125. https://doi.org/10.1007/s10803-016-2742-4

Buhrmester, M., Kwang, T., & Gosling, S. D. (2011). Amazon's mechanical Turk: A new source of inexpensive, yet high-quality, data? *Perspectives on Psychological Science, 6*(1), 3–5.

De Groot, K. (2020). Non-clinical autistic traits correlate with social and ethical but not with financial and recreational risk-taking. *Frontiers in Psychology, 11*. https://doi.org/10.3389/fpsyg.2020.00360

DeMartino, B., Harrison, N. A., Knafo, S., Bird, G., & Dolan, R. (2008). Explaining enhanced logical consistency during decision making in autism. *Journal of Neuroscience, 28*, 10746–10750.

Epstein, S., Pacini, R., Denes-Raj, V., & Heier, H. (1996). Individual differences in intuitive-experiential and analytical-rational thinking styles. *Journal of Personality and Social Psychology, 71*, 390–405.

Frith, U., & Frith, C. (2003). Development and neurophysiology of mentalizing. *Philosophical Transactions of the Royal Society of London – Series B: Biological Sciences, 258*, 459–473.

Gaeth, G. J., Levin, I. P., Jain, G., & Burke, E. V. (2016). Toward understanding everyday decision making by adults across the autism spectrum. *Judgment and Decision making, 11*, 537–546.

Hill, E. L. (2004). Evaluating the theory of executive dysfunction in autism. *Developmental Review, 24*, 189–233. https://doi.org/10.1016/j.dr.2004.01.001

Johnson, S. A., Yechiam, E., Murphy, R. M., Queller, S., & Stout, J. C. (2006). Motivational processes and autonomic responsivity in Asperger's disorder: Evidence from the Iowa gambling task. *Journal of the International Neuropsychological Society, 12*, 668–676.

Kahneman, D. (2003). A perspective on judgement and choice. *American Psychologist, 58*, 697–720.

Levin, I.P. (1999). Why do you and I make different decisions? Tracking individual differences in decision making. Presidential Address, *Societry for Judgment and Decision making*, Los Angeles, CA, November 1999.

Levin, I. P., & Hart, S. S. (2003). Risk preferences in young children: Early evidence of differences in reaction to potential gains and losses. *Journal of Behavioral Decision making, 16*, 397–413. https://doi.org/10.1002/bdm.453

Levin, I. P., Schneider, S. L., & Gaeth, G. J. (1998). All frames are not created equal: A typology and critical analysis of framing effects. *Organizational Behavior and Human Decision Processes, 76*(2), 149–188.

Levin, I. P., Gaeth, G. J., Schreiber, J., & Lauriola, M. (2002). A new look at framing effects: Distribution of effect sizes, individual differences, and independence of types of effects. *Organizational Behavior and Human Decision Processes, 88*(1), 411–429.

Levin, I. P., Bossard, E. A., Gaeth, G. J., & Yan, H. (2014). The combined role of task, child's age and individual differences in understanding decision processes. *Judgment and Decision making, 9*, 274–286.

Levin, I. P., Gaeth, G. J., Foley-Nicpon, M., Yegorova, V., Cederberg, C., & Yan, H. (2015). Extending the study of decision making competence to special populations: A case study of persons on the autism spectrum. *Frontiers in Psychology, 6*, 539. https://doi.org/10.3389/fpsyg.2015.00539

Levin, I.P., Gaeth, G. J., Levin, A.M., Burke, E.V., & Chen, S. (2019). The risky decision making processes of high-functioning adults on the Autism Spectrum. Paper presented at the Subjective Probability, Utility and Decision making Conference, Amsterdam, the Netherlands, November 2019.

Levin, I. P., Gaeth, G. J., Levin, A. M., & Burke, E. V. (2020a). Decision making processes of high-functioning adults on the autism spectrum. In K. Morsanyi & R. M. J. Byrne (Eds.), *Thinking and reasoning in autism*. Psychology Press/Routledge series on Current Thinking and Reasoning.

Levin, A.M., Gaeth, G. J., Levin, I.P., Burke, E., & Chen, S. (2020b). Relating social skills and social media use to consumer behavior: An investigation across the autism spectrum. Manuscript available from authors.

Lindson, J. (2019). 75% of staff at this successful IT company are on the autism spectrum. Fast Company (online). My 7, 2019.

Luke, L., Ring, H., Redley, M., & Watson, P. (2012). Decision making difficulties experienced by adults with autism spectrum conditions. *Autism, 16*, 612–621.

Parker, A. M., & Fischhoff, B. (2005). Decision making competence: External validation through an individual differences approach. *Journal of Behavioral Decision making, 18*, 1–27. https://doi.org/10.1002/bdm.4813054-3067

Payne, J. W., Bettman, J. R., & Johnson, E. I. (1993). *The adaptive decision maker*. Cambridge University Press.

Stanovich, K. E., & West, R. F. (1998). Individual differences in rational thought. *Journal of Experimental Psychology: General, 127*, 161–188.

Tager-Flusberg, H. (2007). Evaluating the theory-of-mind hypothesis of autism. *Current Directions in Psychological Science, 16*, 311–315. 128.255.47.255.

Toplak, M. E., & Weller, J. A. (2016). *Individual differences in judgment and decision making: A developmental perspective*. Psychological Press.
Trope, Y., & Liberman, N. (2003). Temporal construal. *Psychological Review, 110*(3), 403.
Tversky, A., & Kahneman, D. (1981). The framing of decisions and the psychology of choice. *Science, 211*(4481), 453–458.
Van de Cruys, S., Everts, K., Van der Hallen, R., Van Eylen, L., Boets, B., & de-Wit, L., & Wagemans, J. (2014). Precise minds in uncertain worlds: Predictive coding in autism. *Psychological Review, 121*, 649–675. https://doi.org/10.1037/a0037665
Weber, E. U., Blais, A. R., & Betz, N. E. (2002). A domain-specific risk-attitude scale: Measuring risk perceptions and risk behaviors. *Journal of Behavioral Decision making, 15*(4), 263–290.
Weller, J. A., Levin, I. P., Rose, J. P., & Bossard, E. (2012). Assessment of decision making competence in preadolescence. *Journal of Behavioral Decision making, 25*(4), 414–426.
Weller, J. A., Moholy, M., Bossard, E., & Levin, I. P. (2015). Pre-adolescent decision making competence predicts interpersonal strengths and difficulties: A two-year prospective study. *Journal of Behavioral Decision making, 28*, 76–88. https://doi.org/10.1002/bdm.1822
Zeif, D., & Yechiam, E. (2020). Autism is not associated with poor or enhanced performance on the Iowa gambling task: A meta-analysis. *Neuroscience and Behavioral Reviews, 113*, 440–447.

Chapter 15
Cognitive, Emotional, and Moral Decision Making in Adolescents and Adults with Autism Spectrum Disorder

Hidetsugu Komeda

This chapter first reviews cognitive decision making in adolescents and adults with and without autism spectrum disorders (ASD), with a focus on executive function in scenarios such as gambling tasks. A second focus is on emotional decision making in adolescents and adults with and without ASD. Alexithymia and interoception have been found to have an important impact on emotional decision making. Alexithymia is characterized by difficulties in recognizing emotions from internal bodily sensations and frequently co-occurs in as many as 50% of individuals with ASD (Hill, et al., 2004; Shah, et al., 2016b). Effects of comorbidity, such as ASD with alexithymia traits, are also considered, along with neuroimaging and behavioral studies of emotional decision making. A third focus is on moral decision making and individual differences in adolescents with ASD and adolescents with callous and unemotional traits. Finally, a support program for enhancement of decision making in adolescents and adults with ASD is proposed.

Alexithymia is a subclinical phenomenon involving a lack of emotional awareness or difficulty in identifying and describing feelings and in distinguishing feelings from the bodily sensations of emotional arousal (Nemiah et al., 1976; Sifneos, 1973). Self-report is the most widely used approach to assessing alexithymia. The Toronto group developed a 20-item assessment instrument, the Toronto Alexithymia Scale (TAS-20) (Bagby et al., 1994a, b). The TAS-20 assesses three facets of alexithymia: difficulty identifying feelings, difficulty describing feelings, and externally oriented thinking. Different dimensions of alexithymia may play a role in different anxiety disorders; in fact alexithymia dimensions of difficulty in identifying and describing emotions seem more correlated to panic disorder (PD), post-traumatic stress disorder (PTSD), social phobia, and generalized anxiety disorder, whereas externally oriented thinking may be more closely related to obsessive compulsive

H. Komeda (✉)
Education, Psychology, and Human Studies, Aoyama Gakuin University, Tokyo, Japan
e-mail: komeda@ephs.aoyama.ac.jp

I. Khemka, L. Hickson (eds.), *Decision Making by Individuals with Intellectual and Developmental Disabilities*, Positive Psychology and Disability Series,
https://doi.org/10.1007/978-3-030-74675-9_15

disorder (De Berardis et al., 2008). Alexithymia may cause anxiety and sleep-related issues (Tani et al., 2004), and the inability to express and externalize emotions in a healthy way can lead to a variety of psychosomatic manifestations, including immune, gastrointestinal, and circadian disruptions, all of which are frequently seen in ASD (Poquerusse et al., 2018).

Psychopathy is defined as an antisocial disorder in which an individual manifests amoral and antisocial behavior, shows a lack of ability to love or establish meaningful personal relationships, expresses extreme egocentricity, and demonstrates a failure to learn from experience and other behaviors associated with the condition (Hermann, 2017). The two most common ways to assess psychopathic traits are to use expert rater devices, such as the Psychopathy Checklist-Revised (PCL-R, Hare, 1991), and self-report inventories, such as the Psychopathic Personality Inventory (PPI, (Lilienfeld & Andrews, 1996) and the Minnesota Multiphasic Personality Inventory (MMPI, Lilienfeld, 1999).

A personality disorder such as psychopathy, and its likely developmental course, is also relevant to intellectual disabilities (ID) (Lindsay, 2007; Morrissey & Hollin, 2011; Torr, 2003). First, early environmental influences, including poor parenting, neglect, and emotional, physical, and sexual abuse, may play a role in the development of these disorders (e.g., Hatton & Emerson, 2004; Hill, 2003; Marshall & Cooke, 1999). Children with ID are more likely to suffer from environmental disadvantages, which may be related to psychopathology involving abnormal cognitions, behavior, and experiences including psychopathy and conduct disorder (Hatton & Emerson, 2004). Second, population studies indicate that conduct disorder is overrepresented in ID populations (Emerson, 2003), and many conduct disordered children and adolescents have significantly elevated rates of ID (e.g., Moffitt et al., 2008; Vizard et al., 2004). Third, longitudinal research has found that low verbal IQ (FSIQ below 90), poor concentration, restlessness, and high impulsivity at age 8–10 years significantly predict high psychopathy scores at age 48 years (e.g., Farrington, 2004, 2006). Thus, children and adolescents with ID may be likely to present with some of the early risk factors associated with the development of psychopathy.

Empathy can be divided into two types: cognitive empathy, which is to identify the emotions of others, and emotional empathy, which is to share or match one's emotions with another's (De Waal & Preston, 2017). Mencl and May (2009) found that cognitive empathy was more strongly related to principle-based evaluations that placed the individual's own responsibilities toward others and the well-being of others first.

Psychopathy and ASD are two conditions associated with empathy deficits. Psychopathy is predominantly characterized by a reduced capacity for remorse and a propensity for callous or antisocial behavior (e.g., Cleckley, 1976) and has been linked to intact cognitive empathy (i.e., ability to infer the thoughts and feelings of others) and to impaired emotional empathy (i.e., ability to experience vicarious arousal and resonate with others' feelings). On the other hand, ASD is characterized by social communication difficulties (American Psychiatric Association, 2013) and has been linked to impaired cognitive empathy but not necessarily to impaired emotional empathy (e.g., Blair, 2008).

Decision Making

Research in neuroscience and genetics has improved our understanding of the biological underpinnings and the nature of brain functioning in decision making in individuals with ASD and other developmental disorders (Libero & Kana, 2013). The development and implementation of functional MRI (fMRI) techniques have caused an explosion of research in the field. Within the past few decades, fMRI has become a sophisticated neuroimaging tool for in vivo study of the human brain, making possible more convincing investigations of the neurobiological basis of ASD and other developmental disorders such as ID and attention deficit hyperactivity disorder (ADHD).

Cognitive Decision Making

Decision making is one of the basic cognitive processes of human behavior by which a preferred option is chosen from among a set of alternatives based on certain criteria. Cognitive decision making is the cognitive domain of decision making. Cognitive decision making is a choice behavior in risky situations, in which the payoff and the probability of each option are explicit (Mochizuki & Funahashi, 2009).

The Cambridge Gambling Task is a typical task requiring cognitive decision making (Clark et al., 2008; Manes et al., 2002). In this task, the participants make a probabilistic decision in order to get a token hidden in one of the ten boxes. Each box is colored in either red or blue. In every trial, the participants choose the color of the box which they think the token is hidden inside. Also the participants select how much to bet on the color choice from the current balance of the game money. Participants have to make a decision based on a comparison of the payoffs and their probabilities among the options in order to earn as much money as possible.

A gambling task is often used to measure cognitive decision making (Table 15.1). Wu et al. (2018) designed and administered a gambling task to 33 adults with ASD and 47 typically developed (TD) participants who were matched for age and IQ. When participants were presented with choices for which they could make either a risky gamble (e.g., 20% chance of winning £5) or a safe choice (e.g., 100% chance of winning £1), the ASD and TD participants did not differ in their overall risk-taking choices; however, the ASD participants were more consistent in their individual choices from trial to trial. Further, while members of the ASD group were slower to make some decisions (i.e., in the win frame and the first half of the lose frame), by the end of the task their decision times were the same as those of the TD group. These results suggest that the tendency toward repetitive behavior exhibited by individuals with ASD may be demonstrated even in high-level decision-making tasks (Wu et al., 2018).

Fuzzy-trace theory has proposed two types of mental representation, gist and verbatim (Reyna & Brainerd, 1995). Fuzzy-trace theory predicts that children and adolescents will use more precise, verbatim-based processing of risks and rewards. Because they trade off risks and rewards and are less influenced by gist or the context of gains or losses, their choices are more consistent. However, adults show more risky-choice framing biases than children (Reyna et al., 2015). Individuals with ASD show weaker gist-based processing but stronger verbatim processing than TD controls (Reyna et al., 2015), consistent with a cognitive strategy of detailed focus found in individuals with ASD (Happé & Frith, 2006).

In the Ultimatum Game (Table 15.1), two people are randomly matched, one as proposer and one as responder, and told they will play a game exactly one time. The proposer is endowed with an amount of money and suggests a division of that amount between herself or himself and her or his responder. The responder observes the suggestion and decides whether to accept or reject. If it is accepted, then both earn the amount implied by the proposer's suggestion. If it is rejected, then both the proposer and responder earn nothing for the experiment (Houser & McCabe, 2014). The Dictator Game (Table 15.1) has one decision point for player 1 and no decision point for player 2. Unlike the Ultimatum Game, the Dictator Game may also be played as a series of successive rounds. Pairs may remain the same across rounds or change for each round (Kahneman, et al., 1986). Hartley and Fisher (2018) compared how children and young adolescents with ASD and language-matched TD controls shared resources in age-appropriate versions of the Ultimatum Game, which illustrates conflict between fairness and economic utility, and the Dictator Game. In the Ultimatum Game, one participant has a desirable resource and is required to offer a proportion to a partner who has nothing. On acceptance, the resource is split as proposed and both persons keep a share. On rejection, neither person keeps any of the resource. The Dictator Game follows the same format except for one important difference: the partner must always accept whatever share is offered. Previous studies showed that TD adults consistently offer 40–45% of the endowed amount in the Ultimatum Game and 20–25% of the endowed amount in the Dictator Game (Camerer 2003; Henrich et al. 2005; Rigdon 2003). Children and young adolescents with ASD were significantly less likely to reciprocate the offers of a puppet in the Ultimatum Game and much more likely to accept unfair offers in the Ultimatum Game, indicating reduced aversion to inequality (Hartley & Fisher, 2018).

Bibby (2016) found that alexithymia is a precursor to loss-chasing in gambling. Loss-chasing is the tendency of a gambler to amplify their betting in an effort to recoup prior losses (Zhang & Clark, 2020). For example, participants high in alexithymia may bet 22.7% after a win. However, after a loss they often bet significantly more (e.g., 27.0%). In other words, participants high in alexithymia tend to chase their losses (Bibby, 2016). The tendency to loss-chase depends on the need to regain prior losses and the failure to process the emotional consequences of those losses. Two areas of research suggest why there is a relationship between alexithymia and loss-chasing. First, individuals who are high in alexithymia may have difficulty processing information about losses (Bibby & Ferguson, 2011), and second, the neurological structures implicated in loss-chasing (Campbell-Meiklejohn et al., 2008) clearly differ in individuals with and without alexithymia (Berthoz et al., 2002; Kano et al., 2003, 2007; Mantani et al., 2005; Moriguchi et al., 2006).

Zhang et al. (2017) compared performance of participants with alexithymia and a control group on the Iowa Gambling Task and the Game of Dice Task. Participants with alexithymia demonstrated performance deficits relative to the control group on the Gambling Task but not on the Dice Task.

Lösel and Schmucker (2004) assessed 49 male prison inmates with the gambling task of Bechara et al. (1994), the Psychopathy Checklist-Revised (PCL-R, Hare, 1991), and standardized tests of the ability to concentrate and sustain attention. The results revealed no general relationship between psychopathy and gambling task performance. However, the finding that individuals with psychopathy and low attention had more prior convictions than those with high attention suggests that our differentiation has implications beyond an experimental problem situation such as the gambling task.

Mayer et al. (2018) presented healthy individuals (violent offenders and controls) with the Empathic Dictator Game, which extends the classical Dictator Game by inducing empathy. Mayer et al. (2018) measured both self-report data and performance in the Movie for the Assessment of Social Cognition (MASC, Dziobek et al., 2006), a video-based measure sensitive to deficits in cognitive empathy. As for participants, the mean age of male violent offenders ($n = 42$) was 32.79, and the mean age of a male control group ($n = 33$) was 28.82. Participants completed one standard Dictator Game scenario with a hypothetical player. Afterward, participants were introduced to the Empathic Dictator Game. They were told that they would watch the same videos two times, with and without empathy ratings. In the case without empathy ratings, they would be asked to distribute 10 monetary units at any rate between themselves and the person in the respective video. In addition, all participants were informed that the monetary units kept during the task would be converted into real money and added to their reimbursement. Just like in the empathy task, each trial started with the randomized presentation of one of the 44 videos, but was then followed by the question "How many points do you want to give to the person?". Participants indicated on a scale from 0 to 10 how many monetary units they wanted to share with the person in the respective video. Following the response, a feedback screen indicating the payoffs for both parties appeared for 3000 ms (e.g., "You get 3 points; the other person gets 7 points"). The final screen contained information about participants' overall payoffs. Although violent offenders exhibited less altruistic behavior than controls, empathy induction increased prosocial behavior in both violent offenders and the control group, although higher alexithymia scores were associated with less altruistic behavior. Psychopathic traits were associated with lower self-reported empathy, higher alexithymia scores, attenuated affective responding following empathy induction, and less altruistic sharing.

Osumi and Ohira (2010) measured electrodermal responses to fair and unfair offers in the Ultimatum Game to examine the decision making of college students with high and low tendencies for psychopathy. Compared to controls with low psychopathy, individuals with a high tendency toward psychopathy more often chose economic utility by accepting unfair offers. This suggests that the affective deficit of

Table 15.1 Evaluations of individuals with autism, alexithymia, and psychopathy relative to typically developing individuals on cognitive decision-making tasks

	Gambling task	Ultimatum and dictator games
Autism	No difference	Accept more unfair offers
Alexithymia	Greater loss-chasing	Exhibit less altruistic behavior
Psychopathy	No difference	Choose economic utility by accepting more unfair offers

psychopathy may be associated with insensitivity to unfairness, which may in turn contribute to a rational decision to accept unfair offers.

Emotional Decision Making

Emotional decision making is a choice behavior under ambiguous conditions in which the information regarding the payoff and the probability of each option is insufficient (Mochizuki & Funahashi, 2009). The Iowa Gambling Task (Bechara, et al., 1994; Bechara, & Damasio, 2005) is one of the typical behavioral tasks which requires emotional decision making. In the Iowa Gambling Task, the subject needs to accomplish 100 card selections from four options of card decks. Every card selection provides gain of the game money, but sometimes provides loss simultaneously. Four decks have different schedules of gains and losses, but the subject is not informed about these schedules and must learn the optimum choice strategy in a trial-and-error manner. Based on the series of experiments using the Iowa Gambling Task, Damasio (1996) proposed the somatic marker hypothesis, an important theory explaining the roles of emotion in decision making. The somatic marker hypothesis assumes that the decision is biased by the autonomic somatic responses closely linked to emotion (Damasio, 1996). These autonomic somatic responses may be related to emotional empathy.

Some individuals with ASD exhibit atypical emotional processing and moral judgments. Because emotional deficits in ASD may be due to co-occurring alexithymia, atypical moral judgments in ASD may also be attributable to alexithymia (Bird et al., 2010). In Brewer et al. (2015), individuals with and without ASD (i.e., matched for alexithymia) judged the moral acceptability of emotional statements (those evoking happiness, sadness, fear, disgust, and anger) and identified the emotions evoked by the statements. Twenty-five individuals with and 22 individuals without a diagnosis of ASD participated in this study. The task validated by a previous study assessed moral judgments (Marsh & Cardinale, 2012). Participants viewed 100 emotive statements evoking happiness, sadness, fear, disgust, and anger. For example, statements included "I bought you a present" (happiness), "I do not want to be friends any more" (sadness), "I could easily hurt you" (fear), "I never wash my hands" (disgust), and "I broke your phone on purpose" (anger). Each statement was presented once, with order randomized across participants. Participants were required to rate the moral acceptability of saying each statement to another person, ranging from 1 (never acceptable) to 4 (always acceptable). Ability to identify the evoked emotion was assessed by presenting the same statements in a random order and requiring participants to identify their own emotional response to each statement, from happiness, sadness, disgust, anger, and fear. In the control group, the mean of happiness was 3.22, the mean of sadness was 2.34, the mean of disgust was 2.02, the mean of anger was 1.88, and the mean of fear was 1.85. In the ASD group, the mean of happiness was 3.50, the mean of sadness was 2.14, the mean of disgust was 1.99, the mean of anger was 1.80, and the mean of fear was 1.69. The ASD and alexithymia-matched control groups did not differ significantly in individual morality scores. Correlation analyses compared the relationship between emotion identification typicality and moral acceptability judgments in each group. In the control group, emotion identification scores correlated with Global Morality scores (higher scores indicate more severe difficulties in judging moral acceptability), r =.741, $p < .001$, whereas these scores were not correlated in the ASD group, $r = .093$, $p < .657$ (Brewer et al. 2015). Alexithymia predicted moral acceptability judgments only for individuals without ASD, and those with ASD did not base their moral acceptability judgments on emotional information. These results are consistent with evidence that decision making is less subject to emotional biases (distortion in cognition and decision making due to emotional factors) in individuals with ASD (Brewer et al., 2015; Damiano, et al., 2012; De Martino et al., 2008). Because the amygdala plays a role in emotionally biased decision making (De Martino et al., 2006), decision making in ASD may be less subject to emotional information because of reduced activation or atypical connectivity of the amygdala (De Martino et al., 2008).

The way choices are framed influences, and these framing effects emerge, when emotional responses are integrated under uncertainty. Framing effects were believed to be reduced in individuals with ASD because of their lower tendency to incorporate emotional information in the decision-making process. However, recent research suggests that emotional processing impairments in ASD may be attributable to co-occurring alexithymia, which is thought to arise from impaired interoception (the ability to perceive the internal state of one's body).

Interoception is the perception of visceral sensations such as cardiac signals and respiratory volume. It contributes significantly to variability in a range of affective experiences, including emotional lability (Schandry, 1981), arousal focus (Barrett et al., 2004), and emotional decision making (Furman et al., 2013; Harshaw, 2015). Furman et al. (2013) found that decision-making deficits in major depressive disorder are associated with reduced heartbeat perception in interoceptive dysfunction. Poorer interoceptive sensitivity is correlated with alexithymia and involves difficulty identifying and communicating about internal signals and emotional states (e.g., Herbert et al., 2011; Kano et al., 2007; Näring & Van der Staak, 1995). This raises the possibility that emotional signals are not perceived by individuals with ASD. Because decision making is impaired in individuals with alexithymia, reduced framing effects in ASD may be a product of co-occurring alexithymia rather than ASD itself (Shah et al., 2016a). Shah et al. (2016a) compared framing effects in

ASD individuals with TD controls matched for alexithymia. Framing effects were significantly smaller in ASD individuals, and there was no relationship between alexithymia or interoception and decision making in the ASD group. However, in the TD group, framing effects were associated with alexithymia and interoception even after controlling for autistic traits. Thus, although framing effects are associated with interoception and alexithymia in the TD population, emotional and interoceptive signals have less impact upon the decision-making process in ASD (Shah et al., 2016a, b).

In an online study ($N = 541$) and a laboratory study ($N = 55$), Samur et al. (2020) required participants with varying levels of alexithymia to read first- and/or third-person narrated texts and then rate their narrative engagement. Narrative engagement was higher for participants who assumed a first-person (rather than third-person) perspective and for those who were lower (rather than higher) on alexithymia. Narrative perspective interacted with affective facets of alexithymia (i.e., emotionalizing and fantasizing), such that first-person (rather than third-person) stories elicited more narrative engagement at lower (but not at higher) levels of affective alexithymia. These findings suggest that alexithymia is related to difficulties in mentally simulating narrative worlds (Samur et al., 2020).

As noted earlier, Brewer et al. (2015) found that moral acceptability judgments were predicted by alexithymia only for individuals without ASD, and those with ASD did not base their moral acceptability judgments on emotional information. Thus individuals with ASD may rely more on rules to judge moral acceptability. Brewer et al. (2015) also found that although ASD did not affect judgments of moral acceptability, it moderated the relationship between alexithymia and these judgments (Table 15.2). In TD individuals, alexithymia was associated with atypical moral acceptability judgments, such that individuals with more severe alexithymia considered it less acceptable to induce happiness in others and more acceptable to induce sadness, fear, disgust, and anger.

Vyas et al. (2017) examined the relationship between utilitarian decision making (e.g., the participant, a bystander, can pull a lever to divert the train onto another track, where only one worker will die) and two conditions considered to be associated with deficits in empathy: psychopathy and ASD (Table 15.2). Those who scored high for either psychopathic or autistic traits did not exhibit better utilitarian decision making than the low trait groups, although the two high trait groups reported that making decisions that caused harm or distress to others would cause them less discomfort.

Table 15.2 Evaluations of individuals with autism, alexithymia, and psychopathy relative to typically developing individuals on emotional decision-making tasks

	Trolley problem	Story task
Autism	No enhancement of utilitarian decision making	Unreliable judgments of accidental and attempted harm as morally different
Alexithymia	Greater utilitarian tendencies	Association with atypical moral acceptability judgments
Psychopathy	No enhancement of utilitarian decision making	Failure to use prospective regret signals to guide choice behavior

Using a counterfactual decision-making paradigm, Baskin-Sommersa et al. (2016) found that individuals who scored higher on a measure of psychopathy reported negative affect in response to regret-inducing outcomes as often as, or more often than, individuals who scored lower on a measure of psychopathy; however, they did not use prospective regret signals to guide choice behavior. Thus Baskin-Sommersa et al. (2016) identified a specific deficit in the ability of individuals with psychopathic traits to integrate prospective counterfactual signals into decision making.

Moral Decision Making

The broader term moral decision making refers to any decision, including judgments, evaluations, and response choices, made within the *moral domain* (Smetana, 2006, Turiel, 1983) for example, decisions regarding moral issues or principles such as justice, harm, fairness, and care (Garrigan et al., 2018). In the empirical studies in moral decision making, the following tasks were used, for example, paradigms involving semantic judgments of sentences with moral content (Heekeren et al., 2003), judgments of disgust and indignation in response to sentences with moral-emotional connotations (Moll et al., 2005), or moral judgments after participation in game tasks such as the Dictator or Ultimatum Games (Hofmann & Baumert, 2010; Takezawa et al., 2006).

Haidt (2001) proposed an intuitionist model of moral judgment in which moral development does not rely on discursive moral reasoning. According to Dempsey et al. (2020), Haidt's model may account for weak moral reasoning among individuals with autism, in whom moral judgments are generally intact. Investigations of moral reasoning in ASD that use an intuitionist approach may successfully identify both social-cognitive strengths and weaknesses among people with ASD.

In order to investigate the developmental processes of moral decision making, Komeda et al. (2016) examined the information used by early adolescents with and without ASD when they judged story protagonists as good or bad. Tables 15.3 and 15.4 show sample stories used in Experiment 1. Komeda et al. (2016) predicted that adolescents with ASD would use protagonists' behavior when making judgments, while TD adolescents would use protagonists' characteristics. In Experiment 1, sentence by sentence reading times and percentages of good or bad judgments were

Table 15.3 Sample stories with good characteristics and good and bad behavior outcomes in Experiment 1

Good characteristics with good behavior	Good characteristics with bad behavior
Takeru-kun is a nice boy who likes to please his father.	Takeru-kun is a nice boy who likes to please his father.
He said to his father, "Let's go watch your favorite football team play!"	He said to his father, "Let's go see my favorite cartoon movie!" when his father was very busy.
His father smiled when he looked at his son's happy face.	

Table 15.4 Sample stories with bad characteristics and good and bad behavior outcome in Experiment 1

Bad characteristics with good behavior	Bad characteristics with bad behavior
Tomoo-kun is a selfish boy who only thinks of himself.	Tomoo-kun is a selfish boy who only thinks of himself.
He said to his father, "Let's go watch your favorite football team play!"	He said to his father, "Let's go see my favorite cartoon movie!" when his father was very busy.
His father smiled when he looked at his son's happy face.	

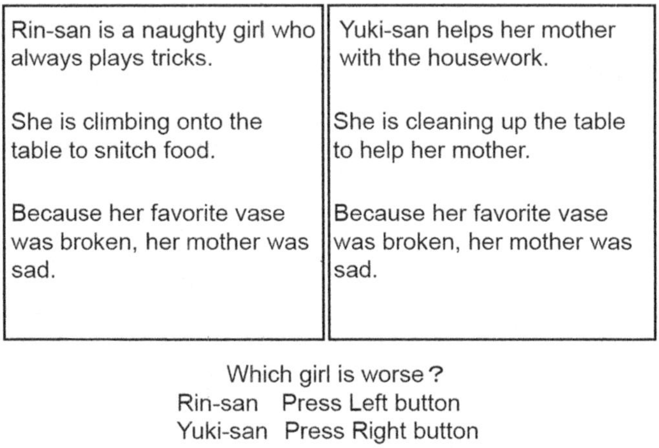

Fig. 15.1 Sample stimuli in Experiment 2 (bad characteristics with bad behavior, bad outcome vs good characteristics with good behavior, bad outcome)

measured. In Experiment 2, two story protagonists were presented, and the participants determined which protagonist was better or worse. Figures 15.1 and 15.2 show the sample stimuli in Experiment 2.

The results of Experiment 1 showed that, in order to judge story protagonists as good or bad, adolescents with ASD used protagonist behaviors and outcomes, whereas TD adolescents used protagonist characteristics, behaviors, and outcomes. In Experiment 2, TD adolescents used protagonist characteristics in determining which protagonist was worse. In situations in which participants could not go back and assess (Experiment 1), and in comparable situations in which all information was available (Experiment 2), adolescents with ASD did not rely on information about individual characteristics when making moral judgments.

People with ID and developmental disabilities often have impaired working memory abilities, and as a consequence, they may have poor decision-making abilities (Caceda et al., 2014). In Komeda et al. (2016), early adolescents with ASD who did not have difficulty with working memory showed altered decision making. The ASD group consisted of 19 participants (two females and 17 males), and the TD group consisted of 20 participants (two females and 18 males). Working memory abilities were measured by the Wechsler Intelligence Scale for Children-Fourth

Rin-san is a naughty girl who always plays tricks. She is cleaning up the table to help her mother. Because her favorite vase was broken, her mother was sad.	Yuki-san helps her mother with the housework. She is climbing onto the table to snitch food. Because her favorite vase was broken, her mother was sad.

Which girl is worse?
Rin-san Press Left button
Yuki-san Press Right button

Fig. 15.2 Sample stimuli in Experiment 2 (bad characteristics with good behavior, bad outcome vs good characteristics with bad behavior, bad outcome)

Edition (WISC-IV). The adolescents with ASD engaged in at least two updating processes during decision making: one to process the congruencies between the characteristics and behaviors and another to process the congruencies between the behaviors and outcomes, whereas TD adolescents appeared to engage in a single updating process for the congruencies between the behaviors and outcomes, when reading stories describing moral situation. As a consequence of these strategic differences, ASD adolescents fail to use characteristics information when making moral judgments about a story protagonist. In Experiment 2, TD adolescents used characteristics information when making moral judgments in a situation where multiple information could be processed at the same time. Taken together, in situations in which participants cannot go back and evaluate (Experiment 1), and in comparable situations in which all information is available at the same time (Experiment 2), adolescents with ASD do not rely on information about individual characteristics when making moral judgments.

Using a story task, Moran et al. (2011) tested whether adults with ASD make atypical moral judgments when they need to consider both the intentions (based on theory of mind) and outcomes of a person's actions. Moran et al. (2011) presented the following story:

Dan is giving a visitor a tour of a laboratory. Before visitors enter the testing room, all test tubes containing disease antigens must be contained in a chamber by flipping a switch. A repairman has just come to fix the switch, which had been broken. The switch has been successfully repaired, so the test tubes are quite safely contained. Thus, anybody who enters the room will be safe and unexposed. Dan believes that the switch is still broken after a conversation with the repairman, so he believes it is not safe for the visitor to enter. Dan tells the visitor to enter the testing room. The visitor does not contract any disease and is fine.

After reading the story, participants used a 7-point scale to judge whether telling the visitor to enter was forbidden (1) to permissible (7). Performance of ASD and TD groups did not differ on the false belief task, but on the moral judgment task, group differences were found for judgments of accidental harm, but not for neutral acts, attempted harm, or intentional harm. The TD group judged accidental harm as less morally wrong than attempted harm, but the ASD group did not find these to be morally different. In judging accidental harm, ASD participants relied less on information about a person's innocent intention and more on negative outcome of the action. To Moran et al. (2011), these results revealed impairments in integrating mental state information for moral judgments in individuals with ASD.

Gleichgerrcht et al. (2013) observed responses to two moral scenarios, one impersonal and one personal moral scenario, as follows:

(a) Impersonal scenario: The trolley dilemma required participants to decide whether to flip a switch to redirect a trolley onto a man and away from a group of five people (utilitarian response) or whether to allow the trolley to hit the five people (deontological response).
(b) Personal scenario: The footbridge dilemma required participants to decide whether to push a man off a bridge so that his body would stop the trolley from hitting five people further down the tracks (utilitarian response) or whether to allow the trolley to hit the five people (deontological response).

Gleichgerrcht et al. (2013) found that individuals with ASD who provided utilitarian responses to moral scenarios demonstrated lower ability to infer other people's mental states and to understand their intentions, as measured both by performance on neuropsychological tests and through dispositional measures. They concluded that greater prevalence of utilitarianism in ASD is associated with difficulties in specific aspects of social cognition.

On the other hand, Patil et al. (2016) investigated moral evaluations in individuals with ASD using a highly emotionally salient moral dilemma task that involved personally carrying out harmful utilitarian behaviors intended to maximize welfare. Individuals with ASD exhibited a normal pattern of moral judgments despite deficits in social cognition and emotional processing. Further, autistic traits were associated with lower utilitarian bias due to elevated personal distress from demanding social situations, while alexithymic traits were associated with greater utilitarian bias due to reduced empathic concern for the victim (Patil et al., 2016).

Patil and Silani (2014) studied responses to emotionally aversive personal moral dilemmas and found that trait alexithymia was associated with greater utilitarian tendencies, due to lower empathic concern for the victim (Table 15.5). These results underscore the importance of empathy in moral judgments in the harm/care domain of morality.

Komeda et al. (2019) investigated the relationship between alexithymia and cognitive empathy in helping motivation. Individuals with ASD and intelligence- and age-matched TD individuals were instructed to read 24 stories (12 which featured protagonists with ASD and 12 which featured TD protagonists) and respond to the following questions: "How did the protagonist feel?" and "Would you help if the protagonist were in trouble?". After controlling for alexithymia and autism-spectrum quotient (AQ) based on multiple regression analyses, individuals with ASD were found to empathize with other people with ASD and were motivated to help other people with ASD.

Further, social skills and attention to detail were associated with decreased helping motivation for story characters with ASD. Social skills among AQ subscales (social skills, attention switching, attention to detail, communication, and imagination) were the dominant predictors of lower helping motivation. These findings suggest that alexithymia and low social skills reduce helping motivation in individuals with ASD (Komeda et al., 2019). Komeda et al. (2019) found that participants with ASD showed greater empathetic responses for people with ASD than did TD participants, whereas TD participants showed greater empathetic responses and greater helping motivation for TD people than did ASD participants (Table 15.5). In the decision-making situation, ASD adolescents are likely to judge other people with ASD more positively than other TD people. The difference between in-group and out-group on decision making should be considered in the social situations.

Patil (2015) showed that trait psychopathy is associated with both reduced outcome aversion (aversion to witnessing harmful outcomes) and action aversion (performing harmful actions), but only action aversion negatively mediates the influence of trait psychopathy on utilitarian moral judgment. Thus, the greater tendency of individuals with psychopathy to make utilitarian moral judgments is in part due to lower aversion to carrying out harmful actions (Table 15.5).

Pletti et al. (2016) reported that participants with high trait psychopathy were more likely to sacrifice one person to save others in sacrificial dilemmas and to pursue a personal advantage in everyday moral situations that caused harm to another. These participants also experienced less unpleasantness during decision making in these situations, compared to participants with low trait psychopathy. But for everyday moral situations that did not entail harm to others, no group differences emerged in choice of action, unpleasantness ratings, or moral judgments. These results suggest that high trait psychopathy affects action choices in sacrificial dilemmas because of reduced emotional reactivity to harmful acts (Pletti et al., 2016).

Koenigs et al. (2012) investigated whether psychopathic subtypes (low-anxious, high-anxious, and non-psychopathic) exhibit significant differences in moral judgment. Three groups of incarcerated participants (low-anxious psychopaths ($n = 12$), high-anxious psychopaths ($n = 12$), and non-psychopaths ($n = 24$) completed a moral judgment test involving hypothetical moral dilemmas that featured *personal* (i.e., involving direct physical harm) or *impersonal* (i.e., involving indirect or remote harm) actions. Participants made judgments on a series of 24 hypothetical moral scenarios (e.g., "pushing one person off a bridge to stop a runaway train car from hitting five people," "pulling a switch to divert a runaway boxcar from hitting five people"), which were selected from a previously published set (Greene et al.,

2001, 2004; Koenigs & Tranel, 2007). Each scenario was presented on a single sheet of paper, followed by a question about a hypothetical action related to the scenario ("Would you ... in order to ...?"). This question format follows previous clinical and prison studies (Koenigs & Tranel, 2007; Cima et al., 2010). Participants chose "yes" or "no, and" "yes" responses always indicated commission of the proposed action. Both low- and high-anxious psychopathic groups were significantly more likely than the non-psychopathic group to endorse the impersonal actions. However, only the low-anxious group was significantly more likely to make the *utilitarian* choice of personal harm when committing the harm would maximize aggregate welfare. The high-anxious and non-psychopathic groups did not significantly differ in their personal moral judgments. In conclusion, the results presented here are broadly consistent with the theoretical perspective that primary (low-anxious) psychopathy may entail a particular affective deficit that is not necessarily present in secondary (high-anxious) psychopathy (Blackburn et al., 2008; Karpman, 1946; Karpman, 1948).

Individuals with psychopathy show antisocial and immoral behavior, but experimental studies have typically failed to identify deficits in their capacities for explicit moral judgment. Young et al. (2012) tested 20 criminal psychopaths and 25 criminal non-psychopaths on a moral judgment task featuring fictional scenarios that systematically varied an actor's intention and the action's outcome. Participants were instructed to assess four classes of actions: accidental harm, attempted harm, intentional harm, and neutral acts. Individuals with psychopathy showed a selective difference, compared with non-psychopaths, in judging accidents in which one person harmed another unintentionally. Specifically, individuals with psychopathy judged these actions to be more morally permissible. Young et al. (2012) suggest that this pattern reflects psychopaths' difficulties to appreciate the emotional factor of the victim's experience of harm. These findings provide experimental evidence of atypical moral judgment in psychopathy (Table 15.5).

Table 15.5 Evaluations of individuals with autism, alexithymia, and psychopathy relative to typically developing individuals on moral decision-making tasks

	Moral dilemma	Story task
Autism	No difference	Reliance on information about individual behaviors rather than individual characteristics
Alexithymia	Associated with increased utilitarian bias	Lower helping motivation in individuals with ASD
Psychopathy	More likely to sacrifice one person to save others in sacrificial dilemmas and to pursue a personal advantage in everyday moral situations entailing harm to another's good	Failure to appreciate emotional aspects of the victim's experience of harm

Support Program for Decision Making

Finally, support programs for decision making are considered based on empirical decision-making studies. The support programs for ASD, alexithymia, and psychopathy are introduced in this section.

Luke et al. (2012) found that participants with ASD reported experiencing problems in decision-making more frequently than TD participants and were also more likely to report decision-making avoidance, as measured by the general decision making style inventory (Scott & Bruce, 1995). The finding suggests that children and adults with ASD could benefit from support during decision making (Luke et al., 2012).

With respect to ASD, training programs have generally focused on children and young adolescents, either targeting behavioral difficulties (e.g., turn-taking or eye contact; Barry et al., 2003) or providing training in cognitive skills (e.g., explicit awareness of others' thoughts or emotions, Gray, 1995).

With respect to alexithymia, Gay et al. (2008) devised an eight-week training program using hypnotic imagery. Thirty-one female college students with alexithymia—defined as scores above 60 on the 20-item TAS-20—were randomly assigned to either an eight-week hypnotic-imagery training program (n = 14) or a control condition (n = 17), which consisted only of attending evaluation sessions. Participants in the hypnotic-imagery condition attended half-hour sessions during which they were read standardized scripts involving a traumatic situation and associated negative feelings, and they were guided to experience different emotions and mental imagery. Alexithymia was measured with the TAS-20. Hypnotic-imagery training resulted in a significant reduction in TAS-20 total score that was independent of changes in mood states. Alexithymia was not found to change significantly in the control group.

With respect to psychopathy, treatment has focused on reducing recidivism in forensic samples and has often been unsuccessful (Pickersgill, 2011). Interviews of a group of neuroscientists revealed that most believed that biological interventions should be used alongside psychotherapeutic strategies, although they were not sure about, or strongly committed to, such interventions. For example, a doctor who believed that neuroscience could help inform psychological interventions and that some kind of psychopharmacological management technique for psychopathy was likely to be necessary was "not sure exactly what that's going to be" (Pickersgill, 2011). Although respondents generally felt that neuroscience research held promise for psychological interventions, they recognized that translating this work into new interventions would not be straightforward, and some saw their work as better suited to identifying interventions that would not be productive.

There are few programs that support lifelong development of decision-making abilities in typically or atypically developing individuals with ASD, alexithymia, and psychopathy. More robust and reliable basic findings will be necessary to develop such training programs. For future advancement of these efforts, it is increasingly important that psychologists, neuroscientists, biologists, medical doctors, counselors, social workers, and school teachers collaborate based on mutual respect.

Conclusion and Recommendations

This chapter reviewed several types of decision making in the individuals with ASD, individuals with alexithymia, and individuals with psychopathy. In the cognitive decision-making tasks, individuals with ASD tend to accept unfair offers, individuals with alexithymia show greater loss-chasing decision making, and individuals with psychopathy choose economic utility by accepting unfair offers. In the emotional decision-making tasks, individuals with ASD do not show enhancement of utilitarian decision making, individuals with alexithymia show greater utilitarian tendencies, and individuals with psychopathy do not use prospective regret to guide choice behavior. In the moral decision-making tasks, individuals with ASD tend to rely on information about individual behaviors rather than individual characteristics, individuals with alexithymia show increased utilitarian bias, and individuals with psychopathy do not appreciate the emotional aspect of victim's experience of harm.

It is important to consider the comorbidity to apply these findings to real-life situations. For example, the incidence of alexithymia is high in the ASD population (40–65% of adults with autism, Bird & Cook, 2013). A recent study shows that alexithymia could be considered a potentially common mechanism underlying psychopathy (Psederska et al., 2019). ID is also comorbid with ASD, alexithymia, and psychopathy. Thus, future studies on decision making should take into consideration the comorbidity of several personality traits and disorders.

Additionally, this chapter classified decision making in terms of its cognitive, emotional, and moral components. These classifications are not always appropriate in real-life settings. For example, moral decision making is associated with cognitive and emotional decision making. It is essential to look carefully at the individual in front of you and to consider individual differences when investigating decision making in special populations. This chapter may provide some evidence of the importance of observing people with different backgrounds from several points of view.

References

American Psychiatric Association. (2013). *Diagnostic and statistical manual of mental disorders* (5th ed.). American Psychiatric Publishing.

Moffitt, T. E., Arseneault, L., Jaffeee, S. R., Kim-Cohen, J., Koenen, K. C., Odgers, C. L., Slutske, W. S., & Viding, E. (2008). Research review: DSM-V conduct disorder: Research needs for an evidence base. *Journal of Child Psychology and Psychiatry, 49*, 3–33.

Bagby, R. M., Parker, J. D., & Taylor, G. J. (1994a). The twenty-item Toronto Alexithymia Scale – I. *Item selection and cross-validation of the factor structure. Journal of Psychosomatic Research, 38*, 23–32.

Bagby, R. M., Taylor, G. J., & Parker, J. D. (1994b). The twenty-item Toronto Alexithymia Scale – II. Convergent, discriminant, and concurrent validity. *Journal of Psychosomatic Research, 38*, 33–40.

Barrett, L. F., Quigley, K., Bliss-Moreau, E., & Aronson, K. R. (2004). Interoceptive sensitivity and self-reports of emotional experience. *Journal of Personality and Social Psychology, 87*, 684–697.

Barry, T. D., Klinger, L. G., Lee, J. M., Palardy, N., Gilmore, T., & Bodin, S. D. (2003). Examining the effectiveness of an outpatient clinic-based social skills group for high-functioning children with autism. *Journal of Autism and Developmental Disorders, 33*, 685–701.

Baskin-Sommers, A. R., Stuppy-Sullivan, A. M., Buckholtz J. (2016) Psychopathic individuals exhibit but do not avoid regret during counterfactual decision making Proceedings of the National Academy of Sciences of the United States of America, 113(50), 14438–14443

Bechara, A., & Damasio, A. R. (2005). The somatic marker hypothesis: A neural theory of economic decision. *Games and Economic Behavior, 52*, 336–372.

Bechara, A., Damasio, A. R., Damasio, H., & Anderson, S. (1994). Insensitivity to future consequences following damage to human prefrontal cortex. *Cognition, 50*, 7–15.

Berthoz, S., Artiges, E., Van De Moortele, P. F., Poline, J. B., Rouquette, S., Consoli, S. M., et al. (2002). Effect of impaired recognition and expression of emotions on frontocingulate cortices: An fMRI study of men with alexithymia. *American Journal of. Psychiatry, 159*, 961–967. https://doi.org/10.1176/appi.ajp.159.6.961

Bibby, P. A. (2016). Loss-chasing, alexithymia, and impulsivity in a gambling task: Alexithymia as a precursor to loss-chasing behavior when gambling. *Frontiers in Psychology, 7*(3), 1–13. https://doi.org/10.3389/fpsyg.2016.00003

Bibby, P. A., & Ferguson, E. (2011). The ability to process emotional information predicts loss aversion. *Personality and Individual Differences, 51*, 263–266. https://doi.org/10.1016/j.paid.2010.05.001

Bird, G., & Cook, R. (2013). Mixed emotions: the contribution of alexithymia to the emotional symptoms of autism. *Translational Psychiatry, 3*, e285. https://doi.org/10.1038/tp.20

Bird, G., Silani, G., Brindley, R., White, S., Frith, U., & Singer, T. (2010). Empathic brain responses in insula are modulated by levels of alexithymia but not autism. *Brain, 133*, 1515–1525.

Blackburn, R., Logan, C., Donnelly, J. P., & Renwick, S. J. (2008). Identifying psychopathic subtypes: combining an empirical personality classification of offenders with the psychopathy checklist-revised. *Journal of Personality Disorders, 22*, 604–622.

Blair, R. J. (2008). The amygdala and ventromedial prefrontal cortex: Functional contributions and dysfunction in psychopathy. *Philosophical Transactions of the Royal Society B: Biological Sciences, 363*, 2557–2565.

Brewer, R., Marsh, A., Catmur, C., Cardinale, E., Stoycos, S., Cook, R., & Bird, G. (2015). The impact of autism spectrum disorder and alexithymia on judgments of moral acceptability. *Journal of Abnormal Psychology, 124*(3), 589–595.

Caceda, R., Nemeroff, C. B., & Harvey, P. D. (2014). Toward an understanding of decision making in severe mental illness. *The Journal of Neuropsychiatry and Clinical Neurosciences, 26*, 196–213.

Camerer, C. F. (2003). *Behavioral game theory*. Princeton University Press.

Cameron, K., Ogrodniczuk, J., & Hadjipavlou, G. (2014). Changes in alexithymia following psychological intervention: A review. *Harvard Review of Psychiatry, 22(3)*, 162–178. https://doi.org/10.1097/HRP.0000000000000036

Campbell-Meiklejohn, D. K., Woolrich, M. W., Passingham, R. E., & Rogers, R. D. (2008). Knowing when to stop: The brain mechanisms of chasing losses. *Biological Psychiatry, 63*, 293–300. https://doi.org/10.1016/j.biopsych.2007.05.014

Cima, M., Tonnaer, F., & Hauser, M. D. (2010). Psychopaths know right from wrong but don't care. *Social Cognitive and Affective Neuroscience, 5*, 59–67.
Clark, L., Bechara, A., Damasio, H., Aitken, M. R., Sahakian, B. J., & Robbins, T. W. (2008). Differential effects of insular and ventromedial prefrontal cortex lesions on risky decision making. *Brain, 131*, 1311–1322.
Cleckley, H. (1976). *The mask of sanity* (4th ed.). Mosby. (Original work published 1941).
Damasio, A. R. (1996). The somatic marker hypothesis and the possible functions of the prefrontal cortex. *Philosophical Transactions of the Royal Society of London (series B), 351*, 1413–1420.
Damiano, C. R., Aloi, J., Treadway, M., Bodfish, J. W., & Dichter, G. S. (2012). Adults with autism spectrum disorders exhibit decreased sensitivity to reward parameters when making effort-based decisions. *Journal of Neurodevelopmental Disorders, 4(1)*, 13. https://doi.org/10.1186/1866-1955-4-13
De Berardis, D., Campanella, D., Serroni, N., et al. (2008). The impact of alexithymia on anxiety disorders: A review of the literature. *Current Psychiatry Reviews, 4*, 80–86.
De Martino, B., Kumaran, D., Seymour, B., & Dolan, R. J. (2006). Frames, biases, and rational decision making in the human brain. *Science, 313*, 684–687.
De Martino, B., Harrison, N. A., Knafo, S., Bird, G., & Dolan, R. J. (2008). Explaining enhanced logical consistency during decision making in autism. *The Journal of Neuroscience, 28*, 10746–10750.
De Waal, F. B. M., & Preston, S. D. (2017). Mammalian empathy: behavioural manifestations and neural basis. *Nature Reviews Neuroscience, 18(8)*, 498–509.
Dempsey, E. E., Johnson, S. A., Stewart, S. H., Smith, I. M., & Moore, C. (2020). Morality in autism spectrum disorder: A systematic review. *Development and Psychopathology, 32*, 1069–1085.
Dziobek, I., Fleck, S., Kalbe, E., Rogers, K., Hassenstab, J., Brand, M., et al. (2006). Introducing MASC: a movie for the assessment of social cognition. *Journal of Autism and Developmental Disorders., 36*, 623–636.
Emerson, E. (2003). The prevalence of psychiatric disorder in children and adolescents with and without intellectual disability. *Journal of Intellectual Disability Research, 47*, 51–58.
Farrington, D. P. (2004). The development of offending and antisocial behaviour from childhood: Key findings from the Cambridge study in delinquent development. *Journal of Child Psychology and Psychiatry, 36*, 929–964.
Farrington, D. P. (2006). Family background and psychopathy. In C. J. Patrick (Ed.), *Handbook of psychopathy* (pp. 229–250). Guilford Press.
Fujino, J., Tei, S., Itahashi, T., Aoki, Y. Y., Ohta, H., Kubota, M., Hashimoto, R. I., Takahashi, H., Kato, N., & Takahashi, H. (2019a). Role of the right temporoparietal junction in intergroup bias in trust decisions. *Human Brain Mapping, 41*, 1677–1688.
Fujino, J., Tei, S., Itahashi, T., Aoki, Y. Y., Ohta, H., Kubota, M., Hashimoto, R. I., Nakamura, M., Kato, N., & Takahashi, H. (2019b). Impact of past experiences on decision making in autism spectrum disorder. *European Archives of Psychiatry and Clinical Neuroscience, 270(8)*, 1063–1071.
Furman, D. J., Waugh, C. E., Bhattacharjee, K., Thompson, R. J., & Gotlib, I. H. (2013). Interoceptive awareness, positive affect, and decision making in Major Depressive Disorder. *Journal of Affective Disorders, 151*, 780–785.
Garrigan, B., Adlam, A. L., & Langdon, P. E. (2018). Moral DecisionMaking and Moral Development: Toward an Integrative Framework. *Developmental Review, 49*, 80. https://doi.org/10.1016/j.dr.2018.06.001
Gay, M., Hanin, D., & Luminet, O. (2008). Effectiveness of an hypnotic imagery intervention on reducing alexithymia. *Contemporary Hypnosis, 25*, 1–13.
Gleichgerrcht, E., Torralva, T., Rattazzi, A., Marenco, V., Roca, M., & Manes, F. (2013). Selective impairment of cognitive empathy for moral judgment in adults with high functioning autism. *Social Cognitive & Affective Neuroscience, 8*, 780–788. https://doi.org/10.1093/scan/nss067

Gray, C. A. (1995). Teaching children with autism to "read" social situations. In K. A. Quill (Ed.), *Teaching children with autism: Strategies to enhance communication and socialization* (pp. 219–241). Delmar.

Greene, J. D., Sommerville, R. B., Nystrom, L. E., Darley, J. M., & Cohen, J. D. (2001). An fMRI investigation of emotional engagement in moral judgment. *Science, 293*, 2105–2108.

Greene, J. D., Nystrom, L. E., Engell, A. D., Darley, J. M., & Cohen, J. D. (2004). The neural bases of cognitive conflict and control in moral judgment. *Neuron, 44*, 389–400.

Haidt, J. (2001). The emotional dog and its rational tail: A social intuitionist approach to moral judgment. *Psychological Review, 108*, 83–95. https://doi.org/10.1037//0033-295X

Happé, F., & Frith, U. (2006). The weak coherence account: Detail-focused cognitive style in autism spectrum disorders. *Journal of Autism and Developmental Disorders, 36*, 5–25.

Hare, R. D. (1991). *The Hare Psychopathy Checklist-Revised*. Multi-Health Systems.

Harshaw, C. (2015). Interoceptive dysfunction: toward an integrated framework for understanding somatic and affective disturbance in depression. *Psychological Bulletin, 141*(2), 311–363.

Hartley, C., & Fisher, S. (2018). Do children with autism spectrum disorder share fairly and reciprocally? *Journal of Autism and Developmental Disorders, 48*(8), 2714–2726.

Hatton, C., & Emerson, E. (2004). The relationship between life events and psychopathology amongst children with intellectual disabilities. *Journal of Applied Research in Intellectual Disabilities, 17*, 109–117.

Heekeren, H. R., Wartenburger, I., Schmidt, H., Schwintowski, H. P., & Villringer, A. (2003). An fMRI study of simple ethical decision making. *Neuroreport, 14(9)*, 1215–1219. https://doi.org/10.1097/01.wnr.0000081878.45938.a7

Henrich, J., Boyd, R., Bowles, S., Camerer, C., Fehr, E., Gintis, H., et al. (2005). "Economic man" in cross-cultural perspective: Behavioral experiments in 15 small-scale societies. *Behavioral and Brain Sciences, 28*, 795–815.

Herbert, B. M., Herbert, C., & Pollatos, O. (2011). On the relationship between interoceptive awareness and alexithymia: Is interoceptive awareness related to emotional awareness? *Journal of Personality, 79*, 1149–1175.

Hermann, H. R. (2017). Chapter 9 - Alternate human behavior. In *Dominance and aggression in humans and other animals* (pp. 139–157). Academic Press.

Hill, J. (2003). Early identification of individuals at risk for antisocial personality disorder. *British Journal of Psychiatry, 182*(Suppl. 44), 11–14.

Hill, E., Berthoz, S., & Frith, U. (2004). Brief report: Cognitive processing of own emotions in individuals with autistic spectrum disorder and in their relatives. *Journal of Autism and Developmental Disorders, 34*, 229–235.

Hofmann, W., & Baumert, A. (2010). Immediate affect as a basis for intuitive moral judgement: an adaptation of the affect misattribution procedure. *Cognition & Emotion, 24(3)*, 522–535. https://doi.org/10.1080/02699930902847193

Houser, D. & McCabe, K. (2014). Experimental Economics and Experimental Game Theory in Neuroeconomics (Second Edition)

Kano, M., Fukudo, S., Gyoba, J., Kamachi, M., Tagawa, M., Mochizuki, H., et al. (2003). Specific brain processing of facial expressions in people with alexithymia: an H215O-PET study. *Brain, 126*, 1474–1484. https://doi.org/10.1093/brain/awg131

Kano, M., Hamaguchi, T., Itoh, M., Yanai, K., & Fukudo, S. (2007). Correlation between alexithymia and hypersensitivity to visceral stimulation in human. *Pain, 132*, 252–263. https://doi.org/10.1016/j.pain.2007.01.032

Karpman, B. (1946). Psychopathy in the scheme of human typology. *Journal of Nervous and Mental Disease, 103*, 276–288.

Karpman, B. (1948). The myth of the psychopathic personality. *American Journal of Psychiatry, 104*, 523–534.

Kahneman, D., Knetsch, J., & Thaler, R. (1986). Fairness as a Constraint on Profit Seeking: Entitlements in the Market. *The American Economic Review, 76*(4), 728–741. Retrieved July 10, 2021, from http://www.jstor.org/stable/1806070

Koenigs, M., & Tranel, D. (2007). Irrational economic decision making after ventromedial prefrontal damage: evidence from the Ultimatum Game. *Journal of Neuroscience, 27*, 951–956.
Koenigs, M., Kruepke, M., Zeier, J., & Newman, J. P. (2012). Utilitarian moral judgment in psychopathy. *Social Cognitive and Affective Neuroscience, 7(6)*, 708–714.
Komaki, G. (2006). Impaired self-awareness and theory of mind: An fMRI study of mentalizing in alexithymia. *Neuroimage, 32*, 1472–1482. https://doi.org/10.1016/j.neuroimage.2006.04.186
Komeda, H., Osanai, H., Yanaoka, K., Okamoto, Y., Fujioka, T., Arai, S., Inohara, K., Koyasu, M., Kusumi, T., Takiguchi, S., & Kawatani, M. (2016). Decision making processes based on social conventional rules in early adolescents with and without autism spectrum disorders. *Scientific Reports, 6*, 37875. https://doi.org/10.1038/srep37875
Komeda, H., Kosaka, H., Fujioka, T., Jung, M., & Okazawa, H. (2019). Do individuals with autism spectrum disorders help other people with autism spectrum disorders? An investigation of empathy and helping motivation in adults with autism spectrum disorder. *Frontiers of Psychiatry, 10*, 376. https://doi.org/10.3389/fpsyt.2019.00376
Libero, L. E., & Kana, R. K. (2013). Advancing our understanding of the brain in autism: Contribution of functional MRI and diffusion tensor imaging. *Imaging in Medicine, 5*, 453–465.
Lilienfeld S. O., & Andrews B. P. (1996) Development and Preliminary Validation of a Self-Report Measure of Psychopathic Personality Traits in Noncriminal Population. *Journal of Personality Assessment, 66*(3), 488–524.
Lilienfeld, S. O. (1999). The relation of the MMPI-2 Pd Harris-Lingoes subscales to psychopathy, psychopathy facets, and antisocial behavior: Implications for clinical practice. *Journal of Clinical Psychology., 55*(2), 241–255.
Lindsay, W. R. (2007). Personality disorders. In N. Bouras & G. Holt (Eds.), *Psychiatric and behavioral disorders in developmental disabilities* (2nd ed., pp. 143–153). Cambridge University Press.
Lösel, F., & Schmucker, M. (2004). Psychopathy, risk taking, and attention: A differentiated test of the somatic marker hypothesis. *Journal of Abnormal Psychology, 113(4)*, 522–529.
Luke, L., Clare, I. C., Ring, H., Luke, L., Clare, I. C., Ring, H., Redley, M., & Watson, P. (2012). Decision making difficulties experienced by adults with autism spectrum conditions. *Autism, 16(6)*, 612–621.
Manes, F., Sahakian, B., Clark, L., Rogers, R., Antoun, N., Aitken, M., & Robbins, T. (2002). Decision making processes following damage to the prefrontal cortex. *Brain, 125*, 624–639.
Mantani, T., Okamoto, Y., Shirao, N., Okada, G., & Yamawaki, S. (2005). Reduced activation of posterior cingulate cortex during imagery in subjects with high degrees of alexithymia: a functional magnetic resonance imaging study. *Biological Psychiatry, 57*, 982–990. https://doi.org/10.1016/j.biopsych.2005.01.047
Marsh, A. A., & Cardinale, E. M. (2012). Psychopathy and fear: Specific impairments in judging behaviors that frighten others. *Emotion, 12*, 892–898.
Marshall, L., & Cooke, D. J. (1999). The childhood experiences of psychopaths. *Journal of Personality Disorders, 13*, 211–225.
Mayer, S., Jusyte, A., Klimecki-Lenz, O., & Schönenberg, M. (2018). Empathy and altruistic behavior in antisocial violent offenders with psychopathic traits. *Psychiatry Research, 269*, 625–632.
Mencl, J., & May, D. R. (2009). The Effects of Proximity and Empathy on Ethical Decision making: An Exploratory Investigation. *Journal of Business Ethics, 85*, 201–226.
Mochizuki, K., & Funahashi, S. (2009). Effect of emotional distracters on cognitive decision making in Cambridge gambling task. *Psychologia, 52(2)*, 122–136.
Moll, J., de Oliveira-Souza, R., Moll, F. T., Ignacio, F. A., Bramati, I. E., Caparelli-Daquer, E. M., et al. (2005). The moral affiliations of disgust – a functional MRI study. *Cognitive and Behavioral Neurology, 18(1)*, 68–78.
Moran, J. M., Young, L. L., Saxe, R., Lee, S. M., O'Young, D., Mavros, P. L., & Gabrieli, J. D. (2011). Impaired theory of mind for moral judgment in high-functioning autism. *Proceedings of the National Academy of Sciences, 108*, 2688–2692. https://doi.org/10.1073/pnas.1011734108

Moriguchi, Y., Ohnishi, T., Lane, R. D., Maeda, M., Mori, T., Nemoto, K., Matsuda, H., & Komaki, G. (2006). Impaired self-awareness and theory of mind: an fMRI study of mentalizing in alexithymia. *NeuroImage, 32*(3), 1472–1482. https://doi.org/10.1016/j.neuroimage.2006.04.186

Morrissey, C., & Hollin, C. (2011). Antisocial and psychopathic personality disorders in forensic intellectual disability populations: What do we know so far? *Psychology, Crime & Law, 17*, 133–149.

Näring, G. W. B., & Van der Staak, C. P. F. (1995). Perception of heart rate and blood pressure: the role of alexithymia and anxiety. *Psychotherapy and Psychosomatics, 63*, 193–200.

Nemiah, J. C., Freyberger, H., & Sifneos, P. E. (1976). Alexithymia: A view of the psychosomatic process. In O. W. Hill (Ed.), *Modern trends in psychosomatic medicine* (Vol. 3, pp. 430–439). Butterworths.

Osumi, T., & Ohira, H. (2010). The positive side of psychopathy: Emotional detachment in psychopathy and rational decision making in the ultimatum game. *Personality and Individual Differences, 49*, 451–456.

Patil, I. (2015). Trait psychopathy and utilitarian moral judgement: The mediating role of action aversion. *Journal of Cognitive Psychology, 27*, 349–366. https://doi.org/10.1080/20445911.2015.1004334

Patil, I., & Silani, G. (2014). Reduced empathic concern leads to utilitarian moral judgments in trait alexithymia. *Frontiers in Psychology, 5*, 501. https://doi.org/10.3389/fpsyg.2014.00501

Patil, I., Melsbach, J., Henning-Fast, K., & Silani, G. (2016). Divergent roles of autistic and alexithymic traits in utilitarian moral judgements in adults with autism. *Scientific Reports, 6*, 23637. https://doi.org/10.1038/srep23637

Pickersgill, M. (2011). 'Promising' therapies: Neuroscience, clinical practice, and the treatment of psychopathy. *Sociology of Health and Illness, 33(3)*, 448–464.

Pletti, C., Lotto, L., Buodo, G., & Sarlo, M. (2016). It's immoral, but I'd do it! Psychopathy traits affect decision making in sacrificial dilemmas and in everyday moral situations. *British Journal of Psychology, 108*, 351–368.

Poquerusse, J., Pastore, L., Dellantonio, S., & Esposito, G. (2018). Alexithymia and autism spectrum disorder: A complex relationship. *Frontiers in Psychology, 9*, 1196. https://doi.org/10.3389/fpsyg.2018.01196

Psederska, E., Savov, S., Atanassov, N., & Vassileva, J. (2019). Relationships Between Alexithymia and Psychopathy in Heroin Dependent Individuals. *Frontiers in Psychology, 10*, 2269. https://doi.org/10.3389/fpsyg.2019.02269

Reyna, V. F., & Brainerd, C. J. (1995). Fuzzy-trace theory: An interim synthesis. *Learning and Individual Differences, 7*, 1–75.

Reyna, V. F., Wilhelms, E. A., McCormick, M. J., & Weldon, R. B. (2015). Development of risky decision making: Fuzzy-trace theory and neurobiological perspectives. *Child Development Perspectives, 9*(2), 122–127.

Rigdon, M. L. (2003). Dictator games. In L. Nadel (Ed.), *Encyclopedia of cognitive science* (pp. 379–382). MacMillan.

Samur, D., Tops, M., Slapšinskaitė, R., & Koole, S. L. (2020). Getting lost in a story: How narrative engagement emerges from narrative perspective and individual differences in alexithymia. *Cognition and Emotion. https://doi.* https://doi.org/10.1080/02699931.2020.1732876

Schandry, R. (1981). Heart beat perception and emotional experience. *Psychophysiology, 18(4)*, 483–488.

Scott, S. G., & Bruce, R. A. (1995). Decision making style: The development and assessment of a new measure. *Educational and Psychological Measurement, 55*, 818–831.

Shah, P., Catmur, C., & Bird, G. (2016a). Emotional decision making in autism spectrum disorder: The roles of interoception and alexithymia. *Molecular Autism, 7*, 43–52.

Shah, P., Hall, R., Catmur, C., & Bird, G. (2016b). Alexithymia, not autism, is associated with impaired interoception. *Cortex, 81*, 215–220.

Sifneos, P. E. (1973). The prevalence of "alexithymic" characteristics in psychosomatic patients. *Psychotherapy and Psychosomatics, 22*, 255–262.

Smetana, J. G. (2006). Social-cognitive domain theory: Consistencies and variations in children's moral and social judgments. In M. Killen & J. G. Smetana (Eds.), *Handbook of Moral Development* (pp. 119–155). Lawrence Erlbaum Associates.

Takezawa, M., Gummerum, M., & Keller, M. (2006). A stage for the rational tail of the emotional dog: roles of moral reasoning in group decision making. *Journal of Economic Psychology, 27(1)*, 117–139. https://doi.org/10.1016/j.joep.2005.06.012

Tani, P., Lindberg, N., Joukamaa, M., Nieminen-von Wendt, T., von Wendt, L., Appelberg, B., Rimón, R., & Porkka-Heiskanen, T. (2004). Asperger syndrome, alexithymia and perception of sleep. *Neuropsychobiology, 49*, 64–70. https://doi.org/10.1159/000076412

Torr, J. (2003). Personality disorder in intellectual disability. *Current Opinion in Psychiatry, 16*, 517–521.

Turiel, E. (1983). *The development of social knowledge: Morality and convention.* Cambridge University Press.

Vizard, E., French, L., Hickey, N., & Bladon, E. (2004). Severe personality disorder emerging in childhood: A proposal for a new developmental disorder. *Criminal Behaviour and Mental Health, 14*, 17–28.

Vyas, K., Jameel, L., Bellesi, G., Crawford, S., & Channon, S. (2017). Derailing the trolley: Everyday utilitarian judgments in groups high versus low in psychopathic traits or autistic traits. *Psychiatry Research, 250*, 84–91.

Wechsler, D. (2003). *WISC-IV technical and interpretive manual*. Psychological Corporation.

Wu, H. C., White, S., Rees, G., & Burgess, P. W. (2018). Executive function in high-functioning autism: decision making consistency as a characteristic gambling behaviour. *Cortex, 107*, 21–36.

Young, L., Koenigs, M., Kruepke, M., & Newman, J. P. (2012). Psychopathy increases perceived moral permissibility of accidents. *Journal of Abnormal Psychology, 121(3)*, 659–667. https://doi.org/10.1037/a0027489

Zhang, K., & Clark, L. (2020). Loss-chasing in gambling behaviour: Neurocognitive and behavioural economic perspectives. *Current Opinion in Behavioral Sciences, 31*, 1–7.

Zhang, L., Wang, X., Zhu, Y., Li, H., Zhu, C., Yu, F., & Wang, K. (2017). Selective impairment of decision making under ambiguity in alexithymia. *BMC Psychiatry, 17(1)*. https://doi.org/10.1186/s12888-017-1537-2

Part IV
Applications of Decision Making

Chapter 16
Decision Making During Transition to Adulthood

Maria P. Mello and Kelli Sanderson

Overview of Transition Planning

Between the ages of 16 and 25 years, adolescents and young adults experience a major life transition when they leave high school and prepare for adulthood. For many typical adolescents and young adults, this may be an exciting time full of welcome changes. However, for individuals with intellectual and developmental disabilities (IDD), this transition can be difficult and overwhelming. According to the American Association on Intellectual and Developmental Disabilities (AAIDD), IDD includes individuals with both ID and DD (see Schalock et al., 2021), with ID being "characterized by significant limitations in both intellectual functioning and in adaptive behavior" and DD defined as having a "severe, chronic disability that is attributable to a mental and/or physical impairment." Additionally, in AAIDD's most recent (2021) classification manual, an IDD diagnosis requires that either or both conditions must be manifested before the age of 22.

For adolescents and young adults with IDD, transition from school to postschool life has been described as *falling off a cliff*. That is, when they leave or graduate from school, they typically have few options available in regard to service, employment, housing/living, community engagement, recreation, social life, and other areas of life. However, this major transition to adult life does not have to be so daunting given proper transition planning and services. In order to support this transition to postschool life for adolescents and young adults with IDD, the Individuals

M. P. Mello (✉)
St. John's University, Jamaica, NY, USA
e-mail: mellom@stjohns.edu

K. Sanderson
California State University, Long Beach, Long Beach, CA, USA

I. Khemka, L. Hickson (eds.), *Decision Making by Individuals with Intellectual and Developmental Disabilities*, Positive Psychology and Disability Series,
https://doi.org/10.1007/978-3-030-74675-9_16

with Disabilities Education Act of 1990 (IDEA) introduced Individualized Transition Plans (ITPs) as a required part of the Individualized Education Program (IEP). Reauthorizations of IDEA in 1997 and 2004 have emphasized the importance of transition. Implemented before students' 16th birthday, the ITP is a required part of the IEP that outlines postsecondary goals in the areas of employment, postsecondary education, and independent living. It is essential for students to have a meaningful transition plan with goals that reflect their interests and preferences.

ITPs specify a set of transition services that are designed to help students succeed in postschool life. Transition planning includes determining student strengths and abilities, interests and preference, needs, and goals. The point of transition planning is to connect those preferences, needs, and goals to analogous transition services. Such services, in turn, support the adolescents and young adults with IDD to reach their postsecondary goals and to learn the necessary skills to succeed in adulthood and meet their postschool needs. As students prepare to exit school, the transition planning process plays a critical role in preparing adolescents and young adults with IDD for their future (Wehman et al., 2015). Transition planning looks forward, envisioning how the adolescents and young adults will engage with their community in terms of work, postsecondary education, and independent living once they leave the high school setting.

Effective transition planning can produce long-term postschool success for adolescents and young adults with IDD. Postschool outcomes for adolescents and young adults with disabilities have notoriously been worse compared to the outcomes of those without disabilities, with those with disabilities often having poor postschool outcomes in regard to employment, independent living, and community engagement. Postschool outcomes are even worse for adolescents and young adults with IDD compared to individuals with high incidence disabilities, such as specific learning disabilities (Newman et al., 2009).

Two important skills that can be addressed during transition planning are self-determination and decision making. Such skills can help students to make decisions about their life and access the necessary supports they need to be successful after high school and to be the director of their own life. The purpose of this chapter is to discuss the decision-making process and how to address decision making during transition planning for adolescents and young adults with IDD. Given the importance of decision making for adolescents and young adults with IDD during transition, this chapter also covers current research-based strategies for decision making and self-direction during the transition process. We begin by providing a brief description of the role of self-determination and decision making in the transition process. Next, we provide an overview of decision making in four key transition areas. We conclude by discussing student-directed IEPs, person-centered planning, and different evidence-based and promising curricula that can improve decision-making skills and self-direction for adolescents and young adults with IDD.

Self-Determination and Decision Making

Decision making is an inherent and central skill within the broader self-determination framework. Wehmeyer et al. (1996) defines self-determination as "acting as the primary causal agent in one's life and making choices and decisions regarding one's quality of life free from undue external influence or interference" (p. 632). Adolescents and young adults with IDD who are self-determined are those who independently make decisions about their own life and, therefore, self-direct their own life (sometimes with self-selected supports).

Self-determination also plays a large role in postschool success and long-term outcomes for adolescents and young adults with IDD. During transition, adolescents and young adults with IDD who have better self-determination skills may have better quality of life, community engagement, employment, and college outcomes and are likely to be more independent (Jameson et al., 2015; Powers et al., 2012; Shogren et al., 2015). Shogren et al. (2015) found that students with IDD with better self-determination skills were more likely to have better outcomes in terms of employment and community access. Similarly, Wehmeyer and Palmer (2003) found that students with IDD who were more self-determined when leaving high school were more likely to be living independently, managing their own money and finances, caring for their own health, and likely to have a job/be employed. Self-determination has also been positively correlated with better work and employment outcomes (Martorell et al., 2008) and a better quality of life (Biggs & Carter, 2016; Lachapelle et al., 2005). Indeed, for transition-aged youth with autism, a higher level of self-determination was positively correlated to quality of life across all domains – physical, psychological, parent, social and peers, and school (Biggs & Carter, 2016). Considering the variety of positive outcomes associated with self-determination, it follows that self-determination skills (including making decisions for one's self) would be an integral set of skills to develop during transition planning.

Self-determination "means having the power to make decisions, to direct one's actions, to dream and take risks, and to exercise rights and responsibilities" (Powers et al., 2012, p. 2182). We can break self-determination down into component parts: (1) goal-setting, (2) problem-solving, (3) evaluating and selecting options, (4) taking action for selected options and decisions made, and (5) self-monitoring progress and adjusting goals and actions as needed. Learning and practicing such skills can help adolescents and young adults become more self-determined, which can lead to greater postschool success on their transition goals (Powers et al., 2012).

Decision Making in Transition Planning

A central part of becoming a self-determined adolescent and adult is decision making. In fact, this is one of the most critical skills an adolescent and young adult with IDD can develop to lead the life they want to have. As adolescents and young adults prepare to leave school, they must make a plethora of decisions about various areas of their life. Decision making can be a complex process – it can involve identifying the available options, weighing the possible consequences of each option, and then selecting an option (Burke et al., 2019; Hickson & Khemka, 2013).

Typical adolescents and young adults between the ages of 16 and 25 years make a lot of decisions about the trajectory of their lives that impact long-term success in adulthood. These decisions include, but are not limited to, deciding where they want to go to college (or even if they want to go), in which sports and hobbies they want to participate, where and with whom they want to live, what job/career they wish to have, financial and money decisions, health decisions, decisions about with whom they want to be friends with or in a relationship with, and other important areas of adult life. For each of these areas, young adults have to consider their options, preferences, and consequences and then select the best option for themselves. These are some of the most important decisions adolescents and young adults with IDD may make, which can frame and impact the rest of their adult life.

When adolescents and young adults turn 18 years old, they reach the "age of majority." At this time, there is a transfer of rights, wherein individuals are granted new legal rights – including such things as the right to vote, marry, open a credit card, and sign legal documents. For adolescents and young adults with IDD, turning 18 also signifies that they can legally assume decision-making power in regard to their special education services (Age of Majority, 2017). In essence, the decision making power is transferred from the parent or guardian to the student, meaning the student is able consent to evaluation or change of placement or even sign their own IEP without their guardians' approval. This Transfer of Rights can lead to struggles between the adolescents and young adults with IDD and their guardians, with disagreements over who should be able to make (or will make the best) decisions regarding the student's future.

In certain cases, when the student has extensive support needs, the transfer of rights may be delayed or denied via a court process in which the parent or guardian assumes power of attorney or conservatorship/guardianship (Millar & Renzaglia, 2002). Guardianship refers to an individual who is appointed by the court to make decisions for the adolescent or young adult with IDD because they have been determined to have limited capacity in decision making. However, guardianship can serve as a barrier for decision making (Devi et al., 2020; Jameson et al., 2015). One of the main issues with guardianship is the restrictive nature of its legal status, "At its core, guardianship is too often sought based on the assumption of a lack of "capacity" and can be seen as a restrictive form of adult support" (Jameson et al., 2015, p. 38). Jameson et al. (2015) conducted a parent survey of guardianship and found that school personnel were not involved in the guardianship process and did not inform the guardianship process. When the transfer of rights does happen, both guardians and adolescents and young adults with IDD must be knowledgeable about this process. IDEA (2004) requires the education agency to notify guardians in advance of the transfer; thus, it is imperative that school personnel be informed about the guardianship process (Reynolds et al., 2018).

Parents and guardians must find ways to meaningfully support their young adults as they reach the age of majority and assume their legal rights. In situations of guardianship, guardians continue to hold the decision-making powers in all areas of life, including special education and transition planning. Even in these cases, parents and guardians can encourage their children to share their interests and preferences and honor them to the greatest extent possible, even if not legally required. With support, practice, and guidance, adolescents and young adults with IDD can make informed decisions that will lead to positive outcomes. Many of the decisions made during the transition planning process can affect students' quality of life after high school (PACER, 2015). Schools and parents should assume that adolescents and young adults with IDD are competent and capable of making decisions and being the leaders of their own life during transition planning (Payne-Christiansen & Sitlington, 2008; Rood et al., 2015). We review two types of decision making that students can engage in during transition planning: independent decision making and supported decision making.

Independent Decision Making

Independent decision making is a process wherein an individual makes decisions on their own. This process may require that the adolescent and young adult consider all options, weigh the risks and benefits, and select and put into action the right decision. Other factors that can play a role in decision making are situational factors (e.g., setting); motivational processes, such as self-esteem beliefs (e.g., self-concept), personal agency beliefs (e.g., locus of control, self-efficacy), and integrated goal processes (e.g., goal identification); environmental factors (e.g., social supports); and personal factors (e.g., age, gender, race; Hickson & Khemka, 2006).

The independent decision-making process can be complex and intricate during transition planning. Students can be involved in creating and deciding upon their own postsecondary goals connected to their needs, strengths, interests, and preferences. Ideally, during transition planning, all components of the student's transition plan should focus on supporting the student's progress toward their selected post-school goals. Postsecondary goals influence a variety of factors, including the courses the student will take, the skills the student will work on over the next school year, and the types of transition services the student will receive (Mazzotti et al., 2009).

In some cases, parents or guardians do not believe the adolescent or young adult with IDD has the ability to make informed decisions alone. While all students should have a say as to what happens in their own lives, some guardians worry that their young adult will make hasty, uninformed decisions that could potentially damage their future prospects. For instance, a student may consent to an inappropriate change of placement, or a frustrated student may decide to drop out of school. There are relatively few studies that examine the independent decision-making abilities and skills of adolescents and young adults with IDD in relation to transition planning. According to Hickson and Khemka (2006), some concerns about the decision-making ability for adolescents and young adults with IDD may be warranted. While the studies described in Hickson and Khemka were not specifically related to making decisions during transition planning, they highlighted how in some instances adolescents with IDD struggled to make appropriate decisions compared to their typical peers, particularly in relation to decisions about interpersonal conflict and issues related to harm, coercion, and abuse. However, they also found that even if adolescents and young adults with IDD struggled with decision making, it is possible to teach them independent decision-making skills (i.e., in relation to peer interactions and decisions related to abuse/harm; Khemka et al., 2005, 2016). Moreover, other studies have found that students who participate in their own transition planning have better self-determination skills (Powers et al., 2012) – of which decision making is a critical attribute.

Thus, there are concerns with both the restrictive nature of guardianship and the ability of adolescents and young adults with IDD to make independent decisions. Additionally, most people do not make decisions alone but often request support in decision making from family or friends. More recently alternatives have been suggested to replace the dichotomy of *student-only* or *parent-only* during decision making in transition planning for adolescents and young adults with IDD. One such practice is supported decision making, which empowers the individual with IDD to make decisions with the support of trusted individuals.

Supported Decision Making

Supported decision making shifts the primary decision-maker role to the adolescent and young adult with IDD but includes support from various stakeholders in making decisions. In this model, adolescents and young adults with IDD work collaboratively with their parents/guardians to prepare themselves for their decision-making responsibilities. Parents and/or trusted others support the individuals with IDD in the decision-making process and encourage them to take an active role in their IEP development and transition planning prior to the transfer of rights at the age of majority (PACER, 2015). Family participation is essential and integral in transition planning during the IEP process. Indeed, family participation in this process improves long-term postschool outcomes of students with IDD, including obtaining postschool employment (Carter et al., 2012; Test et al., 2009). Family involvement in decision making can result in individuals with IDD obtaining more services and better satisfaction with services and providers (Neely-Barnes et al., 2008).

Family involvement may also be a barrier for decision making of adolescents and young adults with IDD during transition. Family members, such as parents or siblings, support staff, or other professionals often times make crucial decisions for the adolescent and young adult with IDD without including them in the decision-making process (Burke et al., 2019; Nonnemacher & Bambara, 2011). When practitioners make decisions for the student, the student often feels powerless or not in control of their life, even when that decision by the practitioner was stated to be in the best interest of the adolescent or young adult (Nonnemacher & Bambara, 2011). Devi et al. (2020) highlight that one of the barriers to supported decision making is the focus on the continuous assessment of mental capacity of the young adult. Rather than continuously assessing mental capacity during transition, the support needs of adolescents and young adults with IDD should be assessed. Assessment of support needs informs the types of assistance necessary for decision making and planning for the individual with IDD and ensures they are provided with these supports during transition planning (Shogren et al., 2017). In a review of supported decision making, Shogren et al. (2017) found that family attitudes about decision-making capacity played a major role on the decision-making process and opportunities in which their child with IDD engaged.

Supported decision making shifts the decision-making responsibility to the adolescent and young adult with IDD, with support from various stakeholders, but the ultimate decision-maker is the adolescent and adult themselves and not parents or siblings. In a supported decision-making model, stakeholders who can help support the student during transition planning can be parents, siblings, friends, professionals, and those who know the student well and whom the student trusts. The adolescent or young adult with IDD should be the one to decide who will provide support at their decision-making and planning meetings – that is, they choose their support system. This support system selected by the student can "help them understand the situations and choices they face, so they may make their own decisions" (Blanck & Martinis, 2015, p. 24).

Jameson et al. (2015) highlighted the primary characteristics of supported decision making: "(a) The individual retains legal decision-making authority, (b) the relationship is freely entered into and can be terminated at will, (c) the individual actively participates in decision making, and (d) decisions made with support are generally legally enforceable" (p. 38).This model acknowledges both the importance of the individual's decision-making capacity and the importance of family participation. To elucidate this point, Miller et al. (2018) states, "it becomes a balancing act as parents walk the tightrope of not undermining their young adult's self-determination and independence, while being strong advocates and negotiating the complex provider system" (p. 268). Supported decision making should be presented as an option and adopted in the transition planning process because it can empower adolescents and young adults with IDD to make decisions about their own life and goals (Blanck & Martinis, 2015).

Areas of Decision Making in Transition Planning

Adolescents and young adults need to make both everyday and long-term decisions in various areas of their life when they transition from school to postschool life. These areas include employment, independent living, postsecondary education, and financial decisions. There are many other areas of decision making during transition planning that are not highlighted in this chapter, but are also important to consider, including health, social friendships and relationships, sexuality, recreation, and faith/religion. This chapter focuses on long-term decisions during transition planning that impact lifelong goals, rather than the daily choices adolescents and young adults with IDD make.

Employment

Employment is one of the major areas of focus during transition planning and decision making. Employment can lead to independence in other life areas such as living independently and financial independence. Unfortunately, employment outcomes for adolescents and young adults with IDD remain dismal (Newman et al., 2009). Many decisions regarding employment are made during transition planning, including (a) type of career, (b) type of work environment, (c) early work experience needed for future career, and (d) the specific skills needed for employment and career options. High parent expectations, social skills, self-determination skills, independent living skills, and early work experience can lead to better employment outcomes for individuals with IDD (Blustein et al., 2016; Carter et al., 2012).

Providing early work experiences during transition seems to be an important factor in better outcomes for postschool employment (Carter et al., 2012). Students can engage in various types of work experiences and access vocational content in various classes, especially during adolescence and emerging adulthood. These should include opportunities to work in various fields to gain both experience and understanding of fields of interest. For example, if a student's postsecondary employment goal is to work in healthcare, their IEP/ITP would include a course of study that supports their future career (e.g., biology or physiology courses), annual IEP/ITP goals that would feature foundational skills needed to eventually work in the healthcare field (e.g., math skills, communication skills, identifying health risks, etc.), and transition experiences that help the student develop in-depth knowledge of health careers should also be included (e.g., job-shadowing various healthcare professionals). By gaining these early work experiences and vocational skills, students can potentially engage in more informed decision making about careers and future employment. Providing students with the opportunities to work and experience different classes and jobs may lead to better decision making about employment.

Independent Living

Young adults with IDD are more likely to live long term with caregivers, parents, siblings, or in community group homes than independently. According to the Residential Information Systems Project (RISP) 2017 report, 60% of adults lived at home with family, and 29% lived in group homes (with 6% in small group homes and 23% in large group homes). Only 12% of adults with IDD were living independently in their own home (Larson et al., 2017). Living at home is even more common for students up to 8 years out of high school, compared to those recently out of school. Two studies based on the National Longitudinal Transition Study-2 (NLTS-2) data reported very high rates of living in the family home for young adults with IDD, between 78% and 97% (Anderson et al., 2014; Bouck, 2014).

Often times living in the family home is assumed to be the norm for adolescents and young adults with IDD by their families and stakeholders, without regard for the preference of the individual. The complex system of adult services and supports is often hard to navigate when considering the potential postsecondary living environments. Multiple factors have to be considered when deciding where to live, such as location, safety, roommates, types of supports available, costs, and risks. The complexity and high stakes of this decision can take time to address (Shogren et al., 2017). Additionally, family context, physical and financial ability to provide those supports, and trust in service providers may play an important role in the decision of where the adolescent and young adult with IDD will live.

However, one important step during transition planning is to discuss with the adolescent or young adult with IDD where and with whom they would like to live in the future and to make plans to work toward that goal. Ensuring that the adolescent and young adult with IDD has a voice in their residential arrangements is integral to supporting their decision-making skills. For example, if the individual with IDD would like to live independently with a roommate in the future, their transition goals, courses, and services should address this desire. Thus, courses will need to address the life skills required to maintain a safe and clean home and how to make financial decisions to pay bills on time (to avoid consequences such as eviction). Furthermore, the adolescent and young adult must acquire the social skills needed to make decisions regarding roommates and sharing space with another individual.

Postsecondary Education

One of the major areas of growth in the past decade in the United States has been the provision of postsecondary education opportunities for adolescents and young adults with IDD after leaving high school (ThinkCollege, 2020). Postsecondary settings provide opportunities for students to improve on their social skills, self-determination skills, employment, friendships, and relationships (Miller et al., 2018). For typical adolescents and young adults, college is a formative experience where they make career-related and independent living decisions that have long-term effects on their life (Arnett, 2000). College is a time to learn about different topics that can inform your interests and preferences in employment, which can further inform decisions about life goals and outcomes (e.g., independent living).

Engaging in college experiences, adolescents and young adults with IDD can gain skills that can improve their decision-making abilities. During college students have to navigate many new types of experiences that can possibly lead to improved decision-making skills. College experiences include opportunities to make decisions regarding social relationships, careers, recreation, health, and other independent living skills. In addition, research suggests that for those students who attended college, parents perceived increased outcomes that allowed them to *let go* of the young adult with IDD. That is, parents stated they felt that the student had many more capabilities and could do things more independently; thus, parents could let go from providing more intensive support (Miller et al., 2018).

Since adolescents and young adults with IDD currently have more postsecondary options available, they now have the ability to decide whether or not they would like to attend college. If attending college is a goal expressed by the adolescent or young adult with IDD, transition plans must include steps toward applying to appropriate college programs. Families and teachers should work together so that the student can visit different programs to decide which college program they like the best and would like to attend. Furthermore, transition goals should work on college readiness skills such as navigating a campus space independently, organizational skills, life skills, and social skills.

Financial Matters

Financial literacy skills are crucial independent living skills for adolescents and young adults with intellectual disabilities (ID) and autism to acquire for more independent decision making in adulthood (Suto et al., 2005). "Financial literacy skills enable individuals to navigate the financial world, make informed decisions about their money and minimize their chances of being misled on financial matters…" (Marcolin & Abraham, 2006, p. 2). A major concern during transition planning is the individual's ability to manage finances and make sound financial decisions. Adolescents and young adults with IDD are often *underbanked* or *unbanked*, which means they often opt to use alternative methods of money management rather than using traditional banking methods used by typical adolescents and adults to manage their money (Goodman & Morris, 2017). In financial matters, there are long-term financial decisions (e.g., savings) and short-term daily financial decisions (e.g., grocery shopping, etc.). Adolescents and young adults with IDD are capable of making both long-term and daily financial decisions when provided with instruction in financial literacy and given opportunities to practice financial decision making and skills (Suto et al., 2005). Similarly, Goodman and Morris (2017) suggest that adults with IDD who participate in vocational rehab services have "a unique opportunity to integrate benefits and financial counseling as part of a pathway to employment with career advancement potential that focuses on short- and longer-term financial goals, informed financial decision making and strategies" (p. 9).

Focusing on financial literacy skills and financial decision making throughout transition planning by creating financial literacy goals is imperative for adolescents and young adults with IDD to build the ability to make independent financial decisions. For example, a postsecondary transition goal can be for the student to budget, save, and manage their own money. If an individual participates in early work experiences in high school, then they can learn how to manage the money they earn from those work experiences. In high school, students can be given bank accounts supported by their parents as an opportunity to learn updated banking methods. Additionally, adolescents and young adults with IDD may receive financial benefits they will have to learn to manage, such as Supplemental Security Income (SSI). In 2014, the Achieving Better Life Experience (ABLE) act was signed into law. This act created ABLE accounts that are like bank accounts for people with disabilities where income can be earned and kept without being subject to income tax and without impacting SSI and Medicaid eligibility. The person with IDD can thus manage and control their own ABLE accounts (Morris et al., 2016). Financial skills are critical because adolescents and young adults with IDD, who are increasingly engaging in competitive employment and earning wages and income, will need to learn how to manage their income, ABLE accounts, and SSI.

Methods for Increasing Decision Making and Self-Direction During Transition

The decisions and goals the students embrace during transition planning impact the services they receive to reach those goals, as well as the long-term outcomes for adolescent and young adults with IDD. Hence, it is essential for the decisions reflecting the preferences and desires of students to be included in transition planning and for students to engage in self-direction. Each of the four areas previously discussed (i.e., employment, independent living, etc.) has some level of complexity in the decision-making process; however, adolescents and young adults with IDD with the right supports can participate in creating goals and deciding the trajectory of their life.

Best practice suggests that student involvement in transition planning is necessary and integral to postschool success. According to Martin et al., (2007, p. 14) "Student-directed learning strategies give students opportunities to become involved in their educational decision making process" (Martin et al., 2007; p. 14). Because postsecondary goals developed during transition planning are so critical to both the student's current and future situations, students should have a say in what these goals entail. In fact, students should be actively involved in every aspect of the transition planning process to the greatest extent possible (Wehman, 2011). There have been various methods to teach self-determination, decision making, and self-direction during transition planning so that adolescents and young adults are empowered, through active participation in transition planning, to become the primary decision-makers of their life. The next section outlines the methods and practices associated with student-directed IEPs and person-centered planning.

Interventions to Promote Self-Direction and Decision Making in the Transition Process

As discussed, self-determination for adolescents and young adults with IDD includes active and ongoing participation in the decision-making process. Decision making skills must be developed early on and practiced regularly in a variety of settings, including the individual's home, community, and school (Connor, 2012). While the home and community settings offer a plethora of decision-making prospects, the school setting offers a unique opportunity for students with disabilities to demonstrate a sense of agency – through participating in the development of their IEPs. In particular, student involvement in the IEP, including the transition component of the IEP, is a major opportunity for students with disabilities to engage in self-determination and decision-making practices (Wehmeyer et al., 2007).

Student-Directed IEPs

Student attendance during transition planning meetings in the IEP process is an important first step toward promoting meaningful participation during transition planning meetings and, ultimately, self-direction (Martin et al., 2004). With recent pushes for student-directed IEPs and student-centered planning in special education (Mazzotti et al., 2015), one might assume that students would play a critical role in the development of their IEPs. However, this is not often the case. In fact, IDEA (2004) provides relatively little guidance on student involvement during IEP meetings, merely stating that the student should attend IEP meetings "when appropriate" [§ 300.321]). *Some* further guidance is provided once students reach transition age and the transition component of the IEP is introduced. To a certain extent, IDEA (2004) requires some student involvement in the transition planning process. Specifically, schools must *invite* the student to attend their IEP meeting if a purpose of the meeting will be the consideration of postsecondary goals and the transition services needed to reach those goals. If the adolescent and young adult with IDD does not attend the IEP meeting, the school must take other steps to ensure that the student's preferences and interests are considered. This includes such practices as administering age-appropriate transition assessments and using results to guide the development of the transition plan. However, there are no hard regulations surrounding student attendance at IEP. Even during transition planning, the student is not required to attend the IEP meeting – rather the school is only required to document that an invitation was extended to the student if transition planning will be discussed. Further, if the student is not present at the meeting, the school must only *consider* the student's preferences. They are not required to actually incorporate them into the ITP.

Perhaps because of the lack of hard rules and regulations, some students do not attend their IEP meetings. Specifically, students with IDD may be less likely to attend and/or participate in educational planning meetings. Shogren and Plotner (2012) found that, compared to other disability groups, students with IDD were among the least likely to attend their transition planning meetings. Further, even when students do attend their IEP meetings, most are inactive participants (Powers et al., 2005; Thoma et al., 2001). Data from the NLTS-2 showed that over half of transition-aged students with ID (57%) or ASD (67.3%) either did not attend their meeting or, if present, participated very little or not at all (Shogren & Plotner, 2012). Wagner et al. (2012) found that less than 1/4 of high school students assumed an active leadership role at their IEP meeting. Martin et al. (2006) observed over 100 educational planning meetings of high school students and found that, on average, students attending IEP meetings spoke for only 3% of the meeting time. In a more recent study, Johnson et al. (2020) found that students with IDD had a limited role in their transition planning and seldom participated in leadership roles in transition planning. The biggest barrier to the participation in transition planning seemed to be limitations in expressive communication and self-advocacy skills (Johnson et al., 2020).

This lack of student participation is unfortunate, given that active involvement in IEP/transition planning meetings is associated with many benefits for adolescents and young adults with IDD. Benz et al. (2000) found that high school students who participated in transition planning during their IEP meetings exhibited a greater ability to set and accomplish goals. Participation in IEP meetings has been positively associated with the development of self-determination skills in students with disabilities (Wehmeyer, 2005), which have been linked to improved postschool outcomes for students with IDD (Shogren et al., 2015; Test et al., 2009).

However, the skills necessary to lead, or even meaningfully participate in, an IEP meeting do not come naturally to most students. "It is naïve to presume that the youth attending their IEP/transition planning meeting will learn how to actively participate and lead the process through serendipity—yet this is precisely what current practice tends to expect" (Martin et al., 2004, p. 4). To address this problem, a variety of self-directed interventions have been developed by special education researchers. These interventions primarily consist of self-advocacy curricula, promoted as tools to enhance both IEP/transition planning participation and the development of self-determination for adolescent and young adults. Many of the curricula share similarities, including teaching students about the IEP and transition planning process using such strategies as mnemonics and role-playing. Table 16.1 provides an overview of interventions designed to increase student participation in IEP transition planning meetings.

Though several such interventions exist, the *Self-Advocacy Strategy* and the *Self-Directed IEP* are two of the most well-known and oft-used self-advocacy curricula. The *Self-Advocacy Strategy* (Van Reusen et al., 1994) is a published curriculum created to explicitly teach middle and high school students with disabilities how to actively participate in IEP meetings. Students are a taught to inventory their strengths, needs, goals, and preferences. Using mnemonic devices, students learn appropriate communication behaviors; self-determination skills, such as identifying and sharing preferences; and IEP-specific skills, such as naming and making

Table 16.1 An overview of interventions used to promote student participation and decision making during transition planning

Intervention	Developers	Purpose	Description
Beyond High School	Wehmeyer et al. (2006)	Promote involvement in adult services and future planning for students with intellectual disability aged 18–21 years	Three stages focused on (1) independent and supported decision making and goal-setting, (2) person-centered planning, and (3) progress monitoring
Charting a Course for the Future	Colorado Dept. of Education (2018)	Help students effectively plan for their future after high school	Eight modules that introduce components of transition planning and support the student in making decisions regarding postschool goals and services (independent and supported decision making)
Charting the LifeCourse	University of Missouri – Kansas City Reynolds et al. (2018)	Designed to help individuals with IDD create a vision and make decisions for their life through person-centered planning and supported decision making	A holistic planning process that recognizes the individual with IDD makes decisions about interconnected life outcomes with supports within the family and community context (supported decision making)
ME! Lessons for Teaching Self-Awareness and Self-Advocacy	Cantley et al. (2010)	Help educators teach students critical transition skills	Ten units, featuring 23 lessons that teach self-determination skills, including self-awareness, decision making, goal-setting, and communications skills (independent decision making)
Next S.T.E.P.	Halpern et al. (1997)	Teach students the skills needed to actively engage in transition planning	Sixteen lessons teaching self-determination skills relevant to transition planning (e.g., independent decision making, goal-setting, self-evaluation, self-directing a meeting)
Project MY VOICE	Van Laarhoven-Myers et al. (2016)	Provide students with disabilities skills and supports to meaningfully participate in transition planning meetings	Students participate in both independent and supported decision making by identifying preferences, setting future goals, and directing a transition planning meeting

(continued)

Table 16.1 (continued)

Intervention	Developers	Purpose	Description
Self-Advocacy Strategy[a]	Van Reusen et al. (1989)	Help students prepare for and participate in education or transition planning meetings by sharing their strengths and needs	Five steps, presented using mnemonics, that teach students effective communication skills, how to inventory strengths and needs, and determine and decide goals (independent decision making)
Self-Directed IEP[b]	Martin et al. (2006)	Teach students the skills needed to lead their IEP and transition planning meetings	Eleven lessons featuring video models, workbook activities, and role-play that teach students self-determination skills, including independent decision making
Student-Directed Summary of Performance	Martin et al. (2007)	Provide students an opportunity to learn about their disability, compile transition assessment data, identify accommodations, and outline postschool goals	Lessons describe transition assessments and guide students in completing their Summary of Performance, including deciding which accommodations best meet their needs (independent decision making)
Student-Directed Transition Planning Curriculum	Center for Disability Resource, U. of South Carolina School of Medicine	Teach students the knowledge needed to actively participate in their transition-focused IEP meetings	Students create a "Summary of Performance Script" to facilitate discussions and decision making at transition planning meetings (independent and supported decision making)
TAKE CHARGE for the Future	Powers et al. (2001)	Teach students transition planning strategies related to achievement, partnership, self-regulation, and meeting participation	Bi-weekly coaching sessions focused on teaching how to determine and decide transition goals, activities, and the supports needed to achieve them (independent decision making)
Whose Future is it Anyway? (2nd ed.)	Wehmeyer et al. (2004)	Teach students the skills needed for a student-directed transition planning process	Thirty-six sessions that provide students with opportunities to practice self-awareness, problem-solving, decision making, goal-setting, and communication skills (independent decision making)

[a]Self-Directed IEP is part of the Choice Maker Self-Determination Curriculum. The Self-Directed IEP is the portion of the curriculum that teaches students how to participate in IEP meetings
[b]The Self-Advocacy Strategy was originally titled IEP Participation Strategy (IPARS)

decisions on goals and learning supports. Skills are practiced in individual and group settings over the course of several weeks, leading up to an actual IEP/transition planning meeting. Teachers can also work with students to generalize skills and behaviors learned through the *Self-Advocacy Strategy* to settings beyond IEP meetings. The *Self-Directed IEP* (Martin et al., 1996) also aims to promote active student engagement in the IEP process. Students learn about the IEP process and the skills needed to actively lead an IEP meeting, such as stating the purpose of the meeting and reviewing past goals and their progress toward the goals. Skills are taught to students over 11 sequential 45-minute lessons. Lessons include a combination of direct instruction, video-modeling, and student workbook activities designed to provide students an opportunity to practice and apply skills, such as self-determination and decision-making skills.

With any of these interventions, the ultimate goal is for students to attend and be active participants at their IEP meetings. Participation may look different depending on a student's individual strengths and areas of need, as well as how much instruction and practice they have received – some students may only share their thoughts on specific areas of the transition plan, while others may take a more active role and lead the entire meeting. A student-led IEP involves the student engaging in the following activities: introducing IEP team members, stating the purpose of the meeting, sharing postschool preferences, presenting annual goals and/or postsecondary goals, stating supports needed to achieve their goals, and sharing preferences and making decisions in regard to courses, educational placement, or learning supports (Mason et al., 2002). Through participating in the IEP meeting, students are practicing a variety of self-determination skills, including problem-solving, decision making, and self-advocacy (Wehmeyer et al., 2007).

Researchers have examined the effects of these self-advocacy curricula on student participation during IEP meetings. In a literature review, Test et al. (2004) identified interventions to increase adolescent student involvement in IEP planning, meeting, and monitoring. Results highlighted the positive outcomes from interventions to promote student participation in IEP/transition planning meetings. Building on the review by Test and colleagues, Griffin (2011) analyzed interventions used to promote IEP participation for high school students from culturally and linguistically diverse backgrounds, finding favorable results. Recently, Sanderson and Goldman (2020) conducted a meta-analysis on this topic. Analyzing data from six group design experimental research studies, they found that adolescents who participated in self-advocacy curriculum interventions made significantly more contributions during IEP meetings compared to students in control groups. Additionally, researchers have also examined the impact of these interventions on the quality of student participation, assessing if students made more relevant (on-topic) contributions during IEP meetings after partaking in a self-advocacy curriculum. Indeed, several of these studies found positive results for students with IDD (Cease-Cook et al., 2013; Kelley et al., 2011; Test & Neale, 2004).

Though the research on these interventions is encouraging, there are some limitations that should be noted. First, the vast majority of studies examining self-advocacy curricula to increase student participation involved students with learning disabilities (LD). Though these results were favorable for students with LD, we are left to wonder how generalizable these results are for students with IDD. Second, most research is limited to students who are able to communicate verbally. We know next to nothing about how nonverbal students or those who communicate with augmentative and alternative communication (AAC) devices participate in their transition planning meetings. One study (Johnson et al., 2020) found that students with limited expressive language are less likely to participate or lead their IEP/transition planning meetings. Future research must examine the effectiveness of self-advocacy curricula on a more diverse group of students, including individuals with extensive support needs and those who use alternative modes of communication.

Even with these limitations, the research is promising. It supports efforts of school districts to go above and beyond, rather than merely complying with IDEA's (2004) loose regulations surrounding involvement in IEP and transition planning meetings. Schools should consider mandatory student attendance policies at IEP/transition planning meetings. By implementing a self-advocacy curriculum and directly teaching students how to effectively participate in their IEP meetings, schools can take steps to ensure adolescents and young adults with IDD are meaningfully included in the decisions made during the transition planning process.

Person-Centered Transition Planning

Person-centered planning is a planning approach and strategy to help improve the quality of life of individuals with IDD. Through a collaborative process, stakeholders in the individuals' circle of support create a *goal-oriented* plan of supports for the adolescent or young adult with IDD (Claes et al., 2010). More recently, person-centered planning has been used extensively as an approach to service delivery in the adult service system in opposition to traditional service-centered approaches. Person-centered planning is used to plan for the direct service support for adults with IDD by including the adult with IDD and their family as the leader of their own plan. "In PCP [person-centered planning] decision making is driven by the individuals themselves and by those who care about them, with particular emphasis on self-determination, choice and autonomy" (Ratti et al., 2016, p. 64).

While person-centered planning has been garnering widespread use in adult services and transition planning, there is a limited research evidence base. However, the results of two meta-analyses (Claes et al., 2010; Ratti et al., 2016) suggest person-centered planning may have a moderate effect that in some areas of adult life. Daily decision making was one of the main areas of improvement following person-centered planning. Across both meta-analyses, 13 studies reported better choice/decision making for adults with IDD who received person-centered planning practices (Claes et al., 2010; Ratti et al., 2016). Robertson et al. (2006) found that person-centered planning had potential positive benefits for community engagement, contact with friends and family, and choice.

Miner and Bates (1997) examined the impact of person-centered planning on transition planning, and the results suggested an increase in parent participation in transition planning after a person-centered planning session meeting. The study also suggests that parents perceived greater participation of their student with IDD during the transition planning, although this was not directly measured. Ratti et al. (2016) also suggests that there may be some positive impact of person-centered planning on community engagement and increase in social networks of friends. However, they caution that the impact of person-centered planning may only be on short-term goals and not on long-term goals for adolescents and young adults with IDD. Person-centered planning can be most effective when applied to important areas of life selected by the individual rather than applied broadly across all areas of life.

In person-centered planning, the adolescent and young adult with IDD should be responsible for identifying and deciding the outcomes that are important for them specifically. Person-centered planning is most effective when the participant is "directly involved in decision making" (Ratti et al., 2016, p. 79). Moreover, the authors suggest that having "active participation in decision making" would result in improved life outcomes for adults with IDD (Ratti et al., 2016, p. 79).

Person-centered planning can support student decision making throughout the transition planning process, either when implemented before the IEP/Transition meeting or during the transition planning meeting itself (Hagner et al., 2014). Transition planning in schools already includes some of the basic ideas of person-centered planning, such as focusing on students' strengths, preferences, and interests (Carter et al., 2015). Person-centered planning itself is an approach but does not need to be conducted as a formalized method. There are various formal models of person-centered planning, such as Essential Lifestyle Planning, Personal Futures Planning, Planning Alternative Tomorrow with Hope (PATH), Picture Method, Life-Lifestyle Planning, and McGill Action Planning System (MAPS; Claes et al., 2010; Ratti et al., 2016). Each of these methods has the goal of creating an individualized service plan based on individuals' strengths, interests, and support needs. However, they differ slightly in how and what pertinent information is gathered and whether they address short- or long-term goals (Robertson et al., 2006). Key features of person-centered planning, such as valuing and planning with the individual, can be implemented during transition planning (Robertson et al., 2006). Additionally, to increase participation in person-centered planning during transition, schools can provide accommodations, such as preparation for meetings, flexible meeting designs, and support for alternative communication methods (Hagner et al., 2014).

In essence, person-centered planning is a collaborative process between the individual with IDD and trusted stakeholders that are selected by the student to be in their circle of support. In addition to the adolescent and young adult with IDD, the other participants in person-centered planning can be parents, siblings, other family members, professionals, teachers, etc. Person-centered planning includes the following tenets: (1) valuing the rights, independence, choice, and inclusion of the individual with IDD; (2) gathering information by asking the individual what is important to them now and what they think will be important for them in the future and what supports the person believes they need to be successful; (3) gathering information on the strengths and capabilities of the individual with IDD and their support needs; (4) may include a facilitator who has knowledge of best practices for planning and collaborating – the facilitator may be a professional staff member or teacher or someone else with extensive knowledge of the student, and (5) a final plan is created that details the actions and who is responsible for those actions (Jameson et al., 2015; Robertson et al., 2006; Sanderson et al., 2006).

Jameson et al. (2015) notes that person-centered planning is not only conducted during transition meetings but continuously used in practice in everyday supports of the adolescent and young adult with IDD. Person-centered planning can include decisions on ongoing changes throughout the individual's life and can be conducted at any time the individual requires. Information gathered and determined in the person-centered planning meeting can be used to address decisions regarding both the daily needs of the adolescent and young adults with IDD and their long-term goals.

Examples of Curricula for Decision Making and Self-Direction

Below we describe three examples of research-based curricula that support decision making and self-direction (also seen in Table 16.1). These are formalized frameworks and tools that can be utilized to implement person-centered planning and self-direction practices and increase self-determination skills and decision-making skills in the transition planning process. Please note that the programs described below are only a few of the current promising practices; there are many other programs and frameworks that exist and can be utilized during transition planning (see Table 16.1).

Charting the LifeCourse *Charting the LifeCourse* (Reynolds et al., 2018) is a framework and set of tools that promotes "a different way of thinking, encouraging high expectations, and having life experiences move the trajectory in the desired direction, and integrating multiple types of supports" (Charting the LifeCourse Nexus, 2020). The framework recognizes that both the individual with IDD and their family and community context are important within the decision-making process. The family and community are defined by the person with IDD during the planning process. Supported decision making and person-centered planning are emphasized throughout this planning process. The framework also provides a process that creates opportunities and plans for life outcomes in key interconnected areas of the individual's life based on the student's interests and needs (Reynolds et al., 2018). This system is designed to help individuals with IDD create a vision and make decisions for their life to "identify how to find or develop supports and discover what it takes to live the lives they want to live" (Charting the LifeCourse

Nexus, 2020). While not transition specific, this framework can be applied to the transition planning process.

ChoiceMaker *ChoiceMaker* (Martin et al., 2006) is a curriculum developed to improve self-determination skills and student participation and decision making in transition planning. *ChoiceMaker* has three sections "(1) Choosing goals, (2) Expressing Goals, and (3) Taking Action" (Martin & Marshall, 1995). Self-directed IEPs as seen in Table 16.1 are a part of the *ChoiceMaker* curriculum. The creators note that IDEA (2004) requires student input during transition planning; thus, they created a curriculum to support students to provide input and make decisions about their goals and supports during their transition planning. *ChoiceMaker* includes self-directed IEP modules to assist students to be leaders in their IEP transition planning and decision-making process during transition (Martin & Marshall, 1995; Martin et al., 2008). This curriculum promotes learning independent decision-making skills.

The Self-Advocacy Strategy The *Self-Advocacy Strategy* (originally called IPARS; Van Reusen et al., 1989) is a five-step process and set of tools to help students with disabilities participate and make decisions during the IEP transition planning process. Studies on the *Self-Advocacy Strategy* found that students with IDD improved the quality of their participation in the IEP meeting and transition planning (Cease-Cook et al., 2013; Test & Neale, 2004). Additionally, they found that students were able to maintain and generalize their participation skills. This curriculum promotes learning independent decision-making skills. The first step in this strategy is for the individual to inventory their strengths, needs, and goals with structured worksheets. The second step is to share the inventory with stakeholders. The third step is to listen during the IEP meeting and respond to questions accordingly. The fourth step is to ask questions when the individuals with IDD do not understand what someone is saying or suggesting. The final step is to share the goals the individual selected. This strategy provides worksheets, communication cards, and other support tools to help students participate in their IEP meetings and plan for the future in all areas of life pertinent to the student.

Conclusion

There is a significant gap in the postschool outcomes of adolescents and young adults with IDD compared to their peers with high incidence disabilities and those without disabilities. Skill sets that have been shown to be effective at improving postschool transition outcomes are self-determination and decision making. In putting together all the approaches detailed in this chapter, several strategies that schools can implement to support decision making and self-determination were identified: (a) including the student in transition planning during the IEP process; (b) ensuring that the interests, preferences, and strengths of the individual with IDD are tied to their postsecondary goals; (c) employing self-directed IEPs and person-centered planning to give the student with IDD the leadership role during transition planning; (d) assessing students' needs in a valid, accurate, and equitable way, that is, not assessing capacity and competence but assessing support needs; (e) ensuring high family expectations and involvement; (f) including support from other trusted stakeholders selected by the individual with IDD; and (g) explicitly teaching and providing opportunities to build self-determination and decision-making skills with evidence-based curricula. These actions are not completed independently of each other but are interconnected to support the adolescent and young adult with IDD during the transition planning process. In addition to these suggestions, school personnel and families should be made aware and informed of all alternative options to guardianship, such as supported decision making, during the transfer of rights for the adolescent and young adult with IDD (Rood et al., 2015).

In agreement with federally mandated transition planning under IDEA, schools first and foremost need to make a concerted effort to include adolescents and young adults with IDD in transition planning meetings. Furthermore, stakeholders in the transition process must acknowledge that adolescents and young adults with IDD can make decisions and lead their own life. Stakeholders can create meaningful transition plans that reflect adolescent and young adults' strengths, preferences, and interests. Person-centered planning and participating in self-directed IEPs during transition planning are effective strategies to center decision making on the adolescent and young adult with IDD rather than teachers, parents, or other stakeholders. Such strategies also support the adolescent and young adult with IDD to practice and build decision-making skills. Effective strategies include specifically teaching self-determination and decision-making skills, such as the TAKE CHARGE program, engaging in person-centered planning, and promoting self-directed IEPs and in school transition planning participation (Martin et al., 1996; Powers et al., 2012; Ratti et al., 2016). These programs teach students how to identify goals and areas of life they would like to improve, how to develop and create plans for improvement, and how to direct and make decisions for their own transition plans. Creating opportunities for adolescents and young adults with IDD to be the director of their own life and make their own decisions in transition planning is essential to improving their outcomes.

Furthermore, providing opportunities for students with IDD to practice decision making and self-determination in schools is important. Such opportunities should be provided beyond just the transition planning meeting, but in all areas of school life. For example, rather than having all students in a transition program engage in the same work experiences, students should be encouraged to make decisions about early work experiences they would like to engage in (Suto et al., 2005). Lastly, even though IDEA (2004) states students must be invited to attend their IEP/transition planning meetings at age 16 and above, it would be better to include students as early as possible. Earlier engagement in IEP meeting can give the student even more opportunities to practice self-determination and decision-making skills.

When the student is the leader in their planning, they are free to make their own decisions and/or decisions with guidance from trusted stakeholders. Including the family in transition planning is essential to improving student life outcomes, but a balance is needed between student and family input in transition planning. Family involvement and supports can assist students to make decisions that improve their life outcomes. In supported decision making models and person-centered planning frameworks, students include trusted family members and stakeholders to support them in decision making during transition planning. Schools must be careful to ensure that family members and professionals are not the final decision-makers but are present to support the students to make their own decisions. Thus, family participation in decision making in transition requires that families act in a guidance role, supporting students rather than making definitive decisions. Families and stakeholders can also assist students with IDD to put their decisions and plans into action.

Finally, in reviewing the research in decision making, self-determination, and self-direction in transition planning, a few gaps were noted. There is limited research on person-centered planning, especially during transition planning. There is also limited research on how person-centered planning can improve decision-making skills, especially in regard to decisions for long-term goals rather than daily choice making. Additionally, self-direction and IEP participation strategies have not been well researched for students with more extensive support needs and alternative communication methods. Finally, there is limited research on how specific decision-making skills and factors during transition planning impact long-term outcomes of adolescents and young adults. Moreover, there is limited research on how decisions are made by adolescents and young adults with IDD within each area of transition during the planning process (e.g., employment, postsecondary, etc.).

For adolescents and young adults with IDD, participating in making decisions about their own lives during transition to adulthood is essential. When students are self-determined and make their own decisions about their lives, they can have better quality of life and increased community engagement, employment skills, independent living skills, money management skills, as well as health outcomes (both physical and mental health). It is essential that adolescents and young adults with IDD are the leaders of their own life during transition planning.

References

Age of Majority. (2017). Center for Parent Information and Resources. https://www.parentcenterhub.org/age-of-majority/

Anderson, K. A., Shattuck, P. T., Cooper, B. P., Roux, A. M., & Wagner, M. (2014). Prevalence and correlates of postsecondary residential status among young adults with an autism spectrum disorder. *Autism, 18*(5), 562–570. https://doi.org/10.1177/1362361313481860

Arnett, J. J. (2000). Emerging adulthood: A theory of development from the late teens through the twenties. *American Psychologist, 55*(5), 469–480. https://doi.org/10.1037/0003-066X.55.5.469

Benz, M. R., Lindstrom, L., & Yovanoff, P. (2000). Improving employment outcomes for students with disabilities: Predictive factors and student perspectives. *Exceptional Children, 66*, 509–529. https://doi.org/10.1177/001440290006600405

Biggs, E. E., & Carter, E. W. (2016). Quality of life for transition-age youth with autism or intellectual disability. *Journal of Autism and Developmental Disorders, 46*, 190–204. https://doi.org/10.1007/s10803-015-2563-x

Blanck, P., & Martinis, J. G. (2015). "The Right to Make Choices": The national resource center for supported decision making. *Inclusion, 3*(1), 24–33. https://doi.org/10.1352/2326-6988-3.1.24

Blustein, C. L., Carter, E. W., & McMillan, E. D. (2016). The voices of parents: Post-high school expectations, priorities, and concerns for children with intellectual and developmental disabilities. *The Journal of Special Education, 50*(3), 164–177. https://doi.org/10.1177/0022466916641381

Bouck, E. C. (2014). The post-school outcomes of students with mild intellectual disability: Does it get better with time? *Journal of Intellectual Disability Research, 58*(6), 534–548. https://doi.org/10.1111/jir.12051

Burke, M. M., Lee, C. E., Hall, S., & Rosetti, Z. (2019). Understanding decision making among individuals with intellectual and developmental disabilities (IDD) and their siblings. *Intellectual and Developmental Disabilities, 57*(1), 26–41. https://doi.org/10.1352/1934-9556-57.1.26

Cantley, P., Little, K., & Martin, J. E. (2010). *ME! Lessons for teaching self-awareness and self-advocacy.* http://www.ou.edu/content/education/centers-and-partnerships/zarrow/trasition-education-materials/me-lessons-for-teaching-self-awareness-and-self-advocacy.html

Carter, E. W., Austin, D., & Trainor, A. A. (2012). Predictors of postschool employment outcomes for young adults with severe disabilities. *Journal of Disability Policy Studies, 23*(1), 50–63. https://doi.org/10.1177/1044207311414680

Carter, E. W., Boehm, T. L., Biggs, E. E., Annandale, N. H., Taylor, C. E., Loock, A. K., & Liu, R. Y. (2015). Known for my strengths: Positive traits of transition-age you with intellectual disability and/or autism. *Research and Practice for Persons with Severe Disabilities, 40*(2), 101–119. https://doi.org/10.1177/1540796915592158

Cease-Cook, J., Test, D. W., & Scroggins, L. (2013). Effects of the CD-ROM version of the Self-Advocacy Strategy on quality of contributions in IEP meetings of high school students with intellectual disability. *Education and Training in Autism and Developmental Disabilities, 48*, 258–268. https://www.jstor.org/stable/23880644

Charting a Course for the Future: A transition tool kit. (2018). Colorado Department of Education. https://www.cde.state.co.us/cdesped/transition_tk

Charting the Life Course Nexus. (2020). *Charting the life course framework.* Curators of the University of Missouri. https://www.lifecoursetools.com/lifecourse-library/lifecourse-framework/

Claes, C., Van Hove, G., Vandevelde, S., Loon, J. V., & Schalock, R. L. (2010). Intellectual and developmental disabilities., *48*(6), 432–453. https://doi.org/10.1352/1934-9556-48.6.432

Connor, D. J. (2012). Helping students with disabilities transition to college: 21 tips for students with LD and/or ADD/ADHD. *Teaching Exceptional Children, 44*(5), 16–25. https://doi.org/10.1177/004005991204400502

Devi, N., Prodinger, B., Pennycott, A., Sooben, R., & Bickenbach, J. (2020). Investigating supported decision making for persons with mild to moderate intellectual disability using institutional ethnography. *Journal of Policy and Practice in Intellectual Disabilities, 17*(2), 143–156. https://doi.org/10.1111/jppi.12314

Goodman, N., & Morris, M. (2017). Banking status and financial behaviors of adults with disabilities: Findings from the 2017 FDIC national survey of unbanked and underbanked households and focus group research. *National Disability Institute*. Retrieved from: https://www.nationaldisabilityinstitute.org/wp-content/uploads/2019/11/ndi-banking-report-2019.pdf

Griffin, M. M. (2011). Promoting IEP participation: Effects of interventions, considerations for CLD students. *Career Development for Exceptional Individuals, 34*(3), 154–164. https://doi.org/10.1177/0885728811410561

Hagner, D., May, J., Kurtz, A., & Cloutier, H. (2014). Person-centered planning for transition-aged youth with autism spectrum disorders. *Journal of Rehabilitation, 80*(1), 4–10.

Halpern, A. S., Herr, C. M., Wolf, N. K., Doren, B., Johnson, M. D., & Lawson, J. D. (1997). *Next S T E P. student transition and educational planning*. Pro Ed.

Hickson, L., & Khemka, I. (2006). The role of motivation in the decision making of adolescents with mental retardation. *International Review of Research in Developmental Disabilities, 31*, 73–115. https://doi.org/10.1016/S0074-7750(05)31003-2

Hickson, L., & Khemka, I. (2013). Problem solving and decision making. In M. Wehmeyer (Ed.), *The Oxford handbook of positive psychology and disability*. Oxford University Press. https://doi.org/10.1093/oxfordhb/9780195398786.001.0001

Individuals with Disabilities Education Improvement Act of 2004, 20 U.S.C. ξ 1440 (2004).

Jameson, J. M., Riesen, T., Polychronis, S., Trader, B., Mizner, S., Martinis, J., & Hoyle, D. (2015). Guardianship and the potential of supported decision making with individuals with disabilities. *Research and Practice for Persons with Severe Disabilities, 40*(1), 36–51. https://doi.org/10.1177/1540796915586189

Johnson, D. R., Thurlow, M. L., Wu, Y., Lavelle, J. M., & Davenport, E. C. (2020). IEP/Transition planning participation among students with the most significant cognitive disabilities: Findings from NLTS 2012. *Career Development and Transition for Exceptional Individuals, 43*(4), 226–239. https://doi.org/10.1177/2165143420952050

Kelley, K. R., Bartholomew, A., & Test, D. W. (2011). Effects of the Self-Directed IEP delivered using computer-assisted instruction on student participation in educational planning meetings. *Remedial and Special Education, 34*, 67–77. https://doi.org/10.1177/0741932511415864

Khemka, I., Hickson, L., & Mallory, S. B. (2016). Evaluation of a decision making curriculum for teaching adolescents with disabilities to resist negative peer pressure. *Journal of Autism and Developmental Disorders, 46*, 2372–2384. https://doi.org/10.1007/s10803-016-2770-0

Khemka, I., Hickson, L., & Reynolds, G. (2005). Evaluation of a decision making curriculum designed to empower women with mental retardation to resist abuse. *American Journal on Mental Retardation, 110*(3), 193–204. https://doi.org/10.1352/0895-8017(2005)110<193:EOADCD>2.0.CO;2

Lachapelle, Y., Wehmeyer, M. L., Haelewyck, M.-C., Courbois, Y., Keith, K. D., Schalock, R., Verdugo, M. A., & Walsh, P. N. (2005). The relationship between quality of life and self-determination: An international study. *Journal of Intellectual Disability Research, 49*(10), 740–744. https://doi.org/10.1111/j.1365-2788.2005.00743.x

Larson, S., Eschenbacher, H., Anderson, L., Pettingell, S., Hewiit, A., Sowers, M., Fay, M. L., Taylor, B., & Agosta, J. (2017). *In-home and residential long-term supports and services for persons with intellectual or developmental disabilities: Status and trends through 2014*. University of Minnesota, Research and Training Center on Community Living, Institute on Community Integration. https://risp.umn.edu/publications

Marcolin, S., & Abraham, A. (2006). Financial literacy research: Current literature and future opportunities. In P. Basu, G. O'Neill, & A. Travaglione (Eds.), *Proceedings of the 3rd international conference on contemporary business*. Faculty of Commerce, Charles Stuart University. https://ro.uow.edu.au/commpapers/223/

Martin, J. E., & Marshall, L. H. (1995). ChoiceMaker: A comprehensive self-determination transition program. *Intervention in School and Clinic, 30*(3), 147–156. https://doi.org/10.1177/105345129503000304

Martin, J. E., Marshall, L. H., & DePry, R. L. (2008). Participatory decision making: Innovative practices that increase student self-determination. In R. W. Flexer, T. J. Simmons, P. Luft, & R. M. Baer (Eds.), *Transition planning for secondary students with disabilities* (3rd ed.). Merrill Prentice Hall.

Martin, J. E., Marshall, L. H., Maxson, L. M., & Jerman, P. L. (1996). *The self-directed IEP*. Sopris West.

Martin, J. E., Marshall, L. H., & Sale, P. (2004). A 3-year study of middle, junior high, and high school IEP meetings. *Exceptional Children, 70*, 285–297. https://doi.org/10.1177/001440290407000302

Martin, J. E., Van Dycke, J., D'Ottavio, M., & Nickerson, K. (2007). The student-directed summary of performance: Increasing student and family involvement in the transition planning

process. *Career Development and Transition for Exceptional Individuals, 30*, 13–26. https://doi.org/10.1177/08857288070300010101
Martin, J. E., Van Dyke, J. L., Christensen, W. R., Greene, B. A., Gardner, J. E., & Lovvett, D. L. (2006). Increasing student participation in IEP meetings: Establishing the self-directed IEP as an evidence-based practice. *Exceptional Children, 72*, 299–316. https://doi.org/10.1177/001440290607200303
Martorell, A., Gutierrez-Recacha, P., Pereda, A., & Ayuso-Mateos, J. L. (2008). Identification of personal factors that determine work outcome for adults with intellectual disability. *Journal of Intellectual Disability Research, 52*(12), 1091–1101. https://doi.org/10.1111/j.1365-2788.2008.01098.x
Mason, C. Y., McGahee-Kovic, M., Johnson, L., & Stillerman, S. (2002). Implementing student-led IEPs: Student participation and student and teacher reactions. *Career Development for Exceptional Individuals, 25*(2), 171–192. https://doi.org/10.1177/088572880202500206
Mazzotti, V. L., Kelley, K. R., & Coco, C. M. (2015). Effects of self-directed summary of performance on postsecondary education students' participation in person-centered planning meetings. *The Journal of Special Education, 48*(4), 243–255. https://doi.org/10.1177/0022466913483575
Mazzotti, V. L., Rowe, D. A., Kelley, K. R., Test, D. W., Fowler, C. H., Kohler, P. D., & Kortering, L. J. (2009). Linking transition assessment and postsecondary goals: Key elements in the secondary transition planning process. *Teaching Exceptional Children, 42*(2), 44–51. https://doi.org/10.1177/004005990904200205
Millar, D., & Renzaglia, A. (2002). Factors affecting guardianship practices for young adults with disabilities. *Exceptional Children, 68*, 465–484. https://doi.org/10.1177/001440290206800404
Miller, K. D., Schleien, S. J., White, A. L., & Harrington, L. (2018). "Letting Go:" Parent perspectives on the outcomes of an inclusive postsecondary education experience for students with developmental disabilities. *Journal of Postsecondary Education and Disability, 31*(3), 267–285.
Miner, C. A., & Bates, P. E. (1997). The effect of person centered planning activities on the IEP/transition planning process. *Education and Training in Autism and Developmental Disabilities, 32*(2), 105–112.
Morris, M., Rodriguez, C., & Blanck, P. (2016). ABLE accounts: A down payment on freedom. *Inclusion, 4*(1), 21–29. https://doi.org/10.1352/2326-6988-4.1.21
Neely-Barnes, S., Graff, J. C., Marcenko, M., & Weber, L. (2008). Family decision making: Benefits to persons with developmental disabilities and their family members. *Intellectual and Developmental Disabilities, 46*(2), 93–105. https://doi.org/10.1352/0047-6765(2008)46[93:FDMBTP]2.0.CO;2
Newman, L., Wagner, M., Cameto, R., & Knokey, A. -M. (2009). The post-high school outcomes of youth with disabilities up to 4 years after high school: A report from the National Longitudinal Transition Study-2 (NLTS2). NCSER 2009–3017. National Center for Special Education Research. https://files.eric.ed.gov/fulltext/ED505448.pdf
Nonnemacher, S. L., & Bambara, L. M. (2011). "I'm supposed to be in charge": Self-advocates perspectives on their self-determination support needs. *Intellectual and Developmental Disabilities, 49*(5), 327–240. https://doi.org/10.1352/1934-9556-49.5.327
PACER. (2015). *Prepare your child for age of majority and transfer of rights*. Retrieved May 14, 2020 from https://www.pacer.org/transition/resource-library/publications/NPC-19.pdf
Payne-Christiansen, E. M., & Sitlington, P. L. (2008). Guardianship: Its role in the transition process for students with developmental disabilities. *Education and Training in Developmental Disabilities, 43*(1), 3–19. https://www.jstor.org/stable/23879740
Powers, K., Gil-Kashiwabara, E., Geenen, S., Powers, L., Balandran, J., & Palmer, C. (2005). Mandates and effective transition planning practices reflected in IEPs. *Career Development for Exceptional Individuals, 28*, 47–59. https://doi.org/10.1177/08857288050280010701
Powers, L. E., Geenen, S., Powers, J., Pommier-Satya, S., Turner, A., Dalton, L. D., Drummond, D., & Swank, P. (2012). My life: Effects of a longitudinal, randomized study of self-determination enhancement on the transition outcomes of youth in foster care and special education. *Children and Youth Services Review, 34*, 2179–2187. https://doi.org/10.1016/j.childyouth.2012.07.018

Powers, L. E., Turner, A., Westwood, D., Matuszewski, J., Wilson, R., & Phillips, A. (2001). TAKE CHARGE for the Future: A controlled field-test of a model to promote student involvement in transition planning. *Career Development for Exceptional Individuals, 24*, 88–103. https://doi.org/10.1177/088572880102400107

Ratti, V., Hassiotis, A., Crabtree, J., Deb, S., Gallagher, P., & Unwin, G. (2016). The effectiveness of person-centered planning for people with intellectual disabilities: A systematic review. *Research in Developmental Disabilities, 57*, 63–84. https://doi.org/10.1016/j.ridd.2016.06.015

Reynolds, M. C., Palmer, S. B., & Gotto, G. S. (2018). Reconceptualizing natural supports for people with disabilities and their families. *International Review of Research in Developmental Disabilities, 54*, 177–209. https://doi.org/10.1016/bs.irrdd.2018.07.006

Robertson, J., Emerson, E., Hatton, C., Elliott, J., McIntosh, B., Swift, P., Krinjen-Kemp, E., Towers, C., Romeo, R., Knapp, M., Sanderson, H., Routledge, M., Oakes, P., Joyce, T., & Felce, D. (2006). Longitudinal analysis of the impact and cost of person centered planning for people with intellectual disabilities in England. *American Journal on Mental Retardation, 111*, 400–416. https://doi.org/10.1352/0895-8017(2006)111[400:LAOTIA]2.0.CO;2

Rood, C. E., Kanter, A., & Causton, J. (2015). Presumption of incompetence: The systematic assignment of guardianship within the transition process. *Research and Practice for Persons with Severe Disabilities, 39*(4), 319–328. https://doi.org/10.1177/1540796915571005

Sanderson, H., Thompson, J., & Kilbane, J. (2006). The emergence of person centered planning as evidence based practice. *Journal of Integrated Care, 14*(2), 18–25. https://doi.org/10.1108/14769018200600014

Schalock, R. L., Luckasson, R., & Tasse, M. J. (2021). *Intellectual disability: Definition, diagnosis, classification, and systems of supports* (12th ed.). American Association on Intellectual and Developmental Disabilities.

Sanderson, K. A., & Goldman, S. E. (2020). A systematic review and meta-analysis of interventions used to increase student IEP meeting participation. *Career Development and Transition for Exceptional Individuals, 43*(3), 157–168. https://doi.org/10.1177/2165143420922552

Shogren, K. A., & Plotner, A. J. (2012). Transition planning for students with intellectual disability, autism, or other disabilities: Data from the national longitudinal transition study-2. *Intellectual and Developmental Disabilities, 50*, 16–30. https://doi.org/10.1352/1934-9556-50.1.16

Shogren, K. A., Wehmeyer, M. L., Lassmann, H., & Forber-Pratt, A. J. (2017). Supported decision making: A synthesis of the literature across intellectual disability, mental health, and aging. *Education and Training in Autism and Developmental Disabilities, 52*(2), 144–157.

Shogren, K. A., Wehmeyer, M. L., Palmer, S. B., Rifenbark, G. G., & Little, T. B. (2015). Relationships between self-determination and postschool outcomes for students with disabilities. *The Journal of Special Education, 48*(4), 256–267. https://doi.org/10.1177/0022466913489733

Suto, W. M. I., Clare, I. C. H., Holland, A. J., & Watson, P. J. (2005). The relationships among three factors affecting the financial decision making abilities of adults with mild intellectual disabilities. *Journal of Intellectual Disability Research, 49*, 210–217. https://doi.org/10.1111/j.1365-2788.2005.00647.x

Test, D. W., Mason, C., Hughes, C., Konrad, M., Neale, M., & Wood, W. M. (2004). Student involvement in individualized education program meetings. *Exceptional Children, 70*, 391–412. https://doi.org/10.1177/001440290407000401

Test, D. W., Mazzotti, V. L., Mustian, A. L., Fowler, C. H., Kortering, L., & Kohler, P. (2009). Evidence-based secondary transition predictors for improving postschool outcomes for students with disabilities. *Career Development for Exceptional Individuals, 32*(3), 160–181. https://doi.org/10.1177/0885728809346960

Test, D. W., & Neale, M. (2004). Using *The Self-Advocacy Strategy* to increase middle graders' IEP participation. *Journal of Behavioral Education, 13*, 135–145. https://doi.org/10.1023/B:JOBE.0000023660.21195.c2

ThinkCollege. (2020). *College search: Find the college that is right for you.* Retrieved from: https://thinkcollege.net/college-search

Thoma, C., Rogan, P., & Baker, S. (2001). Student involvement in transition planning: Unheard voices. *Education and Training in Mental Retardation and Developmental Disabilities, 36*, 16–29. https://www.jstor.org/stable/24481611

Van Laarhoven-Myers, T. E., Van Laarhoven, T. R., Smith, T. J., Johnson, H., & Olson, J. (2016). Promoting self-determination and transition planning using technology: Student and parent perspectives. *Career Development and Transition for Exceptional Individuals, 39*(2), 99–110. https://doi.org/10.1177/2165143414552518

Van Reusen, A. K., Bos, C. S., Schumaker, J. B., & Deshler, D. D. (1994). The self-advocacy strategy for education and transition planning. Edge Enterprises.

Van Reusen, A. K., Deshler, D. D., & Schumaker, J. B. (1989). Effects of a student participation strategy in facilitating the involvement of adolescents with learning disabilities in individualized education program planning process. *Learning Disabilities, 1*, 23–34.

Wagner, M., Newman, L., Cameto, R., Javitz, H., & Valdes, K. (2012). A national picture of parent and youth participation in IEP and transition planning meetings. *Journal of Disability Policy Studies, 23*, 140–155. https://doi.org/10.1177/1044207311425384

Wehman, P. (2011). *Essentials of transition planning*. Brookes Publishing.

Wehman, P., Sima, A. P., Ketchum, J., West, M. D., Chan, F., & Leuking, R. (2015). Predictors of successful transition from school to employment for youth with disabilities. *Journal of Occupational Rehabilitation, 25*, 323–334. https://doi.org/10.1007/s10926-014-9541-6

Wehmeyer, M. (2005). Self-determination and individuals with severe disabilities: Re-examining meanings and misinterpretations. *Research and Practice in Severe Disabilities, 30*, 113–120. https://doi.org/10.2511/rpsd.30.3.113

Wehmeyer, M., Kelchner, K., & Richards, S. (1996). Essential characteristics of self-determined behavior of individuals with mental retardation. *American Journal on Mental Retardation, 100*(6), 632–642.

Wehmeyer, M. L., Garner, N., Yeager, D., Lawrence, M., & Davis, A. K. (2006). Infusing self-determination into 18-21 services for students with intellectual or developmental disabilities: A multi-stage, multiple component model. *Education and Training in Developmental Disabilities, 41*, 3–13. https://www.jstor.org/stable/23879862

Wehmeyer, M. L., Lawrence, M., Kelchner, K., Palmer, S. B., Garner, N., & Soukup, J. H. (2004). *Whose future is it anyway? A student directed transition planning process*. Kansas University Center on Developmental Disabilities.

Wehmeyer, M. L., & Palmer, S. B. (2003). Adult outcomes for students with cognitive disabilities three-years after high school: The impact of self-determination. *Education and Training in Developmental Disabilities, 38*(2), 131–144. https://www.jstor.org/stable/23879591

Wehmeyer, M. L., Palmer, S. B., Soukup, J. H., Garner, N. W., & Lawrence, M. (2007). Self-determination and student transition planning knowledge and skills: Predicting involvement. *Exceptionality, 15*(1), 31–44. https://doi.org/10.1080/09362830709336924

Chapter 17
Financial Decision Making and People with Intellectual Disabilities

Glynis H. Murphy and Isabel C. H. Clare

Introduction

In previous eras, in many countries, people with intellectual disabilities (ID), i.e. those with significant deficits in intellectual skills and adaptive behaviours, were not expected to make their own decisions. There was often no suggestion that they had the right to autonomy, and the attitudes of other people towards them were frequently paternalistic and overbearing, perhaps best illustrated by the fact that they were often referred to as *boys* and *girls*, even when they were adults. Moreover, they did not receive the same education as typically developing children did, and their lives were often very different from other people's. In the UK, for example, universal basic education did not include those with ID until the 1970 Education Act in England and Wales (and this came even later in the rest of the UK). Instead, as children, they attended *training centres* where the emphasis was on gaining basic self-care skills, rather than literacy and numeracy. Even as adults, people with ID were not expected to understand numbers, money or finances more generally, and they were usually not given any autonomy over their money. Many adults with ID lived in institutions until the late 1980s and 1990s. In England, for example, over 50,000 people with ID were living in long-stay hospital settings in 1976, reducing to fewer than 10,000 in 1995 and just over 3500 by 2001 (Emerson, 2004). They received little money, even when they undertook chores that contributed to the life of the institution. Moreover, there was usually nothing more than a tuck shop in the

G. H. Murphy (✉)
Tizard Centre, University of Kent, Canterbury, Kent, UK
e-mail: G.H.Murphy@kent.ac.uk

I. C. H. Clare
NIHR ARC East of England & Department of Psychiatry, University of Cambridge, Cambridge, UK

I. Khemka, L. Hickson (eds.), *Decision Making by Individuals with Intellectual and Developmental Disabilities*, Positive Psychology and Disability Series,
https://doi.org/10.1007/978-3-030-74675-9_17

institutions, and they very rarely visited the community or went into community-based shops, so their experiences of finances were extremely limited.

Nowadays, however, children with ID receive a far better education and are often (at least partially) integrated into mainstream settings, and they may gain some literacy and numeracy skills. As adults, people with ID are gaining experience with handling money for basic shopping, and some of them earn money. Nevertheless, members of this group often struggle with budgeting and are likely to need support in relation to complex finances, such as saving for holidays or other large items of expenditure.

People with ID are rarely wealthy in their own right, and ID is often associated with high levels of deprivation and poverty (Emerson, 2007). In fact, although a minority of adults are now able to obtain employment, they tend to earn low wages and sometimes earn no wages at all (Ridley & Hunter, 2012). So while, as vulnerable adults, people with ID may be at risk of financial abuse, they are rarely targeted by fraudsters on a large scale. Nevertheless, financial abuse sometimes does occur, and even where it is on a small scale, it can be very frightening and inexplicable for people with ID.

Financial Abuse

Financial abuse has been defined as "theft, fraud, exploitation, pressure in connection with wills, property or inheritance or financial transactions, or the misuse or misappropriation of property, possessions or benefits" (Department of Health, 2000, Section 2.7, p. 9). Such abuse is recognised as a common social problem for some people, and it is thought to occur at a prevalence rate of between 0.7% and 14.4% for older people (Davies et al., 2013; Lichtenberg et al., 2015). It is considered to be much more likely to occur when a person has lost their capacity to make financial decisions, for example, as a result of dementia (Davies et al., 2013), and it has also been recognised as more likely to occur when combined with social factors, such as loneliness, social isolation and emotional vulnerability (Fenge & Lee, 2018), conditions which could also apply to those with ID.

Precise figures for financial abuse of people with ID, however, are hard to establish, especially as any such abuse is likely to be hidden in domestic settings. Even in studies of abuse against those with ID, financial abuse is often not mentioned. However, several studies have examined the allegations of abuse against vulnerable groups, and some of these studies provide a guide as to the prevalence of financial abuse. For example, Brown and Stein (1998) found that amongst referrals for adult protection to two Social Services departments in England, 36% of referrals were for the protection of older adults, 34% for people with ID, 16% for those with mental health needs and 14% for people with physical disabilities (those with autistic spectrum conditions were not mentioned). Given that there are numerically far more *older people* in the population than people with ID, it was clear that adult protection issues had been identified in *proportionately* far more people with ID than in any other vulnerable group. However, it was apparent that the types of abuse varied with the vulnerable group: it seemed that financial abuse was much more common amongst older people, but sexual abuse was more common for people with ID. Precise figures were hard to gauge, as the exact incidence of adult protection referrals differed considerably between the two Boroughs, and this was confirmed in a subsequent larger study of 10 areas (Brown & Stein, 2000). The Department of Health's (2000) *No Secrets* guidance, which followed, was in part an attempt to provide a uniform set of definitions and procedures for adult protection to ensure equity across different parts of the country. Thus:

- Abuse was defined as a *violation of an individual's human and civil rights by another person or persons*, taking any of a number of forms: physical, sexual, psychological, financial, discrimination and persistent neglect.
- *Vulnerable adult* was defined as a person who is 'by reason of mental or other disability, age or illness unable to take care of him or herself, or unable to protect him or herself against significant harm or exploitation'.

A later study by Mansell et al. (2009), in two areas of S.E. England, examining over 6000 adult protection referrals, again found some variation between areas. They reported similar figures to those of Brown and Stein (1998) for the percentage of adult protection referrals concerning older people (48%) and people with ID (32%). Again abuse of people with autism spectrum conditions was not mentioned. Of the types of abuse reported, physical abuse and multiple types of abuse were most common (at 24% and 30%, respectively), while approximately 15% concerned financial abuse, and this was more common amongst older people than amongst those with ID. More recent figures from NHS Digital have confirmed these figures: in 2014–2015, 16% of safeguarding referrals overall were for financial abuse, averaged across all care groups, and the equivalent figure was 17% in 2015–2016 (Dalley et al., 2017).

An exploration of referrals relating to people with ID alone (taken from the same study as Mansell et al., 2009) showed an incidence rate of about 20 referrals for adult protection per 100,000 in the general population per year, about 7% of which related to financial abuse (Beadle-Brown et al., 2010). Of course, not all referrals for adult protection proceed to a confirmed outcome, and in this study, overall for people with ID, 41% of cases were confirmed, 21% discounted and 35% recorded as having insufficient evidence. Extremely few were prosecuted by the police.

Decision-Making Capacity: The Legal Basis

Psychological studies of thinking and decision making have suggested that people in the general population make decisions in a variety of ways. Some categorisations of decision-making contrast the rational (or reflective) style with the intuitive style (see Epstein et al., 1996); others describe a number of different decisional styles, such as rational, intuitive, dependent, avoidant and spontaneous decision making (Scott & Bruce, 1995).

Legal views of decision making, on the other hand, tend to assume a rational style, and in recent years, many jurisdictions have sought to ensure autonomous decision making in vulnerable groups. The associated legislation in the various jurisdictions differs somewhat in the exact definitions of decision-making capacity (see Davidson et al., 2016, for some examples). In most cases, however, jurisdictions allow substitute decision making for those who lack the capacity to make their own decisions.

In England and Wales, for example, the Mental Capacity Act (MCA) 2005, which came into force in 2007, was based on five principles for adults (people aged 16 years or over): a presumption of capacity, support for individuals to make their own decisions, the right to make unwise decisions, the *best interests* principle for decisions made on behalf of someone without capacity and the duty to select the less restrictive option. Thus the MCA approach was that those with ID, who might not previously have been permitted to decide for themselves, but who had capacity (to make the specific decision under consideration), could now make their own health and welfare decisions. It was for others to show that they lacked capacity, based on the balance of probabilities. The MCA established a two-stage test for judging mental capacity:

- Stage 1 required the determination of whether the person had an 'impairment or disturbance in the mind or brain' which might affect the ability to make a decision (the so-called *diagnostic* test).
- Then, if relevant, stage 2 required the determination of whether the person could (a) understand information relevant to the decision to be made, (b) retain the information, (c) use or weigh the information and (d) communicate a decision using any means.

This second stage was essentially a functional test of capacity (Murphy & Clare, 2003), and it was similar to that used elsewhere (e.g. in the USA, see below). Its emphasis on the *decision to be made* implied that capacity should not be considered in a generic way, but should be assessed for the specific decision at issue.

The MCA 2005 of England and Wales (Scotland has a different legislation, currently under review) has been criticised by the United Nations in relation to their CRPD, on the grounds that (1) the inclusion of a *diagnostic test* resulted in discrimination against people with disabilities and (2) that it fell short of supported decision making (see https://www.5sah.co.uk/knowledge-hub/articles/2015-11-27/a-gilded-cage-is-a-cage-no-less-the-mental-capacity-act-interfaced-with-the-un-convention-on-human-rights-of-disabled-people-and-the-eu-convention-on-human-rights). However, the House of Lords review in 2014 concluded that the Act was generally robust, though even after 7 years of implementation, it was not as widely understood and implemented as it might have been (see https://publications.parliament.uk/pa/ld201314/ldselect/ldmentalcap/139/13902.htm). Only one part was subjected to major criticism: the Deprivation of Liberty Safeguards (DOLS), a late addition to the MCA and not relevant to financial capacity (DOLS were revised under the

Mental Capacity (Amendment) Act 2019 and are now termed the Liberty Protection Safeguards; they will come into force in April 2022). The House of Lords review concluded that authorities often remained paternalistic and/or risk averse, with capacity assessments frequently not being carried out at all and substitute decision making not being well embedded in professional practice. Research studies have also asserted that mistakes are very common in the way that practitioners assess mental capacity (Wade, 2019). According to Jayes et al. (2020), judges have also criticised practitioners on the grounds that they do not always use a functional test, and/or they assess generic capacity, rather than that required for the specific decision in question, and/or they fail to provide sufficient information about or support to the person being assessed.

The criteria for considering a person to have the mental capacity to make a specific decision vary in different jurisdictions, though many include components similar to those in England and Wales. In the USA these components are usually stated as 'understanding information relevant to the decision; appreciating the information (applying the information to one's own situation); using the information in reasoning; and expressing a consistent choice' (Dunn et al., 2006). The criteria in England and Wales (see above) differ slightly in that they include being able to retain the information relevant to the decision, but do not separately list appreciating that the information applies to oneself (this is considered to be subsumed under *understanding the information*).

A number of instruments for assessing mental capacity have been developed. Dunn et al. (2006), for example, reviewed 15 instruments for assessing capacity to make treatment decisions and 10 for assessing capacity to make a decision to take part in research (2 of the total of 23 instruments were designed to do both). Most had at least some evidence of reliability and validity (such as inter-rater reliability: Cronbach's alpha; comparisons of scores from different groups). About half of the tools were structured clinical assessment tools, such as the MacCAT-T (Grisso et al., 1997), designed to assess capacity in relation to the specific issue in question, whereas the other half used pre-set vignettes. Dunn et al. found only one that was designed for people with ID (Cea & Fisher, 2003). In fact, others for people with ID do exist (see Arscott et al., 1999), but certainly there are far fewer for this care group.

Financial Decision Making and Capacity

Financial capacity has been defined in a number of ways. Two examples of recent definitions are as follows:

- The capacity to manage money and financial assets in ways that meet a person's needs and which are consistent with his/her values and self-interest (Widera et al., 2011, p. 698)
- The ability to manage one's financial affairs in a manner that is consistent with self-interest and personal values (Marson & Hebert, 2008)

A number of instruments have been designed to measure the capacity to make financial decisions, and, according to Pinsker et al. (2010), the methods of assessment include three basic types: clinical interviews, standardised neuropsychological tests and performance-based assessments. In clinical interviews, the person themselves or their family members are asked to report on their financial capacity, but these strategies have disadvantages, in that the person themselves may not realise (or want to admit to) their difficulties, while family members may not always recall difficulties and may be influenced by biases and burdens of care (Pinsker et al., 2010). Neuropsychological assessments on the other hand tend to assess correlates of financial capacity, such as attention, language, memory and executive functioning, which may be correlates of financial capacity but are not direct measures. Lastly, performance-based tests take a more direct approach to assessing ability to manage finances. They may concentrate on simple activities like adding, subtracting and counting coins, but may also cover complex tasks such as the ability to make financial investments. Some are brief measures embedded in longer instruments designed for other purposes, such as the money management subscale of the Independent Living Scales (Loeb, 2003), while others focus only on financial capacity. None of the tools found by Pinsker et al. were designed for people with ID.

In two recent reviews by Sudo and Laks (2017) and Ghesquiere et al. (2019) of structured assessments for evaluating capacity to make financial decisions, 12 different instruments were found altogether, mostly from the USA. These instruments are summarised in Table 17.1. All were developed for older adults in the community or for people with dementia in supported settings, though the FCAI also considered adults with acquired brain injury (n = 36), schizophrenia (n = 29), those with ID (n = 32), as well as those with dementia (n = 22). The majority of assessment instruments consisted of tasks related to financial knowledge and management, though some also considered vulnerability to undue influence and obligations to others. A few included a clinical interview with informants, as well as the main participant, for information (e.g. Cramer et al., 2004). Some can be completed in only 15 minutes; others took 90 minutes. Most had reliability and validity data, though the extent of these varied (see Ghesquiere et al., 2019; Sudo & Laks, 2017, for details).

Since then a new measure has emerged from Italy, the Numerical Activities of Daily Living – Financial (NADL-F) test (Arcara et al., 2019). The measure focuses on seven different, relatively simple, financial tasks (e.g. counting coins, purchasing from a shop, reading bills, understanding banking terms and cheques), and it excluded estate management. It also entailed an interview, asking the participant if they had any difficulty in the tasks they had just done. Comparisons between the general population and those with mild cognitive impairment showed differences on all tasks, and it appeared from the interview that people were not aware of their own difficulties.

Table 17.1 Instruments evaluating capacity to make financial decisions in older people

Researchers	Tool	Participants	Comments and details
Marson et al. (2000)	Financial Capacity Instrument (FCI)	Older people with dementia	Examines declarative knowledge, procedural knowledge and judgement, using a series of 18 tasks (e.g. basic monetary skills, financial knowledge, cash transactions, chequebook management, bank statement management). These are subsumed under 9 domains, producing 2 global scores
Gerstenecker et al. (2015)	Financial Capacity Instrument – Short form (FCI-SF)	Older people with early dementia	37 items covering 4 domains (coins, chequebooks, bank accounts, conceptual knowledge and problem-solving)
Bassett (1999)	Financial Competency Questions (FCQ)	Older people with dementia	Financial relationships and legal obligations
Black et al. (2007)	Financial Assessment & Capacity Tool (FACT)	Geriatric psychiatry patients	46 items covering 9 domains
Kershaw & Webber (2008)	Financial Competence Assessment Inventory (FCAI)	Adults with cognitive impairments: acquired brain injury, schizophrenia, dementia and intellectual disabilities	Approx. 40 items in 6 subscales (everyday finance, financial judgement, estate management, cognitive functioning, debt management, support resources)
Edelstein (1999)	Hopemont Capacity Assessment Interview, finance section (HCAI)	Older adults with cognitive impairments	30 items related to three vignettes about finances. Maps onto understanding, appreciation, rational consideration and choice
Lichtenberg et al. (2015)	Lichtenberg Financial Decision Rating Scale (LFDRS)	Older adults	61 items in 4 domains (financial situation awareness, psychological vulnerability, financial transactions, undue influence)
Lichtenberg et al. (2016)	Lichtenberg Financial Decision Screening Scale (LFDSS)	Older adults	10 items in two domains (intellectual functioning and susceptibility to undue influence)
Cramer et al. (2004)	Measure Assessing Awareness of Financial Skills (MAFS)	Older adults	3 separate questionnaires (34 self-report items for participant, 34 items for informant and 6 functional tasks for participant)

(continued)

Table 17.1 (continued)

Researchers	Tool	Participants	Comments and details
Wadley et al. (2003)	Prior Financial Capacity Form (PFCF) and Current Financial Capacity Form (CFCF)	Older adults with dementia	Covers similar domains to FCI; the PCFC is for carers to measure finance capacity before dementia; the CFCF is for the person with dementia to measure current financial competence
Darzins et al. (2000)	Property and Finance Capacity Assessment (PFCA)	Older adults	4 domains: knowledge of assets debts, income and expenses; obligations to family; understanding choices; appreciating consequences
Marson et al. (2009)	Semi-structured Clinical Interview for Financial Capacity (SCIFC)	Older adults with cognitive impairment	8 items in 8 domains similar to the FCI

Financial Decision Making and Capacity in People with ID

Extremely few studies of everyday financial decision making have been undertaken with people with ID, and there are none with people with autism spectrum conditions (though some studies using laboratory tasks have considered how people with autism spectrum conditions make other decisions, see, e.g. Farmer et al., 2017; Vella et al., 2017). The only instrument examining financial decision making in the reviews above that included people with ID was the FCAI (Kershaw & Webber, 2008), and even then they were not the focus of the study. None included people with autism spectrum conditions.

However, Suto and her colleagues in Cambridge (UK) have completed a series of studies of financial abilities (and the underlying skills necessary), specifically focused on people with ID. In the first study, completed before the MCA of 2005, Suto et al. (2002) investigated the records relating to Enduring Powers of Attorney at the Court of Protection (the body responsible for overseeing the management, by others, of a person's finances). They found that the assessments of financial capacity were frequently completed by a general practitioner (family doctor) and were often simply diagnostic, although functional assessments were recognised as preferable (Murphy & Clare, 2003).

Suto and her colleagues went on to consider the underlying skills necessary for making financial decisions (Suto et al., 2006), arguing that there was very little work on such constructs as quantity, numbers and money in people with intellectual disabilities. They developed five tasks, based on the order in which typically developing children acquire an understanding of quantity and numbers. The tasks were as follows (see Suto et al., 2006, for details):

1. *Ordering quantities*
2. *Number familiarity*
3. *Money familiarity*

4. *Ordering numbers*
5. *Ordering money*

People with ID found that ordering quantity was the easiest task, followed by number familiarity and money familiarity. The hardest tasks were ordering numbers and money. The results suggested that people with ID acquired these skills in the same order as typically developing children.

Suto and colleagues then constructed a measure of financial capacity, consisting of five simple vignettes of increasing complexity (Suto et al., 2005a). In the first vignette, the person (*John*) was going to the supermarket and had to decide whether to buy milk or chocolate, as he had only enough money for one of these; in the second, he was buying jeans and could buy one cheap pair or one more expensive pair; in the third, the person was going to work in an outdoor job and had to decide whether to go or not, because it was raining, but he would lose his payment if he did not go; in the fourth, the person was trying to decide whether it was worth getting an old car repaired or whether it would be better to buy another; in the final task, the person was deciding about shares in a power company. Each scenario was accompanied with pictures, and there were five questions following each (the answers were scored 0, 1, 2), such that a maximum score of 50 could be gained. The questions related to the identification of the problem described, understanding the choices to be made, reasoning about the choices, appreciating who the choices would affect and communication. People with ID ($n = 30$) scored significantly less well than a group of average ability and less well than a group of people with high ability. Nevertheless, 40% of the people with ID were able to make at least one decision (i.e., scored 10/10 on that decision), while none of them could make all five decisions. Members of the general population group were all able to make one decision, but only about 20% were able to make all five. The very able group were also mostly unable to make the most complex decision (about one third scored 10/10 for the shares vignette). The most difficult of the five questions for each vignette, for the ID group, were the understanding and reasoning questions. The results suggested that financial capacity was a continuum and that not everyone in the general population could make the most difficult decisions. It emphasised the need to assess financial capacity carefully, in a decision-specific way.

In a later study, Suto et al. (2005b) examined the extent to which capacity to make financial decisions (as measured by the new instrument) correlated with intellectual ability, with understanding of financial concepts (quantity numbers, money) and experience of making everyday choices in life. All four factors were significantly correlated, but it seemed that intellectual ability affected only financial decision making through its influence on the understanding of financial concepts. Experience of everyday choices on the other hand appeared to directly affect financial decision-making abilities. The implication was that to improve the financial autonomy of people with ID, it would be important both to support them to improve their understanding of financial concepts and to allow them more opportunities for everyday decision making. Based on their studies, Suto and her colleagues produced practical guidance for supporting people with ID to make financial decisions, including the development of the underlying skills (Suto et al., 2007).

Willner et al. (2010b) argued that very little is known about the abilities of people with ID to *weigh up* information, even though this is one of the required stages in decision making in the MCA 2005 (and is equivalent to *reasoning* in other definitions). In one study, they demonstrated that people with ID did not appear to weigh up two variables when deciding whether they wanted a small reward straight away or a larger reward in a few days' time (using the laboratory task of temporal discounting). They found that people with ID sometimes responded to the task in a random way, but when they responded nonrandomly, they rarely used both sources of information (size of reward and degree of delay), unlike people without ID (Willner et al., 2010a). In a second study, using both the temporal discounting task and the measure of financial capacity devised by Suto and her colleagues (with some adjustments to the scoring), together with measures of IQ and executive functioning, Willner et al. (2010b) replicated the Suto et al. (2005a) results on the financial decision-making tool, finding very similar scores for their group of people with ID, though with slightly lower scores on the appreciation aspect. Willner et al. also found that financial decision-making abilities correlated with IQ, as did Suto et al. (at around 0.40), and they documented a significant correlation with executive functioning that they argued was more important to decision making than IQ. As Willner et al. had not measured basic numerical ability (unlike Suto and colleagues), it was not possible for them to say whether executive functioning contributed more than, or less than, numerical ability in determining financial capacity.

Case Example

Ms Lane (not her real name) was 24 years old and had mild ID. She was referred for an assessment of her capacity to make decisions, including her capacity to manage her finances.

Ms Lane had been born very prematurely (at 26 weeks) and had respiratory distress and a large intraventricular haemorrhage, along with other neurological problems, which left her with severe cerebral palsy. She had limited use of her left hand but could not stand, walk or transfer independently. She used a motorised wheelchair and could operate it with her left hand but had no sitting balance, so she needed carers to help her wash, dress and use the toilet, but she could eat and drink independently. Ms Lane was able to communicate verbally (though she was somewhat dysarthric) and could use an iPhone, for making phone calls, and an iPad. Her reading skills were relatively good, and her writing skills were adequate for short emails. She was able to go out independently to the local area, managing small roads, and could withdraw cash from an ATM locally and buy herself coffee at the local café.

A personal injury claim had been made against the hospital where Ms Lane was born, and she obtained an award of £3 million. Some of this had been used to buy her house where she lived with her family (her mother, father and two younger brothers). At first, the personal injury award was managed by Ms Lane's father, but following family arguments, Ms Lane now had a finance deputy (this is a person, often a lawyer, appointed to manage finances, called a guardian or conservator in other jurisdictions). However Ms Lane had recently argued with her finance deputy (refusing to see her or talk on the phone) and said she wanted to manage her own finances. She had also stated recently that she wanted to live alone in her house with a friend and wanted her family to move out. Her family disputed this and felt she did not understand the implications. They argued that they would be unable to buy a house themselves as they had insufficient savings. They were resentful that, after looking after Ms Lane for 24 years, she was rejecting them. Ms Lane was referred for an assessment of her financial capacity and (not considered here) her capacity to decide on her own care arrangements.

As regards her financial capacity, Ms Lane was assessed using Suto et al.'s financial assessment vignettes. Ms Lane was able to score 10/10 for each of the first three vignettes (buying milk in a supermarket, buying jeans, taking a day off work) but only scored very low scores on the other vignettes (getting a car repaired and investing in a power company). She was also assessed for her ability to understand the value of coins and notes and to calculate her change from simple purchases, which she was able to do. Moreover, Ms Lane was able to estimate what she could buy with £10 (five chocolate bars or some food, like bread, milk, butter, cheese) and £100 (a week's food for several people; or clothes like jeans, t-shirts and jumpers; or a coat). She was not sure what £1000 would buy but thought £10,000 would buy a car or a swimming pool. Ms Lane thought £100,000 would buy *a mansion*, and she thought her current house would be worth £40,000. In fact it was probably worth about £400,000, and in her part of the UK, £100,000 would not be enough even for a one-bedroom flat. Ms Lane was asked how much a holiday (in UK) would cost, and she thought about £2000 but added that if she went with her paid carers, it would cost £30,000 (a major overestimate). Ms Lane was shown some monthly bank statements, giving income and expenditures, and was asked to explain them. She was unfamiliar with them and could not do so. However, Ms Lane could understand electricity bills and water bills and understood her phone bill and her phone contract.

It was concluded that, as regards her financial capacity, Ms Lane was able to use coins and make simple purchases, up to a value of about several hundred pounds. She was not able to handle larger amounts of money without support nor did she currently understand the values of expensive items such as houses. She was able to understand utility bills, but not able to understand bank statements, though she could withdraw money from a bank.

Ms Lane had had very few experiences in life involving financial choices, apart from purchasing small items. She had not been shown bank statements before and had not discussed the values of large items of expenditure. It was recommended that at present, she did not have the financial capacity to manage her own money and affairs entirely, but she could manage small amounts of money up to the value of several hundred pounds. It was recommended that she should be provided with further financial education and support and be re-assessed in a year's time. There was also a considerable amount of family work needed, to ensure that the family understood that the personal injury claim was for Ms Lane and not for the family. It transpired that Ms Lane did have the capacity to decide on her own care, but wanted to maintain contact with her family. Clearly there would need to be considerable help and support for Ms Lane and the family to come to amicable arrangements for the future, quite apart from the financial issues at stake.

Following the assessment, Ms Lane had a budgeted amount of her own money, from her personal injury claim, to spend as she liked and a new finance deputy for larger monetary decisions. The new finance deputy lived more locally to Ms Lane and could make face-to-face visits to discuss Ms Lane's wishes. In addition, finance education classes were being sought to assist Ms Lane to understand more about her finances.

Models of Financial Capacity

At first sight, financial decision making might seem to entail a single set of abilities, but many studies have found that it is really a complex network of skills, ranging from simple numerical skills (such as counting coins, understanding quantity, adding and subtracting), through to knowledge of financial processes (how to use a chequebook, how to use a bank, how to read bank statements) and, finally, understanding of expenditure, debts, budgets, assets, interest, inflation, wills and so on. Moreover, the social skills of understanding exploitation by others, and financial abuse, as well as the concept of obligations to others, have been considered relevant in some of the financial capacity instruments for older people considered above.

Marson (2016) has suggested that there are six basic theoretical models underlying the concept of financial capacity – see Table 17.2. Arguably, several of these are extremely similar (e.g. models 2 and 3; models 4 and 5) and are exemplified by slightly different aspects of the same instruments. Nevertheless, from the table, it is clear that different instruments have rather different emphases, and they can all be criticised for different reasons. For example, Pinsker et al. (2010) have criticised the third model as being strong on cognitive factors but not sufficiently strong on social, emotional and cultural factors. Likewise, Marson (2016) has criticised the fourth model as useful only for indicating a person's capacity to make specific decisions, not for indicating a person's overall capacity.

In practice, a combination of approaches to assessing financial capacity is often required, and this is illustrated in the case example above. Moreover, after assessments of financial capacity have taken place, the person's wishes need to be paramount, with any substitute decision-maker ensuring that they understand what the person themselves wants. In fact, this comes very close to the model of supported decision making when properly executed.

Table 17.2 Marson's six models of financial capacity (Marson, 2016)

Model	Examples	Main tenets
1. Instrumental activity of daily living	Lawton & Brody (1969)	Financial capacity as an *instrumental activity of daily living*, like using the telephone, shopping, food preparation, housekeeping, laundering, use of transportation and use of medicine. Financial capacity classified as *dependent* (needing support with all financial tasks), *partially dependent* (being able to handle daily purchases, but otherwise needing support for more complex activities) or *independent* (being able to handled financial transactions at all levels)
2. Clinical model	Marson et al. (2000)	The ability to *perform financial skills* (from basic ones such as counting coins to more complex ones like reviewing bank statements or making investments) and to *make judgements* about what is in one's best interests
3. Cognitive psychological model	Marson et al. (2000)	Considers *declarative knowledge* (such as knowing the values of coins, arithmetic rules, the meaning of financial terms, such as interest rate, loan and bond) and *procedural financial knowledge* (such as calculating change when making purchases, writing cheques) and *judgement* (the ability to combine declarative and procedural financial skills to make personal decisions in line with self- or best interest)
4. Financial capacity as a decision-making capacity (as for decision making about medical treatment)	Grisso & Appelbaum (1998), Lai et al. (2008) instrument; also part of Lichtenberg et al.'s (2015) measure.	Considers the ability to understand information relevant to a financial decision, appreciate the personal effects of the possible choices, reason about the potential outcomes of the various possible choices and make a choice
5. The person-centred model	Lichtenberg et al. (2015)	Intellectual factors (from the decisional model above) are combined with contextual factors (such as likely exploitation and any undue influence) and personal values, to come to a judgement about a person's financial capacity
6. Institute of Medicine model (IOM) (USA)	IOM model was constructed to assist Social Security Administration (see Medicine & National Academies of Sciences, 2016)	Focused on 'financial capability', i.e. the real-world capability of the person to manage their finances. Assessments of financial capacity in the clinic are only part of the person's financial capability, since outside the clinic, they may be less able to exert their skills and knowledge, due, for example, to differing stresses, supports and resources

Summary and Conclusions

People with ID have only been encouraged to manage their own finances in recent decades. They are known to be sometimes subjected to financial abuse, though this is probably less frequently a risk than for older people, partly because people with ID tend to be less wealthy.

In many jurisdictions there are now formal legal arrangements for managing the money of those people with ID who do not have capacity to manage their own finances (replacing the very widespread informal arrangements of the past). However, there has been surprisingly little research on how to assess financial capacity in those with ID and even less into how to promote and support them to acquire these skills, so as to empower people with ID to exercise full autonomy. The Mental Capacity Act in England and Wales does allow and encourage support for people to make decisions (see https://www.5sah.co.uk/knowledge-hub/articles/2015-11-27/a-gilded-cage-is-a-cage-no-less-the-mental-capacity-act-interfaced-with-the-un-convention-on-human-rights-of-disabled-people-and-the-eu convention-on-human-rights). However, it falls short of full *supported decision making* as envisaged by the CRPD.

Meanwhile, for people with autistic spectrum conditions, even less is known about their decision making. In laboratory tasks, intellectually able people with autism seem to make more effortful and logical decisions than the general population (Farmer et al., 2017; Vella et al., 2017), and this seems to extend to some other everyday decisions (Luke et al., 2012), but whether it applies to their everyday financial decision making is not certain.

It is important to note that financial decision making is not the same as decision making in other areas of life. For example, it seems to require specific numerical skills, according to Suto et al. (2005b), something unlikely to be relevant to decisions about other matters, such as medical treatment. It is also clear that experience in making choices is important to financial decision making, suggesting that these are both areas for carers to be aware of and to encourage, in order to maximise the financial capacity of people with ID.

References

Arcara, G., Burgio, F., Benavides-Varela, S., Toffano, R., Gindri, P., Tonini, E., Meneghello, F., & Semenza, C. (2019). Numerical Activities of Daily Living – Financial (NADL-F): A tool for the assessment of financial capacities. *Neuropsychological Rehabilitation, 29*(7), 1062–1084.

Arscott, K., Dagnan, D., & Kroese, B. S. (1999). Assessing the ability of people with a learning disability to give informed consent to treatment. *Psychological Medicine, 29*, 1367–1375.

Bassett, S. S. (1999). Attention: Neuropsychological predictor of competency in Alzheimer's disease. *Journal of Geriatric Psychiatry and Neurology, 12*(4), 200–205.

Beadle-Brown, J., Mansell, J., Cambridge, P., Milne, A., & Whelton, B. (2010). Adult protection of people with intellectual disabilities and challenging behaviour or mental health problems. *Journal of Applied Research in Intellectual Disabilities, 23*, 573–584.

Black, E. L., Ross, T. J., Flanagan, T., Rabheru, K., & Breiter, H. J. (2007). A financial assessment & capacity test (FACT) for a psychogeriatric population: Development and concurrent validity. *Research Insights of the Regional Mental Health Care, London/St. Thomas, 4*, 1–23.

Brown, H., & Stein, J. (1998). Implementing adult protection policies in Kent and East Sussex. *Journal of Social Policy, 27*(3), 371–396.

Brown, H., & Stein, J. (2000). Monitoring adult protection referrals in 10 English local authorities. *Journal of Adult Protection, 2*(3), 19–31.

Cea, C. D., & Fisher, C. B. (2003). Health care decision making by adults with mental retardation. *Mental Retardation, 4*, 78–87.

Cramer, K., Tuokko, H. A., Mateer, C. A., & Hultsch, D. F. (2004). Measuring awareness of financial skills: Reliability and validity of a new measure. *Aging & Mental Health, 8*, 161–171.

Dalley, G., Gilhooly, M., Gilhooly, K., Levi, M., & Harries, P. (2017). Researching the financial abuse of individuals lacking mental capacity. *Journal of Adult Protection, 19*, 394–405.

Darzins, P., Molloy, D. M., & Strange, D. (Eds.). (2000). *Who can decide?: The six step capacity assessment process*. Memory Australia.

Davidson, G., Brophy, L., Campbell, J., Farrell, S. J., Gooding, P., & O'Brien, A. M. (2016). An international comparison of legal frameworks for supported and substitute decision making in mental health services. *International Journal of Law and Psychiatry, 44*, 30–40.

Davies, M. L., Gilhooly, M., Gilhooly, K. J., Harries, P. A., & Cairns, D. (2013). Factors influencing decision making by social care and health sector professionals in cases of elder financial abuse. *European Journal of Ageing, 10*, 313–323.

Department of Health. (2000). *No secrets: Guidance on developing and implementing multi-agency policies and procedures to protect vulnerable adults from abuse*. Department of Health.

Dunn, L. B., Nowrangi, M. A., Palmer, B. W., Jeste, D. V., & Saks, E. R. (2006). Assessing decisional capacity for clinical research or treatment: A review of instruments. *American Journal of Psychiatry, 163*(8), 1323–1334.

Edelstein, B. (1999). *Hopemont capacity assessment interview manual and scoring guide*. West Virginia University.

Emerson, E. (2004). Deinsititutionalisation in England. *Journal of Intellectual and Developmental Disability, 29*(1), 79–84.

Emerson, E. (2007). Poverty and people with intellectual disabilities. *Mental Retardation and Developmental Disabilities Research Reviews, 13*, 107–113.

Epstein, S., Pacini, R., Denes-Raj, V., & Heier, H. (1996). Individual differences in intuitive-experiential and analytical-rational thinking styles. *Journal of Personality and Social Psychology, 71*, 390–405.

Farmer, G. D., Baron-Cohen, S., & Skylark, W. J. (2017). People with Autism Spectrum Conditions make more consistent decisions. *Psychological Science, 28*(8), 1067–1076.

Fenge, L.-A., & Lee, S. (2018). Understanding the risks of financial scams as part of elder abuse prevention. *British Journal of Social Work, 48*, 906–923.

Gerstenecker, A., Eakin, A., Triebel, K., Martin, R., Swenson-Dravis, D., Petersen, R. C., & Marson, D. (2015). Age and education corrected older adult normative data for a short form version of the Financial Capacity Instrument. *Psychological Assessment, 28*, 737–749.

Ghesquiere, A. R., McAfee, C., & Burnett, J. (2019). Measures of financial capacity: A review. *Gerontologist, 59*, 109–129.

Grisso, T., & Appelbaum, P. (1998). *Assessing competence to consent to treatment: A guide for physicians and other health professionals*. Oxford University.

Grisso, T, Appelbaum, P. S., & Hill-Fotouhi, C. (1997). The MacCAT-T: A clinical tool to assess patients' capacities to make treatment decisions. *Psychiatric Services 11*(48), 1415–1419.

Jayes, M., Palmer, R., Enderby, P., & Sutton, A. (2020). How do health and social care professionals in England and Wales assess mental capacity? A literature review. *Disability and Rehabilitation, 42*(19), 2797–2808.

Kershaw, M. M., & Webber, L. S. (2008). Assessment of financial competence. *Psychiatry, Psychology and Law, 15*(1), 40–55.

Lai, J. M., Gill, T. M., Cooney, L. M., Bradley, E. H., Hawkins, K. A., & Karlawish, J. H. (2008). Everyday decision making ability in older persons with cognitive impairment. *American Journal of Geriatric Psychiatry, 16*(8), 693–696.

Lawton, M., & Brody, E. (1969). Assessment of older people: Self-maintaining and instrumental activities of daily living. *The Gerontologist, 9*, 179–185.

Lichtenberg, P. A., Stoltman, J., Ficker, L. J., Iris, M., & Mast, B. (2015). A person-centered approach to financial capacity assessment: Preliminary development of a new rating scale. *Clinical Gerontologist, 38*, 49–67.

Lichtenberg, P. A. P. A., Ficker, L. P., Rahman-Filipiak, A. M., Tatro, R. B., Farrell, C. M., Speir, J. J. M., & Jackman, J. D. J. M. (2016). The Lichtenberg financial decision screening scale (LFDSS): A new tool for assessing financial decision making and preventing financial exploitation. *Journal of Elder Abuse & Neglect, 28*(3), 134–151.

Loeb, P. A. (2003). *Independent living scales (ILS)*. Psychological Corporation.

Luke, L., Clare, I. C. H., Ring, H. A., Redley, M., & Watson, P. (2012). Decision making difficulties experienced by adults with autism spectrum conditions. *Autism, 16*(6), 612–621.

Mansell, J., Beadle-Brown, J., Cambridge, P., Milne, A., & Whelton, B. (2009). Adult protection: Incidence of referrals, nature and risk factors in two English local authorities. *Journal of Social Work, 9*(1), 25–40.

Marson, D. (2016). Conceptual models and guidelines for clinical assessment of financial capacity. *Archives of Clinical Neuropsychology, 31*, 541–553.

Marson, D. C., & Hebert, K. R. (2008). Financial capacity. In B. L. Cutler (Ed.), *Encyclopedia of psychology and the law* (Vol. 1, pp. 313–316). Sage Publications.

Marson, D. C., Sawrie, S. M., Snyder, S., McInturff, B., Stalvey, T., Boothe, A., & Harrell, L. E. (2000). Assessing financial capacity in patients with Alzheimer disease - A conceptual model and prototype instrument. *Archives of Neurology, 57*, 877–884.

Marson, D. C., Martin, R. C., Wadley, V., Griffith, H. R., Snyder, S., Goode, P. S., Kinney, F. C., Nicholas, A. P., Steele, T., Anderson, B., Zamrini, E., Raman, R., Bartolucci, A., & Harrell, L. E. (2009). Clinical interview assessment of financial capacity in older adults with mild cognitive impairment and Alzheimer's disease. *Journal of the American Geriatrics Society, 57*, 806–814.

Murphy, G. H., & Clare, I. C. H. (2003). Adults' capacity to make legal decisions. In D. Carson & R. Bull (Eds.), *Handbook of psychology in legal contexts* (2nd ed., pp. 31–66). Wiley.

Pinsker, D. M., Pachana, N. A., Wilson, J., Tilse, C., & Byrne, G. J. (2010). Financial capacity in older adults: A review of clinical assessment approaches and considerations. *Clinical Gerontologist, 33*, 332–346.

Ridley, J., & Hunter, S. (2012). Employment. In H. L. Atherton & D. J. Crickmore (Eds.), *Learning disabilities: Towards inclusion* (6th ed., pp. 467–484). Churchill Livingstone Elsevier.

Scott, S. G., & Bruce, R. A. (1995). Decision making style: The development and assessment of a new measure. *Educational and Psychological Measurement, 55*, 818–831.

Sudo, F. K., & Laks, J. (2017). Financial capacity in dementia: A systematic review. *Aging & Mental Health, 21*(7), 677–683.

Suto, W. M. I., Clare, I. C. H., & Holland, A. J. (2002). Substitute decision making in England and Wales: A study of the Court of Protection. *Journal of Social Welfare and Family Law, 24*, 37–54.

Suto, W. M. I., Clare, I. C. H., Holland, A. J., & Watson, P. C. (2005a). Capacity to make financial decisions among people with mild intellectual disabilities. *Journal of Intellectual Disability Research, 49*(3), 199–209.

Suto, W. M. I., Clare, I. C. H., Holland, A. J., & Watson, P. C. (2005b). The relationships among three factors affecting the financial decision making abilities of adults with mild intellectual disabilities. *Journal of Intellectual Disability Research, 49*(3), 210–217.

Suto, W. M. I., Clare, I. C. H., Holland, A. J., & Watson, P. C. (2006). Understanding of basic financial concepts among adults with mild learning disabilities. *British Journal of Clinical Psychology, 45*, 261–266.

Suto, W. M. I., Clare, I. C. H., & Holland, A. J. (2007). *Financial decision making: Guidance for supporting financial decision making by people with learning disabilities*. British Institute of Learning Disabilities.

Vella, L., Ring, H. A., Aitken, M. R. F., Watson, P. C., Presland, A., & Clare, I. C. H. (2017). Understanding self-reported difficulties in decision making by people with autism spectrum disorders. *Autism, 22*(5), 549–559.

Wade, D. T. (2019). Determining whether someone has mental capacity to make a decision: Clinical guidance based on a review of the evidence. *Clinical Rehabilitation, 33*(10), 1561–1570.

Wadley, V. G., Harrell, L. E., & Marson, D. C. (2003). Self- and informant-report of financial abilities in patients with Alzheimer's disease: Reliable and valid? *Journal of the American Geriatrics Society, 5*(11), 1621–1626.

Widera, E., Steenpass, V., Marson, D., & Sudore, R. (2011). Finances in the older patient with cognitive impairment: "He didn't want me to take over". *Journal of the American Medical Association, 305*(7), 698–706.

Willner, P., Bailey, R., Parry, R., & Dymond, S. (2010a). Performance in temporal discounting tasks by people with intellectual disabilities reveals difficulties in decision making and impulse control. *American Journal on Intellectual and Developmental Disabilities, 54*, 366–379.

Willner, P., Bailey, R., Parry, R., & Dymond, S. (2010b). Evaluation of the ability of people with intellectual disabilities to 'weigh up' information in two tests of financial reasoning. *Journal of Intellectual Disability Research, 54*(4), 380–391.260.

Chapter 18
Decision Making and Vulnerability to Maltreatment

Linda Hickson and Ishita Khemka

Introduction

Having an intellectual and/or a developmental disability (IDD) is a major risk factor for experiencing maltreatment. According to some studies, up to 90% of women with IDD and up to 86% of men with IDD experience maltreatment sometime during their lives (e.g., Hughes et al., 2011). *Maltreatment* encompasses the full range of abuse involving physical, sexual, or emotional harm to a person. During adolescence, maltreatment research tends to focus on social victimization, consisting of bullying or cyberbullying by peers, which is a common occurrence in schools (Hickson & Khemka, 2021). During adulthood, the focus shifts to domestic violence (DV), intimate partner violence (IPV), and sexual assault (SA) (Hickson & Khemka, 2016). Most troubling is the finding that between 97% and 99% of the abuse of individuals with IDD is perpetrated by persons who are known, and likely trusted, by the person with IDD (Baladerian, 1991; Stevens, 2012). Perpetrators include parents, peers, intimate partners, extended family members, caregivers, teachers, bus drivers, and other service providers.

Our own interest in the role of decision making in the maltreatment of individuals with IDD began about 25 years ago with the highly publicized Glen Ridge Trial of four young men for the SA of a young woman with intellectual disabilities (ID) (see Hickson et al., 1998). Transfixed by the 5-month trial, we became sensitized to the central role of decision making in the vulnerability of the young woman. Decision making played a similarly central role in a more recent event where a

L. Hickson
Teachers College, Columbia University, New York, NY, USA
e-mail: lh76@tc.columbia.edu

I. Khemka (✉)
St. John's University, Queens, NY, USA
e-mail: khemkai@stjohns.edu

I. Khemka, L. Hickson (eds.), *Decision Making by Individuals with Intellectual and Developmental Disabilities*, Positive Psychology and Disability Series,
https://doi.org/10.1007/978-3-030-74675-9_18

30-year-old woman with ID went out with a group of women and men whom she had thought were her friends. Over a period of several days, after humiliating her and brutalizing her, they stabbed and strangled her to death (McKinney, 2018).

An awareness of events such as these, and their all too common variations, has undergirded our focus on the independent decision-making capabilities of individuals with IDD in interpersonal situations that could jeopardize their safety. In one of our early studies on this topic, we asked adults with and without ID what a protagonist should do in a series of vignette situations where there was a threat of harm, including physical danger, involvement in stealing, and sexual assault. Adults without ID gave vigilant decision responses (i.e., responses that appeared to be in the best interest of the protagonist) 91% of the time, while adults with mild ID gave vigilant decision responses only about 50% of the time (Hickson et al., 1998). The situations were based on the real-life experiences of adults with ID living in the community, with one situation based on the initiating event in the Glen Ridge case. The vignette depicted a young woman, "Emily," who was pressured, with the lure of a date, by "Jeff," who had often made fun of her in the past, to go with him and a group of guys to his friends' house where, in the real-life situation, she was sexually assaulted. When presented with this scenario, only one of the study participants without ID said that Emily should go with Jeff. However, about two thirds of the adults with ID said that Emily should go with Jeff.

In another study (Khemka & Hickson, 2000), women and men with ID were presented with videos of decision-making scenarios that placed a key decision-maker at risk of physical, sexual, or verbal/psychological abuse. Overall, participants said that the key decision-maker should take direct prevention-focused action to stop or avoid the abuse only 45% of the time. They recommended actions that were other-dependent and involved seeking help an additional 20% of the time, but in most of the situations, immediate, independent action would have been necessary to stop the abuse. In effect, 35% of the time participants suggested decisions that did not involve stopping the abuse or reporting afterwards.

Around that same time, several studies had reported that individuals with IDD, primarily ID, manifested serious shortcomings with various aspects of decision making. These early research studies tended to assume that people followed a deliberative, stepwise model of decision making encompassing at least four steps: (1) framing the problem, (2) generating alternatives, (3) evaluating consequences, and (4) choosing a course of action (see Hickson & Khemka, 1999 for a review of these studies). The early studies generally pointed toward cognitive limitations as the source of decision-making difficulty for individuals with ID (e.g., Healey & Masterpasqua, 1992; Jenkinson & Nelms, 1994; Short & Evans, 1990; Smith, 1986). However, it soon became apparent that the sources of the difficulty were not limited to cognitive factors or to attempts to apply a stepwise process. There was mounting evidence that decision making involves the complex interplay of cognitive, motivational, and emotional components and that decision making does not always involve a planful, stepwise process (see Hickson & Khemka, 2013, 2014; Khemka & Hickson, 2017b).

It is important to note that our interest in understanding and documenting the decision-making patterns of individuals with IDD is not motivated simply by a desire to provide a detailed description of their limitations. Rather, we have long been committed, in the spirit of positive psychology, to seek ways to strengthen resilience and reduce vulnerability to maltreatment so that individuals with IDD can freely engage "in positive and satisfying interpersonal relationships and improved quality of life" (Hickson & Khemka, 2013, p. 198).

The next section briefly describes the types of assessments that we have used to study interpersonal decision making in our studies. This is followed by reviews of the research literature that links maltreatment to decision making in adolescents and adults with IDD. Risk and protective factors for maltreatment are highlighted.

Assessment of Interpersonal Decision Making

Building on the findings of the early research, we undertook a series of studies aimed at gaining a clearer understanding of the role of decision making, by adolescents and adults with various types of IDD, in interpersonal situations with a risk of victimization or abuse. To assess various aspects of how individuals with IDD approach decision making in those situations, we have typically used brief vignettes to simulate maltreatment situations likely to be encountered by individuals with IDD. Although we have not been in a position to systematically observe their real-life decision making, we have worked closely with teachers and support professionals, in focus groups and informally, to ensure that the vignette situations were drawn from the real-life experiences of individuals with IDD. Table 18.1 lists some of these studies and illustrates the range of measures that we have used in our effort to capture different dimensions of decision making in response to the hypothetical scenarios.

In all of our studies, we have focused on various aspects of independent decision making on the premise that most individuals with IDD will have to handle difficult situations on their own at various points in their lives. This focus feels all the more urgent in the domain of maltreatment because of the frequent occurrence of abuse by caregivers, intimate partners, family members, and other trusted individuals.

In the first of our decision-making studies (the one that included a vignette based on the Glen Ridge case), we measured vigilant decision responding (Hickson et al., 1998). In that study, adults with ID were asked to respond to 12 vignettes that were read aloud to them by an interviewer (adults without ID received a paper-and-pencil version of the same vignettes). Scenarios included interpersonal situations in which a protagonist was called upon to make a decision. Situations typically involved a conflict between seeking a possible social or material gain (e.g., a date) versus the advisability of taking a socially responsible or self-protective action to avoid a negative consequence. Participants were then asked what the (named) protagonist should do and why. Responses reflecting planful decision making that appeared to be in the best interest of the protagonist were classified as vigilant. Responses that were

Table 18.1 Assessments of interpersonal decision making by individuals with IDD

Scale	Reference(s)	Domains of maltreatment	Decision Question(s)
Interpersonal Decision-Making Scale (Adults) (12 vignettes)	Hickson et al. (1998)	Sexual abuse Physical abuse Verbal/psychological abuse	What should ___ do and why?
Social Interpersonal Decision-Making Video Scale (Adults) (12 vignettes)	Khemka (2000) Khemka and Hickson (2000) Hickson et al. (2008)	Sexual abuse Physical abuse Verbal/psychological abuse	Is ___ faced with a problem? What is ___'s problem? What is the best thing for ___ to do? Why is this the best choice for ___?
Self Social Interpersonal Decision-Making Scale (Adults) (6 vignettes)	Khemka (2000) Khemka et al. (2005)	Sexual abuse Physical abuse Verbal/psychological abuse	What would you do if you were ever in this situation?
Decision-Making Scale (Adults) (6 vignettes)	Hickson et al. (2015)	Sexual abuse Physical abuse Verbal/psychological abuse	What is happening in this story? What should ___ do?
Decision Making in Coercive Situations Scale (Adolescents) (12 vignettes)	Khemka et al. (2009)	Coercion with a lure Coercion with a threat General coercive tactic	What should ___ do?
Adolescent Decision-Making Scale-PEER (Adolescents) (8 vignettes)	Khemka et al. (2016)	Negative peer pressure	What is the best thing for ___ to do?
Cyberbullying Decision-Making Scale (Adolescents) (6 vignettes)	Khemka (2016)	Cyber situations of peer pressure or victimization	How do you think ____ might be feeling in this situation?

considered maladaptive included those that were hypervigilant, avoidant, complacent, or other.

In an effort to present decision-making situations that were closer to real-life situations, we used a video measure in three subsequent studies (Khemka, 2000; Khemka & Hickson, 2000; Hickson et al., 2008). This measure, the *Social Interpersonal Decision-Making Video Scale* (Khemka, 2000), contained 12 abuse scenarios (equal numbers of sexual, physical, and verbal/psychological abuse) and 12 filler items involving interpersonal conflict that could be resolved safely through compromise or negotiation. The vignettes consisted of short video clips (less than a minute) with a visual overlay. Each video was immediately followed by four questions read by an interviewer: (1) *Is ___ faced with a problem?* (2) *What is ___'s problem?* (3) *What is the best thing for ___ to do?* (4) *Why is this the best choice for ___?* Independent prevention-focused decision-making scores were based jointly on responses to questions 3 and 4 as follows: (a) independent decision making with a specific rationale (3 points), (b) independent decision making with a general rationale (2 points), independent decision making with no rationale (1 point), or no independent decision making (0 points).

Two studies included a self decision-making verbal measure consisting of 6 vignettes (Khemka, 2000; Khemka et al., 2005). The hypothetical situations included equal numbers of situations of sexual, physical, and verbal abuse. The interviewer read the vignette aloud and asked, *What would you do if you were ever in this situation?* Credit was given for each independent prevention-focused decision suggested by the participant.

The Decision Making in Coercive Situations Scale (Khemka et al., 2009), consisting of 12 verbally presented vignettes, was developed to assess the ability of adolescents with ID to handle situations involving three types of coercion – coercion with a threat, coercion with a lure, and coercion with no specific coercive tactic. After reading each vignette, the interviewer asked the following question: *What should ___do?* The 12 coercion vignettes were interspersed with 12 filler vignettes depicting noncoercive conflict situations. Responses were coded as recommending (1) independent, assertive action, (2) reporting responses, or (3) other (see Table 18.2).

An updated, verbally presented *Decision-Making Scale*, consisting of 6 abuse vignettes (2 sexual, 2 physical, and 2 verbal/psychological) was included in the pretest/posttest battery for the evaluation of ESCAPE-DD, an abuse prevention curriculum for adults with IDD (Hickson et al., 2015). Vignettes were read aloud to participants with IDD and they were asked two questions: (1) *What is happening in this story?* and (2) *What should ___ do?* Responses were scored for overall *effective decision making* (attempts to seek safety now and/or later through independent action or by seeking help) and *safe-now effective decision making* (independent actions aimed at seeking immediate safety) (see Table 18.3).

Table 18.2 Sample decision-making vignettes presented to adolescents with IDD in Khemka et al. (2009)

Coercion with a threat Lisa goes for job training in a hospital where she is learning to do laundry. One day while Lisa is working in the laundry room, Mr. Drake the janitor comes in. Suddenly, Mr. Drake tells Lisa that he will hurt her if she does not kiss him. *What should Lisa do?*
Coercion with a lure Carol goes for job training in a shoe store where she is learning to shelve boxes. One day, while Carol is working in the storage room, Mr. Jones the janitor comes in. Suddenly, Mr. Jones offers to give Carol some money if she will kiss him. *What should Carol do?*
Coercion with no specific tactic Julia goes for job training in a cafeteria where she is learning to wash dishes. One day while Julia is working in the kitchen, Mr. Smith the janitor comes in. Suddenly, Mr. Smith urges Julia to kiss him. *What should Julia do?*

Table 18.3 Sample decision-making vignettes presented to adults with IDD in Hickson et al. (2015)

Sexual abuse Alex has a job at the shoe store. Recently, Jason, a new guy, was hired to also work in the store. One day at work, Alex was buying a Coke™ from the vending machine during his lunch break. Suddenly, Jason, the new guy in the store, comes up behind Alex. Jason then grabs *Alex* from behind and touches his private parts. *What is happening in this story?* *What should Alex do?*
Verbal/psychological abuse Rosie lives with her parents. Rosie has just gotten a job in the neighborhood store. One day, when Rosie was leaving to go to work, her father gives her the finger. Rosie's father laughs loudly at Rosie and tells her, "You are too stupid to hold a job, you loser. You will be fired by the end of the week." *What is happening in this story?* *What should Rosie do?*
Physical abuse Betty and Peter have been married for a year. Betty works at the grocery store and does not get home until late at night. Peter, her husband, stays at home and is supposed to cook the meals. One night, Betty comes home tired after work and finds her husband Peter watching TV. When Betty asks Peter what is for dinner, he does not answer Betty. When Betty asks again, Peter gets very upset and picks up a dish. He throws the dish at Betty and it hits her on the side of her head. *What is happening in this story?* *What should Betty do?*

An adolescent measure was developed for an evaluation of PEER-DM, a decision-making curriculum to teach youth with IDD to resist negative peer pressure (Khemka et al., 2016). After reading each of the 12 vignettes to participants, the examiner asked the following question: *(1) What is the best thing for ___ to do?* Effective decision responses were those that involved immediate, direct attempts to confront the pressuring peer or take action that implied not going along with the negative peer pressure (e.g., *don't do it; seek help*). Additional details about some of these studies are presented, as relevant, elsewhere in this chapter. There is a clear need for continued instrument development to provide more life-like and standardized measures of decision making to be used across studies (see Petitpierre and Tabin in this volume for a discussion of these issues).

Maltreatment and Decision Making During Adolescence

Research has shown that adolescents with IDD have a heightened risk for maltreatment relative to their typically developing (TD) peers (see Blake et al., 2012; Reiter et al., 2007). Although research focuses on peer victimization during adolescence, physical and sexual abuse by family members (usually referred to as child abuse) continues into adolescence, with sexual abuse increasing until age 18.0 (see Hickson & Khemka, 2021). Peer victimization during adolescence most often consists of bullying and related forms of peer aggression or coercion, including cyberbullying using electronic media. Based on a definition set forth by Olweus (1993) in conjunction with his pioneering work in Norway, bullying is typically defined as having three components: (1) peer aggression intended to cause harm, (2) repeated behavior by the same individuals, and (3) an imbalance of power. Peer victimization can consist of various forms of physical or relational peer aggression and bullying as well as other forms of negative peer influence or coercion (see Khemka et al., 2009). Bullying and other forms of peer victimization are quite common, even for adolescents without disabilities, and can cause both immediate and long-term harm, including anxiety, depression, and suicidal ideation for both victims and bullies (Swearer et al., 2010).

Peer Aggression and Bullying

Christensen et al. (2012) compared a group of 13-year-olds with ID with a group of their TD peers. They found higher rates of bullying victimization for the adolescents with ID (62%) than for those without ID (41–42%). Rates of victimization were related to social problems and social withdrawal. Cook et al. (2010) suggested that limitations in social competence and communication skills can have an overall negative impact on the peer relations and friendships of children and youth with autism spectrum disorders (ASD) and ID, leading to peer rejection and loneliness.

In our study on decision making in coercive situations (Khemka et al., 2009), we explored the ability of adolescents with ID to make self-protective decisions when faced with three types of coercive tactics. They were asked to respond to a series of hypothetical decision-making vignette situations depicting coercion with a lure, coercion with a threat, and a similarly urgent (coercive) request, but without the specific coercive tactic of a threat or a lure. Overall, the participants suggested that the protagonist take independent prevention-focused action only about half the time. They were less likely to make decisions suggesting independent action to resist the abuse in the coercion-with-a-threat condition compared with the lure- or no-specific-coercive-tactic conditions. Under the threat condition, they were more likely to suggest reporting the abuse and requesting help than in the other two conditions, suggesting that they did not feel well-prepared to take action to handle the situations on their own (refer to Table 18.2 for samples of the decision-making vignettes used in that study).

Youth with ASD are even more vulnerable to peer victimization than their peers with other disabilities (Humphrey & Hebron, 2015). After reviewing the literature from 2002 to 2013, Schroeder et al. (2014) reported that, relative to adolescents with other special needs and those from the general population, adolescents with ASD were at greater risk for physical, verbal, and relational bullying. Zeedyk et al. (2014) conducted in-depth interviews comparing 13-year-olds with ASD and ID with their TD peers and found that the adolescents with ASD were victimized more often than those in the other two groups. Further evidence that youth with ASD may be more vulnerable to victimization, compared to peers with other disabilities, has been reported by Maïano et al. (2016). Sreckovic et al. (2014) reviewed literature published between 2002 and 2014 and concluded that adolescents with ASD were victimized at very high rates, with yearly rates between 46% and 94%. Similarly, Fisher and Taylor (2016) reported that 73% of their sample of adolescents with ASD reported that they had been victimized.

On the premise that decision-making effectiveness can play a key role in many situations of maltreatment, in one of our studies (Khemka et al., 2013), we compared the performance of adolescents with ASD with that of a group, matched for mean age and IQ, with other disabilities (including ID, LD, and speech/language disorder). Participants were asked to respond to 12 decision-making vignettes depicting negative peer pressure to engage in substance use, stealing, shirking responsibilities, and risking abduction. The adolescents with ASD produced fewer effective decision-making responses than the adolescents with other disabilities, supporting the need to consider factors other than IQ to account for the findings – possibly differences in social understanding. According to the peer interaction model proposed by Humphrey and Symes (2011), a lack of social skills can lead to a reduced frequency of positive peer interactions, conflict with peers, and rejection and isolation for youth with ASD.

Van Roekel et al. (2010) reported that adolescents with ASD who were often bullied tended to interpret bullying situations as nonbullying situations. This tendency was associated with weaker performance on theory-of-mind tasks and higher rates of bullying perpetrator behavior. It has been suggested that youth with ASD experience difficulty with responding to bullying situations because of limitations in their understanding of the dynamics of social situations and because of difficulty overcoming autism-related inflexibility (Garner & Hinton, 2010; Wainscot et al., 2008). They may also be hindered by less ability to employ an intuitive decision-making style and greater reliance on a characteristically slow, deliberative decision-making style – especially in situations with a threat of victimization where a rapid decision and response is called for (e.g., Brosnan et al., 2014; Levin et al., 2015; Luke et al., 2012).

Cyberbullying

Cyberbullying, also referred to as *cyber aggression*, can be extremely harmful, even if it happens only once, because it can be viewed by hundreds of peers and it can remain online permanently (Underwood & Ehrenreich, 2017). Because cyberbullying is a relatively new phenomenon, there have been few studies to date with adolescents with IDD.

In a study with 5th, 8th, and 11th graders without disabilities, it was found that beliefs endorsing bullying were significant predictors of cyberbullying as well as physical and verbal bullying (Williams & Guerra, 2007). Hase et al. (2015) reported that most of the middle and high school students without disabilities in their study who were victims of cyberbullying online were also victims of in-person bullying and that both were associated with negative mental health outcomes. Rice et al. (2015) reported that most cyberbullying by the middle school students in their study occurred on Facebook™ or via text messaging. Frequent victims were likely to report using the Internet for at least 3 hours a day. Experiences of cyberbullying present challenging demands for effective and timely decision making.

In a recent study (Khemka et al., submitted), we examined the decision-making preferences of 6th, 7th, and 8th grade TD girls and boys on how they thought a peer (from a vulnerable group) should handle a situation of physical, verbal, or cyberbullying. Boys were more likely than girls, and older students were more likely than younger students, to endorse decisions involving retaliatory aggression in response to physical aggression scenarios. The findings of this study raise questions about how middle school students' normative beliefs about the acceptability of various types of aggressive behavior might influence their decision making in the presence of peers and their own potential involvement as bullies or as bystanders to bullying in school environments.

Kowalski and Fedina (2011) surveyed 5th to 12th graders with Attention Deficit Hyperactivity Disorder (ADHD) and/or "Asperger syndrome." They found that 57% had been traditionally bullied and 21% had been cyberbullied during the past 2 months. Heiman and Olenik-Shemesh (2013) looked at the cyberbullying involvement of three groups of 12- to17-year-olds in Israel. They found that students with comorbid learning disabilities (LD) and ADHD or other behavioral or communication problems were victims or perpetrators of cyberbullying more often than students with only LD or no disability. Within the comorbid group, girls were more likely than boys to report having been victims of cyberbullying.

In one of the first studies to investigate cyberbullying in adolescents with IDD, Didden et al. (2009) interviewed students, most of whom had ID, who were attending a special education school in the Netherlands. They reported that students were bullied via both cell phones and the Internet and that many of the same students were both victims and bullies. Higher rates of victimization were associated with higher IQs, lower self-esteem, and higher rates of depression and distress. Rose et al. (2015) reported that, overall, secondary students with disabilities experienced more bullying, relational victimization, and online victimization than students without disabilities. More recently, Wright and Wachs (2020) reported that adolescents with IDD are more likely than adolescents without disabilities to be victims of both bullying and cyberbullying. In a pilot study, Khemka (2016) asked 19 transition-aged (18–21 years old) students with IDD how they would handle six hypothetical decision-making situations involving cyberbullying via the Internet or phone. Only 50.83% of their responses indicated effective, self-protective, independent decision making. The remaining responses involved either seeking help from someone or making no attempt to resist the bullying (see Table 18.4).

Table 18.4 Sample cyberbullying vignettes presented to adolescents in Khemka (2016)

Situation 1 One day, Larry is at the home computer with his stepbrother, Javier. Javier forces Larry to sign up for an Internet site that shows videos of people having sex. Javier warns Larry that if Larry tells anyone about this, Larry will get into trouble. *What would you do if you were ever in a situation like Larry?*
Situation 2 One day, Keira checks her Facebook page. Keira's Facebook friend, Danny, has written mean comments under Keira's picture. The comments make fun of the way Keira dresses and Danny calls her "uncool." *What would you do if you were ever in a situation like Keira?*
Situation 3 Antonia and Jared have been dating each other for a long time. Lately, Jared has been pressuring Antonia to drink beer at their friend's house, but Antonia has been saying no. When Antonia goes on Instagram, she sees that Jared has posted a picture with the words, "Antonia, a big loser. Stay far away from her." *What would you do if you were ever in a situation like Antonia?*

Adolescent Risk Factors

As noted earlier, just having IDD is associated with increased risk of maltreatment for adolescents. There is ample evidence that adolescents with IDD experience higher rates of physical, sexual, and emotional victimization than their peers without disabilities (e.g., Christensen et al., 2012; Cook et al., 2010; Reiter et al., 2007). Further, it appears that adolescents with ASD are even more vulnerable to victimization than their peers with other types of IDD (e.g., Humphrey & Hebron, 2015; Zeedyk et al., 2014).

In addition, there is ample evidence that limited decision-making skills constitute a key risk factor for the maltreatment of individuals with IDD. Limitations in this skill set surfaced as an important risk factor at the adolescent life stage in a recent lifespan review of the literature on the maltreatment of individuals with IDD (see Hickson & Khemka, 2021). These findings are consistent with our observations that decision making can play a pivotal role in many situations involving the maltreatment of adolescents with IDD. The following discussion situates decision-making limitations among other risk factors for maltreatment that have been identified in research studies with adolescents with IDD.

Farmer et al. (2015) highlighted additional risk factors for peer victimization that included individual characteristics such as anxiety, depression, social withdrawal, and passivity as well as social isolation and lower levels of social competence. Similarly, Blake et al. (2016), drawing upon a random sample of self-reported responses by 13- to 16-year-old students with disabilities from the National Longitudinal Transition Study-2 (NLTS2), found that respondents with elevated depression and anxiety and those with relationship difficulties and less social support were the most likely to be the victims of bullying. Caucasian respondents in their study were more likely than African-American or Hispanic respondents to be bullied. Reiter et al. (2007) reported that girls were more likely to be victimized than boys.

In addition to the demographic and personal risk factors that have been highlighted in the literature, numerous studies have associated shortcomings in various aspects of the decision-making process with a heightened risk of experiencing bullying or abuse. In particular, it is apparent that adolescents with IDD may encounter difficulty with each of the three basic processes involved in decision making – cognition, motivation, and emotion. Cognitive limitations can constrain memory, comprehension, risk awareness, and processing capacity and, especially, the hypothetical processing needed to consider and select the best course of action from multiple possible options (see Hickson & Khemka, 2013, 2014, 2017; Khemka & Hickson, 2017b).

Motivational goal processes also play an important role in adolescent decision making. To explore the role of self-actualizing/safety goals, we gave a goal selection inventory to 49 adolescents with IDD. Individual items represented either self-actualizing goals prioritizing personal safety (e.g., *How important is it for you to be safe?*) or other-oriented goals (*How important is it for you to go along with what others want?*) (Khemka & Hickson, 2016). Based on their scores, the adolescents were divided into a high self-actualizing goal group and a low self-actualizing goal group. The decision responses of the two groups to a set of negative peer pressure vignettes were then compared. Results indicated that the high-self-actualizing goal group produced more independent, self-protective decision responses, indicating the potential importance of goal orientation in the decisions of youth with IDD. Motivation-related difficulties can also include unclear goal priorities and weak personal agency beliefs (e.g., low self-efficacy).

There is evidence that unregulated emotional processing can also interfere with interpersonal decision making in adolescents. Rieffe et al. (2012) reported that youth with ASD who showed higher dysregulation of anger relative to TD youth were at higher risk for victimization. Lower rates of meaningful friendships and higher rates of peer rejection were also associated with higher rates of peer victimization (Cappadocia et al., 2012). Jawaid et al. (2012) attributed difficulty with peer relations to deficits in emotion recognition, facial processing, and empathy. These emotion processing difficulties are likely to impact decision making in peer relationships. Emotional issues can include high levels of anxiety and distress and dysregulation of anger, resulting in difficulties with peer relationships and a lack of friends that can further increase vulnerability to bullying and cyberbullying (see Khemka, Chapter 11 of this volume for a fuller discussion of these issues).

In summary, our review of the literature on the victimization of adolescents with IDD yielded support for a number of risk factors. Having IDD is itself a major risk factor, showing different patterns for youth with ID and ASD. Two studies showed that girls were at greater risk than boys, and another showed that Caucasian youth were at greater risk for bullying than African-American and Hispanic youth. Limitations in a cluster of decision-making skills were also associated with increased vulnerability to victimization. Many of the risk factors that have been identified as predictors of adolescent victimization can be linked with one or more of the three basic processes involved in decision making – cognition, motivation, and emotion. Furthermore, Hebron et al. (2017) pointed out that the effect of multiple risk factors is cumulative – an important consideration as we seek ways to mitigate the vulnerability to maltreatment.

Adolescent Protective Factors

Wright and Wachs (2020) investigated the buffering effects of parental support on the frequency and sequelae of in-person bullying, cyberbullying, and bystander experiences of adolescents with IDD. Participants completed questionnaires in 7th grade (Time 1) and again in 8th grade (Time 2). The researchers found that higher levels of parental social support at Time 1 were related to (1) lower levels of Time 1 and Time 2 victimization and bystander experiences for face-to-face bullying and cyberbullying as well as (2) lower levels of Time 2 health complaints and suicidal ideation. They recommended follow-up research to better understand the specific nature of buffering parental support for adolescents with IDD.

In conjunction with a special issue of *Remedial and Special Education* on the involvement of adolescents with disabilities in peer victimization, Farmer et al. (2015) proposed several protective strategies. Among their suggestions for protecting individual students against peer victimization was training to strengthen the individual's interpersonal competencies, behaviors, and attitudes in order to reduce the likelihood of their involvement in peer victimization. They also recommended the formation of small friendship groups for high-risk students to promote the acquisition of positive social-emotional skills.

Maltreatment and Decision Making During Adulthood

The focus of the literature on the maltreatment of adults with IDD shifts away from peer victimization in school settings and reflects increasing rates of SA, DV, and IPV (Hickson & Khemka, 2016, 2021). Higher rates of violence against individuals with IDD, relative to the general population, have been consistently reported (e.g., Hughes et al., 2012). The most frequent perpetrators are male intimate partners, other family members, and health care and personal assistance providers (Plummer & Findley, 2012).

In a study comparing the views of DV/SA and IDD support professionals (Hickson et al., 2013), the two groups responded to a series of decision-making vignettes depicting situations in which a key decision-maker with IDD had to decide what to do in an abuse situation. When we asked them what the decision-maker should do, the two groups of support professionals largely agreed in their responses to the three workplace situations. However, they differed in how they thought the decision-maker should handle the three abuse scenarios in home settings. The IDD professionals felt that the decision-maker should strive for better communication and "talk to" the abuser. The DV/SA professionals felt that the decision-maker should be aware that the abuse was a violation of the law or of the person's rights and that law enforcement should be called in. At the time of the study, a *New York Times* article had just reported that law enforcement was being notified in only 5% of abuse allegations in New York State residential facilities for individuals with IDD (Hakim, 2011).

Most studies have found that women with IDD experience higher rates of abuse than men with IDD, who also experience higher rates than women and men in the general population (e.g., Hewitt, 2014; Mitra & Mouradian, 2014). Recognizing the heterogeneity of women survivors of abuse, it has been reported that these women are often women of color living in poverty or women who have dual diagnoses of IDD and mental illness (Ballan et al., 2014; Lightfoot & Williams, 2009). Ballan suggested that this heterogeneity has far-reaching implications for the types of interventions that are needed to address these issues.

Adult Risk Factors

As with adolescents, having IDD is a major risk factor for the maltreatment of adults with IDD. For example, Hughes et al. (2011) found that up to 90% of women and 86% of men with IDD experience maltreatment during their lifetime. Although there are fewer empirical studies in the adult literature that directly link specific risk factors to maltreatment, it is clear that limited decision-making skills play a role. In this section, we draw upon reviews of the adult maltreatment literature (Hickson & Khemka, 2016, 2021; Khemka & Hickson, 2017a) and a previously presented framework for the prevention of maltreatment (see Hickson & Khemka, 2016), to highlight six key risk factors for the maltreatment of adults with IDD. All six of these sources of vulnerability have direct connections with the three basic processes (i.e., cognition, motivation, and emotion) involved in interpersonal decision making.

In order to make effective, self-protective decisions in situations involving abuse, an individual with IDD has to have knowledge about abuse in order to recognize the abuse and assess the risks in the situation. However, there is evidence that individuals with IDD have limitations in both knowledge of abuse (Hickson et al., 2015; Khemka et al., 2005) and risk awareness (Khemka et al., 2016). In fact, Greenspan et al. (2011) proposed that risk unawareness was the essence of ID. In one study (Khemka et al., 2005), women with ID responded correctly, prior to decision-making training, only 38% of the time when asked open-ended questions that probed for their understanding of different types of abuse. In another study (Hickson et al., 2015), women and men were able to articulate the nature of the abuse situation depicted in six decision-making vignettes only 56% of the time prior to training. Because the ability to understand the dynamics of abuse situations and to predict the risks associated with possible actions are integral cognitive components of effective, self-protective decision making, these limitations can have a profound impact.

The role of independent decision-making capabilities in preventing the maltreatment of individuals with IDD is critical because of the fact that the abuser is almost always someone who is known to the person with IDD, often in their close inner circle. Ideally family members, service providers, and friends will play key support roles for individuals with IDD, allowing them to enjoy the benefits of supported and shared decision making. Unfortunately, it is not always safe to assume that familiar individuals will be trustworthy sources of support to individuals with IDD. It is essential that individuals with IDD are able to recognize when someone close to them is not acting in their best interest or is behaving in a way that may be harmful to them. In cases of abuse, individuals with IDD need to be prepared to take decisive actions to recognize and stop, or escape from, the abuse. Although early studies focused on documenting the many decision-making limitations associated with IDD (see Hickson & Khemka, 1999), more recent researchers, with a positive psychology perspective, have shifted toward developing ways to promote effective independent decision-making skills (Fisher 2013; Hickson et al., 2015; Khemka, 2000; Khemka & Hickson, 2017b; Khemka et al., 2005, submitted).

The motivational component of decision making involves both personal agency beliefs and goal processes. In order to take action on a decision, it is necessary to believe in your ability to change the situation (i.e., empowerment and/or self-efficacy). An absence of these personal agency beliefs is manifested in *learned helplessness*, which has often been ascribed to individuals with IDD (e.g., Jenkinson, 1999) and is likely to demotivate attempts at decision making.

The other motivational component in decision making has to do with goal awareness and selection. Individuals with IDD often face a lack of clarity about their goals as well as conflicting goal priorities. Individuals with IDD may be conflicted between the desire to have a romantic partner and a fear of being sexually abused by that same person. In other situations, they may be dependent on the financial or physical support of a caregiver which would be jeopardized if they decided to confront or resist the caregiver's abuse.

Hartley and MacLean (2005) reported elevated stress levels in individuals with mild ID. The participants in their study indicated that a major source of that stress was frequent negative interpersonal interactions. In a later study, we found greater use of passive/avoidant decision-making strategies and higher levels of stress for women with IDD who had a recent history of abuse than for women with no history of abuse (Hickson et al., 2008). Many of the women with a history of abuse also reported unregulated emotional coping strategies that included crying, smashing and breaking things, hitting or punching a friend, and screaming or yelling. Though it is not possible to know whether these behaviors are risk factors or sequelae of abuse, or both, it is clear that they are ineffective coping strategies likely to exacerbate already high levels of stress.

Because individuals with IDD often have limited opportunities for friendships and intimate relationships, they may experience loneliness and isolation, leaving them vulnerable to abusive or exploitative relationships (Ward et al., 2010, 2013). They may lack a sharp awareness of social cues, effective assertiveness skills, or the ability to read the intentions of others (e.g., Eastgate et al., 2012; Nettelbeck et al., 2001), all of which may interfere with their effective decision making.

To summarize, although our most recent reviews of the adult literature did not yield a comprehensive set of empirically validated risk factors for maltreatment, the available research did provide a basis for several *suggested* risk factors for maltreatment (Hickson & Khemka, 2021). These included having ID or ASD, being a woman, limited knowledge and understanding of abuse, difficulty handling relationships, limited goal clarity and empowerment, low levels of emotion regulation and coping, and generally limited decision-making skills.

Adult Protective Factors

Few studies have sought to identify protective factors pertinent to the abuse of adults with IDD. In our study comparing women with IDD with and without a documented history of abuse, we noted that the women without a history of abuse were significantly more likely than those with a history of abuse to have many friends (Hickson et al., 2008). It is possible that these friendships played a protective role for them. However, a clear interpretation of this finding will require further research to rule out the possibility that the social isolation of the women with a history of abuse was a result of the abuse itself.

In one of the few studies that set out to identify protective factors as a way to reduce the high rates of bullying and victimization faced by adults with IDD, Griffin et al. (2019) interviewed 18 women and men with IDD about their bullying experiences. They identified two types of protective factors: (1) support from others (e.g., friends, family, staff members) and (2) a self-determined orientation. Attempts by others to intervene in a supportive way were sometimes effective or helpful, but the authors recommended training for staff and family members to increase the effectiveness of their support. A self-determined orientation was reported by half of the participants and included self-advocacy, decision making, goal setting, and self-regulatory behaviors. These behaviors were often highly effective and could be fortified by targeted training.

In an investigation of a highly publicized "exposé of abuse in residential services for individuals with ID in the UK, Marsland et al. (2015) identified factors that had contributed to the widespread abuse. In summarizing their conclusions and the challenges ahead, they emphasized the role of families as potentially protective, even when the individual with ID was living in a residential facility. They endorsed supporting families "who are willing and able to support themselves and each other" (p. 143) and encouraged the staff to make use of the "natural motivation of family/friends and enable them to safeguard their loved ones more effectively" (p. 143).

Conclusions and Recommendations

There is an urgent need to apply a positive psychology perspective to addressing the alarming vulnerability of individuals with IDD to maltreatment. Engaging with the existing literature on the abuse and victimization of adolescents and adults with IDD suggests at least three possible courses of action that would be consistent with the tenets of positive psychology.

First, risk factors need to be addressed and deactivated. The long-term goal, of course, would be to eliminate IDD as a risk factor by addressing any limitations that can render individuals with IDD vulnerable to maltreatment. The core features of ID and/or ASD do not in and of themselves make these individuals vulnerable. However, some of these core features do impinge upon aspects of interpersonal decision making, limiting the ability of adolescents and adults with IDD to draw upon a repertoire of effective, self-protective decision-making strategies in situations of abuse. To address these limitations, we encourage the further development and evaluation of targeted interventions to strengthen the relevant component skills, including social cognition and understanding, motivational goal priorities and empowerment, and emotion regulation skills.

Second, protective factors need to be nurtured and embraced. While there is a large body of research identifying important risk factors for the maltreatment of individuals with IDD at both the adolescent and adult age levels, there is very little research pointing to protective factors. While identifying further protective factors needs to be a high-priority direction for future research, there are some things that we do already know. For example, we know that connectedness with supportive others is protective. Research has pointed to the importance of parental support and positive peer relationships and friendships as protective. It is also clear that frameworks like self-determination and self-advocacy can play an empowering and protective role, which can be fostered through structured circles of support or friendship groups.

Third, the ongoing development and evaluation of promising interventions to strengthen the self-protective decision-making skills of individuals with IDD is of pivotal importance. Research by ourselves and others (e.g., see Hickson & Khemka, 2021; Hickson et al., 2015; Hughes et al., 2020; Khemka, 2000; Khemka & Hickson, this volume; Khemka et al., 2005; *Petitpierre & Tabin*, this volume; *Uher,* et al., this volume) suggests that a strong repertoire of effective, independent decision-making skills for situations where there is a threat of abuse or victimization can offer a powerful protective shield. At both the adolescent and adult life stages, training to strengthen relevant knowledge and skills, self-efficacy and goal clarity, and emotion regulation skills, as well as strong family supports (see *Burke* et al., this volume; *Vanegas* et al., this volume) and high-quality friendships, are consistently endorsed as protective. In addition, access to the types of supports envisioned by the supported decision making movement can do much to enhance opportunities for safe decision making when trustworthy supporters are available (e.g., *Bigby & Douglas*, this volume; *Shogren* et al., this volume). In addition, evidence-based intervention approaches need to be expanded and fine-tuned to provide individuals with specific subtypes of IDD with a tailored repertoire of effective, self-protective decision-making skills (see *Goscicki* et al., this volume) and to ensure that they will apply them spontaneously in real-life situations where there is a threat of harm or abuse.

In conclusion, it is clear that no one, with or without IDD, would be vulnerable to maltreatment were there not perpetrators with the will to impose their harmful, predatory behaviors on others. Along with all of our collective efforts to strengthen the self-protective decision-making capacity of individuals with IDD, it is imperative that societies work harder to establish cultural climates where people with disabilities are valued and where their safety is ensured. Further, policies and practices must be put in place that effectively eliminate the conditions that engender and tolerate the maltreatment of vulnerable others. Work has begun on many fronts, as reflected by the widespread international adoption of the United Nations Convention on the Rights of Persons with Disabilities (CRPD) (Werner, 2012), as well as the establishment of supports to eliminate abuse and DV and to decrease tolerance for abuse by those in power. The vulnerability of individuals with IDD to maltreatment would disappear entirely if policies of zero tolerance were universally adopted and enforced.

References

Baladerian, N. J. (1991). Sexual abuse of people with developmental disabilities. *Sexuality and Disability, 9*, 323–335.

Ballan, M. S., Freyer, M. B., Marti, C. N., Perkel, J., Webb, K. A., & Romanelli, M. (2014). Looking beyond prevalence: A demographic profile of survivors of intimate partner violence with disabilities. *Journal of Intimate Partner Violence, 29*(17), 3167–3179.

Blake, J. J., Lund, E. M., Zhou, Q., Kwok, O., & Benz, M. R. (2012). National prevalence rates of bully victimization among students with disabilities in the United States. *School Psychology Quarterly, 27*(4), 210–222. https://doi.org/10.1037/spq0000008

Blake, J. J., Zhou, Q., Kwok, O., & Benz, M. R. (2016). Predictors of bullying behavior, victimization, and bully-victim risk among high school students with disabilities. *Remedial and Special Education, 37*(5), 285–295.

Brosnan, M., Chapman, E., & Ashwin, C. (2014). Adolescents with autism spectrum disorder show a circumspect reasoning bias rather than 'jumping-to-conclusions'. *Journal of Autism and Developmental Disorders, 44*(3), 513–520. https://doi.org/10.1007/s10803-013-1897-5

Cappadocia, M. C., Weiss, J. A., & Pepler, D. (2012). Bullying experiences among children and youth with autism spectrum disorders. *Journal of Autism and Developmental Disorders, 42*(2), 266–277.

Christensen, L., Fraynt, R. J., Neece, C. L., & Baker, B. L. (2012). Bullying adolescents with intellectual disability. *Journal of Mental Health Research in Intellectual Disability, 5*(1), 49–65.

Cook, C. R., Williams, K. R., Guerra, N. G., Kim, T. E., & Sadek, S. (2010). Predictors of bullying and victimization in childhood and adolescence: A meta-analytic investigation. *School Psychology Quarterly, 25*(2), 65–83.

Didden, R., Scholte, R. H. J., Korzilius, H., DeMoor, J. M. H., Vermeulen, A., O'Reilly, M., Lang, R., & Lancioni, G. E. (2009). Cyberbullying among students with intellectual and developmental disability in special education settings. *Developmental Neurorehabilitation, 12*(3), 146–151.

Eastgate, G., Scheermeyer, D., van Driel, M. L., & Lennox, N. (2012). Intellectual disability, sexuality and sexual abuse prevention: A study of family members and support workers. *Australian Family Physician, 41*(3), 135–139. Retrieved from https://www.researchgate.net/publication/221683681_Intellectual_disability_sexuality_and_sexual_abuse_prevention_A_study_of_family_members_and_support_workers.

Farmer, T., Wike, T. L., Alexander, Q. R., Rodkin, P. C., & Mehtaji, M. M. (2015). Students with disabilities and involvement in peer victimization: Theory, research, and considerations for the future. *Remedial and Special Education, 36*(5), 263–274.

Fisher, M. H. (2013). Evaluation of a stranger safety training program for adults with Williams syndrome. *Journal of Intellectual Disability Research, 58*(10), 903–914.

Fisher, M. H., & Taylor, J. L. (2016). Let's talk about it: Peer victimization experiences as reported by adolescents with autism spectrum disorder. *Autism, 20*(4), 402–411.

Garner, P. W., & Hinton, T. S. (2010). Emotional display rules and emotion self-regulation: Associations with bullying and victimization in community-based after school programs. *Journal of Community & Applied Social Psychology, 20*(6), 480–496. https://doi.org/10.1002/casp.1057

Greenspan, S., Switzky, H. N., & Woods, G. W. (2011). Intelligence involves risk-awareness and intellectual disability involves risk-unawareness: Implications of a theory of common sense. *Journal of Intellectual and Developmental Disability, 36*, 246–257.

Griffin, M. M., Fisher, M. H., Lane, L. A., & Morin, L. (2019). Responses to bullying among individuals with intellectual and developmental disabilities: Support needs and self-determination. *Journal of Applied Research in Intellectual Disabilities, 32*, 1514–1522.

Hakim, D. (March 12, 2011). At state-run homes, abuse and impunity. *New York Times*, A1.

Hartley, S. L., & MacLean, W. E., Jr. (2005). Perceptions of stress and coping strategies among adults with mild mental retardation: Insight into psychological distress. *American Journal on Mental Retardation, 119*, 285–297.

Hase, C. N., Stuck, A., & Campain, J. (2015). Impacts of traditional bullying and cyberbullying on the mental health of middle school and high school students. *Psychology in the Schools, 52*(6), 607–617.

Healey, K. N., & Masterpasqua, F. (1992). Interpersonal cognitive problem-solving among children with mild mental retardation. *American Journal on Mental Retardation, 96*, 367–372.

Hebron, J., Oldfield, J., & Humphrey, N. (2017). Cumulative risk effects in the bullying of children and young people with autism spectrum conditions. *Autism, 21*(3), 291–300. https://doi.org/10.1177/1362361316636761

Heiman, T., & Olenik-Shemesh, D. (2013). Cyberbullying experience and gender differences among adolescents in different educational settings. *Journal of Learning Disabilities, 20*(10), 1–10.

Hewitt, O. (2014). A survey of experiences of abuse. *Tizard Learning Disability Review, 19*(3), 122–129.

Hickson, L., & Khemka, I. (1999). Decision making and mental retardation. In L. M. Glidden (Ed.), *International review of research in mental retardation* (Vol. 22). Academic Press.

Hickson, L., & Khemka, I. (2013). Problem solving and decision making. In M. Wehmeyer (Ed.), *The Oxford handbook of positive psychology and disability*. Oxford University Press.

Hickson, L., & Khemka, I. (2014). The psychology of decision making. In R. Hodapp (Ed.), *International review of research in developmental disabilities* (Vol. 47, pp. 185–229). Elsevier.

Hickson, L., & Khemka, I. (2016). Prevention of maltreatment of adults with intellectual and developmental disabilities. In J. R. Lutzker, K. Guastaferro, & M. L. Benka-Coker (Eds.), *Maltreatment of people with intellectual and developmental disabilities* (pp. 233–261). American Association on Intellectual and Developmental Disabilities.

Hickson, L., & Khemka, I. (2017). Preventing social victimization of youth with Autism Spectrum Disorder. In H. Chiang (Ed.), *Curricula for teaching students with Autism Spectrum Disorder* (pp. 225–240). Springer.

Hickson, L., & Khemka, I. (2021). Maltreatment of individuals with intellectual and developmental disabilities: A lifespan view. In L. M. Glidden (Ed.), *APA handbook on intellectual and developmental disabilities*. American Psychological Association.

Hickson, L., Golden, H., Khemka, I., Urv, T., & Yamusah, S. (1998). A closer look at interpersonal decision making in adults with and without mental retardation. *American Journal on Mental Retardation, 103*, 209–224.

Hickson, L., Khemka, I., Golden, H., & Chatzistyli, A. (2008). Profiles of women with mental retardation with and without a documented history of abuse. *American Journal on Mental Retardation, 113*(2), 133–142.

Hickson, L., Khemka, I., Golden, H., & Chatzistyli, A. (2013). Views and values of developmental disabilities and domestic violence/sexual assault support professionals regarding the prevention and handling of situations of abuse. *Journal of Policy and Practice in Intellectual Disabilities, 10*(3), 207–214.

Hickson, L., Khemka, I., Golden, H., & Chatzistyli, A. (2015). Randomized controlled trial to evaluate an abuse prevention curriculum for women and men with intellectual and developmental disabilities. *American Journal on Intellectual and Developmental Disabilities, 120*(6), 490–503.

Hughes, R. B., Lund, E. M., Gabrielli, J., Powers, L. E., & Curry, M. A. (2011). Prevalence of interpersonal violence against community-living adults with disabilities: A literature review. *Rehabilitation Psychology, 56*(4), 302–319.

Hughes, K., Bellis, M. A., Jones, L., Wood, S., Bates, G., Eckley, L., … Officer, A. (2012). Prevalence and risk of violence against adults with disabilities: A systematic review and meta-analysis of observational studies. *The Lancet, 379*, 1621–1629.

Hughes, R. B., Robinson-Whelen, S., Davis, L. A., Meadours, J., Kincaid, O., Howard, L., Millin, M., Schwartz, M., McDonald, K. E., & The Safety Project Consortium. (2020). Evaluation of a safety awareness group program for adults with intellectual disability. *American Journal on Intellectual and Developmental Disabilities, 125*(4), 304–317.

Humphrey, N., & Hebron, J. (2015). Bullying of children and adolescents with autism spectrum conditions: A 'state of the field' review. *International Journal of Inclusive Education, 19*(8), 845–862. https://doi.org/10.1080/13603116.2014.981602

Humphrey, N., & Symes, W. (2011). Peer interaction patterns among adolescents with autistic spectrum disorders (ASDs) in mainstream school settings. *Autism, 15*(4), 397–419. https://doi.org/10.1177/1362361310387804

Jawaid, A., Riby, D. M., Owens, J., White, S. W., Tarar, T., & Schulz, P. E. (2012). 'Too withdrawn' or 'too friendly': Considering social vulnerability in two neuro-developmental disorders. *Journal of Intellectual Disability Research, 56*(part 4), 335–350.

Jenkinson, J. C. (1999). Factors affecting decision making by young adults with intellectual disabilities. *American Journal on Mental Retardation, 104*(4), 320–329.

Jenkinson, J., & Nelms, R. (1994). Patterns of decision making behavior by people with intellectual disability: An explanatory study. *Australia and New Zealand Journal of Developmental Disabilities, 19*, 99–109.

Khemka, I. (2000). Increasing independent decision making skills of women with mental retardation in simulated interpersonal situations of abuse. *American Journal on Mental Retardation, 105*, 387–401.

Khemka, I. (2016). Examining decision making and locus of control patterns of transition-aged students with intellectual and developmental disabilities in response to simulated situations of cyberbullying and abuse: Two pilot studies. Unpublished manuscript.

Khemka, I., & Hickson, L. (2000). Decision making by adults with mental retardation in simulated situations of abuse. *Mental Retardation, 38*, 15–26.

Khemka, I., & Hickson, L. (2016). Examining the impact of self-actualized goal orientation on effective, self-protective decision making performance of adolescents with intellectual and developmental disabilities in hypothetical interpersonal situations of peer pressure. Unpublished manuscript.

Khemka, I., & Hickson, L. (2017a). Empowering women with intellectual disabilities to resist abuse in interpersonal relationships: Promising interventions and practices. In A. J. Johnson, J. R. Nelson, & E. M. Lund (Eds.), *Religion, disability, and interpersonal violence* (pp. 67–86). Springer.

Khemka, I., & Hickson, L. (2017b). The role of positive psychology in interpersonal decision making. In K. A. Shogren, M. Wehmeyer, & N. N. Singh (Eds.), *Handbook of positive psychology in intellectual and developmental disabilities: Translating research into practice* (pp. 145–171). Springer.

Khemka, I., Hickson, L., & Reynolds, G. (2005). Evaluation of a decision making curriculum designed to empower women with mental retardation to resist abuse. *American Journal on Mental Retardation, 110*(3), 193–204.

Khemka, I., Hickson, L., Casella, M., Accetturi, N., & Rooney, M. E. (2009). Impact of coercive tactics on the decision making of adolescents with intellectual disabilities. *Journal of Intellectual Disability Research, 53*(4), 353–362.

Khemka, I., Hickson, L., Mallory, S., & Zealand, R. (2013). Social vulnerability of adolescents with autism spectrum disorders and adolescents with other developmental disabilities in situations of negative peer pressure. *Symposium paper presented at the 46th annual Gatlinburg conference on research and theory in intellectual and developmental disabilities.* San Antonio, TX.

Khemka, I., Hickson, L., & Mallory, S. B. (2016). Evaluation of a decision making curriculum for teaching adolescents with disabilities to resist negative peer pressure. *Journal of Autism and Developmental Disorders, 46*(7), 2372–2384.

Khemka, I., Hickson, L., & Gilic, L. (submitted). What is the best thing for a target of peer aggression to do? Views of middle schoolers in Catholic schools (Submitted).

Kowalski, R. M., & Fedina, C. (2011). Cyberbullying in ADHD and Asperger Syndrome populations. *Research in Autism Spectrum Disorders, 5*, 1201–1208.

Levin, I. P., Gaeth, G. J., Foley-Nicpon, M., Yegorova, V., Cederberg, C., & Yan, H. (2015). Extending decision making competence to special populations: A pilot study of persons on the autism spectrum. *Frontiers in Psychology, 6*, 539. https://doi.org/10.3389/fpsyg.2015.00539

Lightfoot, E., & Williams, O. (2009). The intersection of disability, diversity, and domestic violence: Results of national focus groups. *Journal of Aggression, Maltreatment & Trauma, 18*(2), 133–152.

Luke, L., Clare, I. C., Ring, H., Redley, M., & Watson, P. (2012). Decision making difficulties experienced by adults with autism spectrum conditions. *Autism, 16*(6), 612–621.

Maïano, C., Normand, C. L., Salvas, M. C., Moulle, G., & Aimé, A. (2016). Prevalence of school bullying among youth with autism spectrum disorders: A systematic review and meta-analysis. *Autism Research, 9*, 601–615.

Marsland, D., Oakes, P., & Bright, N. (2015). It can still happen here: Systemic risk factors that may contribute to the continued abuse of people with intellectual disabilities. *Tizard Learning Disability Review, 20*(3), 134–146. https://doi.org/10.1108/TLDR-11-2014-0039

McKinney, D. (2018). The invisible hate crime. *Intelligence Report, 165*, 1–17.

Mitra, M., & Mouradian, V. E. (2014). Intimate partner violence in the relationships of men with disabilities in the United States: Relative prevalence and health correlates. *Journal of Interpersonal Violence, 29*(17), 3150–3166.

Nettelbeck, T., Wilson, C., Potter, R., & Perry, C. (2001). The influence of interpersonal competence on personal vulnerability of persons with mental retardation. *Journal of Interpersonal Violence, 15*(1), 46–62.

Olweus, D. (1993). *Bullying at school*. Blackwell Publishing.

Plummer, S. B., & Findley, P. A. (2012). Women with disabilities' experience with physical and sexual abuse review of the literature and implications for the field. *Trauma, Violence & Abuse, 13*(1), 15–29.

Reiter, S., Bryen, D. N., & Shachar, I. (2007). Adolescents with intellectual disabilities as victims of abuse. *Journal of Intellectual Disabilities, 11*(4), 371–387.

Rice, E., Petering, R., Rhoades, H., Winetrobe, H., Goldbach, J., Plant, A., Montoya, J., & Kordic, T. (2015). Cyberbullying perpetration and victimization among middle-school students. *American Journal of Public Health, 105*(3), e66–e72.

Rieffe, C., Camodeca, M., Pouw, L. B. C., Lange, A. M. C., & Stockmann, L. (2012). Don't anger me! Bullying, victimization, and emotion dysregulation in young adolescents with ASD. *European Journal of Developmental Psychology, 9*(3), 351–370. https://doi.org/10.1080/17405629.2012.680302

Rose, C. A., Simpson, C. G., & Moss, A. (2015). The bullying dynamic: Prevalence of involvement among a large-scale sample of middle and high school youth with and without disabilities. *Psychology in the Schools, 52*(5), 515–531.

Schroeder, J. H., Cappadocia, M. C., Bebko, J. M., Pepler, D. J., & Weiss, J. A. (2014). Shedding light on a pervasive problem: A review of research on bullying experiences among children with autism spectrum disorders. *Journal of Autism and Developmental Disorders, 44*, 1520–1534.

Short, E. J., & Evans, S. W. (1990). Individual differences in cognitive and social problem-solving skills as a function of intelligence. In N. W. Bray (Ed.), *International review of research in mental retardation* (Vol. 16, pp. 89–123). Academic Press.

Smith, D. C. (1986). Interpersonal problem-solving skills of retarded and nonretarded children. *Applied Research in Mental Retardation, 7*, 431–444.

Sreckovic, M. A., Brunsting, N. C., & Able, H. (2014). Victimization of students with autism spectrum disorder: A review of prevalence and risk factors. *Research in Autism Spectrum Disorders, 8*, 1155–1172.

Stevens, B. (2012). Examining emerging strategies to prevent sexual violence: Tailoring to the needs of women with intellectual and developmental disabilities. *Journal of Mental Health Research in Intellectual Disabilities, 5*, 168–186.

Swearer, S. M., Espelage, D. L., Vaillancourt, T., & Hymel, S. (2010). What can be done about school bullying? Linking research to educational practice. *Educational Researcher, 39*(1), 38–47.

Underwood, M. K., & Ehrenreich, S. E. (2017). The power and pain of adolescents' digital communication: Cyber victimization and the perils of lurking. *American Psychologist, 72*(2), 144–158.

van Roekel, E., Scholte, R. H. J., & Didden, R. (2010). Bullying among adolescents with autism spectrum disorders: Prevalence and perception. *Journal of Autism and Developmental Disorders, 40*, 63–73.

Wainscot, J. J., Naylor, P., Sutcliffe, P., Tantam, D., & Williams, J. (2008). Relationships with peers and use of the school environment of mainstream secondary school pupils with Asperger syndrome (high-functioning autism): A case control study. *International Journal of Psychology and Psychological Theory, 8*(1), 25–38.

Ward, K. M., Bosek, R. L., & Trimble, E. L. (2010). Romantic relationships and interpersonal violence among adults with developmental disabilities. *Intellectual and Developmental Disabilities, 48*(2), 89–98.

Ward, K. M., Atkinson, J. P., Smith, C. A., & Windsor, R. (2013). A friendships and dating program for adults with intellectual and developmental disabilities: A formative evaluation. *Intellectual and Developmental Disabilities, 51*(1), 22–32.

Werner, S. (2012). Individuals with intellectual disabilities: A review of the literature on decision making since the Convention on the Rights of People with Disabilities (CRPD). *Public Health Reviews, 34*, 1–27.

Williams, K. R., & Guerra, N. G. (2007). Prevalence and predictors of internet bullying. *Journal of Adolescent Health, 41*, S14–S21.

Wright, M. F., & Wachs, S. (2020). Parental support, health, and cyberbullying among adolescents with intellectual and developmental disabilities. *Journal of Child and Family Studies, 29*, 2390–2401.

Zeedyk, S. M., Rodriguez, G., Tipton, L. A., Baker, B. L., & Blacher, J. (2014). Bullying of youth with autism spectrum disorder, intellectual disability, or typical development: Victim and parent perspectives. *Research in Autism Spectrum Disorders, 8*, 1173–1183.

Chapter 19
Criminal Justice Decisions and People with Intellectual Disabilities

Isabel C. H. Clare and Glynis H. Murphy

Introduction

In recent years, there have been encouraging developments in the area of research into offending by people with putative or known intellectual disabilities (ID) as defined in the major psychiatric classification systems (see DSM-5 (American Psychiatric Association, 2013) and ICD-10 (World Health Organization, 1996), ICD-11 (World Health Organization, 2018)). More recently, there has also been growing clinical and research interest involving people who may, or may possibly have, other neurodevelopmental disabilities (such as ADHD and/or autism and/or developmental language disorders). Much of this interest, however, has focussed on individuals *without* concomitant significant intellectual impairments (e.g. Crane et al., 2016; Maras et al., 2018; Underwood et al., 2013; Winstanley et al., 2020;Woodbury-Smith et al., 2005, 2006, 2010) and will not be included here.

Focusing on people with ID, the methodological flaws of a large number of studies, mainly carried out in institutional settings including both 'long-stay' social care hospitals and custodial facilities, that seemed to demonstrate beyond doubt that 'ID' (or its equivalent) was inextricably linked to 'offending' (e.g., Hodgins, 1992; but see Lindsay & Dernevick, 2013) have been exposed by a powerful series of critiques (Lindsay & Taylor, 2018; Murphy & Mason, 2014; Noble & Conley, 1992). These have raised concerns about the conceptualisation and measurement of the 'intellectual' and 'adaptive' domains and the weight to be accorded to each, as well as the developmental criterion of the definition.

I. C. H. Clare (✉)
Department of Psychiatry, University of Cambridge, Cambridge, UK
e-mail: ichc2@medschl.cam.ac.uk

G. H. Murphy
Tizard Centre, University of Kent, Canterbury, Kent, UK

I. Khemka, L. Hickson (eds.), *Decision Making by Individuals with Intellectual and Developmental Disabilities*, Positive Psychology and Disability Series,
https://doi.org/10.1007/978-3-030-74675-9_19

At the same time, there has been greater attention to the meaning of 'offending'. Many people with ID detained under civil or 'criminal' mental health legislation in their jurisdiction are 'alleged' offenders. They may or may not have engaged in anti-social or illegal behaviour, but at some stage as they progress through the criminal justice system from what, in English law, is the putative *actus reus* (the act) to an 'offence', they have been 'filtered out': there may have been no formal report to the police or a similar investigative body; no investigation; no interview or interrogation may have been carried out; no charge may have been made, perhaps because it seems unlikely that the person could form a guilty 'state of mind' (*mens rea* in English (England and Wales) law); or they may have been found unable to take part in any trial, because they have been found 'incompetent to stand trial' (USA) or 'unfit to plead' (Law Commission (England & Wales), 2016) or unable to participate meaningfully; and they may not be found guilty so do not proceed to sentencing.

This more careful approach to the definition of 'offending' has coincided with growing interest in research that extends into the field of ID questions that have long been of concern in mainstream criminology: (1) *Why* might some people be at greater risk than others of engaging in behaviour that leads to the attention of the criminal justice system? (2) *How* are those who have engaged in alleged illegal acts 'transformed' through the criminal justice system into 'offenders'? Both these questions have been addressed in thinking about alleged or actual offenders with ID using decision-making approaches.

Decision Making and 'Offenders'

One of the earliest, and still influential, criminological theories seeking to address the question of why some people commit illegal acts is they are the outcome of a rational decision-making process (Beccaria & Caso, 2013; Bentham, 2017), made by individuals with free will (Beirne, 1993). The proposed answer is that there is some difference (or set of differences, Farrington & West, 1995) between those individuals whose behaviour may lead or has led to contact with the criminal justice system and others. It is assumed (following Gottfredson & Hirschi's (1990) influential reformulation of 'social control' theory) that the particular difference is causative: it is proposed that stronger or weaker 'social bonds' affect the self-control that individuals use to prevent themselves from engaging in illegal behaviour.

For research involving people with ID, the approach to decision making has shifted, over time, from a straightforward 'status' approach (see Murphy & Clare, 2003), in which a diagnosis of 'ID' was considered sufficient to explain an individual's illegal behaviour, to a more sophisticated individual differences approach, often explicitly or implicitly based on social control theory. This stream of research has often focussed on examining variations *within* the group of people with ID, often with the intention of identifying treatment goals (Murphy & Clare, 2012), or assessing risk (see Nicholas et al., 2018). The range of individual variables for which there is some evidence of differences between people (predominantly men) with ID whose illegal behaviour could bring or has brought them into contact with the criminal justice system and their counterparts who do not engage in such behaviour is very broad: it encompasses biological, personality and other psychological, and (so far, limited) social factors. A full account is beyond the scope of this chapter.

A decision-making approach has also been used as a framework to examine the process through which some people with ID make the transition from allegedly engaging in illegal acts to offenders. Based on a pioneering observational study of suspects, none of whom met formal criteria for ID, detained for interviewing at one of two police stations in England, Irving and Hilgendorf (Irving & Hilgendorf, 1980; Hilgendorf & Irving, 1981) conceptualised this part of the criminal justice process as a series of decision-making tasks:

- Whether or not to seek legal advice
- Whether to speak or remain silent
- Whether or not to tell the truth
- Whether to tell the whole or a part of the truth
- Whether or not to make self-incriminating admissions including a confession

At each point, they suggest that the suspect's decision depends on their view, however unrealistic, of (a) the situation and (b) the adverse and beneficial consequences, for the self and/or others, of strategies for coping with this situation.

As Clare (2003) has argued, this conceptualisation might be extended to the entire process from arrest to sentencing (e.g. in England and Wales, a convicted offender cannot receive a community order as punishment unless they have decided to comply).

Over a long period, Gudjonsson and his colleagues (for review, see Gudjonsson, 2003) have adopted an individual differences approach. Through these studies, Gudjonsson (1994) has identified 'psychological vulnerabilities', psychological disadvantages (such as ID) that are likely to compromise suspects' decision making and likely to lead to behaviour during police detention and interviewing that does not best serve their interests. It should be emphasised that Gudjonsson believed that psychological vulnerabilities *on their own* rarely provided a sufficient explanation of a suspect's having made admissions, including confessions, that were misleading or even false. The relevance of a particular vulnerability has always depended on all the circumstances of the case. Nevertheless, this 'individual differences' perspective has drawn on an essentially 'status' approach to decision making.

More recently, in the USA, England and Wales, Scotland, and other jurisdictions, a different approach, a 'functional' approach, has gained increasing influence and has been enshrined in civil legislation in some jurisdictions (e.g. the *Mental Capacity Act (England &Wales)* 2005). With some striking exceptions (Grisso, 1981; Clare, 2003), it has rarely been used in criminal justice system contexts.

Briefly, a functional approach, pioneered by Grisso (1986, 2003) and his colleagues, is based on the proposition than an individual's legal competence (in England and Wales, capacity) to make a decision that furthers their personal goals reflects an interaction between:

- The person's 'functional abilities' (p. 86; Grisso, 1986), that is, what they understand, know, believe, or can do that is directly relevant to a particular context (such as that of a suspect in a police interview rather than as a witness in court)
- The extent to which these functional abilities meet the demands of a specific situation within a given context (e.g. being interviewed for a minor property offence rather than the murder of a child).

A gap between an individuals' functional abilities and the demands of the situation is likely to lead to impairments in decision making. Theoretical analysis of the abilities involved in making valid legally significant decisions has suggested that understanding of information relevant to the specific decision is necessary, but not sufficient (Appelbaum & Grisso, 1988, 1995; Grisso & Appelbaum, 1998). Other abilities are necessary, though these vary a little between jurisdictions: in the USA, for example, there is retaining the information; using or weighing it as part of the process of making the decision; appreciating its personal relevance; and communicating the choice made. In England and Wales, 'appreciation', often referred to as 'insight' and included in 'using or weighing', is omitted (s.3 (1), *Mental Capacity Act (England & Wales)* 2005).

In contrast with a 'status' approach, and paraphrasing Grisso (p. 2, 2003), the concept of legal competencies of any sort recognises (i) individuals have the right to make decisions about their own lives; (ii) that some individuals may lack the necessary abilities and that their well-being and/or that of others may be jeopardised as a result; and (iii) that when incompetence is recognised, then measures need to be taken to protect the individual. While seeking to allow individuals to make for themselves those decisions for which they have the relevant competence/capacity, attention is directed towards the contexts in which those decisions need to be made. Potentially, a functional approach is more radical.

In the following sections, we will consider some studies involving people with ID that have adopted a decision-making approach to the questions: (1) *Why* might some people be at greater risk of engaging in illegal behaviour, leading to the attention of the criminal justice system, than others? (2) *How* are those who have engaged in alleged illegal acts 'transformed' through the criminal justice system into 'offenders'? We will see that studies with participants with ID that have attempted to address the first question have generally adopted an individual differences approach. In contrast, research focussed on the second question has adopted a functional approach.

Why Might Some People Be at Greater Risk than Others of Engaging in Behaviour That Leads to the Attention of the Criminal Justice System?

In a criminological critique of social control theory as formulated by Gottfredson and Hirschi (1990), Wikström (2004 *et seq.*) set out an alternative, and influential, theory of crime causation (https://www.cac.crim.cam.ac.uk/resou/sat for more details). While we cannot do justice to Situational Action Theory here (see Wikström, 2004 *et seq.*), Wikström and Treiber (2007) argue that illegal behaviours, like other actions, result from the way that individuals (a) perceive their alternatives for action and (b) choose between these alternatives. In the same situation, different people will perceive these alternatives differently and will make different choices. *In part*, their perceptions will reflect what Garrigan et al. (2018) refer to as 'moral decision making'. This includes any 'decision, including judgements, evaluations, and response choices' (p. 80, *ibid.*) within the 'moral domain' (Smetana, 2006; Turiel, 1983, cited in Garrigan et., *ibid.*), that is, 'decisions regarding moral issues or principles such as justice, harm, fairness and care' (p. 80, *ibid.*).

Reviewing the empirical literature, carried out primarily within the framework of stage theories based on Piaget's (1932) ideas, such as that of Kohlberg (1969) and, later, Gibbs (2010: see Table 19.1), Langdon et al. (2010a) concluded, with great caution because of a variety of significant methodological problems, that the moral development of children, young people, and adults with ID was likely to be slower than that of their typically developing peers. However, experimentally, this conclusion could be mitigated by matching for 'mental age'. Subsequently, Langdon and his colleagues reviewed the literature addressing the relationship between behaviour and moral development in people with ID (Langdon et al., 2011a) and found no

Table 19.1 Gibbs' sociomoral developmental theory (Gibbs, 2010), from Langdon et al. (2011b)

Level and stage	Description
Level 1: Immature	
Stage 1: Unilateral and physicalistic	Moral justifications are based on unilateral authority and rules or related to punitive consequences of the violation of rules.
Stage 2: Exchanging and instrumental	Moral justifications based on an understanding that has arisen from social interaction. For example, decisions to help others may be justified because that person may help you in the future. However, justifications remain superficial.
Level 2: Mature	
Stage 3: Mutual and prosocial	Moral justifications are characterised by further decentration and are based on a prosocial understanding of emotional states (e.g. empathy), care, and good conduct.
Stage 4: Systemic and standard	Further maturity is indexed by the development of an understanding of the complex social structures in which humans live. Justifications are also based on constructs such as rights, values, and character within society. Other justifications may be based on social justice and responsibility or conscience.

studies of the relationship between moral reasoning, ID, and illegal behaviour in adults.

To examine the relationship between illegal behaviour and moral decision making, Langdon et al. (2011b) carried out a study in the UK with four groups ($n = 20$ in each group) of men with or without ID (see Table 19.2), and with or without histories of arrest, cautions (a formal police warning not to repeat an illegal behaviour, acknowledged by the perpetrator, that remains on their file and can be used in decision making by criminal justice practitioners in relation to any subsequent alleged criminal offence), or convictions. Men were chosen because there is some evidence that the judgements of women differ from those of men (Gilligan, 1982; Walker, 1995, cited in Langdon et al. *ibid.*, but see McDermott & Langdon, 2016). The measure of moral decision making was the Sociomoral Reflection Measure – Short Form (SRM-SF, Gibbs et al., 1992), a measure related to Gibb's developmental stage theory, whose psychometric properties for groups of men with ID had previously been examined and found to be satisfactory (Langdon et al., 2010b). The SRM-SF is a brief, 11-item questionnaire measure, relating to seven constructs: (a) contract, (b) truth, (c) affiliation, (d) life, (e) property, (f) law, and (g) legal justice. Participants were asked, first, to use a three-point forced choice format (from 'very important' to 'not important') to consider the importance of behaving in a particular way, or making a particular decision, and then to explain their response. All the participants responded orally. Scoring is complex (described in more detail in Gibbs et al. (1992) cited in Langdon et al., (2011b)), but a high level of inter-rater agreement with an independent scorer was achieved ($r_I = 0.99$).

The findings indicated that the moral decision making of the two groups of men with ID was at a developmentally earlier Gibbsian stage (Stage 2) than that of their general population (GP) counterparts (Stage 3). Of relevance to this Chapter, the decision making of the 'offender' group with ID was more mature than that of their ID non-offender counterparts. Once intellectual functioning was taken into account, the difference was no longer statistically significant. Nevertheless, men with ID and no history of alleged illegal behaviour were reasoning at Stage 1 in relation to the property, law, and legal justice constructs. In contrast, the ID offenders' decision making was at Stage 2; indeed, their performance was very similar to that of GP

Table 19.2 Participants in Langdon et al. (2011b), drawn from data presented in the paper

	Group A		Group B		Group C		Group D	
	Mean	SD	Mean	SD	Mean	SD	Mean	SD
Chronological age	45.35	16.57	33.60	7.54	38.80	15.20	38.70	12.99
Full scale IQ	58.8	5.87	62.9	5.22	89.5	11.12	103.25	5.77
No. of participants	20		20		20		20	

A Men with ID and no self-reported history of arrests, police cautions, or convictions
B Men with ID and a documented history of at least one Crown Court conviction that led to a custodial sentence
C GP men with a documented history of at least one Crown Court conviction that led to a custodial sentence
D GP men with no self-reported history of arrests, cautions, or convictions

young people referred to as 'delinquents' (Stams et al., 2006, cited in Langdon et al., 2011b).

These findings suggest – and they are not inconsistent with Wikström and Treiber's (2007) account – that there is a group of men with ID who are likely to be law-abiding because their decision making is associated with compliance with authorities' rules and regulations and/or they are fearful of the consequences of violations of such rules and regulations. In contrast, alleged or convicted offenders with ID are at greater risk of illegal behaviour because their decision making in respect of key law-related constructs is more likely to be contingent on their interactions with others: there is no internal cognitive 'barrier' to illegal behaviour. However, the direction of the association between behaviour and moral decision making cannot be known from this study.

Gibbs (2010) does not argue that engaging in illegal behaviour is *only* associated with a developmental delay in moral decision making; this is a distal factor. Proximal factors, such as distorted cognitions and limited social skills, also play a part in such behaviour (see also Palmer, 2003). Drawing on this theoretical basis, Langdon and his colleagues (Langdon et al., 2013) went on to adapt for men with mild ID a programme initially designed by Gibbs and his colleagues (Gibbs et al., 1995, 1996; Potter et al., 2001) for young people whose behaviour brought them into contact with the criminal justice system: the Equipping Youth to Help One Another Programme (EQUIP; see Langdon et al. (2013) for a full description of the amended programme). The programme was piloted using a single-case design with three men (the four other men, with autism but not ID, are not the focus of this chapter), detained in secure provision under mental health legislation following a criminal conviction (sexual offences for two participants; arson for the third). Briefly, over the course of four 1-hour sessions a week for 12 weeks, the group engaged in thirty 'active treatment' meetings of three different sorts: (i) anger management and thinking error correction, involving psychoeducation and skills in better managing feelings of anger; (ii) social skills training involving role-play and other methods of dealing with difficult situations constructively; and (iii) social decision making aimed at enhancing moral development through guided discussion and debate about potentially problematic situations. While the original situations of Gibbs and his colleagues were retained, Langdon et al. (2012) made some modifications to make them culturally appropriate for the UK and suitable for use with adults.

Comparison of the three men's pre- and post-treatment scores were encouraging, with two participants showing improvements on all three 'proximal' measures (cognitive distortions, problem-solving, anger). Importantly, on the distal measure of decision making (using the Sociomoral Reflection Measure, see Langdon et al., 2012), the scores of all three participants improved. While there were a number of methodological limitations, such as the absence of a follow-up, and the design did not allow causality to be established, the study suggested a promising addition to interventions for individuals with criminal convictions. A randomised controlled trial of EQUIP has recently been completed, but the findings have not yet been submitted for publication (Langdon, personal communication, February, 2021).

An aspect of moral decision making that may be important, because it relates to the promotion of social relationships (Pojman, 2004), which strengthen 'social bonds' according to social control theory, but has rarely been examined among adults with ID, is prosocial decision making. In her doctoral thesis, Bolton (2006) used five hypothetical vignettes initially devised for children (Eisenberg-Berg & Hand, 1979; Eisenberg et al., 1995), adapted slightly for adult participants (p. 58, Bolton, 2006). In each vignette, the protagonist was placed in a dilemma. They were asked to decide whether or not to help another character who was in need, at a potential cost to themselves in terms of money, social commitments, health, or food. Using the prosocial moral judgement interview devised by Eisenberg-Berg and her colleagues (Eisenberg-Berg & Hand, 1979; Eisenberg et al., 1995), participants were first asked to suggest what they thought the protagonist should do.

While there was an overlap between the performance of the two groups of men, the ID group, recruited from community settings and with no self-reported offending ($N = 20$; mean Full Scale IQ: 58.55 (SD 5.72; range: 52–67), was statistically significantly *less* likely (at $p \leq 0.05$) to act with a 'helping' response than the GP participants ($N = 20$; mean Full Scale IQ on the WASI (Wechsler, 1999): 106.5 (SD: 17.00; range: 80–127)).

However, Bolton demonstrated that, using a 'staged' version, in which the text of the vignettes was simplified and illustrated with cartoons, and understanding was checked after the presentation of each vignette element, the proportion of participants with ID who provided a 'helping' response increased; there was no longer any statistically significant difference between the two groups.

Bolton then compared the participants' reasoning about their decisions, again using standard and 'staged' versions of the vignettes. Once more, there was an overlap between the groups: the only statistically significant difference (at $p \leq 0.05$) was that the prosocial reasoning of the participants with ID was less likely to be 'internal', that is, it did not show the most sophisticated perspective-taking ability and abstract reasoning skills.

Subsequently, Bolton used her 'staged' version of the vignettes to compare her sample of men with ID living in the community ('Community') with a group of 20 similar men detained in hospital under mental health legislation and with an existing or previous conviction by a court ('Detained'). There was no statistically significant difference (at $p \leq 0.05$) between the number of 'help' responses provided by the two groups. Figure 19.1 shows the frequency of different types of prosocial reasoning used by each group. Again, there was no statistically significant difference between the groups: among both, accounts consistent with a 'needs-oriented' stage predominated.

These are preliminary findings, based on a small sample of individuals not all of whom, even using the 'staged' version, appeared to understand the vignettes fully. Nevertheless, they suggest that people with mild ID may be less inclined to act prosocially because they simply fail to realise the benefits for the other person. Such difficulties may be even more salient in everyday life where there are rarely opportunities to replay situations to establish what is happening.

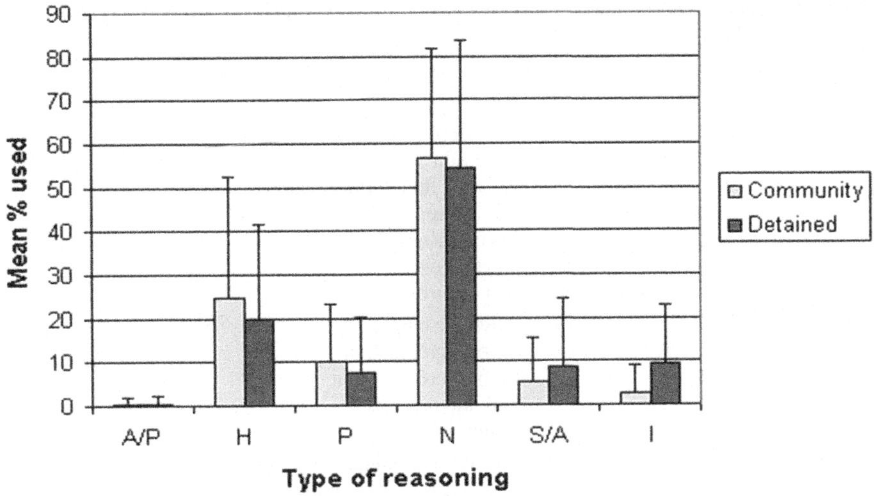

Fig. 19.1 Comparison of the prosocial reasoning profiles used by the 'Community' and 'Detained' groups of men with ID. (From Bolton, 2006). Key: A/P Authoritarian/punishment based, H Hedonistic, P Pragmatic, N Needs-oriented, S/A Stereotypic/approval based, I Internal. Note: error bars = + 1 SD

Still, Bolton's (2006) findings have some implications for the treatment of people with ID who engage in illegal behaviour. In common with frameworks for addressing problematic behaviour in people with ID (e.g. LaVigna & Willis, 1995) and the 'Good Lives Model' (Ward & Brown, 2004) for GP convicted offenders, her findings suggest that the promotion of positive functioning deserves a level of attention similar to that given to the minimisation of illegal behaviour.

Recently, Garrigan and her colleagues (Garrigan et al., 2018) have attempted to bring together the many different components of moral decision making from the research literature into a Social Information Processing-Moral-Decision-making (SIP-MDM) framework. The framework comprises a number of elements (Table 19.3).

The model is complex (see Fig. 19.2, p. 92, Garrigan et al., 2018). Nevertheless, while they acknowledge the risks of creating a 'theory of everything' (p. 95, Garrigan et al., 2018), the challenges of reconciling different theories and underlying philosophies, and the problems of incorporating social neuroscientific research into an area traditionally dominated by developmental psychology, the SIP-MDM is an ambitious attempt to bring together moral decision making, moral development, and behaviour. Though, no doubt, the framework will be refined, it provides an agenda for further research into moral decision making that may lead to better outcomes for 'offenders' and alleged offenders with ID. As yet, however, the relationship of SIP-MDM to current criminological theories is uncertain.

Table 19.3 Social information processing-moral-decision-making (SIP-MDM) framework. (Adapted from Garrigan et al., 2018)

Component type	Component
Cognitive	Working memory Perspective-taking Attention Abstract thought/reasoning Logical reasoning Schemas/scripts Attributions Self-control
Affective	Affective empathy Emotion regulation Emotion recognition Somatic markers Intuition
Social	Social functioning/competence/skills Peer interaction/socialisation Socioeconomic status Culture Parenting/family function
Other	Brain development and integrity Temperament/personality Social information processing

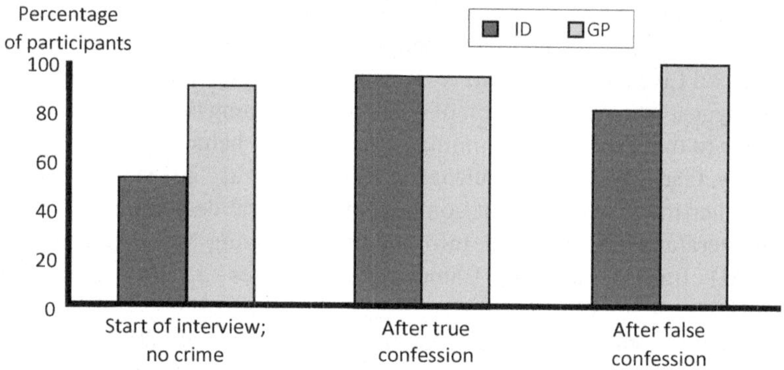

Fig. 19.2 Perceived need for legal advice for the suspect at different stages of the police interview. (From Clare, 2003)

How Are Some People Who May or Not Have Committed Illegal Behaviour 'Transformed' into 'Offenders'?

It is a criminological truism that the 'transformation' of behaviours into criminal 'offences' depends on a series of informal and formal decisions. Irving and Hilgendorf's (Irving & Hilgendorf, 1980; Hilgendorf & Irving, 1981) pioneering study led, at least in the UK, to a focus on a previously neglected area: the decision making of suspects. In this section, we consider what is known for people with ID about their understanding of the *Miranda* rights, the caution, and other legal documents; appreciation of the significance of making a confession; and understanding of magistrates' courts (lower Courts) in England and Wales.

Understanding of the Miranda Rights (USA) and the Caution (England and Wales)

According to a legal competencies or capacity approach, individuals cannot make a valid decision if they have not understood relevant information. In the context of this Chapter, the first important study was carried out by Grisso (1981, cited in Grisso, 2003). He developed and used a set of well-standardised and operationalised methodologies (the *Comprehension of Miranda Rights, CMR*; see Grisso, 2003) to examine the extent to which young people (compared with adults) understood the *Miranda* rights (*Miranda v Arizona*, 384, U.S.436 (1966). Before being interviewed by the police (and at other stages, not of concern here), suspects must be advised of (i) the right to silence, (ii) the potential use of any statement as evidence in Court, (ii) the right to legal advice, and also (iv) that this legal advice is free of charge. Confessions are only admissible in court if (a) the suspect has received this information and (b) has 'waived', or relinquished, the safeguards that it provides 'voluntarily, knowingly, and intelligently' (para.10). Similar provisions exist in a range of jurisdictions.

Briefly, examining performance on one of the *CMR* tasks involving paraphrasing each of the four *Miranda* items among a large sample of adult convicted and non-offenders ($n = 260$, mean prorated IQ: 89.9 (s.d.13.0; test not spceified)) and young people ($N = 431$, mean prorated IQ: 86 (s.d.16.3)) held in custody following alleged or proven 'felonies' (serious offences): (i) performance was significantly related to IQ. The most intellectually disadvantaged persons (IQ $\leq$ 70) demonstrated the poorest level of understanding; (ii) understanding was not related to experience of arrest. In a second task (the *Comprehension of Miranda Rights*, see Grisso, 2003) in which the participants were asked to explain the meaning of key words, performance was also related to intellectual ability. The word 'right' appears both in the *Miranda* items and the information presented orally in the caution and in written format in the 'Notice to Detained Persons' in England and Wales. This was presented as:

> 'Right. You have the right to vote' (Grisso, 1981, p. 238)

followed by a question 'Can you tell me more about what right means' (*ibid.*, p238). Adequate explanations of the word 'right' were provided by fewer than half (43.1%) of the adults and only a quarter (26.7%) of the young people. They did not understand that it was a safeguard. Consistent with the *CMR* results, understanding of a 'right' was significantly related to intellectual ability, with the most intellectually disadvantaged participants performing worst.

Grisso's (1981) methodology was later used in the USA by Fulero and Everington (1995) to examine understanding of the *Miranda* rights among two groups: first, 'non-offenders', diagnosed with 'mild to moderate mental retardation' (*ibid.*, p. 536; no data on intellectual ability provided) and attending specialist day services; 10 percent had convictions for minor offences. The second group ($N = 25$) comprised 'offenders' (mean Full Scale IQ on the WAIS-R (Wechsler, 1981): 65, no details provided) who had all received probation orders following a conviction for a criminal offence; 96 percent had previous convictions.

As expected, on the measures of understanding, the performance of both the groups of people with ID was poorer than that of both samples in Grisso's (1981) study. The extent of their impairment was very striking. On the *Comprehension of Miranda Rights*, the overwhelming majority (90%) of the 'non-offenders' and two-thirds (68%) of the 'offenders' were unable to demonstrate adequate understanding of one or more of the rights. Similarly, the majority of both the 'non-offenders' (83%) and the 'offenders' (56%) were unable to offer any reasonable explanation of the 'rights' item of the *Comprehension of Miranda Vocabulary*. Further studies have been carried out on the equivalent rights in other jurisdictions, with a range of participants (e.g. England & Wales: Clare & Gudjonsson, 1991, 1992; Clare et al., 1998; Scotland: Cooke & Philip, 1998; Canada: Olley & Ogloff, 1993), including people with ID (reported in Clare, 2003).

While most of the studies have had small samples, and there may be arguments about the variety of measures of 'understanding' used (paraphrased recall of the information presented in its entirety and/or as discrete elements (Grisso, 1981), paraphrased recall of single key words (Grisso, 1981), identification of sentences as the 'same' or 'different' from those in the information (*ibid.*), and answering 'yes'/'no' questions about the material (Clare & Gudjonsson, 1991)), the main findings are all consistent. Regardless of the extent of their experience of contact with the criminal justice system, understanding of the Miranda Rights (or their equivalent, e.g. the caution, and written Notice to Detained Persons in England and Wales) is negatively associated with intellectual ability. Nor can it be assessed by self-report (e.g. by asking a suspect 'do you understand?'). Supporting the findings of Clare (2003) with participants with ID and an earlier version of the caution, Fenner and her colleagues (Fenner et al., 2002) demonstrated in England and Wales that while the overwhelming majority (96%, $N = 54$) of both unemployed participants attending a 'job centre' and suspects detained by the police for interviewing claim to understand the caution, their reports cannot be relied on. Worryingly, however, Fenner and her colleagues (Fenner et al., 2002) found that only six participants (11%, $N = 52$) understood the caution in its entirety, even when presented sentence by sentence rather than in its entirety as it would be in real life, and under experimental conditions.

As Gudjonsson (2003) points out, the implication of these studies is that practice needs to change so that:

- Each sentence of the material relating to suspects' rights should be presented in turn.

- The police and their legal advisors or any other person independent from the police, such as parent or (England and Wales) 'appropriate adult', should ask suspects to explain the meaning of their rights in their own words.

He also notes that a third change is required: police officers should be able to explain material about the *Miranda* rights or their equivalents themselves. There is evidence, albeit limited, that in jurisdictions in which the wording of the caution is complex (e.g. England & Wales; Republic of Ireland), it is not completely understood by police officers (Clare et al., 1998), let alone the general public. Unfortunately, as far as we know, no published research has examined the extent to which this simple guidance is being followed in any jurisdiction.

These findings suggest that suspects, and particularly suspects with ID, detained by the police for interviewing may be disadvantaged in their decision making: in particular, they may not understand the 'right to silence' so they inadvertently make admissions; and they may not understand that they can ask for legal advice, even if they are indigent. Such an outcome seems far from the intention of the Supreme Court's majority decision in their judgement in *Miranda* (*Miranda v Arizona*, 384, U.S.436 (1966)).

Decision Making During Police Detention and Interviewing

Unfortunately, while there are case studies (see Gudjonsson, 2003), so far, there has been very little experimental or observational research that has examined decision making by people with ID during police detention and interviewing. While it is now more than 25 years old, the study by Clare and Gudjonsson (1995; Clare, 2003) is therefore included in this Chapter.

In a small study in the UK, Clare and Gudjonsson (1995; Clare, 2003) sought to investigate the decision making of people with ID, focusing on their perception of the significance and consequences of particular courses of action. It was expected that, compared with GP adults, the participants with mild ID would be less likely to appreciate the implications of making a false confession to a major offence.

For the study, a short film was developed and shot in an actual police station. Briefly, it showed a suspect (an actor) being interviewed by a police officer (a serving experienced senior police officer) about a house burglary during which it is alleged that the householder was killed. When confronted with eyewitness evidence, the suspect rescinds his initial denial and admits that he burgled the house ('true confession'). The police officer then suggests to the suspect that he 'knows when people are telling the truth' and that what the suspect has told him is 'half the truth'. The police officer assures the suspect that he will feel 'greatly relieved' about his role in the householder's death when 'he gets it off his chest'. The suspect then admits to killing the householder ('false confession').

Understanding of the context of the police interview was assessed before it was shown to participants in two groups, matched for chronological age. The groups comprised 21 adults (mean Full Scale IQ score on the WAIS-R: 68 (SD 5.2; range: 60–75), all of whom were attending designated community or in-patient facilities for people with ID, and 20 GP participants (mean WAIS-R Full Scale IQ score: 102 (SD 9.02; range: 90–118). All the participants seemed to understand the context of the film, and its ecological validity appeared satisfactory.

As the film was shown, it was paused at pre-set intervals so that different parts of a semistructured interview could be presented to participants. To avoid acquiescent responding to 'yes'/'no' questions (Clare & Gudjonsson, 1993), multiple-choice formats were used, as far as possible, for the presentation of direct questions. Adapting van Someren et al.'s (1994) 'think aloud' technique, those taking part were encouraged to explain their responses. The responses were written down (normally verbatim) for analysis.

Four issues were examined by Clare and Gudjonsson (1995; Clare, 2003). The first related to the perceived need for legal advice at different stages of the interview: at the start, if the suspect had not committed any crime, after the 'true' confession, and after the 'false' confession. Figure 19.2 shows the number of participants in the ID and GP groups who thought the suspect in the film should have legal advice at the start of the police interview, even if he had not, in fact, committed any crime.

Consistent with the findings of the studies referred to earlier in this chapter, significantly fewer of the ID group perceived a need for legal advice at this stage, but understanding of the *benefits* of legal advice at the start of the interview was *not* significantly poorer in the ID group. However, their response typically indicated that legal advice was not required by a suspect who had not committed the offence about which he was being questioned.

Figure 19.2 also shows the proportions of participants in the two groups who thought that the suspect should seek legal advice after, first, making a true confession to the burglary and, then, a false confession to killing the householder. While there were no significant differences between the two groups, the pattern of responses differed slightly. There was a very small *increase* in the percentage of GP participants who perceived the need for legal advice after the suspect had confessed to killing the householder. In contrast, there was a small *decrease* among the ID group.

Three participants with ID who reported that the suspect should have legal advice after the true confession had changed their minds. Unfortunately, only one of the three could give an account of his response. He explained that:

> 'He doesn't need a solicitor because he's making it all up about the killing'.

Second, Clare and Gudjonsson (1995; Clare, 2003) examined the perceived consequences of confessing to a serious crime. Figure 19.3 shows the proportions of participants in the ID and GP groups who selected each of a series of possible 'outcomes' as the most likely consequence of the suspect's (false) confession to killing the householder. Significantly *more* of the ID than the GP group reported that the suspect would be allowed home until his case went to Court. Conversely, significantly *fewer* of the ID group reported that the suspect would be remanded in prison.

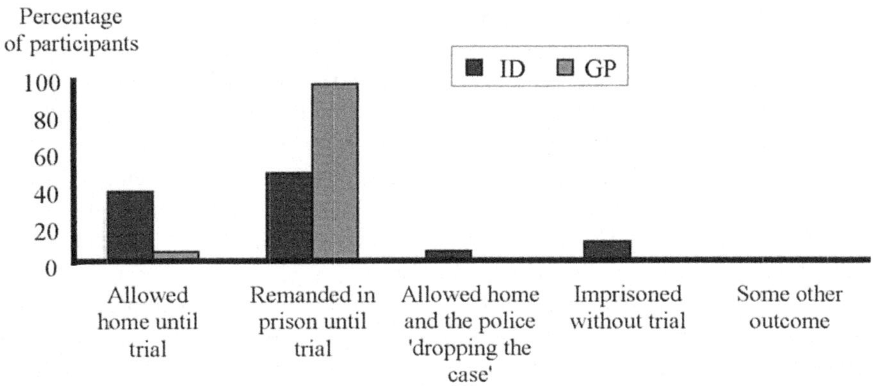

Fig. 19.3 The proportions of participants selecting each of a series of possible outcomes following confession to a serious crime. (From Clare, 2003)

Thirdly, since many suspects later retract admissions made during a police interview, Clare and Gudjonsson (1995; Clare, 2003) examined the likely response to the suspect attempting to withdraw his confession, first to the police officer and then to a court. Again, participants with ID were more likely than their GP counterparts to report that a retracted confession would be believed by the police officer (ID group: 24%, $N = 21$; GP group: 5%, $N = 20$; not statistically significant at $p \leq 0.05$).

Of the four participants with ID who were able to explain why they thought a retraction would be believed, two suggested that the police officer would not have been convinced by the initial confession; the remaining two indicated that it would be convincing because it reflected what had actually happened, that is, that the suspect 'didn't do it'.

Clare and Gudjonsson (1995; Clare, 2003) then examined the participants' perception of the likelihood of a court being convinced by a false confession that was maintained. Again, similar proportions of the two groups reported that they were uncertain whether or not the suspect would be found guilty if he stated his false confession in court. The most frequent explanation of this response was that the outcome would depend on the strength of the other evidence against him. However, a greater proportion of the participants with ID than their counterparts (ID: 29%; n-21; c.f. GP: 10%, $n = 20$; not statistically significant at $p \leq 0.05$) reported that the court would acquit him.

Of the six participants with ID who believed that the court would not be convinced by the suspect's false confession, only one person's explanation (that the suspect was clearly a liar because he had denied and then admitted the burglary during the audiotaped interview) drew on evidence that might be available during a trial. The other participants implied that the court had access to special knowledge and would know that the defendant was not guilty.

The findings suggest that, compared with their GP peers, people with ID may be more sanguine about the significance of making a confession to a very serious offence. They are at increased risk both of appearing to think that suspects might the police would have no problem in beht be problematic retracted and, almost incredibly, that a confession would not be perceived as evidence of guilt even if it were repeated in court. This would matter less if they understood the practical importance of the caution and their legal rights so that they could seek to exercise them at the first opportunity. However, only about half of the sample of people with ID reported that legal advice would be needed from the start of the interview, even if a suspect was innocent of the illegal behaviour about which they were being questioned. Worryingly, only four in five participants thought that legal advice should be sought after a false confession to a very serious offence.

This study has a number of methodological shortcomings: the sample sizes were very small; for ethical reasons, there was no sense for the participants of personal threat; the participants' responses were not audio-recorded so no measure of inter-rater agreement could be carried out; despite simplification during piloting, the interview schedule remained complex and was not accompanied by any visual material to assist the participants. For all that, the findings suggested that the perceptions of police interviewing and its consequences among people with ID might differ from those of their 'general population' counterparts so that they would be disadvantaged in protecting themselves from making admissions or other self-incriminating statements. In terms of the framework used in the USA, the participants with ID did not 'appreciate' the possible personal significance of making a false confession in court. In England and Wales, there were impairments in the ability to 'use and weigh' the information.

In terms of its practical implications, the findings strongly speak to the importance of legal advice. There is overwhelming evidence that, even when legal advisers do not speak, suspects who receive such help are less likely to make self-incriminating admissions, including confessions, and more likely to exercise their right to silence (Gudjonsson & Petursson, 1991; McConville & Hodgson, 1993; Pearse et al., 1998).

Understanding of Magistrates' Courts in England and Wales

As noted previously, according to a functional approach to decision making, individuals cannot make valid decisions if they do not understand relevant information. In England and Wales, 95% of trials (involving approximately 1.5 million defendants a year) begin and end in magistrates' courts, where they are heard by a district judge, or more commonly, three magistrates. In contrast with Crown courts, where the most serious alleged offences are heard, there is no jury.

Until now, though there have been a number of potentially relevant studies, little systematic information has been available about the extent to which aspects of their functioning are understood by anyone (though see Souza & Kemp, 2009; Talbot, 2008; McLeod et al., 2010), let alone people with ID.

In his doctoral thesis, McCombie (2017) used an established video that was developed to train lawyers (Hannibal & Mountford, 2013) in three experimental studies. In turn, these studies focussed on (i) understanding of legal terminology, (ii) the identification of court actors (such as the prosecution and defence lawyers) and understanding of their role, and (iii) understanding of the court's proceedings. The three studies involved the same groups of 30 adults with mild ID (Full Scale IQ on the WASI-II (Wechsler, 2011): 60.2 (s.d. = 10.6; range: 45–83). For ethical reasons, McCombie was not able to ask about whether or not they had any experience of a Magistrates' court, as a defendant, alleged victim, and/or another kind of witness. The performance of this group was compared with that of a control group of 30 GP individuals (mean FSIQ: 112.0; s.d. = 14.28; range: 83–142).

The findings indicated that the group of people with ID had a more limited understanding of Magistrates' courts than their GP counterparts. First, they had many more (often statistically different at $p < 0.001$) misconceptions about the legal terminology used. They were best able to explain terms that may be heard outside a courtroom (e.g. 'burglary', 'stole') but found great difficulty with 'technical' language (e.g. 'aggravating feature', 'either-way offences'). Of very considerable concern, only one person (out of 15 in this part of the study) and consistent with other studies in the ID group (such as Ericson & Perlman, 2001) was able to provide the meaning of a key term – 'prosecuted' – correctly: they thought it meant 'guilty'.

Second, the group of people with ID were less able (mainly significantly less able $p < 0.001$) than their GP counterparts to identify from the film the roles of court actors such as 'defendant', 'defence solicitor' (lawyer), 'Magistrates'. Table 19.4 compares the number and percentage of correct identifications of court actors between the ID and GP groups and their general population (GP) counterparts.

The participants were also worse at describing the roles of court actors, often referring instead to what the person was physically doing (sitting, talking etc.). Of concern, the participants with ID showed very limited understanding of the defence solicitor's role in helping defendants present their best case. Such a lack of awareness is consistent with the findings of Clare and Gudjonsson (1995; Clare, 2003) regarding the need for legal advice during police detention and interviewing.

Finally, McCombie (2017), as part of his investigation of participants' understanding of Magistrates' courts proceedings, carried out a content analysis of the misunderstandings demonstrated by participants. Compared with their GP

Table 19.4 The number and percentages of each participant group who correctly identified court actors. (Adapted from McCombie, 2017)

Court actor	ID group $N = 30$ (%)	GP group $N = 30$ (%)
Defendant	2 (6.7)	26 (80.0)
Prison officer	5 (16.7)	19 (63.3)
Legal advisor	0 (0.0)	9 (30.0)
Magistrate	5 (16.7)	21 (70.0)
Prosecution solicitor	2 (6.7)	24 (80.0)
Defence solicitor	3 (10.0)	21 (70.0)

counterparts, the participants with ID were much more likely to attribute guilt to the defendant in the film, even though the charges against him had not been proved.

The findings of McCombie's (2017) set of studies suggested that Magistrates' court trials are too complicated for individuals with ID to understand, potentially placing them at a disadvantage, compared with their counterparts, in making decisions that are in their best interests. This does not imply, however, that people with ID should not attend court if they are charged with a criminal offence; however, they should have access to support. Strikingly, some of the GP participants, including the most intellectually able participants, struggled with one or more of the three domains investigated in the study. The implication is that Magistrates' court trials are too complicated, not only for people with ID but also for the general population. Following a functional approach to decision making (Grisso, 1986, 2003), the findings support the need for the reform of these courts.

Conclusions

In this Chapter, we have described studies relating to decision making by people with ID that have sought to contribute to two important questions in criminological research: (1) *Why* might some people be at greater risk of engaging in illegal behaviour, leading to the attention of the criminal justice system, than others? (2) *How* are those who have engaged in alleged illegal acts 'transformed' through the criminal justice system into 'offenders'?

Our account has highlighted some challenges for further research. First, and as frequently happens in applied research, the investigators have had to develop their own methodologies in an attempt to improve ecological validity. With few exceptions, such as Fulero and Everington's (1995) follow-up to Grisso's (1981) study, the methodologies have rarely been used on more than one occasion; replication is virtually unknown. Second, study designs have been unsophisticated: for example, the associations between performance on different measures have not been investigated. Thirdly, almost always, sample sizes have been very small. Power calculations, used to minimise the likelihood of false positive and false negative findings, are virtually unknown. Rather, convenience samples, differing across studies, have been used. From a practical perspective, this has meant that it has not been possible to examine whether there are important intersections between different characteristics, including those of ethnicity and self-reported gender; instead, the population of people with 'ID' has been treated as homogeneous.

Nevertheless, the findings of the studies have practical implications for 'alleged' and convicted offenders (Langdon et al., 2010a *et seq.*; Bolton, 2006) and for systemic reform (Clare & Gudjonsson, 1995; Clare, 2003; McCombie, 2017). The next task will be to implement and evaluate the impact of these changes on the lives of people with ID whose behaviour has brought them, or may bring them, into contact with the criminal justice system.

Acknowledgements We are grateful to Dr Clare Bolton and Dr Andrew McCombie for permission to include data from their doctoral theses and for their comments on the text. We are also grateful to Dr Peter Langdon for information about his RCT of EQUIP.

References

American Psychiatric Association. (2013). *Diagnostic and statistical manual of mental disorders* (5th ed.). American Psychiatric Association.

Appelbaum, P. S., & Grisso, T. (1988). Assessing patients' capacities to consent to treatment. *The New England Journal of Medicine, 319*, 1635–1638.

Appelbaum, P. S., & Grisso, T. (1995). The MacArthur Treatment Competence Study - I: Mental illness and competence to consent to treatment. *Law and Human Behavior, 19*, 105–126.

Beccaria, C., & Caso, A. (2013). *On crimes and punishments*. Branden Books.

Beirne, P. (1993). *Inventing criminology: Essays on the rise of "Homo Criminalis"*. State University of New York.

Bentham, J. (2017). *An introduction to the principles of morals and legislation*. Anados Books.

Bolton, C. S. (2006). Pro-social decision making in men with learning disabilities. PhD Thesis. Cambridge: University of Cambridge.

Clare, I. C. H. (2003). *'Psychological vulnerabilities' of adults with mild learning disabilities: Implications for suspects during police detention and interviewing.* PhD Thesis. London: King's College.

Clare, I. C. H., & Gudjonsson, G. H. (1991). Recall and understanding of the caution and rights in police detention among persons of average intellectual ability and persons with mild mental handicap. In *Proceedings of the 1st Annual Conference of the Division of Criminological and Legal Psychology, 1*, pp. 34–42. Leicester: British Psychological Society.

Clare, I. C. H., & Gudjonsson, G. H. (1992). Devising and piloting an experimental version of the "Notice to Detained Persons". *Royal Commission on Criminal Justice, Research Study No. 7.* H.M.S.O.

Clare, I. C. H., & Gudjonsson, G. H. (1993). Interrogative suggestibility, confabulation, and acquiescence in people with mild learning disabilities (mental handicap): Implications for reliability during police interrogations. *British Journal of Clinical Psychology, 32*, 295–301.

Clare, I. C. H., & Gudjonsson, G. H. (1995). The vulnerability of suspects with intellectual disabilities during police interviews: A review and experimental study of decision making. *Mental Handicap Research (Journal of Applied Research in Intellectual Disabilities), 8*, 110–128.

Clare, I. C. H., Gudjonsson, G. H., & Harari, P. M. (1998). Understanding of the current police caution (England and Wales*). Journal of Community & Applied Social Psychology, 8*, 323–329.

Cooke, D. J., & Philip, L. (1998). Comprehending the Scottish caution: Do offenders understand their right to remain silent? *Legal and Criminological Psychology, 3*, 13–27.

Crane, L., Maras, K., Hawken, T., Mulcahy, S., & Memon, A. (2016). Experiences of autism spectrum disorder and policing in England and Wales: Surveying Police and the Autism Community. *Journal of Autism and Developmental Disorders, 46*(6), 2028–2041.

Eisenberg, N., Carlo, G., Murphy, B., & Van Court, B. (1995). Prosocial Development in Late Adolescence: A longitudinal study. *Child Development, 66*(4), 1179–1197.

Eisenberg-Berg, N., & Hand, M. (1979). The relationship of preschoolers' reasoning about prosocial moral conflicts to prosocial behavior. *Child Development, 50*, 356–363.

England & Wales in the Act not England and Wales Mental Capacity (England & Wales) 2005. https://www.legislation.gov.uk/ukpga/2005/9/contents.

Ericson, K. L., & Perlman, N. B. (2001). Knowledge of legal terminology and court proceedings in adults with developmental disabilities. *Law and Human Behavior, 25*(5), 529–545.

Farrington, D. P., & West, D. J. (1995). Effects of marriage, separation and children on offending by adult males. In Z. S. Blau & J. Hagan (Eds.), *Current perspectives on aging and the life cycle* (Vol. 4: Delinquency and disrepute in the life course, pp. 249–281). JAI Press.

Fenner, S., Gudjonsson, G. H., & Clare, I. C. H. (2002). Understanding of the current police caution (England and Wales) among suspects in police detention. *Journal of Community and Applied Social Psychology, 12*, 83–93.

Fulero, S. M., & Everington, C. (1995). Assessing competency to waive *Miranda* rights in defendants with mental retardation. *Law and Human Behavior, 19*, 533–543.

Garrigan, B., Adlam, A. L. R., & Langdon, P. E. (2018). Moral decision making and moral development: Toward an integrative framework. *Developmental Review, 49*, 80–100.

Gibbs, J. C. (2010). Moral development and reality: *Beyond the theories of Kohlberg and Hoffman* (2nd ed.). Sage.

Gibbs, J. C., Basinger, K. S., & Fuller, D. (1992). *Moral maturity: Measuring the development of Sociomoral reflection*. Erlbaum Associates.

Gibbs, J. C., Potter, G., & Goldstein, A. P. (1995). *The EQUIP programme: Teaching youth to think and act responsibly through a peer-helping approach*. Research Press.

Gibbs, J. C., Potter, G., Barriga, A. Q., & Liau, A. K. (1996). Developing the helping skills and prosocial motivation of aggressive adolescents in peer group programs. *Aggression and Violent Behavior, I*, 283–305.

Gilligan, C. (1982). *In a different voice: Psychological theory and women's development*. Harvard University Press.

Gottfredson, G. H., & Hirschi, T. (1990). *A general theory of crime*. Stanford University Press.

Grisso, T. (1981). *Juveniles' waiver of rights. Legal and psychosocial competence*. Plenum Press.

Grisso, T. (1986). *Evaluating competencies. Forensic assessments and instruments*. Plenum Press.

Grisso, T. (2003). *Evaluating competencies. Forensic assessments and instruments* (2nd ed.). Kluwer Academic/Plenum Publishers.

Grisso, T., & Appelbaum, P. S. (1998). *Assessing competence to consent to treatment: A guide for physicians and other health professionals*. Oxford University Press.

Gudjonsson, G. H. (1992). *The psychology of interrogations, confessions and testimony*. John Wiley.

Gudjonsson, G. H. (1994). Psychological vulnerability: Suspects at risk. In D. Morgan & G. Stephenson (Eds.), *Suspicion and silence: The right to silence in criminal investigations*. Blackstone Press.

Gudjonsson, G. H. (2003). *The psychology of interrogations and confessions. A handbook*. John Wiley.

Gudjonsson, G. H., & Petursson, H. (1991). Custodial interrogation: Why do suspects confess and how does it relate to their crime, attitude and personality? *Personality & Individual Differences, 12*, 295–306.

Hannibal, M., & Mountford, L. (2013). *Criminal litigation handbook*. Oxford University Press.

Hilgendorf, E. L., & Irving, B. (1981). A decision making model of confessions. In S. M. A. Lloyd-Bostock (Ed.), Psychology in legal contexts. Palgrave Macmillan.

Hodgins, S. (1992). Mental disorder, intellectual deficiency and crime: Evidence from a birth cohort. *Archives of General Psychiatry, 49*, 476–483.

Irving, B., & Hilgendorf, L. (1980). *Police interrogation: The psychological approach. Research Studies No. 1*. HMSO.

Kohlberg, L. (1969). Stag and sequence: The cognitive development approach to socialization. In D. A. Goslin (Ed.), *Handbook of socialization theory and research* (pp. 347–480). Rand McNally.

Langdon, P. E., Clare, I. C. H., & Murphy, G. H. (2010a). Developing an understanding of the literature relating to the moral development of people with intellectual disabilities. *Developmental Review, 112*, 629–649.

Langdon, P. E., Murphy, G. H., Clare, I. C. H., & Palmer, E. (2010b). The psychometric properties of the Socio-Moral Reflection Measure – Short Form and the Moral Theme Inventory for men with and without intellectual disabilities. *Research in Developmental Disabilities, 31*, 1204–1215.

Langdon, P. E., Clare, I. C. H., & Murphy, G. H. (2011a). Moral reasoning theory and illegal behaviour by adults with intellectual disabilities. *Psychology, Crime & Law, 17*, 101–115.

Langdon, P. E., Murphy, G. H., Clare, I. C. H., Steverson, T., & Palmer, E. J. (2011b). Relationships among moral reasoning, empathy, and distorted cognitions in men with intellectual disabilities and a history of criminal offending. *American Association on Intellectual and Developmental Disabilities, 116*, 438–456.

Langdon, P. E., Murphy, G. H., Clare, I. C. H., Palmer, E. J., & Rees, J. (2013). An evaluation of the EQUIP Treatment Programme with men who have intellectual or other developmental disabilities. *Journal of Applied Research in Intellectual Disabilities, 26*, 167–180.

LaVigna, G. W., & Willis, T. J. (1995). A positive behavioural support model for breaking the barriers to social and community inclusion. *Learning Disability Review, 10*, 16–23.

Law Commission. (2016). Unfit to plead? Volumes 1 and 2. (Report no. 364). H.M.S.O. http://www.lawcom.gov.uk

Lindsay, W. R., & Dernevick, M. (2013). Risk and offenders with intellectual disabilities: Reappraising Hodgins (1991) classic study. *Criminal Behaviour and Mental Health, 23*, 151–157.

Lindsay, W. R., & Taylor, J. L. (2018). Historical and theoretical approaches to offending in people with intellectual and developmental disabilities. In W. R. Lindsay & J. L. Taylor (Eds.), *Offenders with intellectual and developmental disabilities* (pp. 3–20). Wiley.

Maras, K., Mulcahy, S., Crane, L., Hawken, T., & Memon, A. (2018). Obtaining best evidence from the autistic interviewee: Police-reported challenges, legal requirements and psychological research-based recommendations. *Investigative Interviewing: Research and Practice, 9*, 52–60.

McCombie, A. J. L. (2017). *Understanding magistrates' courts: Implications for individuals with intellectual disabilities*. PhD Thesis. Cambridge: University of Cambridge.

McConville, M., & Hodgson, J. (1993). *Custodial legal advice and the right to silence*. The Royal Commission on Criminal Justice Research Study No. 16. HMSO.

McDermott, E., & Langdon, P. E. (2016). The moral reasoning abilities of men and women with intellectual disabilities who have a history of criminal offending behaviour. *Legal and Criminological Psychology, 21*(1), 25–40.

McLeod, R., Philpin, C., Sweeting, A., Joyce, L., & Evans, R. (2010). *Court experience of adults with mental health conditions, learning disabilities and limited mental capacity. Report 3: At Court*. Ministry of Justice.

Murphy, G., & Clare, I. C. H. (2003). Adults' capacity to make legal decisions. In D. Carson & R. Bull (Eds.), *Handbook of psychology in legal contexts* (2nd ed.). Wiley.

Murphy, G., & Clare, I. C. H. (2012). Working with offenders or alleged offenders with intellectual disabilities. In E. Emerson, C. Hatton, K. Dickson, R. Gone, A. Caine, & J. Bromley (Eds.), Clinical psychology and people with intellectual disabilities (2nd. ed.). Wiley.

Murphy, G. H., & Mason, J. (2014). People with intellectual disabilities and offending behaviours. In E. Tsakanikos & J. McCarthy (Eds.), *Handbook of psychopathology in intellectual disability*. Springer Science.

Nicholas, S., Gray, N. S., & Snowden, R. J. (2018). In W. R. Lindsay & J. L. Taylor (Eds.), *Offenders with intellectual and developmental disabilities* (pp. 3–20). Wiley.

Noble, J. H., & Conley, J. W. (1992). Towards an epidemiology of relevant attributes. In R. W. Conley, R. Luckasson, & G. N. Bouthilet (Eds.), *The criminal justice system and mental retardation* (pp. 17–53). Paul H. Brookes.

Olley, M. C., & Ogloff, J. R. P. (1993). *Competency to understand charter cautions: A preliminary investigation*. Paper presented at the Annual Meeting of the Canadian Psychological Association, Montreal.

Palmer, E. J. (2003). *Offending behaviour: Moral reasoning, criminal conduct and the rehabilitation of offenders*. Willan Publishing.

Pearse, J., Gudjonsson, G. H., Clare, I. C. H., & Rutter, S. C. (1998). Police interviewing and psychological vulnerabilities: Predicting the likelihood of a confession. *Journal of Community and Applied Social Psychology, 8*, 1–21.

Piaget, J. (1932). *The moral judgement of the child*. Routledge and Kegan Paul.

Pojman, L. P. (2004). *Introduction to philosophy: Classical and contemporary readings* (3rd ed.). Oxford University Press.

Potter, G. B., Gibbs, J. C., & Goldstein, A. P. (2001). *The EQUIP implementation guide: Teaching youth to think and act responsibly through a peer-helping approach.* Research Press.
Smetana, J. G. (2006). Social-cognitive domain theory: Consistencies and variations in children's moral and social judgements. In M. Millen & J. G. Smetana (Eds.), *Handbook of moral development* (pp. 119–155). Laurence Erlbaum Associates.
Souza, K., & Kemp, V. (2009). *Study of defendants in magistrates' courts.* Ministry of Justice.
Stams, G. J., Brugman, D., Dekovic, M., van Rosmalen, L., van der Laan, P., & Gibbs, J. C. (2006). The moral judgement of juvenile delinquents: A meta-analysis. *Journal of Abnormal Child Psychology, 34*, 697–713.
Talbot, J. (2008). *No one knows: Report and final recommendations. Prisoners' voices: Experiences of the criminal justice system by prisoners with learning difficulties and disabilities.* Prison Reform Trust.
Turiel, E. (1983). *The development of social knowledge: Morality and convention.* Cambridge University Press.
Underwood, L., Forrester, A., Chaplin, E., & McCarthy, J. (2013). Prisoners with neurodevelopmental disorders. *Journal of Intellectual Disabilities and Offending Behaviour., 4*(1/2), 17–23.
van Someren, M. W., Barnard, Y. F., & Sandberg, J. A. C. (1994). *The think aloud method: A practical guide to modeling cognitive processes.* Academic Press.
Walker, L. J. (1995). Sexism in Kohlberg's moral psychology. In W. M. Kurtines & J. L. Gerwirtz (Eds.), *Moral development: An introduction* (pp. 83–108). Allyn and Bacon.
Ward, T., & Brown, M. (2004). The good lives model and conceptual issues in offender rehabilitation. *Psychology, Crime & Law, 10*, 243–257.
Wechsler, D. (1981). Wechsler adult intelligence scale-revised (WAIS-R). The Psychological Corporation.
Wechsler, D. (1999). Wechsler abbreviated scale of intelligence (WASI). Pearson Assessment.
Wechsler, D. (2011). *Wechsler abbreviated scale of intelligence – 2nd edition (WAIS-II).* NCS Pearson.
Wikström, P.-O. (2004). Crime as alternative: Towards a cross-level situational action theory of crime causation. In J. McCord (Ed.), *Advances in Criminological Theory 13, Beyond empiricism: Institutions and intentions in the study of crime* (pp. 1–37). Transaction.
Wikström, P.-O., & Treiber, K. (2007). The role of self-control in crime causation. *European Journal of Criminology, 4*(2), 237–264.
Winstanley, M., Webb, R. T., & Conti-Ramsden, G. (2020). Developmental language disorders and risk of recidivism among young offenders, *The Journal of Child Psychology and Psychiatry*, https://doi.org/10.1111/jcpp.13299
Woodbury-Smith, M. R., Clare, I. C. H., Holland, A. J., Kearns, A., Staufenberg, E., & Watson, P. (2005). A case-control study of offenders with high functioning autistic spectrum disorders. *Journal of Forensic Psychiatry & Psychology, 16*, 747–763.
Woodbury-Smith, M. R., Clare, I. C. H., Holland, A. J., & Kearns, A. (2006). High functioning autistic spectrum disorders, offending and other law-breaking: Findings from a community sample. *Journal of Forensic Psychiatry & Psychology, 17*, 108–120.
Woodbury-Smith, M. R., Clare, I. C. H., Holland, A. J., Watson, P., Bambrick, M., Kearns, A., & Staufenberg, E. (2010). Circumscribed interests and 'offenders' with autism spectrum disorders: A case-control study. *Journal of Forensic Psychiatry & Psychology, 21*, 366–377.
World Health Organization. (1996). *International statistical classification of diseases and related health problems (ICD-10).* World Health Organization.
World Health Organization. (2018). *International statistical classification of diseases and related health problems (ICD-11).* World Health Organization. https://icd.who.int/browse11/l-m/en

Part V
Positive Psychology Interventions for Decision Making

Chapter 20
From Social Vulnerability Assessment to Active Prevention Measures: A Decision-Making Perspective

Geneviève Petitpierre and Mireille Tabin

Many aspects of abuse prevention involve decision making. In this chapter, we highlight the role played by decision making within the broader domain of abuse prevention. Abuse prevention takes many forms, follows various paths (passive and active), and operates at many levels (environmental and individual). Several typologies have been created in order to classify the forms of abuse prevention according to their characteristics. One such typology aims to classify the various prevention measures according to the degree of involvement and responsibility they require from their recipients. This binary typology has been proposed by the French epidemiologist Anne Tursz, who distinguishes between active and passive preventive actions. According to Tursz (Tursz, 2002; Tursz & Gerbouin-Rérolle, 2002), passive prevention refers to primary prevention measures aimed at increasing environmental safety. It refers to all measures and/or procedures that protect people from risks without requiring their assistance or mobilization. Whereas the purpose of passive prevention is to protect people without them being either informed or aware of it, active prevention implies explicit involvement of people in their own protection, often through their decision making.

G. Petitpierre (✉) · M. Tabin
Department of Special Education, University of Fribourg, Fribourg, Switzerland
e-mail: genevieve.petitpierre@unifr.ch; mireille.tabin@unifr.ch

I. Khemka, L. Hickson (eds.), *Decision Making by Individuals with Intellectual and Developmental Disabilities*, Positive Psychology and Disability Series,
https://doi.org/10.1007/978-3-030-74675-9_20

Active Prevention

Active Prevention Goals and Strategies

Risk-free environments do not exist, and new risks are constantly emerging that require the updating of passive prevention measures. In the area of abuse, risks on the Internet present, for instance, a new threat and require an updating of prevention measures. Passive prevention cannot pretend to completely eradicate risks (see White et al., 2003). It is able only to limit them below an acceptable threshold. In most cases, ensuring the effectiveness of standards, laws, or regulations requires individuals' active participation (Tursz, 2002). In the area of abuse prevention, for instance, people's knowledge of the laws is a very important protective factor. The combination of active and passive prevention measures, therefore, appears not only reasonable but highly necessary (Kosher & Ben-Arieh, 2020). However, some people are reluctant to shift the responsibility for protection measures onto the vulnerable person (Van Gijseghem, 1999). They think that active prevention could be a source of unintended harm (Allen-Scott et al., 2014). They fear that the person with intellectual and developmental disabilities (IDD) might become suspicious of those around them, think that they are living in a malicious world, or become unnecessarily anxious. Over the past decade, however, attitudes have changed under the influence of self-advocacy movements. Most people with IDD express the desire to be informed and to take an active part in decisions that impact their lives, even on sensitive issues, such as abuse prevention (Hughes et al., 2018; Masse et al., 2009). Research has also shown that negative side effects from attending active abuse prevention programs are either nonexistent (Haseltine & Miltenberger, 1990; Lee & Tang, 1998; Mazzucchelli, 2001; Miltenberger & Olsen, 1996;) or rare (Egemo-Helm et al., 2007) as long as certain precautions are taken during implementation.

Active prevention refers to a set of programs that enable people to understand the risks by empowering and teaching them how to cope with the source of danger in order to further reduce their level of risk. Active prevention does not replace passive prevention, but complements it. Moreover, the combination of passive and active prevention is consistent with Sobsey's (1994) ecological explanatory model of abuse which considers abuse to be explained by the interaction of four factors: (1) the potential victim, (2) the potential offender, (3) the immediate environment, and (4) the culture. Factors relating to any of these four systems can increase the likelihood of abuse (Hickson et al., 2013). This chapter focuses on active prevention in relation to the factors that put the victim at risk. Since decision making is a learned skill, which depends in part on opportunities for learning and practice, the roles and responsibility given to people with IDD and/or their proxy in prevention are also discussed.

The active prevention approach is relatively recent. It states that it is necessary to involve the person, at least partially, in making decisions regarding their own protection. Active prevention uses strategies that differ from one another in terms of recipients' degree of empowerment, recognition of their self-determination abilities, and level of shared responsibility (Bury, 1988; Petitpierre, 2009); for instance, it is possible to consider three potential active prevention strategies:

- *Information.* The goal of this strategy is to make the person aware of the existence of the risk and enable him or her to access a more or less complex body of information, including ways to protect and/or safeguard oneself against the risk (Cherbonnier, 2003). Prevention through *information* is an *open* form of prevention that leaves the recipient free to decide whether or not to take up the message (Bury, 1988). It requires the individual to take some responsibility for managing the risk without offering a lot of guidance. It presumes, rightly or wrongly, that the individual in contact with the information will take ownership of it and act accordingly.
- *Risk education.* This strategy consists of providing the person with thorough knowledge about the target danger or threat. It aims to maximize the person's self-protection skills by teaching him or her to understand the preventive content and/or to develop various ways of acting on the key determinants of their security (Trefois, 2003). This kind of action aims to transform the individual, as well as their relationship to the risk situation, by teaching them new knowledge or skills in order to maximize their abilities, and help them to develop various ways of acting, including making self-protective decisions, to increase their safety and promote their empowerment. The risk education programs vary according to the degree of reflectiveness they aim to elicit in the learners.
- *Persuasion.* This strategy is the third form of active prevention. It plays on seduction or fear through messages or images that encourage or repel. This strategy favors implicit channels to communicate its message. Persuasive strategies intend to elicit physical responses such as fighting off the danger, distancing, or getting away. Enabling the person to understand why the situation is dangerous is not part of the strategy. The path used by persuasive strategies contradicts the respect due to the individual's dignity, freedom, and self-determination. Indeed, "the fact of influencing [even] skillfully a person to make him or her think and act as one wishes corresponds exactly to the definition of the word 'manipulate'" (Trefois, 2003, p. 25).

These three active prevention strategies differ considerably from one another. Knowing their respective objectives, modi operandi, advantages, and disadvantages is critical for choosing a prevention strategy that is respectful of ethical principles. We chose to focus on risk education programs, in particular on programs aimed at decision-making skills improvement, because they value, recognize, and emphasize the person's abilities, which is consistent with both the social-ecological paradigm and the paradigm of social participation. Risk education programs do not just aim to improve the person's functioning, but to empower them, a long-term benefit, and to increase their freedom. These programs challenge the perceptions which view people with IDD as dependent on professionals or proxies for guidance and protection. Decision-making learning programs are opportunities to improve the person's self-protection abilities and consequently to increase both their autonomy and independence under conditions of controlled risk. They allow learners to become aware of their strengths and weaknesses in order to better cope with social risks.

Risk Education Programs and Abuse Prevention in IDD

The risk education strategy aims, through educational intervention, to equip the person with knowledge and self-protection skills against danger. Currently, abuse prevention programs specifically designed or adapted for people with IDD are rare. In 2008, a literature review conducted by Mechling, based on articles published between 1976 and 2006 in 18 English-language journals, reported only 6 intervention studies on the topic. A few years later, Doughty and Kane (2010) systematically reviewed relevant empirical investigations published between 1997 and 2007. Six other studies were identified. More recently, in a meta-analysis of publications published between 2001 and 2018, Park (2020) identified five studies, including Mazzucchelli's (2001) already identified by Mechling (2008). Quantitatively, the results of these reviews show that over time abuse prevention is becoming a more dominant theme (6 studies out of 36 concerning other types of risk in Mechling's, 2008 review; 5 studies out of 12 in Park's, 2020 meta-analysis). The methodology has been strengthened; there has been an increase in sample size and an improvement in the quality of methodological designs. The target audience for abuse prevention programs is predominantly adults. The only exceptions are the studies conducted by Gast et al. (1993) and Watson et al. (1992), which both involved children under 10 years, and Lee and Tang (1998) and Thomas et al. (2018), respectively, which involved adolescents between 11 and 18 years.

Challenges Faced by Abuse Prevention Programs in IDD

Although all identified programs aimed to teach individuals with IDD how to protect themselves from victimization, they differed in terms of main objectives, behaviors targeted by the intervention, program content, teaching methods (e.g., individual versus group training, parent/teacher versus expert/researcher instruction, simulation conditions versus in vivo conditions) as well as concrete organizational modalities (e.g., duration of training, number and frequency of sessions, presence/absence of consolidation or recall phase).

Three categories of programs can be identified: (1) programs focused on learning motor scripts or procedural knowledge, (2) multicomponent programs, and (3) programs focused on teaching problem-solving and decision-making skills. Table 20.1 presents the multicomponent programs and programs focused on problem-solving and decision-making skills identified in the abovementioned reviews. The programs in these two categories are designed to support the development of an active prevention approach more directly by focusing on cognitive strategy processes and generalized applications, rather than on learning specific behavioral responses.

Table 20.1 Training of individuals with IDD: Characteristics of primary studies, ordered by date

Author (year)	Risk type	Participants' characteristics	Design	Outcomes	Setting	Intervention/ instructor	Time and duration	Results
Khemka (2000)[2]	Sexual, physical, and verbal abuse	36 women[a], Mild ID, Ages 21–40 yearrs N (G. exp1) = 12 N (G. exp2) = 12 N (G. ctr) = 12	RCT	Social interpersonal video scale Self social interpersonal decision-making scale Nowicki-Strickland internal-external scale	Community agency	PRE-ESCAPE curriculum Sessions for the group trained in a cognitive decision-making strategy (G. exp1). Sessions for the group trained in a cognitive-motivational strategy decision making (G. exp2). / author +graduate students	10 sessions (45′) over several weeks in small group (n = 2–3)	General improvement in exp. groups in comparison to control one Decision-making: Gexp1 and Gexp2 > Gctrl Cognitive-motivational decision making: Gexp2 > Gexp1; Gexp2 > Gctrl; Gexp1 = Gctrl Internal perceptions of control: Gexp2 > Gexp1 > Gctrl *Generalization in real life:* No report

(continued)

Table 20.1 (continued)

Author (year)	Risk type	Participants' characteristics	Design	Outcomes	Setting	Intervention/ instructor	Time and duration	Results
Mazzucchelli (2001) [1, 3]	Unsafe potentially abusive Situations	20 participants[a] (15 women, 5 men) Mild ID Mean age 34 years N Exp group = 10 N ctrl group = 10	Quasi	Feel safe questionnaire (FSQ) Protective behaviour skills evaluation (PBSE) Comprehensive quality of life scale-ID	–	Feel safe curriculum	One 3-hour session a week, over 4 weeks + a booster session	FSQ: Performance increased in the experimental group from pretest to posttest and from pretest to follow-up after 5 weeks, but no significant interaction between groups and time. PBSE: Performance increased in the experimental group from pretest to follow-up after 5 weeks. The size of this effect was large
Khemka et al. (2005)[2]	Sexual, physical, and verbal abuse	36 women (18 after attrition)[a], Mild to moderate ID Mean age 34 years N (G. exp) = 18 N (G. ctr) = 10	RCT	Decision-making video scale Knowledge of abuse concepts scale Empowerment scale Stress management survey Self decision-making scale	Facilities	ESCAPE curriculum/ members of the research team	18 sessions (curriculum lessons +6 maintenance sessions), once or twice a week, over a 6- to 12-week period in small group ($n = 3$)	Significant differences in favor of the intervention group in decision making, knowledge of abuse and empowerment abilities, but not stress management ones *Generalization in real life:* No report *Maintenance (1 week after the twelve lessons):* Gains are maintained on all three measures

Author (year)	Risk type	Participants' characteristics	Design	Outcomes	Setting	Intervention/ instructor	Time and duration	Results
Hughes et al. (2010) [3]	Sexual, physical, and verbal abuse	7 women With diverse disabilities Ages 40–62 years	One-group design	Recognition of abuse questionnaire Safety self-efficacy questionnaire Safety promoting behavior questionnaire	Centers for independent living	Curriculum ASAP for women Instruction in class sessions, information, activities, handbook/staff of the center	Twice weekly over 4 weeks (8 sessions of 2.5 hours each)	Significant increases from baseline to post-intervention were found on measures of self-efficacy and safety skills *Generalization:* No report *Maintenance:* No report
Ward et al. (2013) [3]	Violence in dating	31 adults[a], Gender: 14 women, 17 men	One-group design	Social networks measure (SNM) Interpersonal violence interview (IVI)	Community agencies	Friendships and dating program/ personnel working in community Agencies	20 sessions (1.5 hrs) twice per Week over a 10-week period in group ($n = 3–7$)	Increase of the participants' social network size Decrease of interpersonal violence incidents *Maintenance (after 10 weeks):* Gains are maintained

(continued)

Table 20.1 (continued)

Author (year)	Risk type	Participants' characteristics	Design	Outcomes	Setting	Intervention/ instructor	Time and duration	Results
Hickson et al. (2015) [3]	Sexual, physical, and verbal abuse	58 participants[a] Mild to moderate ID Mean age 38 years N (G. exp) = 30 N (G. ctr) = 28 Gender: 29 men, 29 women	RCT	Decision-making scale On 6 vignettes	Facility	ESCAPE-DD/ special education graduate students	Session on a weekly or bi-weekly basis for an average of about 7 weeks In small group (n = 2–4)	Problem awareness: Gexp = Gctrl Overall effective decision making: Gexp > Gctrl Safe-now effective decision making: Gexp > Gctrl

Note. Publications identified by [1]Mechling (2008), [2]Doughty & Kane (2010), and [3]Park (2020)
[a]Asterisks indicate the total number of participants (the number of participants in the intervention group is half)

Multicomponent Programs

The acquisition of general knowledge (vocabulary, cultural concepts, rules and practices, etc.) is necessary to recognize, identify, and deal with situations of abuse. Many programs combine the teaching of responses or scripts (how to act) with teaching how to recognize conditions when a decision has to be made (when to act). The teaching content is quite similar from one program to another even though concrete materials may vary. It generally includes information on safety rules, appropriate and inappropriate situations, acceptable and unacceptable interactions, safe and unsafe secrets, sexuality, gender differences, words required to understand abuse concepts, and feelings, types of relationships, personal boundaries, first impressions, as well as relaxation and/or self-assertion skills.

Multicomponent programs mainly focus on the information to be communicated rather than on the learner and the way he or she reasons or processes social information. For example, Foxx et al. (1984) designed a board game in which the person had to imagine solutions for social situations involving confrontation (e.g., "You are at a party, and a stranger keeps touching you. What should you say to him?"). However, despite the fact that the task requires significant socio-cognitive processes in order to propose adequate solutions with respect to these critical situations, Foxx et al. (1984) focused on procedural knowledge or skills and did not analyze the responses from a socio-cognitive or decision-making perspective. Furthermore, the generalization of acquired skills remains a challenge for most multicomponent programs.

Programs Focused on Decision-Making and Problem-Solving Skills

Perceiving and being aware of the danger, assessing the risks and benefits inherent in each of the possible courses of action, and choosing an appropriate self-protection path are decisive cognitive abilities that can push back the danger on a permanent basis. They include recent preventive programs that emphasize decision making (Hickson et al., 2015; Khemka, 2000; Khemka et al., 2005). These programs encourage people with ID to identify a possible problem in situations involving interactions with others, analyze situations, and make decisions. ESCAPE, a curriculum created by Khemka et al. (2005), is based on theories of decision making. This curriculum was developed in response to women with ID being underprepared to face the risks of abuse to which they are frequently exposed to in their lives. With the availability of ESCAPE and ESCAPE-DD and the more recent versions of this curriculum, people with IDD, irrespective of gender, are given the opportunity to learn decision-making-based skills so as to not react in a generalized way to risky situations, but rather to analyze them in depth and to make strategic decisions that combine caution and autonomy (Hickson et al., 2015; Khemka et al., 2005). In social-cognitive programs based on decision making, people learn a generalizable strategy for managing an unlimited number of problems. The effectiveness of the curriculum has been evaluated in randomized controlled trial studies. ESCAPE and ESCAPE-DD have proved to be effective in empowering people with ID to protect themselves from social threats (Hickson et al., 2015; Khemka et al., 2005), by improving their decision-making skills.

Another interesting program is The Safety Class program, an abuse prevention group program initially developed for women with diverse disabilities (Robinson-Whelen et al., 2014) which has been adapted, through a community-based participatory process, to the specific needs and wishes of people of any gender with ID (Hughes et al., 2018). The program's objective is identification of warning signs of abuse, safety relationships, safety planning, communication, and help-seeking skills, along with key knowledge about healthy boundaries, nature and types of abuse, relaxation training, respect in relationships, and disability rights. The effectiveness of the program has been evaluated using a participatory research approach involving 12 Centers for Independent Living and 170 adults with ID (Hughes et al., 2020). The results show that, in comparison to the control group, participants in the training group significantly improved their posttest knowledge about healthy relationships, but no longer differed at follow-up. The improvement on key facts about abuse and safety depended on the training site, as far as abilities relating to recognizing warning signs, safety and communication skills, and safety-related self-efficacy feelings were concerned. The improvement noted could not be explained by training, as both groups improved their abilities between the pretest and the posttest. On the contrary, the experimental group showed significantly greater improvement in its risk-planning skills compared to the control group. Differences at follow-up approached significance.

The results of these recent studies show promising effects of active prevention programs, particularly those targeting socio-cognitive and decision-making skills, on improving people's safety knowledge and skills relating to the ability to protect themselves. However, they also show that the gains remain fragile and are not always maintained over time. Some authors (Hughes et al., 2020; Mazzucchelli, 2001) suggest planning booster sessions to improve the fluidity of new skills remobilization. In our view, an overall analysis would be required to understand the reasons why the individuals do not maintain the newly acquired skills on a long-term basis. It is not clear whether the decline in skills is due solely to personal factors. In a previous study which included respect for privacy (Masse & Petitpierre, 2011), we found that practices and lifestyles within families and institutions often contradicted the principles of respect expected and taught in programs. Parents and professionals said that it was not uncommon for them to enter the person's bathroom while the person was there, even though they knew that they should not do so ("at home I do that all the time [going into the bathroom when my son or daughter is there]" (a parent), "Yes, it often happens to me with my kid, ... it's true that I come in from time to time" (another parent), "It's something that happens every day" (a professional), "I mean, it's lived experience" (another professional)). Problem awareness is a key ability in the decision-making process. The person's analysis is influenced by mnemonic counterproductive traces (latent mental structure, Crick & Dodge, 1994) of multiple previous and/or current experiences. An *intrusion tolerant* mnemonic trace can inhibit the reflective process as it is usually only when the person finds it difficult to draw a ready-made answer from their memory that he or she engages in a thoughtful and deliberate thought process (Kahneman, 2011). This also raises the issue of the effect on the decision-making process of being exposed to conflicting cues. Perhaps it would be necessary to train not only cognitive but also metacognitive processes of people with ID to get around this obstacle. With respect to training programs, we must keep in mind that prior representations need to be disclosed, sometimes deconstructed, before new learning and new skills can be developed; otherwise they remain ephemeral and fragile with the risk of a rapid return to previous ways of thinking and behaving (Turiel, 1983). Another concern is the potentially disruptive impact of stress on real-time decision-making processes (Brown, 2011). Although advances have been made in the field of active prevention, future research should continue to evaluate and compare methods for improving self-protective skills.

Social Vulnerability

In our society, the precautionary approach and the need to think in terms of prevention and protection have led to the designation of certain groups as vulnerable and intrinsically fragile (Parliamentary Assembly Council of Europe [PACE], 2009). People with IDD, as well as children, elderly or migrant persons, etc., are generally among the groups considered vulnerable. Their situation is particularly scrutinized with regard to the judicial system, education, employment, housing, scientific research, as well as with regard to racism, discrimination, and, of course, abuse (Brown, 2003). In its broadest sense, the notion of *social vulnerability* refers to "the set of disadvantages that an individual may face when he or she attempts to take their place as a productive member of society" (Jawaid et al., 2012, p.335). In a more limited sense, it refers to victimization (Fisher et al., 2013).

Collective and Personal Social Vulnerability

We propose to distinguish between collective and personal vulnerability. Collective social vulnerability (CSV) provides information on the degree of exposure of a group of individuals to a given category of risk. The assessment of a population's social vulnerability is based on statistical standards, informed by prevalence and/or epidemiological studies that examine the links between the characteristics of this group (including those of its environment) and the risk of being victimized. Prevalence data are valuable in situating the extent of risk for populations with respect to etiology, age, gender, and presence or absence of associated impairments (e.g., Chan et al., 2018; Sobsey et al., 1997; Sullivan & Knutson, 2000). However, "the fact of being recognized as disabled does not make a person certainly, definitively, and in all cases *vulnerable*" (Juilhard & Blanc, 2003, p. 132). Because of interpersonal differences, knowledge about collective social vulnerability is not sufficient to account for the degree of risk incurred by an individual even if he or she is part of a population considered *at risk*.

Personal social vulnerability (PSV) refers, on the other hand, to the risk incurred by a single person given their individual characteristics. By insisting on the distinction between CSV and PSV, the French Senate warned against the limits of the indications conveyed by the indices of CSV and stressed the importance of an individual assessment of vulnerability (Juilhard & Blanc, 2003). This refers to the person's abilities and experience. It is likely to vary according to their life contexts and life stages. The French Senate highlighted that the consideration of PSV should not be seen as a defect in the expression of solidarity and responsibility of the community towards the person with a disability designated as vulnerable, but rather as a tangible sign that aims to express the need for proportionality of protection in order not to compromise the autonomy of the person (Juilhard & Blanc, 2003).

Factors of Personal Social Vulnerability

Factors of PSV can be extracted from various sources: prevalence studies; comparative studies (e.g., Fisher et al., 2012, 2013, 2018; Jawaid et al., 2012); empirical studies which have gathered the views of people with IDD themselves, professionals, or families (Hickson et al., 2013; Masse & Petitpierre, 2011); reviews of research already completed on the topic (Fisher et al., 2008, 2016); and, finally, theoretical models (Greenspan et al., 2001; Nettelbeck & Wilson, 2002) that provide sets of factors and suggest relationships between them. PSV factors, identified from some of these key sources, are presented in Table 20.2.

Many factors determine PSV. However, due to the quality of the data, the respective weights of factors determining individual vulnerability could not be calculated at this stage, even for gender and/or age (Hughes et al., 2012; Jones et al., 2012). With regard to cognitive factors, particularly those underlying cognitive and socio-cognitive functioning, their influence is being clarified thanks to studies focused on fine-tuned modeling of specific processes, such as decision making (Hickson & Khemka, 2014) or problem-solving skills (Greenspan et al., 2001). Further developments are required to prioritize abilities that need to be assessed and to target the goals of preventive actions.

Social Vulnerability Assessments

The evaluation of social vulnerability aims to assess an individual's risk of social victimization (e.g., physical and sexual assault, financial abuse, psychological abuse). In current intervention practices, social vulnerability assessment is usually conducted by professionals as clinical observations. This evaluation format leaves, however, a lot of room for the subjectivity of the observers. In this context, standardized tools (e.g., self-reported or/and informant-rated tests), are welcome as they offer more uniform evaluation criteria. Together with clinical observations, they

Table 20.2 Personal social vulnerability factors identified in key sources

	Prevalence studies, meta-analyses & reviews	People with IDD & proxies' points of view	Nettelbeck & Wilson's model (2002)	Greenspan's model (2001)
Ontogenetic development, history, actual life conditions, lifestyle & routines, activities	Gender, age, etiology, presence of associated impairments	Time spent in institutional contexts; limited opportunities to learn autoprotective skills; past history of abuse; financial dependance; social isolation	Socio-demographic variables, socio-psychological influences, limited opportunistic influences, alcoholism	
Appearance			Attractiveness	
Personality		Passivity/submissiveness, learned helplessness, aggressiveness; asking to be alone when they would need monitoring, introversion; need to be liked or wanted; fearfulness	Aggressiveness, limited self-esteem, gullibility, etc.	Motivation, goals, efficacy beliefs, affect/attention
Interaction & communication		Low assertiveness and self-empowerment skills, limited ability to verbalize, say no, refuse, ask for help; no friends	Limited communication, interpersonal or conflict resolution abilities	Low assertiveness
Everyday intelligence (social & practical)		Limited ability to understand situations	Limited problem solving ability	Fluid intelligence, e.g., perspective taking
		Limited knowledge of the rules/rights (e.g., privacy, dignity, etc.)		Crystallized intelligence, e.g., credulity
Physical competence		Limited motor abilities (e. g., ability to walk away, to take shelter)		Physical state, e.g., fatigue; trait, e.g., strength

provide a more complete and nuanced picture of a person's strengths and limitations. Over the last two decades, several tools aimed at assessing the social vulnerability of people with IDD have been developed in English-speaking countries: the Test of Interpersonal Competence and Personal Vulnerability, the Social Vulnerability Scale, and the Social Vulnerability Questionnaire.

The Test of Interpersonal Competence and Personal Vulnerability

The Test of Interpersonal Competence and Personal Vulnerability (TICPV, Wilson et al., 1996) is a self-reported 20-item measure designed to assess the ability of people with ID to choose the most appropriate response in interpersonal situations where there may be a risk of victimization. The measure explores four risk categories: theft, physical or verbal aggression, inappropriate requests or attempts at manipulation, and sexual assault.

The TICPV aims to evaluate the way in which adults with ID analyze social information present in so-called analogous situations (e.g., hypothetical situations mimicking real-life situations). The person being assessed has to imagine themself facing an inappropriate action or request from someone in their environment (a relative or a stranger). Three alternative answers are proposed (see Table 20.3 for an example). The person has to indicate which one is the *most cautious*. Responses that are overly conciliatory with the aggressor or may increase the degree of hostility shown by the aggressor are considered incorrect (0 points). The expected (*most cautious*) response is the one that has the potential to end the interaction, reduce the harmfulness of the consequences for the victim, or decrease the degree of hostility shown by the aggressor. It is scored 1 point. A high score indicates a low level of social vulnerability, with a maximum score of 20.

The TICPV was first developed and validated in 1996 with a sample of 40 adults with ID. Scores on the TCIPV significantly distinguished individuals with ID who had suffered from victimization (assault, sexual assault, robbery, financial exploitation, break-in) and individuals with ID who had not been victimized. More precisely, lower scores on the TICPV in part indicated that victims had problems discriminating strangers or acquaintances from friends, especially when limits are set on what others can reasonably ask for (Nettelbeck & Wilson, 2001). Results indicated that neither age nor adaptive behavior was significant covariates influencing performance on the TICPV. While performance on the TICPV did correlate in the predicted direction with IQ (i.e., low social vulnerability and high IQ), social vulnerability appears to involve some skills over and above adaptive behavior and IQ, successfully identifying victims from nonvictims (Wilson et al., 1996).

Table 20.3 Example of a TICPV item (Wilson et al., 1996)

Item number 3, "theft" risk	If you are sitting on a bus and a woman says she likes your watch and asks you for it, you should		
Response options	**(a)** Tell her where she can buy one like it	**(b)** Yell at her	**(c)** Give the watch to her

The Social Vulnerability Scale

The Social Vulnerability Scale (SVS) was originally an 18-item measure which aimed to assess the gullibility of people with IDD (Greenspan & Stone, 2002; Stone & Sofronoff, 2006). This informant-rated measure was afterwards adapted and validated among various populations: neurotypical children (Seward et al., 2018), children with Asperger syndrome (Sofronoff et al., 2011), and older adults with neurological conditions (Pinsker et al., 2006). The SVS has a two-factor structure, with some items focusing on credulity (tendency to believe something questionable) and others on gullibility (vulnerability to being tricked) that can lead to vulnerability in daily life (Sofronoff et al., 2011). A third-party respondent indicates how often the person being assessed exhibits the target behaviors. Respondents (e.g., family members, caregivers) are asked to rate the items of the SVS on a 5-point Likert-type scale (see Table 20.4 for an example). A high score indicates a high level of social vulnerability, with a maximum score between 90 and 140 according to the number of items of the SVS version.

Building upon Greenspan and Stone's (2002) work, Pinsker et al. (2006) developed a modified version of the Social Vulnerability Scale (SVS), a 28-item measure, which aimed to assess the gullibility and credulity of older adults with and without a neurological condition. Among older adults, social vulnerability is not related to advancing age, but rather to neurological conditions (Pinsker et al., 2006). Results of the SVS suggest that proficiencies in language and higher-order reasoning may aid with detecting deceptive cues and with formulating strategies to avoid victimization, thus decreasing social vulnerability.

Likewise, the Gullibility Scale, a self-report measure for neurotypical adults, has a two-factor structure: insensitivity to cues of untrustworthiness (e.g., "I'm pretty good at working out when someone is trying to fool me") and persuadability (e.g., "If anyone is likely to fall for a scam, it's me") (Teunisse et al., 2020). Results suggest that the acceptance of the false premise, which may motivate gullible behavior, is likely to play an important role in gullibility (Teunisse et al., 2020).

The Social Vulnerability Questionnaire

The Social Vulnerability Questionnaire (SVQ) is a 28-item measure which aims to assess the potential risk of social victimization in people with IDD (Fisher et al., 2018). This informant-rated measure was validated with a sample of 428 parents or

Table 20.4 Example of an SVS item (Pinsker et al., 2006)

Item number 4, subscale "credulity"	[(S)he] believes everything (s)he reads, e.g., in newspapers, magazines, books, and advertisements		
Response options	Never 0	Sometimes 1 2 3	Always 4

Table 20.5 Example of an SVQ item (item #28, credulous) (Fisher et al., 2018)

Item number 28, factor credulous	[(S)he is] likely to believe false claim		
Response options	Not true/never 1	Sometimes 2 3	Very true/always 4

caregivers of individuals with IDD (Fisher et al., 2018). Respondents (i.e., parents or caregivers) are asked to rate the 28 items of the SVQ on a 4-point Likert-type scale (see Table 20.5 for an example). A high score indicates a high level of social vulnerability, with a maximum score of 112 (Fisher et al., 2018). An exploratory factor analysis provides support for a six-factor structure that explains 61% of the variance. Factors are (1) risk awareness (e.g., recognizing potentially dangerous situations), (2) parental independence (e.g., likely to be left alone for an extended period of time), (3) social protection (e.g., part of a social peer group), (4) credulous (e.g., likely to believe false claims), (5) vulnerable appearance (e.g., others consider him/her to look different from his/her peers), and (6) emotional abuse (e.g., people make fun of him/her) (Fisher et al., 2018).

Individuals with IDD who had a higher level of education and had one or more friends were more aware of potential risks (risk awareness factor) and were given more independence (parental independence factor). As such, individuals having lower levels of education and no friends experienced more social victimization because they did not recognize risk and experienced more social isolation (Fisher et al., 2018).

Social Vulnerability and Decision Making

Scores on the TCIVP successfully distinguished victims from nonvictims in a sample of adults with ID (Wilson et al., 1996). However, among adults with Williams syndrome, autism spectrum disorder and Down syndrome, no distinct pattern of victimization was found. Fisher et al. (2018) found that having no friends was a significant predictor of social vulnerability. On the other hand, some adults, with and without IDD, display a high level of social vulnerability, suggesting that with respect to exposure to social situations, other factors are involved. Overall, research studies involving youths and adults with IDD seem to agree that social vulnerability is not linked to gender and appears to involve some skills over and above IQ, social intelligence, and adaptive behavior (Fisher et al., 2018; Sofronoff et al., 2011; Wilson et al., 1996), thus supporting the overall concept of social vulnerability as a dynamic, fluid, and multidimensional construct.

All three measures of social vulnerability the Test of Interpersonal Competence and Personal Vulnerability (TICPV), the Social Vulnerability Scale (SVS), and the Social Vulnerability Questionnaire (SVQ) assess credulity and/or gullibility, either through vignettes implying being tricked (TICPV) or by asking if the person is easily tricked and similar questions (SVS, SVQ). Greenspan's (2009) proposed model for gullible behavior comprises four factors: situational (e.g., time pressure), cognitive, affective, and personality. As such, Greenspan's defines gullibility, not as a personality trait, but rather as an outcome affected by the four factors. The TICPV, the SVS, and the SVQ each assesses, in a different way, decisional efficiency involving cognition, motivation, and emotional components in social situations. For example, the TICPV (Wilson et al., 1996) aims to assess social vulnerability, through evaluating interpersonal problem-solving skills in hypothetical situations. Asking the person to indicate the most cautious answer in each vignette implies that he or she has to think about the possible consequences of the three answers presented as a, b, or c. This hypothetical thinking is an important, likely pivotal, component in decision making, which simultaneously involves demands on cognitive, motivational, and emotional processes (Hickson & Khemka, 2013). The SVQ (Fisher et al., 2018) measures both sociocultural components (e.g., parental independence) and individual components of social vulnerability, such as credulity and risk awareness. Cognitive limitations may limit comprehension of the nature of the situation requiring a decision (e.g., threatening) (Khemka et al., 2013). Hickson and Khemka (2013, p.6) assume that "at the start of the decision-making process, identifying and defining the problem (and hence recognizing that a decision has to be made) are essential components that must occur before solution strategies can be engaged." Risk awareness is thus a key ability required at a very early stage of the decision-making process, emphasizing that decisional processes are at the core of social vulnerability.

Significant gaps remain in the conceptualization of social vulnerability and its assessment in research and clinical settings. The differences in the SVS, SVQ, and TICPV in conceptualizing and assessing social vulnerability raise the issues about conceptualizing vulnerability more generally (for a scoping review, see Enang et al., 2019). What all these measures of social vulnerability share, however, is a view of an interactive schema of victimization, with the underlying hypothesis that "some individual risk variables cannot be changed (e.g., intellectual disabilities) or are difficult to change (e.g., living situation)"; they are nevertheless of the opinion that other victim characteristics reflecting social vulnerability should be amenable to intervention (Nettelbeck & Wilson, 2002, p. 289). These authors consider social vulnerability not as a stable state across the lifespan, but rather as a holistic construct. Further research should try to bridge this gap by establishing a clear definition of social vulnerability, which will allow effective assessments and intervention models to be crafted.

Our Own Research in the Field

Research in the Field of Social Vulnerability Assessment

In French-speaking countries, measures assessing vulnerability are scarce, if not nonexistent. The development of the TV-22 (Test of Social Vulnerability, 22 items), the French, enhanced, and accessible version of TICPV, was intended to fill this gap. This section presents the characteristics of the TV-22 and its relevance for practice and research.

Initially developed and validated in English by Wilson et al. (1996), the TICPV was translated into French by Petitpierre and colleagues in Petitpierre et al., 2011. TICPV has several strengths: it is a self-report measure which studies vulnerability to several sub-categories of risks; it assesses victimization risks from familiar people versus unknown people; its psychometric properties are satisfactory. Several research studies were conducted to adapt the TICPV and led to the TV-22 (Beaufort, 2013; De Palma, 2018; Lopez, 2013; Tabin et al., 2020). The TV-22 is a socio-cognitive test consisting of 22 illustrated vignettes mimicking a social risk. It explores four categories of social risk: theft, physical or verbal aggression, inappropriate requests or attempts at manipulation, and sexual abuse. In each item (vignette), the person assessed is asked to advise a third person (Pierre or Marie, according to the gender of the respondent) facing a risk. The risk may come from known persons (friends, family members, partner) or unknown persons. The test, which is administered on a computer, consists of two parts. In the first part (part A), the vignette is introduced, and then the person being evaluated has to suggest a decision that Pierre/Marie should make (see Fig. 20.1). In part B, the same vignettes as from part A are introduced, but three possible answers (a, b, c) are presented (see Fig. 20.2). A minimum of two testing sessions is required for the administration of the TV-22: one to administer part A of the test, the other for part B, with ideally an interval of 1 day to 2 weeks between sessions.

Part A comprises open-ended questions. For each vignette, the person is first asked to rephrase the situation presented. This is to ensure that the person has understood the situation presented and whether he or she identifies the risk present. Next, he or she is asked to formulate one or more strategies and, finally, to justify the proposed strategies.

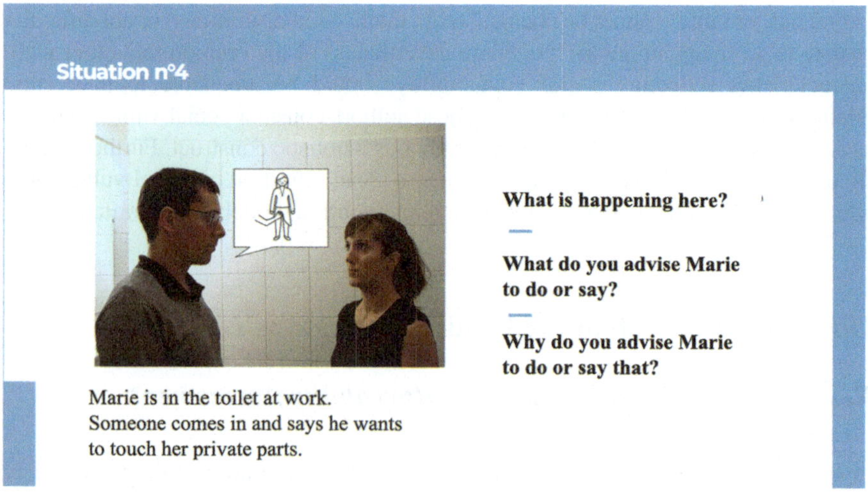

Fig. 20.1 TV-22, female version, part A, situation # 4. (Petitpierre et al., 2020a, translated and reproduced with permission; the pictograms used in the test material come from ARASAAC (http://www.arasaac.org) and are distribuited under the Licence Creative Commons BY-NC-SA)

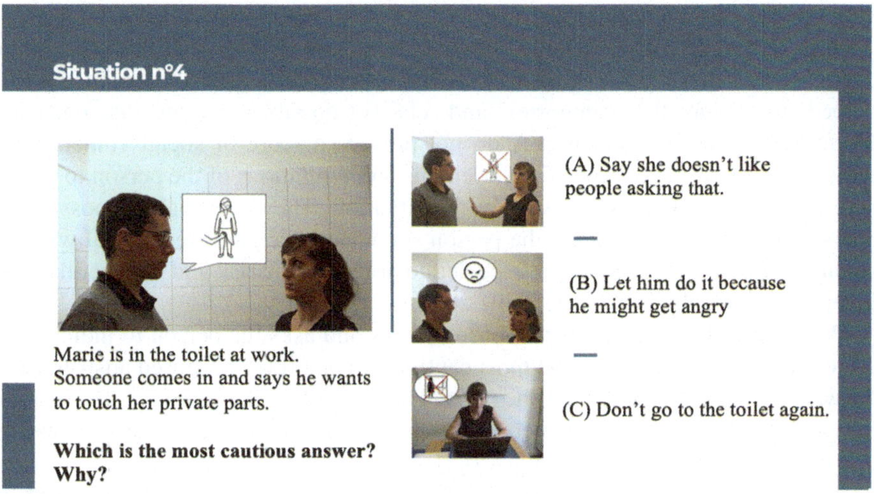

Fig. 20.2 TV-22, female version, part B, summary of the situation # 4. (Petitpierre et al., 2020b, translated and reproduced with permission; the pictograms used in the test material come from ARASAAC (http://www.arasaac.org) and are distribuited under the Licence Creative Commons BY-NC-SA)

Table 20.6 Types of strategies (Petitpierre et al., 2020c, translated and reproduced with permission)

Strategy		Definition
Noneffective decision responses: Strategies that do not minimize or eliminate the risk	No answer	Gives no advice, says "I don't know," or answers nothing
	Submission	Advises to accept, to say "yes," to give the money, etc.
	Emotional	Advises to react aggressively (e.g., by physical or verbal aggression) and to respond to a threat with a threat, or advises to be overly empathetic (e.g., proposing to give a massage to a stranger to calm him down or credulously), advises to ask "why" (e.g., "why are you in my house?" "why are you asking that?"), etc.
	Other non- cautious response	Advises an imprudent response that is neither submissive nor emotional (e.g., walking home late at night, hitchhiking, not going to the toilet anymore, etc.)
Effective decision responses: Strategies that minimize or eliminate risk	Indirect protection	Advises to call on others to resolve the situation (e.g., a trusted person, the police, a doctor, etc.)
	Self-protection	Advises to refuse calmly, to defend oneself verbally by saying "no," to ignore the interlocutor, to leave or to continue on one's way, to close the door, etc.

The types of strategy spontaneously proposed by the person can be grouped into six categories which reflect either effective or noneffective decision responses (see Table 20.6). Two members of the research team separately coded 22 interviews on part A of the TV-22. Cohen's Kappa (κ) was calculated to determine agreement between the two coders on just over 20% of these interviews ($n = 6$). Inter-judge agreement ranged from 0.70 to 1, ranging from satisfactory to perfect agreement (Landis and Koch, 1977). A detailed analysis of the differently coded statements helped to refine the categories and clarify possible coding discrepancies. Nevertheless, results from part A do not provide a score of social vulnerability, rather a qualitative overview of the strengths and limitations of the person to recognize and describe the risk present in the situation. They also make it possible to analyze what types of strategy the person spontaneously proposes to deal with the situation, whether these strategies accelerate or reduce the risk and whether they are independent or submissive (see Table 20.6).

Part B introduces the three response options and asks the person to indicate the option that she or he feels is the most cautious among three proposed answers. The answers are presented one after the other. Then, based on the summary of the situation and the three options, the person is asked to choose the most cautious answer and to justify their answer (see Fig. 20.2).

The results from part B of the TV-22 reflect the individual's overall ability to select appropriate protection strategies that reduce exposure to different categories of social risks. The *most cautious* answer is scored 1, whereas the other options are scored 0. The higher the score (22 points maximum), the more capable the person is of choosing strategies that tend to protect the protagonist (Pierre/Marie).

Psychometric properties of part B of the test were evaluated in a sample of 29 French-speaking adults with ID. The findings provide preliminary support for the use of the TV-22 (Tabin et al., 2020). They show that the tool has very good internal consistency ($\alpha = 0.89$, $\Omega = 0.93$) and good test-retest reliability (r_s (29) = 0.81, $p < 0.01$). As far as external validity is concerned, there is no gender or age effect. As expected, the TV-22 scores correlate with support needs. Participants who appear to be less socially vulnerable (i.e., with higher self-protective skills) are those with lower support needs. There is also a positive relationship between self-protection skills and logical reasoning measured by Raven's Colored Progressive Matrices. Contrary to expectations, however, TV-22 scores do not correlate with scores on the adaptive behavior assessment measured using ABAS-3 (Harrison & Oakland, 2015; see Tabin et al., 2020, for interpretation).

Table 20.7 presents an excerpt of an interview conducted using the TV-22 with an adult with ID, who shows signs of vulnerability.

In part A, the fact that the person answers directly "No!" seems to indicate that he immediately perceives a risk, which tends to be confirmed by his further answer and justification, "Say no. You don't touch private parts." With this self-protection strategy, the interviewee proposes an effective decision response that minimizes the risk. Nevertheless, later on, when he faces three possible answers in part B of the test, he chooses to not go to the toilet again. This choice provides an insight into his difficulty to think about the possible consequences of the answers presented and to assess the adequacy of the chosen solution in the long term. The differences in the answers of this participant between part A and part B of the test also highlight that, for this person, risk awareness may not be his main area of social vulnerability. This example, and particularly the dimension concerning risk awareness, shows the value of combining vulnerability assessment measures. In this case, if we had the results

Table 20.7 Excerpt of an interview conducted with Mr. T (20 years old) using the TV-22

Part A – item #4 – male version	Part B – item # 4 – male version
Psychologist: Situation N°4. Pierre is in the toilet at work. Someone comes in and says she wants to touch his private parts **Interviewee:** No! **Psychologist:** What do you advise Pierre to do or say? **Interviewee:** Say no. you don't touch private parts. **Psychologist:** So he should say no? **Interviewee:** You don't touch private parts. **Psychologist:** Why? **Interviewee:** No because you don't touch private parts	**Psychologist:** Situation N°4. Pierre is in the toilet at work. Someone comes in and says she wants to touch his private parts. I will show you 3 answers, and you will tell me which one is the most cautious. (A) Say he doesn't like people asking that; (B) let her do it because she might get angry; (C) don't go to the toilet again. So, which is the most cautious answer? **Interviewee:** (C) don't go to the toilet again. **Psychologist:** Don't go to the toilet again, why? **Interviewee:** Because private parts should not be shown or touched

Note: English translation

of the SVQ, we could – or not – confirm the fact that the person has more difficulty in assessing consequences than in the risk awareness dimension. Thus, TV-22 data should not be considered in isolation, but rather integrated into a comprehensive assessment of the social vulnerability of individuals with IDD. The score and analysis of the answers help to better understand their risk awareness and self-protection abilities (in hypothetical scenarios). The professionals' clinical observations still remain essential to complete the results of the evaluation. For clinicians, results from the TV-22 could guide the objectives of the individualized assessment if they indicate the need for abuse prevention training, or they could be used as a tool for monitoring the person's progress. For researchers, results from the TV-22 could be part of the assessment of the effectiveness of a decision making-based abuse prevention intervention (e.g., ESCAPE-NOW, Khemka & Hickson, 2015). The participant could answer the TV-22 at baseline (pretest) and after the intervention (posttest). Comparing results from part A between pre- and posttest would allow changes in the participant's decision-making skills for self-protection to be assessed. More precisely, analysis of the participant's decision responses would make it possible to identify if the participant proposes more effective decision responses (e.g., refuse calmly) and/or fewer noneffective decisions (e.g., let him do it because he might get angry) after having experienced the intervention. Results from part B would allow assessment of the effect of a decision making-based abuse prevention intervention on the participant's hypothetical thinking skills and their ability to choose the most self-protective option from the three proposed answers.

Nevertheless, further research is definitely needed on the one hand to ensure the psychometrics properties of part B of the TV-22 and, on the other hand, to further deepen and clarify the definition of social vulnerability. Comparing SVQ and TV-22 results, for example, could provide valuable information into social vulnerability and help to improve its definition.

Implementation of ESCAPE-DD in the Swiss Context: Some Considerations

The ESCAPE-DD curriculum (Khemka & Hickson, 2008) has been translated into French by Petitpierre et al. (2011). The following conditions have been put in place to facilitate implementation of the program in facilities offering residential services and sheltered workshops: ensure that the management of the facility supports the introduction of training; identify an on-site trainer (if possible a person who already provides training at the facility); constitute a pair of trainers (the first round of training has always been delivered by a pair composed of an on-site trainer and an academic collaborator); and identify the necessary adaptations. Adaptations are based on three small exploratory studies carried out in three implementation sites by three master's students (Borloz, 2015; Jonin, 2012; Noir, 2011).

The ESCAPE-DD curriculum consists of 12 small group instructional lessons and six support group sessions. Lessons 11 and 12 are crucial in the program. In Lesson 11, participants apply the 4-step decision-making strategy independently (between learners). As in the following lesson, Lesson 12, the instructor's help should gradually be reduced. Clinical observations suggest that Lesson 11 should not be rushed. All learners must be able to function as leaders of the team before implementing the 4-step decision-making strategy on their own as they will have to do later in Lesson 12. It has been noted that it was energetically and cognitively too demanding for most participants to take the lead for the entire duration of the activity. Asking two or more of them to take it in turns could be a facilitative option. As Hickson et al. (2015) have already pointed out, the third step of the procedure, which involves verifying that the self-protection solutions were considered to meet the requirements of immediate and long-term safety, as well as independence, is the most demanding part of the procedure. In this step, participants also tended to lose their train of thought. An alternative might be to go through steps 2 and 3 in sequence for one solution and then the next, in other words to ask them to generate a solution and then evaluate it right away.

In any case, an important feature of Lesson 11 lies in the fact that it enables participants to hear each other's point of view and to challenge it. For the instructor, access to the learners' position statements in the context of the group is very valuable because it gives him or her a window into the characteristics of the person's reasoning (i.e., their flexibility, independence of reasoning, and so on). It also enables him or her to look at the strength and relevance of the arguments put forward or, on the contrary, the weakness or absence of arguments. In general, over the course of the program, it is also interesting to observe gestures that participants exhibit in conjunction with language (i.e., participants showing awareness of the problem by waving up and down and saying "ouch ouch ouch ouch" or symbolizing the weighing of solutions with their hands). The literature shows that people with ID often use gestures to structure their thinking and support their cognitive activity (Goldin-Meadow, 2003; Lacombe et al., 2020). The "think" gesture (finger rotation movement at temple level) was spontaneously used by some learners in the first step. Gestures expressing doubt or difficulty in choosing were also manifested, particularly in step 3.

Table 20.8 summarizes the adaptations that were found to be necessary when implementing the program in the Swiss context. Some adaptations (in bold in the table) are similar to considerations noted by Hughes et al. (2018) in their feasibility study.

Perspectives and Conclusion

Over the past 5 years, significant progress has been made in identifying individual and environmental risk and protective factors that affect exposure to abuse. Tools have been created to enable the measurement of a person's decision-making skills and/or vulnerability in social situations. They offer an objective measurement and provide useful information on how people reason in certain situations similar to those they might encounter in real life. Neither the strengths nor the weaknesses of these tools are yet known.

With regard to decision-making processes, the initial procedure that is typically attempted by an individual in a given situation is to intuitively search for recollections of past decisions. It is an automatic procedure which does not involve thinking about what the person should do. Rather, when the person finds it difficult to draw a ready-made answer from their memory, he or she engages in a deliberate thought process (Kahneman, 2011). The intuitive procedure works by association and relies on memories derived from past experiences. This implies that being confronted with situations that should not occur, but which nevertheless do occur sporadically in the person's daily environments, can lead them to implicitly record these facts as *ordinary* and can make it difficult for them to think deliberately about them, i.e., be aware of them; distance, analyze, and consider them critically; and develop motivation for change and explore the possibilities for it. Deconstructing the habits of the entourage and the participant's implicit knowledge may require at least five adaptations of the active prevention programs: (1) better integrate the metacognitive dimension to enable the learner to discover their own cognitive strengths and weaknesses; (2) be interested in the participant's previous representations, knowledge, and real-life experiences in order to identify those that could be a potential barrier to new learning; (3) consider working with the immediate entourage (families or professionals) by making them aware of the need for greater congruence between their practices and the principles of protection instilled in the prevention programs; (4) inform the family and caregivers of the contents of the program and ask them to use natural situations that may arise to coach the participant's new decision-making skills; and (5) remember that the best safety net is based on the activation of both sets of measures, active and passive.

Table 20.8 *Adaptations excerpted from French translation of ESCAPE implementation logbooks*

Course of the training	Adaptations	When
A participant discloses abuse during training	Define a clear procedure before the beginning of the training. It should spell out what trainers should do if a participant reports abuse	Before training
The participants' entourage believes that the training will immediately "protect" the participant	Remind those around that learning is a process	
A participant uses a specific means of communication or requires specific conditions to understand	Inform oneself about the way participants communicate and function (fatigability, attentional difficulties, etc.)	
Rename the program with a title that can be understood by Francophones	ESCAPE-DD, French version → "building skills to protect oneself" ("Construire des compétences pour se protéger")	
A participant shows signs of anxiety or stress during the lesson (or training)	It is imperative that the program be facilitated by two instructors present throughout the program. One of them should concentrate on participants' emotional reactions Provide a worry box to collect participants' concerns and remind them of its existence at each session Counterbalance the potentially threatening nature of risk information with activities aimed at identifying participants' strengths and opportunities of supports (cf. Morgan, 2013)	During the training
Participants are afraid to speak up in front of others or of not giving the "right" answer to the questions	Create a respectful learning space for everyone In addition to speaking rules in the instruction manual, give a reminder that learning is the main goal, so everyone has the right not to know, the right to make mistakes, the right to express disagreement, or the right not to understand	During the training
Participants forget or don't use some of the skills they have learned	Review the content of the previous session at the beginning of each new session Explicitly summarize key information after each activity Suggest that participants build up an individual portfolio and update it throughout the sessions: This file should enable them to keep reviewing the notions from one session to the next. It also constitutes a "toolbox" that can be used after the training Take pictures or video of the results of group activities Schedule "refresher" knowledge sessions Inform the family and caregivers of the contents of the program and ask them to use natural situations that may arise to reinforce learning Train proxies in the decision-making process	Before, during, and after the training

(continued)

Table 20.8 (continued)

Course of the training	Adaptations	When
Content of the program		
Module 1 was felt to be very dense, very conceptual, and not easily accessible by some participants.	Increase the number of sessions in the module to allow more time per theme/activity Lighten the module program Translate material into easy to read and understand format	Content adaptation
Tackling the topic of sexual abuse too quickly offends participants' feelings	Reverse the themes by first addressing verbal abuse, then physical abuse, and finally sexual abuse	
Participants generate solutions that are not part of those proposed in the source material	Provide "blank" material to express unanticipated proposals	

References

Allen-Scott, L. K., Hatfield, J. M., & McIntyre, L. (2014). A scoping review of unintended harm associated with public health interventions: Towards a typology and an understanding of underlying factors. *International Journal of Public Health, 59*(1), 3–14. https://doi.org/10.1007/s00038-013-0526-6

Beaufort, N. (2013). *Analyse du traitement de l'information sociale permettant la prise de décision efficace face à Une situation à risques d'abus [analysis of the processing of social information allowing effective decision making in a situation at risk of abuse]*. Unpublished Master Thesis. University of Liège.

Borloz, M. (2015). *Construire des compétences pour se protéger–implémentation du programme Escape DD [Building skills to protect oneself – implementation of the Escape DD program]. Mémoire de maîtrise non publié [Unpublished Master Thesis]*. Université de Fribourg, Faculté de lettres, Département de Pédagogie spécialisée.

Brown, H. (2003). *Safeguarding adults and children with disabilities against abuse*. Council of Europe Publishing.

Brown, H. (2011). The role of emotion in decision making. *The Journal of Adult Protection, 13*(4), 194–202. https://doi.org/10.1108/14668201111177932

Bury, J. A. (1988). *Éducation pour la santé: concepts, enjeux, planifications [Health education: concepts, issues, planning]*. De Boeck-Wesmael.

Chan, K. L., Lo, C. K. M., & Ip, P. (2018). Associating disabilities, school environments, and child victimization. *Child Abuse & Neglect, 83*, 21–30. https://doi.org/10.1016/j.chiabu.2018.07.001

Cherbonnier, A. (2003). Peur, culture et prévention [Fear, culture and prevention]. Bruxelles Santé, no spécial, pp. 42-49. http://www.questionsante.org/03publications/charger/peuretprevention.pdf

Crick, N. R., & Dodge, K. A. (1994). A review and reformulation of social information-processing mechanisms in Children's social adjustment. *Psychological Bulletin, 115*(1), 74–101.

De Palma, N. (2018). TCIVP–FEAI. In *Test de Compétences Interpersonnelles et de Vulnérabilité Personnelle – version Française, Enrichie, Accessibilisée et Informatisée [Test of Interpersonal Competences and Personal Vulnerability–French, streamlined and digitalized version]*. Unpublished Master Thesis. University of Fribourg.

Doughty, A. H., & Kane, L. M. (2010). Teaching abuse-protection skills to people with intellectual, disabilities: A review of the literature. *Research in Developmental Disabilities, 31*(2), 331–337. https://doi.org/10.1016/j.ridd.2009.12.007

Egemo-Helm, K. R., Miltenberger, R. G., Knudson, P., Finstrom, N., Jostad, C., & Johnson, B. (2007). An evaluation of in situ training to teach sexual abuse prevention skills to women with mental retardation. *Behavioral Interventions, 22*(2), 99–119. https://doi.org/10.1002/bin.234

Enang, I., Murray, J., Dougall, N., Wooff, A., Heyman, I., & Aston, E. (2019). Defining and assessing vulnerability within law enforcement and public health organisations: A scoping review. *Health & Justice, 7*(1), 2. https://doi.org/10.1186/s40352-019-0083-z

Fisher, M. H., Baird, J. V., Currey, A. D., & Hodapp, R. M. (2016). Victimisation and social vulnerability of adults with intellectual disability: A review of research extending beyond Wilson and Brewer: Victimisation of adults with intellectual disability. *Australian Psychologist, 51*(2), 114–127. https://doi.org/10.1111/ap.12180

Fisher, M. H., Burke, M. M., & Griffin, M. M. (2013). Teaching young adults with disabilities to respond appropriately to lures from strangers: Responding to strangers. *Journal of Applied Behavior Analysis, 46*(2), 528–533. https://doi.org/10.1002/jaba.32

Fisher, M. H., Hodapp, R. M., & Dykens, E. M. (2008). Child abuse among children with disabilities: What we know and what we need to know. *International Review of Research in Mental Retardation*, 251–289. https://doi.org/10.1016/s0074-7750(07)35007-6

Fisher, M. H., Moskowitz, A. L., & Hodapp, R. M. (2012). Vulnerability and experiences related to social victimization among individuals with intellectual and developmental disabilities. *Journal of Mental Health Research in Intellectual Disabilities, 5*(1), 32–48. https://doi.org/10.1080/19315864.2011.592239

Fisher, M. H., Moskowitz, A. L., & Hodapp, R. M. (2013). Differences in social vulnerability among individuals with autism spectrum disorder, Williams syndrome, and down syndrome. *Research in Autism Spectrum Disorders, 7*(8), 931–937. https://doi.org/10.1016/j.rasd.2013.04.009

Fisher, M. H., Shivers, C. M., & Josol, C. K. (2018). Psychometric properties and utility of the social vulnerability questionnaire for individuals with intellectual and developmental disabilities. *Journal of Autism and Developmental Disorders*. https://doi.org/10.1007/s10803-018-3636-4

Foxx, R. M., McMorrow, M. J., Storay, K., & Rogers, B. M. (1984). Teaching social/sexual skills to mentally retarded adults. *American Journal of Mental Deficiency, 89*, 9–15.

Gast, D. L., Collins, B. C., Wolery, M., & Jones, R. (1993). Teaching preschool children with disabilities to respond to the lures of strangers. *Exceptional Children, 59*, 301–311.

Goldin-Meadow, S. (2003). *Hearing gesture: How our hands help us think*. Harvard University Press.

Greenspan, S. (2009). Foolish action in adults with intellectual disabilities: The forgotten problem of risk-unawareness. *International Review of Research in Mental Retardation, 36*, 147–194. https://doi.org/10.1016/S0074-7750(08)00005-0

Greenspan, S., Loughlin, G., & Black, R. S. (2001). Credulity and gullibility in people with developmental disorders: A framework for future research. In L. M. Glidden (Ed.), *International review of research in mental retardation (Vol. 24)* (pp. 101–135). Elsevier. https://doi.org/10.1016/S0074-7750(01)80007-0

Greenspan, S., & Stone, V. (2002). *Gullibility Rating Scale* [unpublished manuscript].

Harrison, P., & Oakland, T. (2015). *ABAS-3* (3rd ed.) Western Psychological Services.

Haseltine, B., & Miltenberger, R. (1990). Teaching self-protection skills to persons with mental retardation. *American Journal on Mental Retardation, 95*, 188–197.

Hickson, L., & Khemka, I. (2013). Problem solving and decision making. *Oxford University Press*. https://doi.org/10.1093/oxfordhb/9780195398786.013.013.0015

Hickson, L., & Khemka, I. (2014). The psychology of decision making. In *International review of research in developmental disabilities (Vol. 47)* (pp. 185–229). Elsevier. https://doi.org/10.1016/B978-0-12-800278-0.00006-3

Hickson, L., Khemka, I., Golden, H., & Chatzistyli, A. (2013). Views and values of developmental disabilities and domestic violence/sexual assault support professionals regarding the prevention and handling of situations of abuse: Views and values of professionals. *Journal of Policy and Practice in Intellectual Disabilities, 10*(3), 207–214. https://doi.org/10.1111/jppi.12040

Hickson, L., Khemka, I., Golden, H., & Chatzistyli, A. (2015). Randomized controlled trial to evaluate an abuse prevention curriculum for women and men with intellectual and developmental disabilities. *American Journal on Intellectual and Developmental Disabilities, 120*(6), 490–503. https://doi.org/10.1352/1944-7558-120.6.490

Hughes, K., Bellis, M. A., Jones, L., Wood, S., Bates, G., Eckley, L., McCoy, E., Mikton, C., Shakespeare, T., & Officer, A. (2012). Prevalence and risk of violence against adults with disabilities: A systematic review and meta-analysis of observational studies. *The Lancet, 379*(9826), 1621–1629. https://doi.org/10.1016/S0140-6736(11)61851-5

Hughes, R. B., Robinson-Whelen, S., Davis, L. A., Meadours, J., Kincaid, O., Howard, L., Millin, M., Schwartz, M., McDonald, K. E., & The Safety Project Consortium. (2020). Evaluation of a safety awareness group program for adults with intellectual disability. *American Journal of Intellectual and Developmental Disabilities, 125*(4), 304–317. https://doi.org/10.1037/a0019916

Hughes, R. B., Robinson-Whelen, S., Goe, R., Schwartz, M., Cesal, L., Garner, K. B., Arnold, K., Hunt, T., & McDonald, K. E. (2018). "I really want people to use our work to be safe"…using participatory research to develop a safety intervention for adults with intellectual disability. *Journal of Intellectual Disabilities, 24*(3), 309–325., 174462951879346. https://doi.org/10.1177/1744629518793466

Hughes, R. B., Robinson-Whelen, S., Pepper, A. C., Gabrielli, J., Lund, E. M., Legerski, J., & Schwartz, M. (Shell). (2010). Development of a safety awareness group intervention for women with diverse disabilities: A pilot study. *Rehabilitation Psychology, 55*(3), 263–271. https://doi.org/10.1037/a0019916

Jawaid, A., Riby, D. M., Owens, J., White, S. W., Tarar, T., & Schulz, P. E. (2012). 'Too withdrawn' or 'too friendly': Considering social vulnerability in two neuro-developmental disorders: Social vulnerability in neuro-developmental disorders. *Journal of Intellectual Disability Research, 56*(4), 335–350. https://doi.org/10.1111/j.1365-2788.2011.01452.x

Jones, L., Bellis, M. A., Wood, S., Hughes, K., McCoy, E., Eckley, L., Bates, G., Mikton, C., Shakespeare, T., & Officer, A. (2012). Prevalence and risk of violence against children with disabilities: A systematic review and meta-analysis of observational studies. *The Lancet, 380*(9845), 899–907. https://doi.org/10.1016/S0140-6736(12)60692-8

Jonin, C. (2012). *Prévention de la maltraitance à l'encontre des personnes avec une déficience intellectuelle. Administration, validation et adaptations du programme ESCAPE-DD [Prevention of abuse against persons with intellectual disabilities. Administration, validation and adaptations of the ESCAPE-DD program].* Mémoire de maîtrise non publié [unpublished master thesis]. Université de Genève, Faculté de psychologie et des sciences de l'éducation.

Juilhard, J.-M. & Blanc, P. (2003). *Maltraitance envers les personnes handicapées : briser la loi du silence [Abuse of people with disabilities: breaking the code of silence].* www.senat.fr/rap/r02-339-1/r02-339-1.html

Kahneman, D. (2011). *Thinking, fast and slow*. Penguin.

Khemka, I. (2000). Increasing independent decision making skills of women with mental retardation in simulated interpersonal situations of abuse. American Journal on Mental Retardation, 110(3), 387–401. https://https://doi.org/10.1352/0895-8017(2000)105<0387:IIDSOW>2.0.CO;2

Khemka, I., & Hickson, L. (2008). *ESCAPE-DD: An Effective Strategy-Based Curriculum for Abuse Prevention and Empowerment for Adults with Developmental Disabilities.* Center for Opportunities and Outcomes for people with disabilities, Teachers College, Columbia University, New York, NY.

Khemka, I., & Hickson, L. (2015). ESCAPE-NOW: An effective strategy-based curriculum for abuse prevention and empowerment for individuals with developmental Disabilites – NOW. Teachers College, Columbia University, New York, New York.

Khemka, I., Hickson, L., Mallory, S., & Zealand, R. (2013). Social vulnerability of adolescents with autism spectrum disorders and adolescents with other developmental disabilities in situations of negative peer pressure. In *Symposium paper presented at the 46th annual Gatlinburg conference on research and theory in intellectual and developmental disabilities*. TX.

Khemka, I., Hickson, L., & Reynolds, G. (2005). Evaluation of a decision making curriculum designed to empower women with mental retardation to resist abuse. *American Journal on Mental Retardation, 110*(3), 12.

Kosher, H., & Ben-Arieh, A. (2020). Social workers' perceptions of children's right to participation. *Child & Family Social Work, 25*, 294–303. https://doi.org/10.1111/cfs.12685

Lacombe, N., Petitpierre, G., & Dias, T. (2020). Observer les gestes pour analyser les habiletés spatiales des élèves ayant Une déficience intellectuelle [observing gestures to analyze the spatial abilities of students with an intellectual disability]. *Revue suisse de pédagogie spécialisée [Swiss journal of special education], 2*, 47–53.

Lee, K. Y., & Tang, S. C. (1998). Evaluation of a sexual abuse prevention program for female Chinese adolescents with mild mental retardation. *American Journal on Mental Retardation, 103*(2), 105–116.

Lopez, K. (2013). *Adaptation du Test de compétence interpersonnelle et de vulnérabilité personnelle [Adaptation of the Interpersonal Competency and Personal Vulnerability Test].* [unpublished master thesis. University of Geneva]. http://archive-ouverte.unige.ch/unige:30678.

Masse, M., & Petitpierre, G. (2011). *La maltraitance en institution. Les représentations comme moyen de prévention [Institutional abuse. Representations as a means of prevention.].* Editions IES.

Masse, M., Petitpierre, G., Jossevel, J. D. & Vidon, C. (2009). Prévention de la maltraitance en institution: collaboration autour de l'élaboration d'un support pédagogique [Prevention of abuse in institutions: collaboration on the development of an educational support material]. In V. Guerdan, G. Petitpierre, J. P. Moulin & M. C. Haelewyck (Eds). *Participation et responsabilités sociales: un nouveau paradigme pour l'inclusion des personnes avec une déficience intellectuelle [Participation and Social Responsibilities: A New Paradigm for the Inclusion of Persons with Intellectual Disabilities]* (pp. 660–669). Peter Lang.

Mazzucchelli, T. G. (2001). Feel safe: A pilot study of a protective behaviours programme for people with intellectual disability. *Journal of Intellectual & Developmental Disability, 26*(2), 115–126. https://doi.org/10.1080/13668250020054431

Mechling, L. (2008). Thirty year review of safety skill instruction for persons with intellectual disabilities. *Education and Training in Developmental Disabilities, 43*(3), 311–323.

Miltenberger, R. G., & Olsen, L. A. (1996). Abduction prevention training: A review of findings and issues for future research. *Education and Treatment of Children, 19*, 69–82.

Nettelbeck, T., & Wilson, C. (2001). Criminal victimization of persons with mental retardation: The influence of interpersonal competence on risk. In *International review of research in mental retardation, (Vol. 24)* (pp. 137–169). Elsevier. https://doi.org/10.1016/S0074-7750(01)80008-2

Nettelbeck, T., & Wilson, C. (2002). Personal vulnerability to victimization of people with mental retardation. *Trauma, Violence, & Abuse, 3*(4), 289–306. https://doi.org/10.1177/1524838002237331

Noir, S. (2011). La maltraitance envers les personnes avec Une déficience intellectuelle: Administration du 1er module du programme ESCAPE-DD [abuse prevention of persons with intellectual disabilities: Administration of the 1st module of the ESCAPE-DD program]. Mémoire de maîtrise non publié [unpublished master thesis]. Université de Genève, Faculté de psychologie et des Sciences de l'éducation.

Park, E.-Y. (2020). *Meta-analysis on the safety skill training of individuals with intellectual disabilities* (pp. 1–15). International Journal of Disability. https://doi.org/10.1080/1034912X.2020.1761540

Parliamentary Assembly Council of Europe [PACE] (2009). *Une stratégie intégrée contre la violence. Lignes directrices du Conseil de l'Europe sur les stratégies nationales intégrées de protection des enfants contre la violence [Council of Europe Guidelines on integrated national strategies for the protection of children against violence].* https://rm.coe.int/CoERMPublicCommonSearchServices/DisplayDCTMContent?documentId=090000168046eb80.

Petitpierre, G. (2009). Maltraitance et déficience mentale: Des démarches de prévention passives à la prévention participative [From passive to participatory prevention approaches]. In V. Guerdan, G. Petitpierre, J. P. Moulin, & M. C. Haelewyck (Eds.), *Participation et responsabilités sociales: un nouveau paradigme pour l'inclusion des personnes avec une déficience intellectuelle [Participation and Social Responsibilities: A New Paradigm for the Inclusion of Persons with Intellectual Disabilities]* (pp. 660–669).

Petitpierre, G., Noir, S., & Jungo, M. (2011). *Test de compétence interpersonnelle et de vulnérabilité personnelle (TCIVP) [Test of Interpersonal Competences and Personal Vulnerability]* [unpublished manuscript].

Petitpierre, G., Noir, S., Jungo, M. & Bruttin, C. (2011). *Un programme d'apprentissage de stratégies de prévention des abus et de développement de l'empowerment destiné à des adultes ayant une déficience intellectuelle – traduction française du programme ESCAPE DD [A program to teach abuse prevention decision making and empowerment strategies to adults with intellectual disabilities. French translation of the Escape DD program].* [unpublished manuscript].

Petitpierre, G., Tabin, M., Diacquenod, C., & de Palma, N. (2020a). *Test d'évaluation de la Vulnérabilité Sociale–22 items, version féminine, partie A [Social Vulnerability Assessment Test–22 items, female version, part A].* University of Fribourg.

Petitpierre, G., Tabin, M., Diacquenod, C., & de Palma, N. (2020b). *Test d'évaluation de la Vulnérabilité Sociale–22 items, version féminine, partie A [Social Vulnerability Assessment Test–22 items, female version, part B].* University of Fribourg.

Petitpierre, G., Tabin, M., Diacquenod, C., & de Palma, N. (2020c). *Test d'évaluation de la Vulnérabilité Sociale–22 items, manuel de passation [Social Vulnerability Assessment Test–22 items, instruction manual].* University of Fribourg.

Pinsker, D. M., Stone, V., Pachana, N., & Greenspan, S. (2006). Social vulnerability scale for older adults: Validation study. *Clinical Psychologist, 10*(3), 109–119. https://doi.org/10.1080/13284200600939918

Robinson-Whelen, S., Hughes, R. B., Gabrielli, J., Lund, E. M., Abramson, W., & Swank, P. (2014). A safety awareness program for women with diverse disabilities: A randomized controlled trial. *Violence Against Women, 20*(7), 846–869. https://doi.org/10.1177/1077801214543387

Seward, R. J., Bayliss, D. M., & Ohan, J. L. (2018). The Children's social vulnerability questionnaire (CSVQ): Validation, relationship with psychosocial functioning, and age-related differences. *International Journal of Clinical and Health Psychology, 18*(2), 179–188. https://doi.org/10.1016/j.ijchp.2018.02.001

Sobsey, D. (1994). *Violence and abuse in the lives of people with disabilities: The end of silent acceptance?* Brooks Publishing.

Sobsey, D., Randall, W., & Parrila, R. K. (1997). Gender differences in abused children with and without disabilities. *Child Abuse and Neglect, 21*, 707–720.

Sofronoff, K., Dark, E., & Stone, V. (2011). Social vulnerability and bullying in children with Asperger syndrome. *Autism, 15*(3), 355–372. https://doi.org/10.1177/1362361310365070

Stone, V., & Sofronoff, K. (2006). *The social vulnerability scale: SVS*. Unpublished questionnaire.

Sullivan, P. M., & Knutson, J. F. (2000). Maltreatment and disabilities: A population-based epidemiological study. *Child Abuse & Neglect, 24*(10), 1257–1273. https://doi.org/10.1016/S0145-2134(00)00190-3

Tabin, M., Diacquenod, C., De Palma, N., Gerber, F., Straccia, C., Wilson, C., Kosel, M., & Petitpierre, G. (2020). Cross-cultural preliminary validation of a measure of social vulnerability in people with intellectual disabilities. *Journal of Intellectual & Developmental Disability*, 1–13. https://doi.org/10.3109/13668250.2020.1793450

Teunisse, A. K., Case, T. I., Fitness, J., & Sweller, N. (2020). I should have known better: Development of a self-report measure of gullibility. *Personality and Social Psychology Bulletin, 46*(3), 408–423. https://doi.org/10.1177/0146167219858641

Thomas, N., Nattala, P., & Seshadri, S. P. (2018). Effectiveness of behavioral skills training (BST) on knowledge of sexual abuse and resistance ability among children with intellectual disability: A pilot study., *6*(1), 9.

Trefois, P. (2003), *L'usage de la peur – une approche légitime en promotion de la santé ? [The use of fear - a legitimate approach to health promotion?]* http://www.questionsante.org/03publications /charger/peuretprevention.pdf.
Turiel, E. (1983). *The development of social knowledge: Morality and convention*. Jossey-Bass.
Tursz, A. (2002). A quels obstacles spécifiques se heurte la prévention des accidents domestiques intéressant les enfants ? [what are the specific obstacles to the prevention of domestic accidents involving children?] [Oral presentation]. MIRe-DREES Congress on The Prevention Factory.
Tursz, A., & Gerbouin-Rérolle, P. (2002). L'expérience de la prévention des accidents domestiques: des connaissances applicables aux accidents de la route ? [Experience in the prevention of domestic accidents: knowledge applicable to road accidents?]. *Actualité et dossier en santé publique [Public Health News and Features], 41*, 35–38. https://www.hcsp.fr/explore.cgi/Adsp?clef=78
Van Gijseghem, H. (1999). Us et abus de la mise en mots en matière d'abus sexuels [uses and abuses of the discourse touching upon sexual abuse].. Les Éditions du Méridien.
Ward, K. M., Smith, C., & Windsor, R. (2013). A Friendship & Dating Program for adults with intellectual and developmental disability: A formative evaluation. *Research in Developmental Disabilities, 51*, 22–32. https://doi.org/10.1352/1934-9556-51.01.022
Watson, M., Bain, A., & Houghton, S. (1992). A preliminary study in teaching self-protective skills to children with moderate and severe mental retardation. *The Journal of Special Education, 26*, 181–194.
White, C., Holland, E., Marsland, D., & Oakes, P. (2003). The identification of environments and cultures that promote the abuse of people with intellectual disabilities: A review of the literature. *Journal of Applied Research in Intellectual Disabilities, 16*, 1–9. https://doi.org/10.1046/j.1468-3148.2003.00147.x
Wilson, C., Seaman, L., & Nettelbeck, T. (1996). Vulnerability to criminal exploitation: Influence of interpersonal competence differences among people with mental retardation. *Journal of Intellectual Disability Research, 40*(1), 8–16. https://doi.org/10.1111/j.1365-2788.1996.tb00597.x

Chapter 21
Behavioral Approaches to Teaching Decision Making to Individuals with Intellectual and Developmental Disabilities

Alyssa M. Uher, Cynde K. Josol, and Marisa H. Fisher

Introduction

Advocacy movements of the past 50 years have resulted in significant changes and opportunities for individuals with intellectual and developmental disabilities (IDD). The photographic exposé, "Christmas in Purgatory," by Blatt and Kaplan (1974), along with progress in the disability rights movement, prompted deinstitutionalization and the move of thousands of individuals with IDD from institutions into group homes, halfway houses, and independent living facilities (Buchanan & Walmsley, 2006; Mansell, 2006; Nielsen, 2012). With this move, individuals with IDD experienced new possibilities for the development of personal control and self-determination, allowing them to participate in both major life decisions and minor daily living decisions for the first time (Smith et al., 2005).

When physical integration did not lead to significant improvements in the quality of life or well-being of individuals with IDD, self-advocates in the 1980s began to speak and stand up for their rights, aiming to reduce the inequalities and discrimination they experienced and to increase their own self-determination (Buchanan & Walmsley, 2006; Fenn & Scior, 2019). With declarations and publications from the United Nations, World Health Organization, World Bank, and other international organizations, the rights of individuals with IDD have taken center stage, giving them increased control of their own lives. The United Nations Convention on the

A. M. Uher · C. K. Josol · M. H. Fisher (✉)
Michigan State University, East Lansing, MI, USA
e-mail: fishermh@msu.edu

I. Khemka, L. Hickson (eds.), *Decision Making by Individuals with Intellectual and Developmental Disabilities*, Positive Psychology and Disability Series,
https://doi.org/10.1007/978-3-030-74675-9_21

Rights of Persons with Disabilities (2006), for example, highlighted as global priorities that individuals with disabilities should have their rights protected and be enabled to participate fully within their societies. The key tenets of the disability human rights movement emphasize autonomy, choice, independence, equality, and participation for all individuals with disabilities, providing them with the rights to make their own life decisions.

Today, individuals with IDD increasingly live and successfully participate in the community, and many residential, vocational, leisure, and other initiatives have been developed to promote their social inclusion (TASH, 2006). While inclusion leads to more social interactions, job opportunities, and community participation, it also allows individuals with IDD to make their own decisions about how they want to live their lives (Abbott & McConkey, 2006; Kampert & Goreczny, 2007; Simplican et al., 2015). Indeed, individuals are increasingly provided with the opportunity to make decisions related to employment, independent living, and community access (Shogren et al., 2020; Wehmeyer & Garner, 2003).

Unfortunately, these increased opportunities can also lead to increased risks of victimization. Compared to those without disabilities, individuals with IDD are more socially vulnerable or at risk of experiencing adverse events that could affect their emotional, physical, or financial well-being (Fisher et al., 2016; Jawaid et al., 2012; Ticoll, 1994). Individuals with IDD experience higher rates of physical and sexual assault, personal or property crimes, psychological abuse (e.g., bullying, persuasion), discrimination, neglect, and financial victimization (Fisher et al., 2012; Petitpierre et al., 2013; Ticoll, 1994; White et al., 2003). Such increased rates of victimization may be related to the inability to recognize and respond appropriately to potentially harmful situations (Fisher et al., 2013).

With this backdrop, it is important to consider the decision-making capacity of individuals with IDD and to ensure that they are equipped with the skills to make appropriate decisions when faced with potentially dangerous situations. In this chapter we will first briefly describe why it is critically important for individuals with IDD to have the ability to make informed decisions. We will specifically discuss the importance of effective decision making when there is an increased risk of victimization or abuse. Next, we will discuss how behavioral intervention approaches have been effectively developed to build and support decision-making behaviors in response to high-risk decision making situations for individuals with IDD. Specifically, we will describe behavior skills training (BST) and then provide examples of how BST has been used to teach individuals with IDD to make appropriate decisions in response to situations involving potential abduction, sexual assault, bullying, or other forms of social victimization. We will end with a call for future research to evaluate additional behavioral interventions and suggestions for practice to increase the decision-making capacity of individuals with IDD in response to high-risk decision-making situations.

Importance of Decision Making by Individuals with IDD

Being able to engage in autonomous decision making, or the ability to independently make a decision and communicate that decision to others, is an essential part of living a self-determined life for individuals with IDD (Werner, 2012). Indeed, self-determination is achieved when a person is the primary causal agent in their life and is free to make decisions without undue external influence or pressure (Wehmeyer, 1997). For all people, decisions range from low-risk decisions, such as deciding what to eat for dinner each night, to high-risk decisions, such as deciding whether to live independently or to consent to an intimate relationship.

Historically, individuals with IDD were afforded opportunities to make only low-risk decisions related to activities of daily living and were less often provided opportunities to make high-risk decisions (Brown & Brown, 2009). As notions of self-determination have gained in importance, however, there has been a push for adults with IDD to live as independently as possible and to make their own decisions, even in high-risk situations (Bannerman et al., 1990; Sandjojo et al., 2019). Given the wide range of decision-making contexts and the complexity of the decision making process, special considerations may be needed for supporting individuals with IDD in making their own decisions.

Various decision-making frameworks and models have been developed specific to the decision-making abilities of those with IDD. For example, to assist individuals who do not have the capacity to make decisions with full independence, the *Supported Decision Making Inventory* (SDMI) (Shogren et al., 2017a, b, 2020; Shogren & Wehmeyer, 2015, 2016) was developed to identify an individual's decision-making capacity, environmental demands, and supports needed for individuals with IDD to make effective decisions (Shogren, et al., 2017b). Other frameworks for individuals with IDD include the *Pathways Model of Decision Processing* (Hickson & Khemka, 2014), which synthesizes other decision-making models and outlines possible pathways that may be used by individuals with IDD to make independent decisions on their own.

Such frameworks and models can assist in the design of assessments and interventions to promote successful decision making for individuals with IDD in various social contexts and situations. They also allow for the identification of specific considerations to account for when teaching individuals with IDD how to make informed decisions. For example, although individuals with IDD may be provided with opportunities to make decisions, they face numerous barriers that limit their decision-making abilities (Stang et al., 2009; Werner, 2012) and that may lead to difficulties in making effective decisions (Hickson & Khemka, 2014). Such barriers include poor self-awareness, learned helplessness, low self-esteem, self-deprecation, and lack of recognition of strengths and weaknesses (Field, 1996; Field & Hoffman, 1994; Sparks et al., 2016). The barriers will not only impact decision-making abilities, but will also inform how to begin teaching individuals essential decision-making skills. Another critical consideration regarding decision making by individuals with IDD is the social vulnerability of this population in comparison to those without IDD, thus increasing their susceptibility to victimization.

Social Vulnerability and the Importance of Effective Decision Making

Compared to those without disabilities, individuals with IDD experience increased rates of victimization, including child abuse, bullying, sexual and physical assault, and property and financial crimes (Fisher et al., 2016; Jawaid et al., 2012; Ticoll, 1994). These increased rates of victimization are often attributed to both demographic and societal variables that lead to social vulnerability, indicating that individuals with different forms of IDD may be more or less socially vulnerable (Fisher et al., 2013).

Depending on the type of IDD, individuals may experience verbal and memory deficits, difficulties with problem-solving, problems with abstract thinking and overly concrete thought processes, or a tendency toward acquiescence and suggestibility that impede successful decision making (Greenspan et al., 2001; Werner, 2012). For example, individuals with intellectual disability (ID) may be unable to recognize potentially abusive situations, or they may be more gullible, lacking the skills to evaluate whether certain claims are true (Fisher et al., 2013; Greenspan et al., 2001). On the other hand, due to poor social communication skills, individuals with autism spectrum disorder (ASD) may have difficulty recognizing and rejecting certain forms of social victimization (Jawaid et al., 2012; Nettelbeck & Wilson, 2002; Wilson et al., 1996). Finally, because of their extreme sociability and trust in others (e.g., see Jones et al., 2000), individuals with Williams syndrome may experience impaired judgment that could lead to an increased likelihood of complying with questionable requests (Jawaid et al., 2012; Lough & Fisher, 2016; Thurman & Fisher, 2015).

Others have attributed the social vulnerability and victimization of individuals with IDD to certain societal practices that limit the abilities of individuals with IDD to learn to make effective decisions. For example, many individuals with IDD are taught to be compliant and cooperative, decreasing their ability to recognize and stop unwelcome advances (Rosen, 2006; Westcott & Jones, 1999). Further, individuals with IDD often experience decreased privacy and increased dependency, increasing their vulnerability to physical and sexual abuse, as well as personal and financial victimization (Hughes et al., 2012). In essence, when individuals with IDD are not taught and given the opportunity to lead their own lives and to make their own decisions, they are at increased risk of experiencing abuse and victimization.

Given this increased risk, when providing opportunities for self-determination and planning for increased decision making, it is important to teach individuals how to recognize and respond to potentially high-risk decision-making situations. Learning to independently make high-risk decisions will not only provide individuals with IDD the autonomy they deserve, but they will also gain the skills required to keep themselves safe from victimization. Thus, it is critical to design effective decision-making interventions to increase opportunities for successful high-risk decision making and decrease risk of social victimization.

Teaching Effective Decision Making to Decrease Victimization

To provide opportunities for decision making while ensuring the safety of individuals with IDD, it is important to evaluate how we can improve the decision-making skills of individuals with IDD, especially when they are faced with high-risk situations that could lead to negative consequences. In the case of decision making where there is a heightened risk of social victimization, the decision-making process may be non-iterative, and there may be minimal chances for correction (Hickson & Khemka, 2014). That is, there are times when individuals with IDD are faced with difficult decisions that require fast action and less deliberation, particularly in potentially dangerous, high-risk situations such as potential abduction or sexual abuse. Previous research indicates that individuals with IDD often fail to anticipate possible negative consequences of their decisions and often select inappropriate courses of action (Hickson & Khemka, 1999). Further, Hickson and Khemka (2014) reported that individuals with IDD are less likely to apply a systematic decision-making process; rather they rely on their past experiences to arrive at solutions for current problems. Therefore, individuals with IDD may face unique circumstances or barriers that can inhibit successful decision making – particularly in the context of social vulnerability and victimization.

Thus, for high-risk situations, individuals with IDD may need to be taught the safest, quickest, and most appropriate response rather than employing a prolonged decision-making strategy. In these specific circumstances, behavioral intervention approaches such as BST may be the most appropriate method for designing decision-making interventions that are not only tailor-made for specific high-risk circumstances prevalent in the IDD population such as abuse and victimization but also consist of clear manageable steps.

Behavior Skills Training

BST is an applied behavior analytic strategy that is used to not only teach new skills or behavior but to also ensure individuals are able to perform their newly acquired skill in appropriate situations. Applying the principles of behavior analysis, BST has been used to effectively teach new skills to a wide range of individuals, including teaching individuals with and without disabilities a variety of skills, including bullying and abduction prevention (Poche et al., 1988; Stannis et al., 2019), firearm safety (Miltenberger et al., 2004; Morgan & Miltenberger, 2017), home accident prevention (Dancho et al., 2008: King & Miltenberger, 2017), pedestrian/street crossing safety (Rossi et al., 2017), and reporting inappropriate staff-to-resident interactions (Bollman et al., 2009). Before describing the research using BST to teach safety skills to individuals with IDD, we first clearly describe the systematic BST procedures.

BST Procedures

Although BST is continually expanding to teach different groups and individuals under various contexts, the general procedures for implementing BST remain consistent across applications. Instruction is typically conducted first in a classroom/clinical setting and then moved into the applied setting where the skill is most likely to be performed. Although sessions vary in duration depending on time constraints, the complexity of the skill being taught, and the individual's ability to attend to instruction, all sessions are delivered frequently, and performance is monitored until the individual is able to display the behavior independently during rehearsal in the classroom/clinical setting and in the applied setting.

Preparation for BST Sessions

Pre-planning steps are completed prior to beginning a BST session. First, the context of when and where the behavior is expected to be performed is determined. Next, the behaviors for the specific context are operationally defined in measurable and observable terms. For example, the operational definition for appropriately responding to a stranger's lure could be, "when out alone in public, approached by a stranger, and presented with a request to leave with the stranger, the individual will (a) immediately say "no," (b) walk or run at least 10 steps away from the stranger, and (c) tell an adult what the stranger said within 5 minutes of the lure." This definition provides a description of three actions that are observable and can be measured by independent observers with fidelity. The trainer then works with the individual and/or their caregivers to determine in which community locations the individual is most likely to be alone and in need of potentially using this skill.

Components of BST

Setting it apart from other teaching strategies, BST incorporates the principles of applied behavior analysis (ABA) and consists of four components including (1) systematic instruction, (2) modeling, (3) rehearsal, and (4) feedback (Miltenberger, 2015). The order of these components is essential to its effective delivery.

Instruction

BST begins with systematic instruction. The goal of instruction is to teach the desired behavior to the individual, providing details about how to correctly perform each behavior and in which circumstances the behavior is expected to be performed.

During instruction, the trainer describes the desired behavior with very specific details so that the individual knows exactly what the desired behavior looks like as well as what the behavior does not look like. Thus, it is important to stop and check for understanding throughout instruction. For example, when teaching the steps to responding to a lure from a stranger, an appropriate question for the trainer to ask would be "What are the three steps if a stranger approaches you?" with "Say no, walk away, tell an adult" as the correct response.

Modeling

Once instruction has been delivered, the desired behaviors are then modeled either by the trainer or through video modeling (Miltenberger, 2015). The model is used to demonstrate a clear example of how the behaviors are to be performed. The scope of the model includes all the expected behaviors and nothing beyond the scope of the instruction. For example, when modeling a response to a stranger lure, the model includes only an example of a lure and an individual displaying the three-step response. During modeling, no other situations or scenarios are shown.

Rehearsal

The third component of a BST session is rehearsal, during which the individual is provided the opportunity to practice performing the behavior. Rehearsal not only provides opportunities for practice, but it also allows the trainer to observe the individual's performance and to correct any errors (Miltenberger, 2015). Essentially, this step ensures that the individual has acquired the skill and has it within their behavioral repertoire.

While observing the individual during rehearsal, the trainer takes data on the individual's performance during each role-play scenario. Acquisition is typically determined when the individual completes all steps accurately and independently in 80%–100% of role-plays across 2–3 BST sessions consecutively.

Feedback

The final component of BST is feedback, in which the instructor delivers immediate feedback on the individual's performance after rehearsal of the behavior. Feedback consists of praise for correct performance of the behavior or parts of the behavior, identification of incorrect responses, and instructions on how to improve the performance (Miltenberger, 2015). Feedback is given either during the rehearsal when the individual begins to perform a component incorrectly or immediately following each role-play.

In Situ Assessment

Once the individual has demonstrated the ability to perform the behavior through rehearsal in the BST setting, it is critically important to ensure the skill generalizes to the natural environment (i.e., the applied setting) through in situ assessment. To replicate a realistic lure from a stranger, the trainer arranges a situation to assess if the individual will display the behavior without knowing they are being evaluated. This is done by observing the individual in a community setting where the target behavior is expected to be performed (e.g., a park after training in the classroom) and a stranger approaches them and attempts to get the individual to leave with them. The individual is unaware that the trainer is present and watching from a distance, and the stranger is known by the trainer.

If the individual does not make the correct response, there are two different training options. The stranger can terminate the interaction (e.g., "Oh never mind! I see my puppy over there!" and walks away from the individual), or the trainer can appear and provide immediate correction (e.g., "Remember, we do not leave with a stranger. We say no, walk away, and go find an adult"). In situations where the skill does not generalize from the training context to the applied setting, in situ training is used to teach the desired skill in the natural setting.

In Situ Training

In situ training includes the same components of BST, but the behavior is practiced in the natural setting. This training can either be initiated after the trainer determines that the skill did not generalize, or it can be proactively planned as an additional phase in the training procedures. Regardless of method, in situ training helps to improve the likelihood that the skills taught in the training environment generalize to the natural environment. After in situ training, it is critically important to conduct an in situ assessment to ensure the behavior is displayed in the perceived absence of the trainer.

Using BST to Teach Decision Making

Given the systematic process of instruction and evaluation of individual skill acquisition in BST, this teaching strategy is an effective method for teaching individuals with IDD how to respond in high-risk decision-making situations. High-risk decision-making situations are those in which the decision will be either a safe or dangerous response. Given the complexities of high-risk situations, BST simplifies the factors that often lead to difficulties in decision making for individuals with IDD, such as the need to evaluate the situation or to recall all of the information that is necessary to make an effective decision. BST provides individuals with the skills they need to recognize potentially dangerous situations and to make quick and effortless decisions that keep themselves safe. Thus, BST can be used to teach skills that will help individuals with IDD navigate their environment and maintain their safety in an efficient and effective manner.

A small but growing body of research has evaluated the effectiveness of BST to teach individuals with IDD effective and safe responses to potentially dangerous situations. Two common targets of BST intervention research are to teach abduction prevention and abuse prevention skills to individuals with IDD. More recently, research has expanded to teaching decision making related to other safety skills.

Abduction Prevention

In 2019, over 32,000 individuals with disabilities were reported missing (FBI, 2020). Prevention education provides strategies for how to identify and respond to dangerous high-risk situations, such as potential abduction. Identifying if the situation is dangerous, if it is safe to comply with the requests given by the potential abductor, and how they can terminate the interaction safely are just a few examples of decisions that are necessary. Although several curricula have been developed to teach abduction prevention to school-aged children, including Protect Yourself Rules, Second Step, Michigan Model, and KidSmartz (Moore & Bongiovanni, n.d.), only KidSmart includes a section specific to students in special education. As individuals with IDD are more vulnerable to being abducted (FBI, 2020), even into adulthood, education is needed for these individuals as well. In an effort to reduce risk or occurrences of abduction, then, BST has been used to teach individuals with IDD how to recognize and appropriately respond to lures from strangers.

Using BST and in situ simulations with adults, Haseltine and Miltenberger (1990) delivered a self-protection skills curriculum to teach eight adults with IDD how to identify and safely respond to abduction and sexual abuse situations. Seven of the eight participants learned the skills and maintained them at a 6-month follow-up. In another example, Fisher et al. (2013) taught five adults with IDD to respond appropriately to lures from strangers using BST and in situ training. In response to a lure, participants were taught to (a) say "no" within 3 seconds, (b) move away within 3 seconds of the refusal, and (c) to report the event to a trusted adult. The participants were able to perform these responses during classroom and in situ training, but variability was seen during in situ assessments. Despite the variation, all participants' performance increased from baseline during weekly maintenance probes and at follow-up assessments for up to 3 months after training, and no participant agreed to leave with the stranger following BST.

Finally, Fisher (2014) used BST to teach adults with Williams syndrome to respond to lures from strangers. The 21 participants were divided into four groups, ranging from four to six participants per group, and received group-based BST across 3 days. When given potential lure situations, such as "I will buy you ice cream if you come with me," "can you help me carry this to my car?", and "will you go on a walk with me?", participants were taught to (1) say "no," (2) immediately walk at least five steps away, and (3) tell a trusted adult about the stranger. The participants with Williams syndrome rapidly acquired the safety skills during role-play and use of the safety skills increased after BST, although not consistently across participants.

The results of these studies highlight that adults with IDD can be taught important high-risk decision-making skills in response to stranger lures. These skills increase the ability of these individuals to remain safe when out in the community. Having the ability to identify and terminate potential lure attempts increases the individual's independence and ability to safely navigate in community settings.

Sexual Abuse Prevention

BST has also been used to teach sexual abuse prevention to individuals with IDD. Perpetrators of sexual abuse may view individuals with IDD as easy targets, believing that they can be easily manipulated or unable to testify in court (Shapiro, 2018). Because of this increased risk, sexual abuse prevention training for individuals with IDD is essential. As sexual abuse can also occur in private, sexual abuse prevention education and training is used to teach individuals with IDD how to independently identify potential sexual abuse situations.

Sexual abuse can be prevalent within group homes; some estimates indicate that as many as 41% of individuals with IDD living in group homes experience sexual abuse (Baladerian et al., 2013). Several researchers have addressed this concern by using BST to teach adults with IDD to recognize and respond to potential abuse situations. Bollman et al. (2009) used BST and in situ assessment to teach two women with IDD how to report inappropriate staff-to-resident interactions. The reporting behavior included making a refusal or self-advocacy response, walking away, and reporting the interaction. For example, the individual was taught to tell the person to stop, leave the situation, and tell a different staff member that the direct care staff member touched them inappropriately. In their study, both participants learned the safety response, maintained this behavior at 2- and 4-week follow-ups, and generalized the behavior to novel situations within a group home setting.

In another example, Miltenberger et al. (1999) used BST and in situ training and assessment to teach sexual abuse prevention skills to five women with IDD. The skills included (a) not agreeing to engage in or begin to comply with the requested behavior, (b) saying "no" or verbally refusing, (c) leaving the situation or telling staff to leave, and (d) reporting the incident to a different staff person. In three consecutive role-play assessments, participants met performance criteria (i.e., 4 out of 4 correct responses), while verbal report scores were more variable during training, ranging from two to four correct responses. For in situ assessment, post-training scores ranged from zero to four correct responses. After four to eight sessions of in situ training, all participants met criterion on in situ assessments, excluding one who moved away from their residence during in situ training. The results of this study demonstrate the effectiveness of in situ training to increase individual performance to the desired level, particularly when performance does not initially generalize to the applied setting.

Finally, Egemo-Helm et al. (2007) taught sexual abuse prevention skills to five women with IDD using only in situ training within a community setting. Similar to the procedures used by Miltenberger et al. (1999), participants were taught to (a) not agree to engage in or begin to comply with the requested behavior, (b) state "no" or verbally refuse, (c) leave the situation or tell staff to leave, and (d) report the incident to a staff person. Skills were assessed with self-report, role-play, and in situ assessments. Training occurred simultaneously in their residence and in situ. For training within their residence, all five participants performed at criteria within three sessions. For in situ training, two participants met criteria with five BST sessions, one participant met criteria within seven sessions, one participant required 12 sessions, and the last participant dropped out of the study after scoring three out of four consistently during assessments. Additionally, while in situ training was being conducted, performance in the residential training setting continued to be maintained at criterion across all participants, including the participant who left the study.

As these studies illustrate, BST can be highly effective in teaching individuals with IDD the skills required to prevent sexual abuse in their individual environments, such as group home settings. There are limitations to BST, however, as the findings from the previous research demonstrate that not all participants were able to acquire the skills through BST and required other interventions or adaptations to training. Nevertheless, sexual abuse prevention remains an important area of focus given that individuals with IDD may be unable to identify such situations without prior training and a majority of abusers are already known by the individual (Murray et al., 2014) perhaps making it more difficult to identify sexual abuse.

Safety Skills

Along with abduction and sexual abuse prevention, BST can also be an effective approach to teach other safety skills to individuals with IDD, such as bullying victimization prevention. Bullying victimization is quite prevalent in the IDD population, with recent numbers indicating that as many as 40% of individuals with IDD experienced bullying (Pfeffer, 2016). Those with ID can be at even greater risk of experiencing bullying victimization with 56.7% reporting bullying victimization compared to 10.6% of those without disabilities (Sterzing et al., 2012).

Given the prevalence of bullying victimization within the IDD population, researchers have used BST to teach individuals with IDD certain safety skills that may prevent such outcomes. For instance, Stannis et al. (2019) used BST and in situ training to teach a response to bullying to four adults with IDD experiencing bullying victimization. The participants were taught to (a) refrain from retaliating against the bully by avoiding physical contact or vocal statements; (b) state a short comment of disapproval, such as "I don't like that"; (c) walk away from the bully; and (d) tell a staff member about the interaction. BST alone was successful in teaching the response to two participants, BST plus in situ training was effective for one participant, and BST with in situ training plus an incentive was effective for the final participant.

The response to bullying was recently extended to work settings by Peterson (2020), who replicated the procedures to teach four young adults with IDD to recognize and respond to coworker victimization. Using BST and in situ training following assessment, three of the four participants were able to display the four-step safety response during the researcher and confederate in situ assessments.

As the research demonstrates, BST can be incredibly effective at improving the decision making skills of individuals with IDD, especially for dangerous or high-risk situations like potential abduction, sexual abuse, and bullying victimization. More specifically, BST can teach individuals with IDD how to identify such contexts and the specific steps required to successfully make a decision that will result in a positive and safe outcome. The use of specific safety skills is crucial, particularly in high-risk situations that require a quick and efficient response. And, perhaps most importantly, the research demonstrates that BST has the potential to teach individuals with IDD how to successfully make decisions in high-risk situations *independently*, which is critical given that dangerous high-risk situations may likely occur without the presence of others.

Considerations for BST

For individuals with IDD, the ability and opportunity to make decisions within a wide variety of situations is crucial for living an independent and self-determined life. Given that individuals with IDD may be socially vulnerable and therefore at heightened risk for social victimization, the development of decision-making interventions that focus on such contexts is crucial. BST may be an effective intervention approach for building decision-making skills for high-risk situations as it teaches individuals to recognize specific potentially dangerous situations and to make appropriate and safe decisions quickly and effectively.

Despite the small body of research supporting the effectiveness of BST for teaching responding to potentially dangerous situations, there are still limitations to these procedures that will need to be further addressed in future research. First, because BST procedures teach the individual a discrete "all or nothing response," it could be argued that the individual is not making a decision at all. Rather, it may seem that the individual is being taught a rule to always "say no" to situations that could impose danger. Such training could appear to directly contradict the principles of self-determination. However, we argue that the decision point in BST is when the individual must decide to use or not use the acquired safety response in a risky situation. Like other decisions in life, the individual's decision to use or not use the learned safety response may result in an ideal outcome or an unintended negative consequence. The latter, while not ideal, is considered a critical factor in not only developing autonomous decision making but self-determination throughout the lifespan (Sheppard & Unsworth, 2011; Wehmeyer, 2003).

Indeed, individuals with IDD have the right to take risks and make mistakes (Bannerman et al., 1990; Nirje, 1976) in order to live a self-determined life. At the same time, if individuals with IDD are not given the skills necessary to appropriately weigh and understand the risks that are involved in making a high-risk decision, then we have not adequately prepared them for independence. As Bannerman et al. (1990) argued nearly three decades ago, "all people have the right to eat too many doughnuts and take a nap. But along with rights come responsibilities. Teaching clients how to exercise their freedoms responsibly should be an integral part of the habilitation process. While learning, clients should be encouraged to make as many choices as their abilities allow, as long as these choices are not detrimental to the client or to others" (pp. 86).

To reconcile the need to ensure safety but to provide opportunities for risk and independence, we argue that BST is a first step in teaching and preparing individuals with IDD for a life of independence and self-determination. Additionally, BST may also address other barriers to autonomous decision making and self-determination for individuals with IDD. Previous research has found that one of the largest barriers to independence is related to parental concern for the individual's safety. For example, Griffin et al. (2010) found that in a survey of 108 parents of individuals with IDD, their biggest concern regarding sending their child to a post-secondary education program was related to the individual's safety. When parents are concerned about safety, they are more likely to supervise their child's actions and less likely to provide the individual with opportunities for independence (Fisher et al., 2012) including chances for autonomous decision making.

On the other hand, previous research also highlights that when individuals are afforded opportunities for autonomous decision making, the results are not always favorable. Fisher et al. (2012) reported that individuals with IDD who were afforded less parental supervision – despite their ability to identify risky situations – were just as socially vulnerable and likely to experience victimization as those who were less aware of risks. The authors concluded that these findings are consistent with evidence that although individuals with IDD may be able to say what they are expected to do in a risky situation, they are not always able to perform the correct behavior in a real situation (Fisher et al., 2012; Fisher et al., 2014; Mechling, 2008).

Thus, to ease parental concerns about the individual with IDD's safety and to provide more opportunities for safe and independent decision making (including abduction or sexual abuse prevention), BST may be an important first step. Providing individuals with a safety response and observing and reinforcing their decision to use that response in situ may go a long way in easing parental concerns and providing individuals with the safety skills they need to independently navigate their communities and various social situations. Not only does BST teach the individual the safety skill, but training also continues until the individual is able to independently display the skill in community settings. As such, parents can be assured that the individual cannot only *say* but can *do* what they should do in high-risk situations. Doing so can assist individuals with IDD to live a self-determined life in which they are encouraged to make their own life decisions.

Directions for Future Research

Whereas research has shown that BST can be effective to teach safety responding in certain high-risk situations, additional research is needed to evaluate ways to teach or expand other decision-making abilities. Thus far, we have discussed the use of BST to teach individuals with IDD to make appropriate decisions in dangerous high-risk situations such as potential abduction, sexual abuse, and bullying victimization. However, compared to the general population, individuals with IDD are also at more risk for financial victimization or exploitation, particularly adults presenting with more severe ID (Cummings et al., 2006; Horner-Johnson & Drum, 2006).

In fact, in their study of 122 adults with severe intellectual and social limitations, Claycomb et al. (2013) observed that 70% of their participants had experienced at least one type of financial victimization within the past month, with 35% of participants reporting complete loss of their financial funds. Earlier research also suggests that female adults with ID who use personal assistance services may be vulnerable to financial victimization, such as personal assistants stealing their money, forging checks or misusing credit, and even pressuring them for money (Powers et al., 2002). Given these extreme risks for financial victimization, further research is warranted to evaluate the effectiveness of BST for teaching individuals with IDD successful decision-making skills when faced with potential financial victimization.

Another important consideration for future research is an exploration of for whom BST is most successful. In the research described above, not all participants were able to acquire the safety response, even when in situ training was introduced. It would be important to closely examine the characteristics of participants for whom BST is most and least successful. Once these characteristics are identified, research can then be conducted to determine whether the procedures of BST can be adapted to meet the needs of these certain individuals. For example, it is possible that the procedures for BST may need to be adapted to meet the unique needs or considerations of individuals with moderate to severe ID or those who do not communicate verbally. Due to its step-by-step procedure, BST may be able to readily address such considerations to ensure effective teaching of safety responses to individuals with IDD.

Given that individuals with IDD may be more vulnerable to a wide range of victimization scenarios, additional research is warranted to further explore the use of BST to teach safety responses to individuals with IDD with varying degrees of severity (i.e., mild to severe symptomatology) for different types of victimization. Doing so will inform future practice of the effectiveness and limitations of BST in teaching decision-making skills to individuals with IDD in a variety of contexts.

Directions for Future Practice

While the research literature supports the use of BST to teach effective decision-making skills to individuals with IDD, the use of BST in practice remains unclear. That is, not much is known about the application of BST by various professions in different fields.

While this chapter has focused on published research, the procedures used for BST easily translate into practice. The methods used to create instruction are designed in such a way that the creation and implementation of a training program is straightforward and quick. As BST is grounded in Applied Behavior Analysis (ABA), a core principle of ABA is that the methods used are *technological* (Baer et al., 1968). For a practice to be considered technological, its procedures are able to be replicated by providing sufficient guidance for another person to reproduce the work. The methods provided in the report can be seen as an instruction guide for practitioners to adopt. Alongside with a technological component, the principles also emphasize that a practice be *applied* (Baer et al., 1968). That is, the behaviors and skills of interest are of social significance and have real-world applications. For example, in this chapter we provided multiple examples of BST being used to teach individuals how to identify and remove themselves from dangerous situations (i.e., abduction, abuse, social victimization). Given the feasibility of implementing BST, it has promising potential to be readily incorporated into practice across different professional fields.

Overall, to teach efficient and effective decision-making skills to individuals with IDD, it is crucial to use methods and approaches that are not only promising but are grounded in the research literature. BST has the potential to assist individuals with IDD in high-risk decision-making situations by breaking the decision-making process into more manageable steps. More importantly, such strategies can help highlight important factors associated with the decision-making process to individuals with IDD, raising self-awareness of how cognitive processes and behavior can combine to influence the decisions we make in our everyday lives.

References

Abbott, S., & McConkey, R. (2006). The barriers to social inclusion as perceived by people with intellectual disabilities. *Journal of Intellectual Disabilities, 10*(3), 275–287. https://doi.org/10.1177/1744629506067618

Baer, D. M., Wolf, M. M., & Risley, T. R. (1968). Some current dimensions of applied behavior analysis. *Journal of applied behavior analysis, 1*(1), 91. https://doi.org/10.1901/2Fjaba.1968.1-91

Baladerian, N., Coleman, T., Stream, J., (2013). *Abuse of people with disabilities victims and their families speak out: A report on the 2012 national survey on abuse of people with disabilities.* Disability and Abuse. http://www.disabilityandabuse.org/survey

Bannerman, D. J., Sheldon, J. B., Sherman, J. A., & Harchik, A. E. (1990). Balancing the right to habilitation with the right to personal liberties: The rights of people with developmental disabilities to eat too many doughnuts and take a nap. *Journal of Applied Behavior Analysis, 23*(1), 79–89. https://doi.org/10.1901/2Fjaba.1990.23-79

Blatt, B., & Kaplan, F. (1974). *Christmas in Purgatory: A photographic essay on mental retardation*. Human Policy Press.

Bollman, J. R., Davis, P. K., & Zarcone, J. (2009). Teaching women with intellectual disabilities to identify and report inappropriate staff-to-resident interactions. *Journal of Applied Behavior Analysis, 42*(4), 813–817. https://doi.org/10.1901/jaba.2009.42-813

Brown, I., & Brown, R. I. (2009). Choice as an aspect of quality of life for people with intellectual disabilities. *Journal of Policy and Practice in Intellectual Disabilities, 6*(1), 11–18. https://doi.org/10.1111/j.1741-1130.2008.00198.x

Buchanan, I., & Walmsley, J. (2006). Self-Advocacy in Historical Perspective. *British Journal of Learning Disabilities, 34*, 133–138. https://doi.org/10.1111/j.1468-3156.2006.00410.x

Claycomb, M., Black, A. C., Wilber, C., Brocke, S., Lazar, C. M., & Rosen, M. I. (2013). Financial victimization of adults with severe mental illness. *Psychiatric Services, 64*(9), 918–920.

Cummings, J. G., Pepler, D. J., Mishna, F., & Craig, W. M. (2006). Bullying and victimization among students with exceptionalities. *Exceptionality Education Canada, 16*(3), 193–222.

Dancho, K. A., Thompson, R. H., & Rhoades, M. M. (2008). Teaching preschool children to avoid poison hazards. *Journal of Applied Behavior Analysis, 41*(2), 267–271. https://doi.org/10.1901/jaba.2008.41-267

Egemo-Helm, K. R., Miltenberger, R. G., Knudson, P., Finstrom, N., Jostad, C., & Johnson, B. (2007). An evaluation of in situ training to teach sexual abuse prevention skills to women with mental retardation. *Behavioral Interventions: Theory & Practice in Residential & Community-Based Clinical Programs, 22*(2), 99–119. https://doi.org/10.1002/bin.234

Federal Bureau of Investigation. (2020). *2019 National crime information center (NCIC) missing person and unidentified person statistics.* https://www.fbi.gov/file-repository/2019-ncic-missing-person-and-unidentified-person-statistics.pdf

Fenn, K., & Scior, K. (2019). The psychological and social impact of self-advocacy group membership on people with intellectual disabilities: A literature review. *Journal of Applied Research in Intellectual Disabilities, 32*(6), 1349–1358. https://doi.org/10.1111/jar.12638

Field, S. (1996). Self-determination instructional strategies for youth with learning disabilities. *Journal of Learning Disabilities, 29*(1), 40–52. https://doi.org/10.1177/002221949602900107

Field, S., & Hoffman, A. (1994). Development of a model for self-determination. *Career Development for Exceptional Individuals, 17*(2), 159–169. https://doi.org/10.1177/088572889401700205

Fisher, M. H. (2014). Evaluation of a stranger safety training programme for adults with Williams syndrome. *Journal of Intellectual Disability Research, 58*(10), 903–914. https://doi.org/10.1111/jir.12108

Fisher, M. H., Baird, J. V., Currey, A. D., & Hodapp, R. M. (2016). Victimization and social vulnerability of adults with intellectual disability: A review of research extending beyond Wilson and Brewer. *Australian Psychologist, 51*, 114–127. https://doi.org/10.1111/ap.12180

Fisher, M. H., Burke, M. M., & Griffin, M. M. (2013). Teaching young adults with disabilities to respond appropriately to lures from strangers. *Journal of Applied Behavior Analysis, 46*(2), 528–533. https://doi.org/10.1002/jaba.32

Fisher, M. H., Mello, M. P., & Dykens, E. M. (2014). Who reports it best? A comparison between parent-report, self-report, and the real life social behaviors of adults with Williams syndrome. *Research in developmental disabilities, 35*(12), 3276–3284. https://doi.org/10.1016/j.ridd.2014.08.011

Fisher, M. H., Moskowitz, A. L., & Hodapp, R. M. (2012). Vulnerability and experiences related to social victimization among individuals with intellectual and developmental disabilities. *Journal of Mental Health Research in Intellectual Disabilities, 5*, 32–48. https://doi.org/10.1080/19315864.2011.592239

Greenspan, S., Laughlin, G., & Black, R. S. (2001). Credulity and gullibility in people with developmental disorders: A framework for future research. In L.M. *Glidden (Ed.) International Review of Research in Mental Retardation, 24*, 101–135.

Griffin, M. M., McMillan, E. D., & Hodapp, R. M. (2010). Family perspectives on post-secondary education for students with intellectual disabilities. *Education and Training in Autism and Developmental Disabilities, 45*(3), 339–346. https://www.jstor.org/stable/23880108

Haseltine, B., & Miltenberger, R. G. (1990). Teaching self-protection skills to persons with mental retardation. *American journal of mental retardation: AJMR, 95*(2), 188–197.

Hickson, L., & Khemka, I. (1999). Decision making and mental retardation. *International Review of Research in Mental Retardation, 22*, 227–265.

Hickson, L., & Khemka, I. (2014). The psychology of decision making. *In International Review of Research in Developmental Disabilities, 47*, 185–229. https://doi.org/10.1016/B978-0-12-800278-0.00006-3

Horner-Johnson, W., & Drum, C. E. (2006). Prevalence of maltreatment of people with intellectual disabilities: A review of recently published research. *Mental Retardation and Developmental Disabilities Research Reviews, 12*, 57–69. https://doi.org/10.1002/mrdd.20097

Hughes, K., Bellis, M. A., Jones, L., Wood, S., Bates, G., Eckley, L., McCoy, E., Mikton, C., Shakespeare, T., & Officer, A. (2012). Prevalence and risk of violence against adults with disabilities: a systematic review and meta-analysis of observational studies. *The Lancet, 379*(9826), 1621–1629. https://doi.org/10.1016/S0140-6736(11)61851-5

Jawaid, A., Riby, D. M., Owens, J., White, S. W., Tarar, T., & Schulz, P. E. (2012). 'Too withdrawn' or 'too friendly': Considering social vulnerability in two neurodevelopmental disorders. *Journal of Intellectual Disability Research, 56*, 335–350. https://doi.org/10.1111/j.1365-2788.2011.01452.x

Jones, W., Bellugi, U., Lai, Z., Chiles, M., Reilly, J., Lincoln, A., & Adolphs, R. (2000). II. Hypersociability in Williams syndrome. *Journal of Cognitive Neuroscience, 12*(Supplement 1), 30–46. https://doi.org/10.1162/089892900561968

Kampert, A. L., & Goreczny, A. J. (2007). Community involvement and socialization among individuals with mental retardation. *Research in Developmental Disabilities, 28*(3), 278–286. https://doi.org/10.1016/j.ridd.2005.09.004

King, S., & Miltenberger, R. (2017). Evaluation of video modeling to teach children diagnosed with autism to avoid poison hazards. *Advances in Neurodevelopmental Disorders, 1*(4), 221–229. https://doi.org/10.1007/s41252-017-0028-2

Lough, E., & Fisher, M. H. (2016). Parent and self-report ratings on the perceived levels of social vulnerability of adults with Williams syndrome. *Journal of Autism and Developmental Disorders, 46*(11), 3424–3433. https://doi.org/10.1007/s10803-016-2885-3

Mansell, J. (2006). Deinstitutionalisation and community living: Progress, problems and priorities. *Journal of Intellectual and Developmental Disability, 31*, 65–76. https://doi.org/10.1080/13668250600686726

Mechling, L. C. (2008). Thirty year review of safety skill instruction for persons with intellectual disabilities. *Education and Training in Developmental Disabilities, 43*(3), 311–323.

Miltenberger, R. G. (2015). *Behavior modification: Principles and procedures* (6th ed.). Cengage Learning.

Miltenberger, R. G., Flessner, C., Gatheridge, B., Johnson, B., Satterlund, M., & Egemo, K. (2004). Evaluation of behavioral skills training to prevent gun play in children. *Journal of applied behavior analysis, 37*(4), 513–516. https://doi.org/10.1901/jaba.2004.37-513

Miltenberger, R. G., Roberts, J. A., Ellingson, S., Galensky, T., Rapp, J. T., Long, E. S., & Lumley, V. A. (1999). Training and generalization of sexual abuse prevention skills for women with mental retardation. *Journal of applied behavior analysis, 32*(3), 385–388. https://doi.org/10.1901/jaba.1999.32-385

Moore, M., & Bongiovanni, A., (n.d.). *Child abduction prevention education.* [Power Point slides]. Erie 2-Chautauqua-Cattaraugus BOCES. https://www.e2ccb.org/files/1991674/child%20abduction%20prevention%20webinar%20slides.pdf

Morgan, K., & Miltenberger, R. G. (2017). Evaluation of video modeling and in situ training to teach firearm avoidance skills to individuals with autism spectrum disorder. *Advances in Neurodevelopmental Disorders, 1*(3), 122–128. https://doi.org/10.1007/s41252-017-0024-6

Murray, L. K., Nguyen, A., & Cohen, J. A. (2014). Child sexual abuse. *Child and adolescent psychiatric clinics of North America, 23*(2), 321–337. https://doi.org/10.1016/j.chc.2014.01.003

Nettelbeck, T., & Wilson, C. (2002). Personal vulnerability to victimization of people with mental retardation. *Trauma, Violence, & Abuse, 3*(4), 289–306. https://doi.org/10.1177/1524838002237331

Nielsen, K. E. (2012). A disability history of the United States (Vol. 2). Beacon Press.

Nirje, B. (1976). 14. The Normalization Principle. *Changing patterns in residential services for the mentally retarded, 76*(21015), 231–240.

Peterson, A. M. (2020). *Teaching young adults with intellectual and developmental disabilities how to recognize and respond to coworker victimization scenarios*. Master's thesis, Michigan State University.

Petitpierre, G., Masse, M., Martini-Willemin, B., & Delessert, Y. (2013). A complementarity of social and legal perspectives on what is abusive practice and what constitutes abuse. *Journal of Policy and Practice in Intellectual Disabilities, 10*, 196–206. https://doi.org/10.1111/jppi.12038

Pfeffer, R. D. (2016). Childhood victimization in a national sample of youth with autism spectrum disorders. *Journal of Policy and Practice in Intellectual Disabilities, 13*(4), 311–319. https://doi.org/10.1111/jppi.12203

Poche, C., Yoder, P., & Miltenberger, R. (1988). Teaching self-protection to children using television techniques. *Journal of Applied Behavior Analysis, 21*(3), 253–261. https://doi.org/10.1901/jaba.1988.21-253

Powers, L. E., Curry, M. A., Oschwald, M., & Maley, S. (2002). Barriers and strategies in addressing abuse: A survey of disabled women's experiences. *Journal of Rehabilitation, 68*(1), 4–13.

Rosen, D. B. (2006). Violence and exploitation against women and girls with disability. In F. L. Denmark, H. H. Krauss, E. Halpern, S. Esther, & A. Jeri (Eds.), *Violence and exploitation against women and girls* (pp. 170–177). Blackwell Publishing.

Rossi, M. R., Vladescu, J. C., Reeve, K. F., & Gross, A. C. (2017). Teaching safety responding to children with autism spectrum disorder. *Education and Treatment of Children, 40*(2), 187–208. https://doi.org/10.1353/etc.2017.0009

Sandjojo, J., Gebhardt, W. A., Zedlitz, A. M., Hoekman, J., den Haan, J. A., & Evers, A. W. (2019). Promoting independence of people with intellectual disabilities: A focus group study perspectives from people with intellectual disabilities, legal representatives, and support staff. *Journal of Policy and Practice in Intellectual Disabilities, 16*(1), 37–52. https://doi.org/10.1111/jppi.12265

Shapiro, J. (2018). *The sexual assault epidemic no one talks about*. Abused and Betrayed. https://www.npr.org/2018/01/08/570224090/the-sexual-assault-epidemic-no-one-talks-about

Sheppard, L., & Unsworth, C. (2011). Developing skills in everyday activities and self-determination in adolescents with intellectual and developmental disabilities. *Remedial and Special Education, 32*(5), 393–405. https://doi.org/10.1177/0741932510362223

Shogren, K. A., Rifenbark, G. G., Wehmeyer, M. L., Dean, E. E., Killeen, M. B., & Karsevar, J. (2020). Refining the supported decision making inventory. *Journal of Policy and Practice in Intellectual Disabilities, 17*(3), 195–206. https://doi.org/10.1111/jppi.12335

Shogren, K. A., & Wehmeyer, M. L. (2015). A framework for research and intervention design in supported decision making. *Inclusion, 3*(1), 17–23. https://doi.org/10.1352/2326-6988-3.1.17

Shogren, K. A., & Wehmeyer, M. L. (2016). *Supported decision making inventory system (SDMIS) [Pilot Version]*. Kansas University Center on Developmental Disabilities/Beach Center on Disabilities.

Shogren, K. A., Wehmeyer, M. L., Lassman, H., & Forber-Pratt, A. J. (2017a). Supported decision making: A synthesis of the literature to guide assessment and intervention development. *Education and Training in Autism and Developmental Disabilities, 52*, 144–157. https://www.jstor.org/stable/26420386

Shogren, K. A., Wehmeyer, M. L., Uyanik, H., & Heidrich, M. (2017b). Development of the supported decision making inventory system. *Intellectual and Developmental Disabilities, 55*, 432–439. https://doi.org/10.1352/1934-9556-55.6.432

Simplican, S. C., Leader, G., Kosciulek, J., & Leahy, M. (2015). Defining social inclusion of people with intellectual and developmental disabilities: An ecological model of social networks and community participation. *Research in Developmental Disabilities, 38*, 18–29. https://doi.org/10.1016/j.ridd.2014.10.008

Smith, R. B., Morgan, M., & Davidson, J. (2005). Does the daily choice making of adults with intellectual disability meet the normalisation principle? *Journal of Intellectual and Developmental Disability, 30*(4), 226–235. https://doi.org/10.1080/13668250500349359

Sparks, S. L., Pierce, T., Higgins, K., Miller, S., & Tandy, R. (2016). Increasing choice making in students with intellectual disability. *Education and Training in Autism and Developmental Disabilities, 51*(4), 331–343. https://www.proquest.com/docview/1843276145/DCB23827D1F34647PQ/1

Stang, K. K., Carter, E. W., Lane, K. L., & Pierson, M. R. (2009). Perspectives of general and special educators on fostering self-determination in elementary and middle schools. *The Journal of Special Education, 43*(2), 94–106. https://doi.org/10.1177/0022466907313452

Stannis, R. L., Crosland, K. A., Miltenberger, R., & Valbuena, D. (2019). Response to bullying (RTB): Behavioral skills and in situ training for individuals diagnosed with intellectual disabilities. *Journal of Applied Behavior Analysis, 52*(1), 73–83. https://doi.org/10.1002/jaba.501

Sterzing, P. R., Shattuck, P. T., Narendorf, S. C., Wagner, M., & Cooper, B. P. (2012). Bullying involvement and autism spectrum disorders: Prevalence and correlates of bullying involvement among adolescents with an autism spectrum disorder. *Archives of Pediatric and Adolescent Medicine, 166*(11), 1058–1064. https://doi.org/10.1001/archpediatrics.2012.790

TASH. (2006). TASH resolution on postsecondary education [Resolution statement]. https://tash.org/about/resolutions/tash-resolution-post-secondary-education/

Thurman, A. J., & Fisher, M. H. (2015). The Williams syndrome social phenotype: Disentangling the contributions of social interest and social difficulties. In R. M. Hodapp & D. J. Fidler (Eds.), *International review of research in developmental disabilities, Vol. 49* (pp. 191–227). Academic Press.

Ticoll, M. (1994). Violence and people with disabilities: A review of the literature. In *National clearinghouse on family violence*. Ontario. https://www.ncjrs.gov/App/Publications/abstract.aspx?ID=159729

United Nations. (2006). *Convention on the Rights of Persons with Disabilities*. United Nations.

Wehmeyer, M. L. (1997). Self-determination as an educational outcome: A definitional framework and implications for intervention. *Journal of Developmental and Physical Disabilities, 9*(3), 175–209. https://doi.org/10.1023/A:1024981820074

Wehmeyer, M. L. (2003). A functional theory of self-determination: Definition and categorization. In M. L. Wehmeyer, B. H. Abery, D. E. Mithaug, & R. J. Stancliffe (Eds.), *Theory in self-determination: Foundations for educational practice* (pp. 174–181). Charles C Thomas.

Wehmeyer, M. L., & Garner, N. W. (2003). The impact of personal characteristics of people with intellectual and developmental disability on self-determination and autonomous functioning. *Journal of Applied Research in Intellectual Disabilities, 16*(4), 255–265. https://doi.org/10.1046/j.1468-3148.2003.00161.x

Werner, S. (2012). Individuals with intellectual disabilities: a review of the literature on decision making since the Convention on the Rights of People with Disabilities (CRPD). *Public Health Reviews, 34*(2), 1–27. https://doi.org/10.1007/bf03391682

Westcott, H. L., & Jones, D. P. H. (1999). Annotation: The abuse of disabled children. *Journal of Child Psychology and Psychiatry, 40*(4), 497–506. https://doi.org/10.1111/1469-7610.00468

White, C., Holland, E., Marsland, D., & Oakes, P. (2003). The identification of environments and cultures that promote the abuse of people with intellectual disabilities: A review of the literature. *Journal of Applied Research in Intellectual Disabilities, 16*, 1–9. https://doi.org/10.1046/j.1468-3148.2003.00147.x

Wilson, C., Seaman, L., & Nettelbeck, T. (1996). Vulnerability to criminal exploitation: Influence of interpersonal competence differences among people with mental retardation. *Journal of Intellectual Disability Research, 40*, 10–19. https://doi.org/10.1111/j.1365-2788.1996.tb00

Chapter 22
Strategy-Based Interventions for Effective Interpersonal Decision Making

Ishita Khemka and Linda Hickson

Introduction

The human and legal rights of individuals with intellectual and developmental disabilities (IDD) to have agency over their own lives are well established. Widespread support for these rights has been achieved through strengths-based initiatives like the positive psychology and self-determination movements and the passage of the United Nations Convention on the Rights of Persons with Disabilities (CRPD) (Werner, 2012). However, in spite of these expanded rights, the quality of life of individuals with IDD is too often compromised by heightened rates of abuse and victimization which far exceed the rates in the general population (e.g., Horner-Johnson & Drum, 2006; Hughes et al., 2020).

It is clear that limited decision-making skills play a pivotal role in the vulnerability of individuals with IDD to maltreatment (see Hickson & Khemka in this volume). Because we believe that effective decision making is a necessary component for fostering independent and safe navigation of interpersonal relationships by individuals with IDD, we have focused our efforts on the development of targeted decision-making-based interventions. In keeping with this goal, our interventions are designed to draw upon increased understanding of the decision-making processes needed to bolster the decision-making competence of individuals with IDD.

In the interest of providing individuals with IDD with repertoires of effective independent, self-protective decision-making skills, we have conducted a series of intervention studies. With an initial focus on adults with IDD, primarily intellectual disabilities (ID), we have extended our decision-making training efforts to

I. Khemka (✉)
St. John's University, Queens, NY, USA
e-mail: khemkai@stjohns.edu

L. Hickson
Teachers College, Columbia University, New York, NY, USA

I. Khemka, L. Hickson (eds.), *Decision Making by Individuals with Intellectual and Developmental Disabilities*, Positive Psychology and Disability Series,
https://doi.org/10.1007/978-3-030-74675-9_22

adolescents with IDD, including ID and autism spectrum disorders (ASD), developing a line of intervention research aimed at understanding and strengthening their decision processing. These intervention efforts have focused on designing and evaluating curriculum materials to enhance the ability of individuals with IDD to protect themselves from abuse and peer victimization, using a systematic decision-making strategy approach.

Interdependent Processes in Decision-Making Training

Earlier efforts to improve decision-making skills were centered on addressing the cognitive deficits that were implicated as the primary source of decision-making difficulty for individuals with ID (see Hickson & Khemka, 1999). Clearly, decision making is associated with many cognitive processing requirements (e.g., comprehension, working memory, consequential reasoning, inhibitory control) that are needed to generate alternative solutions and evaluate consequences to support effective deliberative decision making. Limitations in these areas of cognitive functioning have long been documented (e.g., Healey & Masterpasqua, 1992; Jenkinson & Nelms, 1994; Short & Evans, 1990; Smith, 1986), and training targeting the cognitive components of decision making has focused on improving the application of a stepwise model of decision making involving framing a problem, generating alternatives, evaluating consequences, and choosing a course of action (see Hickson & Khemka, 1999 for a review of these studies). However, the cognitive focus of the early training efforts began to change with growing evidence that decision making does not always proceed in a planful, stepwise manner and that decision making also involves motivational and emotional factors. This has led to a more comprehensive understanding of decision-making processing, with an emphasis on noncognitive factors interacting with cognitive influences.

Decision-Making Curriculum Development and Evaluation: Focus on Adults with IDD

Khemka (2000) made an important shift in examining both cognitive and noncognitive factors, as integral elements of the decision-making process, in a training study aimed at improving decision-making skills in the context of simulated interpersonal situations involving a threat of abuse. Khemka (2000) targeted both cognitive and motivational components of decision making in an intervention designed to improve self-protective decision-making skills of adult women with ID. Khemka randomly assigned 36 women with ID to three treatment conditions: cognition- and motivation-based decision-making training, cognition-based decision-making training, and a no-treatment control that received only the agency's usual abuse prevention program. Although both training approaches were effective relative to the control condition, the combined cognitive and motivational training approach was superior to the cognitive-only training approach. Participants in the enhanced cognitive and motivation group gave significantly more independent prevention-focused decision responses when asked what a protagonist should do on a decision-making posttest that presented video vignettes of simulated interpersonal situations of abuse. The increased positive effect of the combined training approach was also reflected on a verbally presented generalization task requiring participants to respond to a decision-making situation involving abuse from their own perspective and on a locus of control scale that measured their perceptions of control. In sum, the study provided strong support for focusing on both cognitive and motivational aspects in decision making during training, particularly in response to decision situations where establishing goal clarity and personal agency may be especially important to determining the decision outcome, such as handling decisions in abusive situations.

ESCAPE (Effective Strategy-Based Curriculum for Abuse Prevention and Empowerment)

Based on the efficacy findings of Khemka's (2000) intervention study, the initial version of the ESCAPE curriculum (Khemka & Hickson, 2002) was developed to offer a systematic, structured approach for abuse prevention utilizing decision-making skills instruction as its core emphasis. The decision-making scripts used in Khemka (2000) were expanded and developed into a structured 12-session curriculum designed to provide explicit instruction and practice in using a stepwise strategy for decision making in situations of abuse and to reinforce motivational processes, involving goals clarification and self-empowerment beliefs.

Khemka et al. (2005) conducted a randomized controlled trial to test the efficacy of the ESCAPE curriculum for teaching self-protective decision-making skills to women with ID. Results of that study indicated that women with ID who received the ESCAPE curriculum performed significantly better than women in a randomly assigned wait-list control group on measures tapping knowledge of abuse concepts, empowerment, and decision making. However, despite the significant superiority of the intervention group, it was noted that there was variability in the extent to which the women had benefited from their exposure to the ESCAPE curriculum.

ESCAPE-DD (Effective Strategy-Based Curriculum for Abuse Prevention and Empowerment – Developmental Disabilities)

We broadened the scope of this line of research by developing ESCAPE-DD (Khemka & Hickson, 2008), a version of the original ESCAPE curriculum that was modified to be appropriate for both men and women. The inclusion of men with

IDD was significant, as despite accumulating evidence of their vulnerability (e.g., Doughty & Kane, 2010), men with IDD had been largely excluded from research in this area. The curriculum was updated to increase the ecological validity of the decision-making scenarios of sexual, physical, and verbal (psychological) abuse used for instruction and assessment by basing the scenarios on actual incidents drawn from agency records.

The ESCAPE-DD curriculum included 12 small-group instructional lessons that presented a multifaceted view of decision making by clearly articulating the cognitive, motivational, and emotional processes needed for effective decision making in situations of abuse. The five lessons in Unit 1 were modeled on ESCAPE, but with training scenarios and examples that were appropriate for both women and men. The cognitive components of decision making were designed to instill a broad-based understanding of healthy and abusive interpersonal situations which enabled the necessary problem awareness for decision making. Identification of a situation as abusive indicated the need to apply the stepwise decision-making strategy, whereas identification of a situation as healthy allowed for a more spontaneous, intuitive decision-making approach. The motivational components included the prioritization of self-protective goals associated with personal safety and empowered personal agency beliefs. Specifically, three goals (*be independent*, *be safe now*, and *be safe later*) were identified as a basis for the evaluation of options and consequences of possible actions. The goal of being *safe now* prompted consideration of options that involved immediate independent action to stop or prevent the abuse, while the *safe later* goal prompted decision choices that typically involved reporting the abuse to someone in authority, who could help ensure their long-term safety. Finally, the emotional components of the decision-making curriculum involved recognizing feelings likely to occur in healthy vs. abusive situations and regulating emotions to enable effective decision making. Unit 2 of the curriculum, comprised of a set of seven lessons, focuses on acquisition and application of a stepwise, deliberative, decision-making strategy for situations of abuse, including four specific steps that integrate the learning from Unit 1: (1) identifying a problem situation, (2) generating alternatives, (3) considering possible consequences of each alternative, and (4) choosing a course of action (see *Petitpierre and Tabin* in this volume for a discussion of the implementation of a French translation of ESCAPE-DD in Switzerland).

Hickson et al. (2015) conducted a study to test the effectiveness of ESCAPE-DD, at multiple sites of an adult day agency in New York City where the decision-making curriculum was delivered in small groups of three to five participants. In a randomized controlled trial, 58 women and men with IDD were randomly assigned to an intervention group (n = 30) or a to a wait-list control group (n = 28) with a mean chronological age of 38.81 years (SD = 13.85) and mean IQ of 56.78 (SD = 9.01) for the entire sample. The findings of this study indicated that women and men with IDD who received the ESCAPE-DD curriculum recommended overall effective decision-making responses (for how best a key protagonist should handle a hypothetical scenario of physical, sexual, and verbal abuse, over a set of 12 scenarios) 84% of the time on the posttest, whereas participants in the control group recommended effective strategies only 63% of the time. In addition, participants in the intervention group scored significantly higher on a subset of effective decision responses that indicated taking immediate action to be safe now (35% of the time) compared with the control group (15% of the time).

Although significant differences were seen on the decision-making scores at posttest between the intervention and control group, the groups did not differ on problem awareness, their ability to recognize an abuse situation as a problem situation requiring a self-protective decision. Further, significant correlations were found between pretest problem awareness scores and posttest decision-making scores (both overall effective and safe-now decision-making responses) in the intervention group and between pretest problem awareness scores and posttest overall effective decision-making scores in the control group. These findings suggested that problem awareness, paralleling the first step in the four-step decision-making strategy and determining whether the stepwise strategy was warranted, was an important cognitive appraisal step in effectively setting up the remainder of the decision-making process and in determining the final decision outcome. Although, problem awareness is largely shaped by cognitive input (e.g., comprehension or knowledge about the type of situation, in this case abuse), it might also be linked to noncognitive factors, such as motivation or emotion, underlying the initial identification of the decision problem. This finding presented the need to teach the emotional components of decision making at the initial step of the decision-making strategy, especially when decisions are to be made in highly emotional or pressured situation. This issue was addressed in the most recent revision of the curriculum called ESCAPE-NOW.

ESCAPE-NOW (Effective Strategy-Based Curriculum for Abuse Prevention and Empowerment – NOW)

The most recently updated version of the ESCAPE curriculum, ESCAPE-NOW, draws upon the wealth of research on decision making. Accordingly, it has an added focus on the emotional components underlying decision making, incorporating the effect of emotions on the mental processes, starting with problem identification, and leading up to selecting a decision action. The function of emotions in decision making is emphasized to highlight the different ways in which emotions can interact with cognitive and motivational processes during the course of decision processing. Some of the ways in which the role of emotions in decision making is made evident in the curriculum include providing information about feelings associated with abusive versus healthy situations as well as option generation, enabling (or hindering) rapid choices under conditions of time pressure or uncertainty, determining the focus of attention on relevant aspects of a decision problem, and generating the commitment (or avoidance) needed to reach a decision solution.

ESCAPE-NOW was developed and field-tested with the help of a Project Advisory Board that included individuals with IDD, parents, and IDD support professionals. Staff and participants with IDD at the field-test sites conducted ESCAPE-NOW sessions and provided ongoing feedback and trainer logs, which were the basis for the final modifications of the curriculum. New features in ESCAPE-NOW include the following:

- A wider range of updated and ecologically validated sexual, physical, and verbal abuse vignette situations that include cyber abuse, financial abuse, disability-specific abuse, as well as healthy interpersonal interactions
- Individual participant binders with activities and lesson-by-lesson assessments, which participants can keep for future reference
- An updated, user-friendly decision-making chart with a streamlined four-step strategy for decision making in abuse situations (see Fig. 22.1 for an image of the ESCAPE-NOW decision-making chart)
- Technology-based options, including as follows:
  - A video version of the Vignette Booklet consisting of lesson-by-lesson video files with audio narration
  - A PowerPoint™ version of the Vignette Booklet consisting of a lesson-by-lesson presentation of the visual vignette illustrations accompanied by the written text
  - An alternative format option consisting of SMARTBoard™ applications for Activity Sheet Activities

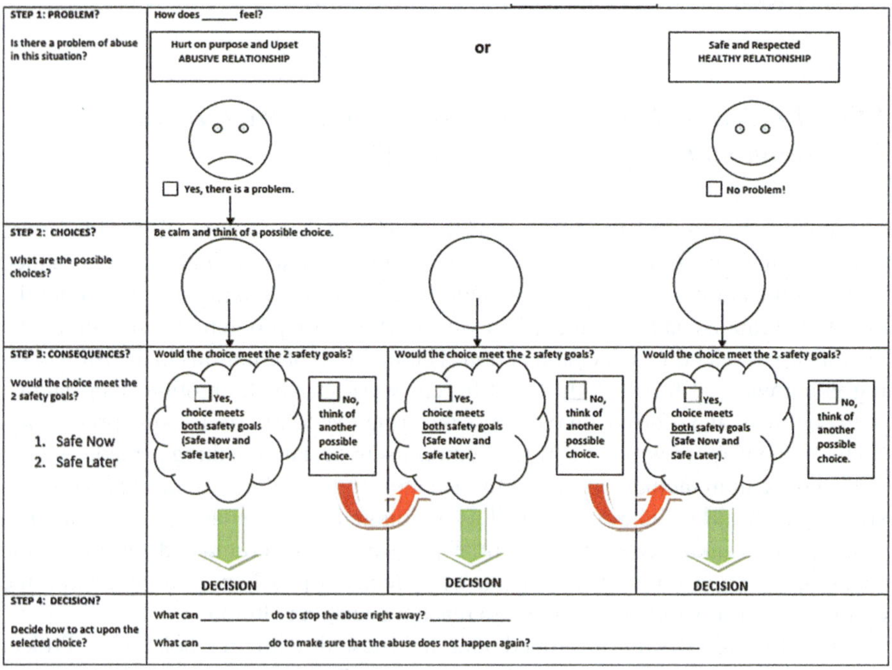

Fig. 22.1 Decision-making chart (Khemka & Hickson, 2015)

ESCAPE-NOW is available electronically free of charge and includes an instructor/staff training module which provides an overview of the curriculum components and general tips for trainers, including a video demonstration of how to teach the four-step decision making strategy.

Decision-Making Training Support Groups

A primary goal of our decision-making curricula for adults with IDD has been to provide explicit instruction in the use of a systematic decision-making strategy, along with a targeted focus on key motivational and emotional factors, as a way of increasing capability for effective decision making. This structured instruction provides a basis for developing a strong foundation for utilizing effective decision-making skills in response to interpersonal decision-making situations. However, we regard this phase of explicit knowledge and skill building provided in these curricula as a starting point to the vital preparation for generalizing the use of a learned decision-making strategy to individualized contexts, informed by various other factors such as personal and environmental influences, and most importantly past decision-making experiences. We therefore intentionally offer guidance and procedures for forming support groups and allocate structured and supervised time and activities (e.g., six support group sessions following the 12 instructional lessons) to allow participants to share and reflect on their individual life experiences and discuss potential applications of their acquired repertoire of decision-making skills in novel, personal situations. The support groups offer a meaningful and normalized space allowing training group members to share and support each other to promote healthy interpersonal relationships in their lives and to bolster feelings of empowerment to make self-protective decisions. Rehearsing ways to act upon different decision choice options through role-playing in these group sessions also provides an opportunity for participants to explore different pathways to decision making and to gradually develop expertise in managing affective cues and practicing their deliberative skills during decision making. The interactive shared processes in these group experiences can also provide opportunities for participants to receive pertinent feedback from others on the potential impact of their decisions. Although measuring decision-making outcomes in real-life encounters, especially in situations of abuse or threat of coercion, has been hard to achieve, we believe that successful participation in both the structured instructional phase and the support group sessions afterward can substantially prepare individuals with IDD to face real-life situations and to apply effective decision-making skills to the maximum extent possible. It is our hope that trained participants with IDD will go on to find ways to continue and form their own support groups in their local communities and that they will become their own powerful voice in making sense of their individual and collective experiences and effecting change in their own lives through their self-directed decision making.

Decision-Making Curriculum Development and Evaluation: Focus on Adolescents with IDD

Building upon the findings of our intervention studies with adults with IDD, we expanded our decision-making research and curriculum development efforts to include adolescents with IDD. Given that adolescence, developmentally, is a challenging time when youth may engage in risk-taking behaviors and/or acquiesce to negative peer influences (Gardner & Steinberg, 2005), the need for timely intervention and supports is urgent, particularly for adolescents with IDD who show heightened vulnerability to peer victimization and bullying (see Chapter 18 by Hickson & Khemka in this volume). Other than the general risk-taking behaviors and the peer victimization witnessed during this period, adolescents with IDD often have difficulty in sustaining positive peer relationships. These difficulties might also be associated with limited ability to articulate one's personal goals (especially when goals are in conflict with peer-induced or other social goals), to generate and evaluate alternative courses of action, and to act in their own best interest when handling a peer situation. Based on data from adolescent risk-taking studies, researchers have recommended targeting decision-making training as a promising approach for addressing the vulnerabilities of these adolescents (e.g., Bexkens et al., 2016; Wagemaker et al., 2020).

PEER-DM: Peers Engaged in Effective Relationships – Decision Making

PEER-DM, a decision-making-based program (Khemka & Hickson, 2013), was developed to teach decision-making skills to adolescents with IDD so that they could be more independent and safe in navigating interactions with peers. The curriculum was designed to teach concepts of positive (*when your peers encourage and support you to do things that are right for you and that you want to do*) and negative peer pressure (*when peers push or bully you to do things that get you into trouble or hurt you)* along with a four-step strategy for making effective decisions. PEER-DM incorporates the cognitive, motivational, and emotional components of decision making (see Hickson & Khemka, 2014; Khemka & Hickson, 2017). At the core of the curriculum is a four-step decision-making strategy, taught using instructional strategies that include modeling and fading, guided practice with interactive games and activities, and repeated opportunities to apply the strategies in a broad range of negative peer pressure situations. The six-weekly-session curriculum is delivered in small groups of three to four participants. The curriculum was designed to provide numerous opportunities to distinguish between a wide range of situations involving positive (no problem) and negative peer pressure (problem) and to apply the four-step strategy in the face of negative peer pressure.

An evaluation of PEER-DM was conducted using a randomized controlled design in which adolescents with IDD were randomly assigned either to a group who received PEER-DM ($n = 22$) or to a wait-list control group ($n = 20$; Khemka et al., 2016). Participants who had received PEER-DM produced significantly more effective decision-making responses and correct risk-perception scores on posttests relative to participants in the control group. The findings in the larger sample held up for a subgroup of participants with ASD.

Although PEER-DM provided clear empirical support in favor of decision-making training to increase understanding and handling of peer relationships, the curriculum's applicability in a school setting as part of a social skills class was limited as it had been designed for small groups. To develop a version that could be readily used in a larger classroom setting, PEER-DM was modified for classroom use.

PEER-DM-C: Peers Engaged in Effective Relationships – Decision Making: Classroom

PEER-DM-C is a modified version of the PEER-DM curriculum described in the Khemka et al. (2016) study, adapted for whole-class instruction and peer-supported learning. The decision-making scenarios covered in the curriculum were expanded from the original PEER-DM curriculum to allow for explicit instruction with situations involving a potential threat of cyberbullying, which adolescents with IDD are known to increasingly encounter (e.g., Kowalski & Toth, 2018). The first four lessons of the six-lesson curriculum focus on building knowledge of peer relationships and distinguishing concepts of negative peer pressure and positive peer pressure. The motivational/emotional aspects of decision making are enhanced with a focused discussion on goal prioritization, especially when competing goals are present (e.g., pleasing a friend vs. staying out of trouble), and an emphasis on effective ways of coping with decisional stress in peer pressure situations.

The remaining two lessons provide guided support for individualized and generalized use of strategies for making effective decisions across a range of hypothetical situations involving positive and negative peer pressure in school, vocational, or community settings, thereby providing situations that are as authentic as possible for practicing the application of the learned decision-making strategy and related skills. The decision-making strategy at the core of PEER-DM-C is essentially the same as the decision-making strategy in our adult curricula, but it is tailored to address adolescent peer pressure situations. The four steps exemplified are as follows: (1) identifying a situation as a problem (i.e., as a situation involving negative peer pressure), (2) generating alternatives (i.e., coming up with possible options for how to handle the situation), (3) considering possible consequences of each alternative and evaluating them in relation to previously established base goals (e.g., health and safety and not getting into trouble), and (4) choosing a course of action (i.e., selecting an independent and empowered option for the final decision; "Is the decision-maker acting on his/her own?") (Hickson & Khemka, 2013; Khemka et al.,

2016). Implemented as a whole-class lesson, follow-up activities and review were done in small groups with peer participation and feedback.

A pilot study was conducted to evaluate the impact of PEER-DM-C with a small group of 20 students with IDD between the ages of 16 and 19 years from a private, self-contained, specialized high school in New York City (Khemka et al., submitted). A quasi-experimental group design was used in this study to evaluate the effects of the decision-making strategy-based curriculum. Two intact classes of adolescents and young adults with IDD were randomly assigned to an intervention group who received PEER-DM-C or a comparison group who did not receive the curriculum. The students in the intervention group received the curriculum, delivered by whole-group instruction, during their group counseling period of 45 minutes once a week for a total 6 weeks.

The posttest performance of students in the intervention group was significantly superior to that of those in the comparison group on measures of decision making, awareness of peer influence, and knowledge about peer relationships. Group differences in effective decision-making skills were measured in response to hypothetical scenarios, including both positive and negative forms of peer pressure. This study provides preliminary support for incorporating PEER-DM-C into a social skills curriculum for adolescents with IDD. Due to lack of random assignment and the small group of participants involved, further research and replication will be needed to establish the efficacy of the whole-group decision-making curriculum.

Considerations in Decision-Making Training

Consistent with the *Pathways Model of Decision Processing* (described by Khemka in Chapters 6 and 11 of this volume), each of our decision-making interventions has been intentionally designed to address the three basic processes of cognition (e.g., *What is the problem? Is it a healthy or abusive situation?*), motivation (e.g., *What are my goals? I believe that I can stop the abuse.*), and emotion (e.g., *What are my feelings in a healthy situation? What are my feelings in an abusive situation?*) that we believe to underlie effective decision making. Figure 22.2 illustrates the overlapping domains and the interplay of these three key processes that are featured in the *Pathways Model.* The figure also shows that interaction with the neurodevelopmental/personal characteristics of the decision-maker can impact decision outcomes which are further regulated by inputs that are specific to situational/environmental factors which influence how the stepwise decision-making strategy in our interventions can operate to determine decision outcomes. The situational and environmental influences are defined by the nature of the specific decision task (e.g., a positive vs. negative peer pressure situation) and other contextual variables, such as the presence of peers, urgency of the decision, cultural norms for decision rules, etc.

After completing one of our decision-making curricula (e.g., ESCAPE-NOW or PEER-DM-C), individuals with IDD should be prepared to draw upon the four decision processing pathways described in the *Pathways Model* (see Chapter 6 by

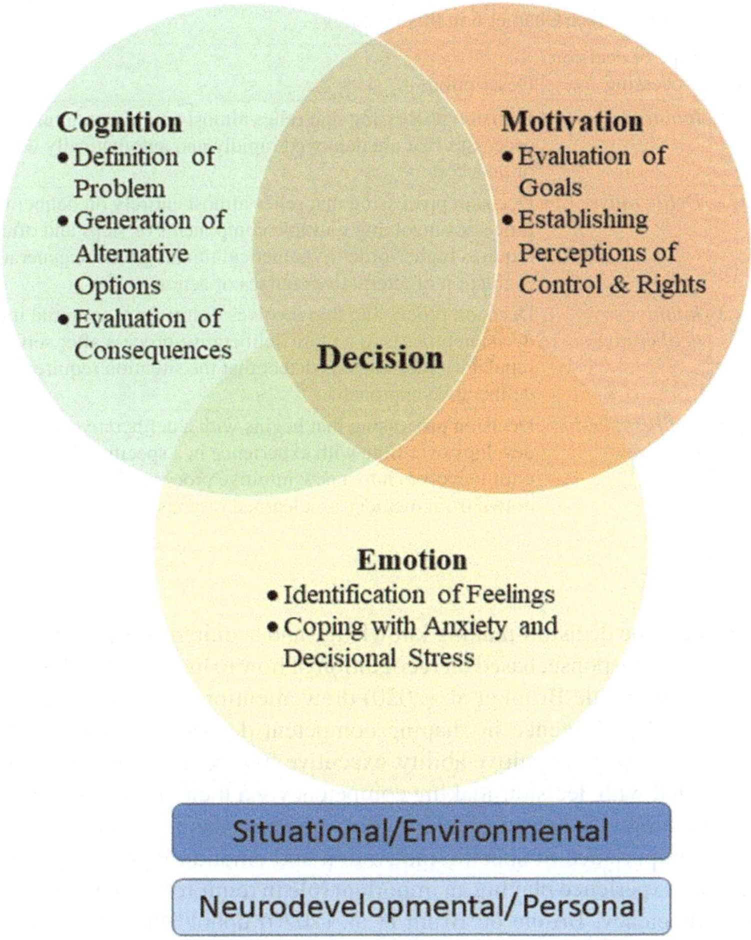

Fig. 22.2 The interplay of the three basic processes of cognition, motivation, and emotion in interpersonal decision making

Khemka in this volume) to pursue a decision action that is consistent with their personal safety goals. The four decision pathways, summarized in Table 22.1, include (1) an *intuitive* pathway (*Pathway 1*) that can be used with little effort in situations that do not carry a threat of abuse or negative peer pressure, (2) a *deliberative* pathway guided by the stepwise strategy taught in the curricula to resist abuse or negative peer pressure (*Pathway 2*), and (3) an *intuitive-to-deliberative* pathway for unstable situations that require a shift from an intuitive strategy to a self-protective, deliberative strategy (*Pathway 3*) and a *deliberative-to-intuitive* strategy that can take shape over time (*Pathway 4*). *Pathway 4* can become available with experience, acquired either through training or over time through practice,

Table 22.1 Descriptions of the four decision processing pathways based on the *Pathways Model of Decision Processing* (see Chapter 6 in this volume)

Pathway	Type of decision processing	Description
Pathway 1	*Intuitive*	Decision processing that relies almost entirely on intuitive processes that are deployed rapidly and automatically with little or no effort
Pathway 2	*Deliberative*	Decision processing that relies almost entirely on deliberative processes involving multiple components or steps and often requires higher-order hypothetical thinking for the generation and evaluation of alternative courses of action
Pathway 3	*Intuitive-to-deliberative*	Decision processing that involves a shift from an initial intuitive decision process to a more deliberative process after self-regulatory mechanisms detect that the situation requires a more deliberative approach
Pathway 4	*Deliberative-to-intuitive*	Decision processing that begins with a deliberative process and develops over time with experience in a specific decision domain until it evolves into a new intuitive process whereby a decision is drawn from memory as a learned response

which can bind the decision-making knowledge and acquired strategy expertise into an intuitive-type response, based on recognition of how to handle a familiar situation.

Recently, Bruine de Bruin et al. (2020) drew attention to the roles of both fluid and crystallized intelligence in shaping competent decision behavior. Aspects of fluid intelligence (e.g., cognitive ability, executive functioning, attention) have long been associated with decision-making competency via their emphasis on the cognitive underpinnings of decision making. According to them, newer interpretations point to the importance of specific motivation and emotion inputs during decision making, with experience playing an important role in reinforcing increased decision-making competency. Bruine de Bruin et al. (2020) uphold that "Individuals who have more experience with specific decisions may not need to deliberate as much about those decisions, because they have acquired crystallized intelligence and have already learned what to do" (p. 188). Emotions that help set the initial feeling response to a decision process also play an essential role in the end to interpret the emotions experienced as a result of a decision, thereby adding to the individual's experience base to inform future preferences and decision effort.

The situational considerations involved in selecting the vignette scenarios for teaching the decision-making strategies in our curricula are two-fold. First, the decision making strategy is taught in the context of a specific type of situation and then practiced over many examples of the same type of situation. For example, in ESCAPE-NOW, the main situation domain is abuse, and various prototype situations reflecting abuse are used in the training. The curriculum covers a range of situations – situations of generic sexual, physical, and verbal abuse as well as situations involving financial abuse, internet/technology-related abuse, and disability-specific abuse. Within the category of sexual abuse, a range of situational variations are covered – setting of the abuse (e.g., home or work place), perpetrator type (e.g., family member, coworker), gender of the perpetrator, individual or group perpetration, etc. In Step 1, definition of the problem and features of the situational context are examined to define whether there is a problem (or not) that requires a deliberative decision-making process. For example, in PEER-DM, definition of the problem requires knowing whether the situation involves negative or positive peer pressure before a decision can be attempted. The demands for decision making therefore vary by situation and are a significant factor in the decision process.

The second situational consideration is that the scenarios presented as hypothetical decision tasks were selected for their relevance to real-world outcomes. For example, in selecting the situations for ESCAPE-NOW, we drew upon input from individuals with diverse perspectives. These included members of our Project Advisory Board, respondents to an online needs survey, and the views of support professionals in the fields of domestic violence/sexual assault and IDD (e.g., Hickson et al., 2013). Consistent with the Bruine de Bruin et al.'s (2020) recommendation that interventions should provide people with the experience they need to master decision-making principles so that, with practice, the application of the decision-making strategy in protected settings will improve and transfer to real-world settings, we build in sufficient practice and maintenance probes in our training to support this growth in decision-making competency.

Interpersonal decisions by nature tend to be open-ended and have outcomes that are often unknown at the start of the process and get derived in the moment as the decision situation unfolds. They tend to be influenced by the social demands of a situation and, in situations of abuse or bullying, can involve the added complexities of a power dynamic. Therefore, these decisions are highly sensitive not only to differences in decision-making ability and individual preferences but also to a host of outside influences relating to the social and cultural bases of the decision-making situation. In the interest of providing individuals with IDD with repertoires of effective, independent, self-protective decision-making skills, we promote and support the use of a well-thought-through, deliberative decision process as detailed in this chapter but recognize that relying on one's intuition for decision making might be a reasonable and satisfactory way to reach an effective decision outcome in certain situations. Further, we know that a new intuitive pathway leading to a positive decision outcome can be formed when repeated use of the same strategy provides an experience advantage and a new pathway over time.

The decision-making strategy taught in our curricula is presented as a visual graphic overlay with the steps clearly marked in a stepwise sequence. The visual sequencing of the steps signals the process of decision making and the sequence of its related sub-components. The use of visuals as a tool to enable cognitive functioning and attention to information needed during the process of decision making has been supported by research. In a study by Bailey et al. (2011), use of visual aids to support integration of information from different sources for reaching a decision was found to be beneficial for individuals with ID (mean Full-Scale IQ = 59.8). Participants trained to use a visual aid to support decision making showed improved quality of decision making on a temporal discounting task and a scenario-based financial decision-making task, suggesting that the visuals helped the participants to not only successfully weigh up information needed for reasoning the decisions but that they were able to provide more information to justify their decisions. The potential of visuals as an aid to decision-making interventions was thus clearly supported by the findings of this study.

The strategy for decision making in the ESCAPE and PEER-DM curricula is intended to be taught via the use of a scripted Instructor's Manual with effective instructional strategies tied to an explicit model of teaching beginning with a sequence of teacher modeling and moving gradually to guided practice in interactive small group activities, fading to semi-independent practice and increasingly to independent application of the strategy to new situations.

Curricular Adaptations for Singular Groups

The increase in etiology-based understanding of specific types of IDD and their decision-making processing patterns has implications for how best to plan and differentiate training by addressing etiology-specific needs (see Chapter 7 by Goscicki et al. in this volume). Although it can be argued that etiology-based interventions, specified for each disability type, may have a larger impact on the learning outcomes for individuals in those specific groups, it may, in practicality, not be possible to provide a totally differentiated approach for each type of IDD. In inclusive settings, due to low numbers of individuals with specific types of IDD (e.g., Williams syndrome (WS), Prader-Willi syndrome (PWS)), it may not be feasible to implement etiology-specific decision-making interventions for each disorder. In addition, research on the efficacy of such approaches will likely be limited due to the general rarity of the disorders (Hodapp & Fisher, 2017). It may be worthwhile to examine whether interventions that have been shown to be successful for a specific type of IDD could benefit individuals with other types of IDD, based on certain commonalities, such as those observed between WS and other developmental disorders, such as Down syndrome (DS) and ASD (Brown et al., 2003; Klein-Tasman et al., 2009). For example, although individuals with DS and WS differ in their reading difficulties, with WS-related reading difficulties resulting from visual-spatial processing difficulties and DS-related reading difficulties connected to delayed phonological awareness and vocabulary; there is research that shows that vocabulary is a significant longitudinal predictor of reading growth for both groups (Steele et al., 2013). Therefore, the *Reading and Language Intervention for Children with Down Syndrome* (2021), a targeted, evidence-based approach to support the development of reading and language skills in individuals with DS, might also benefit the reading skills of individuals with WS. For decision-making interventions, we therefore suggest a focus on the overlap of certain characteristics among individuals with various types of IDD and that best practice for effective programming might involve developing an overall intervention approach with the option of differentiated inputs on an individual or small group basis.

An effort to tailor ESCAPE to meet the needs of 12 women, who were attending a residential program for individuals with WS, was assessed informally in an exploratory pilot study (Hickson & Khemka, 2014). Individuals with WS, with mild to moderate ID, are known to have a distinct social-emotional and behavioral profile with a tendency to be overfriendly and to indiscriminately approach strangers. They have been described as lacking social judgment and having problems with attention (see Doyle et al., 2004; Dykens, 2003). Along with socio-communication difficulties, they tend to face increased risk for victimization. Based on this WS profile and an initial series of observations and assessments conducted at the start of the pilot study, we identified several areas as a basis for adapting the ESCAPE curriculum to match key phenotype requisites. Originally designed to provide women with IDD with an effective repertoire of self-protective decision-making skills, ESCAPE offers systematic instruction in applying a stepwise strategy for decision making. Individual turn taking was incorporated as a compensatory strategy to address attention deficits and a tendency for individual women to dominate the discussion. It also served to increase the engagement of all participants. Enhancing the explicitness of instructions with more concrete examples helped address comprehension weaknesses. Adaptations to address visuospatial issues included replacing wall charts with individual worksheets. These modifications were incorporated into the ESCAPE-WS version of the curriculum (Khemka & Hickson, 2005) which was tailored to the specific needs of women with WS. The small group of participating women responded positively to the adaptations in the ESCAPE-WS curriculum, leading to the conclusion that the instructional implications of distinct decision-making patterns with genetically based syndromes associated with ID merit further attention. Some of the adaptations made for ESCAPE-WS were subsequently incorporated into ESCAPE-NOW to the benefit of participants with other types of IDD (e.g., turn taking and individual participant binders).

Other examples of etiology-specific differentiations that could be used to support decision-making interventions for specific groups of individuals with IDD can be drawn from the phenotype patterns that have been identified for the groups. For instance, those with ASD may not easily show understanding of others' emotions which typically requires the spontaneous, automatic encoding of socially relevant information. Senju et al. (2009) recommend that a comparable understanding can be achieved through the application of explicit verbally mediated reasoning strategies. For individuals with ASD, the information gathering process during decision making can be overwhelming and time-consuming due to their tendency to analytically examine all available information and be overly deliberative before settling on a decision action (Brosnan et al., 2016; Luke et al., 2012; Volkmar et al., 2005). Therefore, it might be helpful to plan for additional strategies that might help them to streamline information so that they are able to focus on the most relevant aspects of a decision task required to evaluate options for decision making. In interpersonal situations where unpredictability is a given, a satisficing tool, such as generating a limited number of options, might need to be added to the four-step strategy for decision making. A longer response time in decision making has been observed among individuals with ASD, which can be adaptive to decision making in well-reasoned, analytical decision tasks. However, to meet more spontaneous, on-the-spot decision demands, typical of interpersonal decision situations, a self-prompting strategy for better allocation of time and resources might prove beneficial. The *Pathways Model* specifies different ways for approaching decision making, and although individuals with ASD most naturally rely on a deliberative, carefully thought-out *Pathway 2*, the model offers alternative mechanisms to decision making that can be considered during training to reduce rigidity in decision-making patterns and to have a wider repertoire of skills that can be flexibly applied depending on the situational context of the decision.

Research has shown that when individuals with ASD encounter quick-paced, unexpected social situations, they tend to search from memory for a similar experience that they can use as a model for their decision (Allman et al., 2005; Volkmar et al., 2005). It is possible that decision-making training can help optimize this preference for lawful, predictable responses in an adaptive way to create experience (leading to memory) with newer more effective ways of responding. Over time, learned responses can serve as regularities or formalistic rules that individuals with ASD could rely on easily to make decisions in social environments. In this way, they can intuitively access expert decision responses to a situation akin to that suggested for *Pathway 4*.

The emotional involvement in decision making for individuals with ASD can be complex, with studies (e.g., Luke et al., 2012; Vella et al., 2018) reporting that individuals with ASD might find the overall experience of decision making overwhelming and exhausting, to the extent that they can get *locked up* under the pressures of their own inflexibility and overly logical thinking during decision making. Therefore, additional strategies for managing anxiety and improving organization and planning skills for increased attention to the decision task on hand may be required. In sum, it will be helpful to design training tools in ways that are consistent with the ways that are known to be effective for individuals with ASD. Finally, self-regulatory mechanisms and ability to exercise inhibitory controls to monitor highly emotional or affective responding will need to be maintained and supported during the course of decision processing.

New Directions in Decision-Making Training and IDD

Application to Broader Domains of Decision Making

In our research, we have tested the application of a systematic decision-making strategy in the context of social interpersonal situations involving some form of maltreatment. Although our study has been limited to social contexts, we believe that the decision-making strategy and tools developed in this specific context provides a generalized structure for making decisions in other domains (e.g., financial or medical). We conceptualize the four-step decision-making strategy as providing a general plan for how to simplify the demands of any decision task with serious long-term consequences and how to navigate a decision toward a final outcome. We believe that this decision approach can be used for scaffolding decision making in novel situations with minor adjustments.

Methodological Challenges in Decision-Making Intervention Research

Although, to date, we have consistently found significant positive treatment effects for our decision-making-based interventions, we acknowledge that the sample sizes have been small. Larger sample size evaluation studies and systematic studies involving more representation from diverse samples of individuals with various types of IDD are required to replicate and aggregate the findings across individual studies to establish the overall effectiveness of these interventions. In our own research studies evaluating the effectiveness of ESCAPE-DD or PEER-DM, the samples have largely included participants across a broad range of classifications within the overall categories of ID and IDD. Participants with ADHD, or those having singular disorders within IDD (e.g., PWS) have rarely been included in interpersonal decision-making program development and evaluation studies. Effective program development for them will require in-depth needs assessment and adaptations to existing program components based on the specific training requirements of these individual disability groups.

In addition, the outcomes of our intervention studies have been associated with indirect measures (e.g., knowledge about peer pressure, defining a problem, suggesting a decision action for a protagonist in a hypothetical situation) rather than direct measures (e.g., frequency of attempts to stop victimization). Future research must develop measures of real-life outcomes associated with the decision-making interventions, to establish their effectiveness in improving decision outcomes in real-life situations.

Future Areas of Decision-Making Intervention Research

In establishing evidence-based practices for improving decision making among individuals with ASD and ID, it is essential that we also consider secondary and tertiary conditions (e.g., comorbidities) in that these conditions can have a measurable impact on how the disability manifests itself over time and interacts with the efficacy of any treatment. Therefore, best practices across fields (e.g., personality/behavior management and mental health services) might need to be combined with any decision-making training efforts to address all aspects of an individual's functioning to accomplish the most effective life outcomes for them.

In line with this emphasis, future interventions should further explore the role of emotions in decision making, since emotions connect with the psychobiological nature of the many different groups within IDD. Although we commonly relegate the primary function of emotion to be a direct cause of behavior (e.g., as in the emotion of fear causing fleeing), there is evidence that (see Baumeister et al., 2007) emotion also influences behavior indirectly via a feedback system. Emotional states promote (or deter) future behavior by providing feedback and appraisal of actions. Rapid, automatic affective responses due to the intensity of the emotional experience inform cognitive and behavioral choices to create a strong memory to guide future decisions. The automatic affective responses in a decision situation might remind the person of past emotional outcomes and provide useful guides as to what emotional outcomes may be anticipated in the present, thereby creating a feedback cycle. Drawing from such emotional inputs, motivational efforts are impacted, fostering strong interdependence between these factors during decision making. This impact of experience on decision competencies has been identified by Bruine de Bruin et al. (2020) as an important variable in how decisions are remembered or become automatic over time with more experience.

Many of the training efforts in decision making have involved individuals with mild to moderate levels of ID. However, decision-making skills would be important intervention aims for individuals with IDD who have severe to profound intellectual disability (S/PID). Decision-making interventions that may improve decision making skills in this population will need to include a focus on communication and behavioral skills, along with more structured teaching interventions and accommodations to address the developmental performance levels of the individuals.

The four-step decision-making strategy that has been key to our intervention work has had a focus on improving decision-making preparedness of individuals with IDD in situations involving different forms of maltreatment. However, the decision-making strategy provides a generalized plan for systematically teaching a decision-making process and related skills for situations with serious consequences. The strategy addresses the overall needs of any deliberate, effortful decision-making process, including planning (cognitive and metacognitive structures), personal resources (emotional and motivational resources), and a range of executive functions, including self-regulation. The situational context can be easily extended to other novel decision-making tasks or situations, and future research should examine the effectiveness of this approach in different domains of decision making.

Given that much of the time people make decisions by routine, habit, or automatic processes (Kay, 2002), it is our belief that exposure to deliberative, strategy-based training in decision making will provide effective tools to individuals with IDD, so that if they are not able to apply a deliberative decision-making process, they will proceed with intuitive decision responding in a more cautious and controlled manner. This view has most relevance for individuals, such as those with ADHD and similar disorders, for whom difficulties with executive function, impulse control, and/or elevated anxiety can lead them to more impulsive or hasty forms of decision responding. More research from a neurodevelopmental perspective is needed to investigate which pathways of our *Pathways Model* are likely to be more adaptive for a particular type of IDD and which resources need to be developed to best individualize training and experience within the broad parameters of the *Pathways Model* to maximize individual decision outcomes.

We are aware that although most of our decision-making training efforts have focused on preparing individuals with IDD to effectively handle situations of coercion or abuse, there is need to extend decision-making training to focus on building positive, healthy emotional and motivational states that promote prosocial behaviors and adaptive social functioning and adjustment. This is in accord with Fredrickson's (2001) notion of the positive psychology of emotions, whereby the experience of positive emotions can enable individuals to broaden their momentary thought-action repertoires and establish positive affective states that over time increase one's physical and psychological resources toward personal well-being. Experiences of positive affect, also according to Isen (1990), produces a "broad, flexible cognitive organization and ability to integrate diverse material" (p. 89), resulting in thought patterns that allocate broader resources to functions of attention, cognition, and action. From a decision-making perspective, such psychological (and resulting social) growth can be fundamental in improving cognitive and intrinsic motivation when engaged in decision-making activities and the ability to endure through difficult decisions.

Conclusion

Our concerted efforts at exploring evidence-based strategies and supports to improve the decision-making capabilities of individuals with IDD have resulted in significant progress in our overall understanding of decision making in individuals with IDD and the development of the *Pathways Model of Decision Processing* to serve as a useful explanatory framework for the study of decision making for this group. Although much progress has been made in understanding the nature of decision-making processes for both adolescents and adults with IDD, the emphasis on training and skills development for improved decision making is still nascent. Research substantiates the multidimensional and complex nature of decision making among those with IDD, with influences of underlying mechanisms spanning cognitive, motivational, and emotional factors across different types of decision tasks. The range of these underlying mechanisms and the extent to which these factors differentially impact different types of IDD needs further investigation. Growing understanding of these aspects of decision processing will provide a more informed basis for designing differentiated decision-making interventions in the future and support improved training outcomes by considering the full breadth of difficulties posed by the different types of IDD in response to different types of decision-making situations.

Empirical evidence for the effectiveness of decision making-based curricula in improving knowledge and decision-making skills of adults and adolescents with IDD as summarized in this chapter points to the possibility of narrowing ability-related differences in decision making over time for this group. In recognizing the critical role of decision making, especially in the social context for preventing maltreatment, we advocate for continuing research for the advancement and evaluation of decision-making strategies for strengthening the interpersonal decision-making skills of individuals with IDD.

To conclude, it is well recognized that individuals with IDD must make significant life decisions as they navigate interpersonal situations with family members, friends, romantic partners, coworkers, and people in their communities. Indeed, the ability to manage personal and interpersonal affairs partly defines what it means to function as an adult in our society. Research examined in this chapter validates strategies to better understand and strengthen interpersonal decision making and offers ways to enable people with IDD to have personal agency of choice and control in their lives. We believe that enabling them to acquire a repertoire of effective independent interpersonal decision-making skills can provide them with a powerful tool for building positive, satisfying friendships and intimate relationships as well as for ensuring their own personal safety and preventing harm. With continuing development of effective evidence-based interventions, and their successful translation into practice, increases in the capabilities of individuals with IDD that allow them to be more independent and effective in their daily decision making are imminent.

References

Allman, J. M., Watson, K. K., Tetreault, N. A., & Hakeem, A. Y. (2005). Intuition and autism: A possible role for Von Economo neurons. *Trends in Cognitive Sciences, 9*(8), 367–373. https://doi.org/10.1016/j.tics.2005.06.008

Bailey, R., Willner, P., & Dymond, S. (2011). A visual aid to decision making for people with intellectual disabilities. *Research in Developmental Disabilities, 32*(1), 37–46. https://doi.org/10.1016/j.ridd.2010.08.008

Baumeister, R. F., Vohs, K. D., DeWall, C. N., & Zhang, L. (2007). How emotion shapes behavior: feedback, anticipation, and reflection, rather than direct causation. *Personality and social psychology review: an official journal of the Society for Personality and Social Psychology, Inc, 11*(2), 167–203. https://doi.org/10.1177/1088868307301033

Bexkens, A., Jansen, B. R., Van der Molen, M. W., & Huizenga, H. M. (2016). Cool decision making in adolescents with behavior disorder and/or mild-to-borderline intellectual disability. *Journal of Abnormal Child Psychology, 44*(2), 357–367.

Brosnan, M., Lewton, M., & Ashwin, C. (2016). Reasoning on the autism spectrum: A dual process theory account. *Journal of Autism and Developmental Disorders, 46*(6), 2115–2125. https://doi.org/10.1007/s10803-016-2742-4

Brown, J. H., Johnson, M. H., Paterson, S. J., Gilmore, R., Longhi, E., & Karmiloff-Smith, A. (2003). Spatial representation and attention in toddlers with Williams syndrome and Down syndrome. *Neuropsychologia, 41*(8), 1037–1046. https://doi.org/10.1016/s0028-3932(02)00299-3

Bruine de Bruin, W., Parker, A. M., & Fischhoff, B. (2020). Decision-making competence: More than intelligence? *Current Directions in Psychological Science, 29*, 186–192.

Doughty, A. H., & Kane, L. M. (2010). Teaching abuse-protection skills to people with intellectual, disabilities: A review of the literature. *Research in Developmental Disabilities, 31*(2), 331–337. https://doi.org/10.1016/j.ridd.2009.12.007

Doyle, T. E., Bellugi, U., Kronerberg, J. R., & Graham, J. (2004). "Everybody in the world is my friend" hypersociability in young children with Williams syndrome. *American Journal of Medical Genetics, Part A, 124A*(3), 263–273.

Dykens, E. M. (2003). Anxiety, fears, and phobias in persons with Williams syndrome. *Developmental Neuropsychology, 23*, 291–316.

Fredrickson, B. L. (2001). The role of positive emotions in positive psychology. *American Psychologist, 56*(3), 218–226.

Gardner, M., & Steinberg, L. (2005). Peer influence on risk taking, risk preference, and risky decision making in adolescence and adulthood: An experimental study. *Developmental Psychology, 41*, 625–635.

Healey, K. N., & Masterpasqua, F. (1992). Interpersonal cognitive problem-solving among children with mild mental retardation. *American Journal on Mental Retardation, 96*, 367–372.

Hickson, L., & Khemka, I. (1999). Decision making and mental retardation. In L. M. Glidden (Ed.), *International review of research in mental retardation* (Vol. 22). Academic.

Hickson, L., & Khemka, I. (2013). Problem solving and decision making. In M. Wehmeyer (Ed.), *The Oxford handbook of positive psychology and disability*. Oxford University Press.

Hickson, L., & Khemka, I. (2014). The psychology of decision making. In R. Hodapp (Ed.), *International review of research in developmental disabilities* (Vol. 47, pp. 185–229). Elsevier.

Hickson, L., Khemka, I., Golden, H., & Chatzistyli, A. (2013). Views and values of developmental disabilities and domestic violence/sexual assault support professionals regarding the prevention and handling of situations of abuse. *Journal of Policy and Practice in Intellectual Disabilities, 10*(3), 207–214.

Hickson, L., Khemka, I., Golden, H., & Chatzistyli, A. (2015). Randomized controlled trial to evaluate an abuse prevention curriculum for women and men with intellectual and developmental disabilities. *American Journal on Intellectual and Developmental Disabilities, 120*(6), 490–503.

Hodapp, R. M., & Fisher, M. H. (2017). Using genetic etiology to intervene with students with intellectual disabilities. In S. Bouregy, E. L. Grigorenko, S. R. Latham & M. Tan Jenkinson, J.C., & Nelms, R. (1994). Patterns of decision making behavior by people with intellectual disability: An exploratory study. *Australia and New Zealand Journal of Developmental Disabilities*

Horner-Johnson, W., & Drum, C. E. (2006). Prevalence of maltreatment of people with intellectual disabilities: A review of recently published research. *Mental Retardation and Developmental Disabilities Research Reviews, 12*(1), 57–69. https://doi.org/10.1002/mrdd.20097

Hughes, R. B., Robinson-Whelen, S., Davis, L. A., Meadours, J., Kincaid, O., Howard, L., Millin, M., Schwartz, M., McDonald, K. E., & The Safety Project Consortium. (2020). Evaluation of a safety awareness group program for adults with intellectual disability. *American Journal on Intellectual and Developmental Disabilities, 125*(4), 304–317.

Isen, A. M. (1990). The influence of positive and negative affect on cognitive organization: Some implications for development. In N. Stein, B. Leventhal, & T. Trabasso (Eds.), *Psychological and biological approaches to emotion* (pp. 75–94). Erlbaum.

Jenkinson, J., & Nelms, R. (1994). Patterns of decision making behavior by people with intellectual disability: An explanatory study. *Australia and New Zealand Journal of Developmental Disabilities, 19*, 99–109.

Kay, J. (2002, August 20). Beware the pitfalls of over-reliance on rationality: Attempting to shoehorn complex decisions into the framework of classical theory can be a mistake. *The Financial Times*, p. 9.

Khemka, I. (2000). Increasing independent decision making skills of women with mental retardation in simulated interpersonal situations of abuse. *American Journal on Mental Retardation, 105*, 387–401.

Khemka, I., & Hickson, L. (2002). *ESCAPE: An effective strategy-based curriculum for abuse prevention and empowerment*. Teachers College, Columbia University.

Khemka, I., & Hickson, L. (2005). *ESCAPE - WS: An effective strategy-based curriculum for abuse prevention and empowerment – Williams syndrome*. Teachers College, Columbia University.

Khemka, I., & Hickson, L. (2008). *ESCAPE-DD: An effective strategy-based curriculum for abuse prevention and empowerment for adults with developmental disabilities*. Center for Opportunities and Outcomes for People with Disabilities, Teachers College, Columbia University.

Khemka, I., & Hickson, L. (2013). *PEER-DM: Peers engaged in effective relationships: A decision making approach*. Teachers College, Columbia University.

Khemka, I., & Hickson, L. (2015). *ESCAPE-NOW: An effective strategy-based curriculum for abuse prevention and empowerment for individuals with developmental disabilities-NOW*. Teachers College, Columbia University.

Khemka, I., & Hickson, L. (2017). The role of positive psychology in interpersonal decision making. In K. A. Shogren, M. Wehmeyer, & N. N. Singh (Eds.), *Handbook of positive psychology in intellectual and developmental disabilities: Translating research into practice* (pp. 145–171). Springer.

Khemka, I., Hickson, L., & Reynolds, G. (2005). Evaluation of a decision making curriculum designed to empower women with mental retardation to resist abuse. *American Journal on Mental Retardation, 110*(3), 193–204.

Khemka, I., Hickson, L., & Mallory, S. B. (2016). Evaluation of a decision making curriculum for teaching adolescents with disabilities to resist negative peer pressure. *Journal of Autism and Developmental Disorders, 46*(7), 2372–2384.

Khemka, I., Hickson, L., & Gilic, L. (submitted). *What is the best thing for a target of peer aggression to do? Views of middle schoolers in Catholic schools.*

Klein-Tasman, B. P., Phillips, K. D., Lord, C., Mervis, C. B., & Gallo, F. J. (2009). Overlap with the autism spectrum in young children with Williams syndrome. *Journal of Developmental and Behavioral Pediatrics: JDBP, 30*(4), 289–299. https://doi.org/10.1097/DBP.0b013e3181ad1f9a

Kowalski, R. M., & Toth, A. (2018). Cyberbullying among youth with and without disabilities. *Journal of Child & Adolescent Trauma, 11*, 7–15. https://doi.org/10.1007/s40653-017-0139-y

Luke, L., Clare, I. C., Ring, H., Redley, M., & Watson, P. (2012). Decision-making difficulties experienced by adults with autism spectrum conditions. *Autism, 16*(6), 612–621.

Reading and Language Intervention for Children with Down Syndrome (RLI). (2021). Dse. https://WWW.down-syndrome.org/en-us/resources/teaching/rli/

Senju, A., Southgate, V., White, S., & Frith, U. (2009). Mindblind eyes: An absence of spontaneous theory of mind in Asperger syndrome. *Science (New York, N.Y.), 325*(5942), 883–885. https://doi.org/10.1126/science.1176170

Short, J. E., & Evans, W. S. (1990). Individual differences in cognitive and social problem-solving skills as a function of intelligence. *International Review of Research in Mental Retardation*, Academic Press, *16*(1990), 89–123. https://doi.org/10.1016/S0074-7750(08)60093-2

Smith, D. C. (1986). Interpersonal problem-solving skills of retarded and nonretarded children. *Applied Research in Mental Retardation, 7*, 431–442.

Steele, A., Scerif, G., Cornish, K., & Karmiloff-Smith, A. (2013). Learning to read in Williams syndrome and Down syndrome: Syndrome-specific precursors and developmental trajectories. *The Association for Child and Adolescent Mental Health, 54*(7), 754–762. https://doi.org/10.1111/jcpp.12070

Vella, L., Ring, H. A., Aitken, M. R., Watson, P. C., Presland, A., & Clare, I. C. (2018). Understanding self-reported difficulties in decision making by people with autism spectrum disorders. *Autism : The International Journal of Research and Practice, 22*(5), 549–559. https://doi.org/10.1177/1362361316687988

Volkmar, F. R., Paul, R., Klin, A., & Cohen, D. J. (2005). *Handbook of autism and pervasive developmental disorders* (Vol. 1, 3rd ed., p. 1283). Wiley.

Wagemaker, E., Huizenga, H. M., Dekkers, T. J., Collot d'Escury-Koenigs, A. L., Salemink, E., & Bexkens, A. (2020). When do those "risk-taking adolescents" take risks? The combined effects of risk encouragement by peers, mild-to-borderline intellectual disability and sex. *Journal of Abnormal Child Psychology, 48*(4), 573–587. https://doi.org/10.1007/s10802-020-00617-8

Werner, S. (2012). Individuals with intellectual disabilities: A review of the literature on decision making since the Convention on the Rights of People with Disabilities (CRPD). *Public Health Reviews, 34*, 1–27.

Part VI
New Directions in Decision-Making Research

Chapter 23
Decision Making and Intellectual and Developmental Disabilities: Future Directions for Research and Practice

Ishita Khemka

Viewing decision making as a natural part of the life experience of individuals with intellectual and developmental disabilities (IDD) promotes positive psychology and a path forward to their full participation and self-determination as envisioned in the UN Convention on the Rights of Persons with Disabilities (CRPD) (UN General Assembly, 2007). Over the past year, as this edited volume was taking shape, individuals with IDD have faced extraordinarily challenging times in response to the global pandemic of COVID-19 and a divisive national election in the United States. This has brought to the forefront their legal right to protect and advocate for self-directed decision-making participation in every aspect of life as endorsed by Article 12 of the CRPD. In the context of the COVID-19 pandemic, decisions to access safe care, supports, and services, many with serious and long-term consequences, have had to be made swiftly. The disproportionate burden of the pandemic on people with IDD (Nygren & Lulinski, 2020) has stimulated inquiry and thought in the disability field, with efforts to safeguard their decision-making rights being paramount. According to Luckasson and Schalock (2020), "The COVID-19 pandemic is the recent example of a crisis that places people with IDD at risk for lopsided societal reactions and threats to them or their wellbeing" (p. 1). During the same time period, many individuals with IDD marched the streets nationwide in support of the Black Lives Matter Movement in the United States, leading up to their actively petitioning in large numbers to exercise their right to vote in the 2020 National Elections, securing their voice to have a choice and establish personal agency.

It is in response to these life events that individuals with IDD have recently witnessed the integral role of their independent and supported decision-making rights

I. Khemka (✉)
St. John's University, Queens, NY, USA
e-mail: khemkai@stjohns.edu

I. Khemka, L. Hickson (eds.), *Decision Making by Individuals with Intellectual and Developmental Disabilities*, Positive Psychology and Disability Series,
https://doi.org/10.1007/978-3-030-74675-9_23

and the impact of their informed choices in determining outcomes relating to their inclusion, equity, and personal freedom. They have also seen that these rights are not always guaranteed and that they can be diverted to others (through substitute decision making) or their rights can be maintained with additional supports, especially in times of crises and major life events. In this context we remind ourselves to continue to value and uphold the principles of the CRPD and to find a balanced approach to decision making, in keeping with the will and preferences of individuals with disabilities. It pushes us, as a field, to examine our stance for how to interpret the procedural safeguards required to protect the legal capacity rights of all persons with disabilities as outlined in the CRPD and how best to allocate our resources and efforts toward training and support building to better prepare to adhere to the mandates.

The coming together of this volume to provoke a thoughtful discussion on factors that facilitate or impede the realization of the decision-making rights of individuals with IDD, while recognizing the importance of the interaction between a person and their context, could not have been more timely. The global sharing of ideas and concepts in this volume, from research to best practices that examine individual characteristics and environments for the purpose of improving the decision-making capacity of individuals with IDD, offers a meaningful opportunity to further reflect upon and align to the principles of the CRPD.

Our (Khemka & Hickson) efforts have been directed toward presenting far-reaching insights based on the findings of decision-making research with individuals with IDD, which we recognize to be a fast-moving field where much exciting progress has been made in recent years. Here, we draw attention to topics that were not covered in the volume but deserve consideration in future discussions, both in research and practice. The first part of this chapter extends the thinking on a few topics covered in the volume to feature certain important implications for practice. Next, future areas for research are identified that have the potential to strengthen our understanding of the decision-making processes of individuals with IDD and to inform the further development of evidence-based approaches for training and for the identification of needed supports.

Implications for Practice

Decision-Making Models

Although supported decision making is well on its way to becoming an effective and recognized practice as a less restrictive alternative to guardianship arrangements (see Shogren et al., 2019), there is still limited empirical research on the challenges and practical strategies for providing decision supports, including training and monitoring of the roles of decision supporters, both formal and informal. Bigby and Douglas (2020) call for a closer look at the availability of capacity building resources for both supporters and those whom they support. Empirically validated practices for providing support for decision making are just beginning to emerge, and continued evaluation of provisions for supported decision making and similar arrangements is essential (Browning et al., 2020). *Wehmeyer*, *Shogren* et al., and *Bigby and Douglas* in this volume provide a strong foundation for understanding supported decision-making models, both as theory and in application. Their chapters set forth valuable next steps to defining systems for assessing and planning environments and supports for individuals with IDD across a wide range of functioning.

We value the freedom of individuals with IDD to always make decisions on their own, even decisions that might result in adverse consequences. In keeping with Perske's (1972) notion of *dignity of risk*, we contend that the underlying assumption when measuring a person's decision capacity should be one of full independence with complete control over their personal agency. Research and practice should be centered on this right.

In building capacity for independent decision making, we have directed our efforts in research and practice to focus on the development of self-initiated, independent decision-making skills, free from other dependence, in the context of interpersonal and social situations involving a potential threat of coercion or abuse. The *Pathways Model of Decision Processing*, updated by *Khemka* in Chapter 6 in this volume, captures our current theoretical understanding of decision making, and interrelated influences in persons with IDD, and has applicability in research and practice across the lifespan. We encourage researchers to consider the components outlined in the *Pathways Model* in future studies of decision making by individuals with IDD. We believe the *Pathways Model* represents a promising avenue to understanding the various mechanisms underpinning the different forms of decision processing and that it can be applied to disentangle the intricate mechanisms as a basis for further interventions. By focusing on differentiated training and support, it promotes an individual differences approach to examining decision-making competency.

The development of validated measures to assess decision-making functioning and the outcomes of effective evidence-based training programs in persons with IDD have been slow, and additional research in this area is a priority. Further, we note that there is a gap in our understanding of how decision performance that is typically measured on artificial or hypothetical decision tasks in controlled environments, will translate into how people make decisions in their own lives. Although many decision tasks in research have begun to include simulations of real-life situations, measurement of decision making in vivo or in naturalized settings is remarkably low. Significant challenges in research design and methodology exist across studies, particularly in social, interpersonal decision situations. New research must find ways to strengthen the research methodologies for assessing decision-making-performance in real-life settings. *Uher* et al. in this volume summarize current work in this direction and reflect on the challenges of conducting naturalistic studies of decision making. It is also essential to conduct model demonstration and field implementation studies to help translate research into practice more effectively and to render intervention studies pertinent to improving real-world decision outcomes.

We recognize that along with greater emphases on supported and independent decision making, there will still need to be, as an exception to the norm, options available for individuals with IDD who may need some assistance with their decision making. Therefore, continuing guidance and education in the form of shared decision making as a collaborative partnership model, or, in extreme cases, involving a substitute decision-maker, must proceed with the best research evidence and in keeping with the rights and preferences of the individual.

Developmental View of Independent Decision Making

Making competent decisions is predictive of better real-life outcomes. In the general decision-making literature, higher overall decision-making competence has been consistently associated with fewer negative life outcomes, such as juvenile delinquency and drug use (Parker & Fischhoff, 2005) and health problems and bankruptcy (see Bruine de Bruin et al., 2007). Moreover, Weller et al. (2015) demonstrate that poor decision scores at the age of 10–11 years predicted interpersonal problems 2 years later. Such evidence compels us to adopt a lifespan view in identifying best practices to support improvements in decision-making abilities of individuals with IDD in natural settings. *Murphy and Clare* and *Clare and Murphy* in this volume use real-life accounts and court case studies to illustrate the role of individual and contextual factors in determining decision outcomes for individuals with ID in situations that involve complex understandings of cause and consequences of actions, such as during financial decision making or in navigating the criminal justice system.

In this volume, we chose to focus on adolescents and adults with IDD as most extant literature in decision making and IDD pertains to these age groups, including our own research. As in the general literature, many studies involving adolescents and young adults examine decision making in the context of heightened risk-taking and reward seeking, characteristic of this developmental period. There is a high focus on decision making and self-determination for individuals with IDD as they make the difficult transition to adulthood and engage in a host of decisions in areas pertaining to living, education, employment, etc. *Mello and Sanderson* in this volume have provided helpful insights into specific training needs during this period of transition. Transition to adolescence, and subsequently adulthood, requires navigation of varied social interpersonal environments, with an increasing need to stay protected from negative peer influences and threats of coercion or social victimization. Several chapters in this volume (*Bexkens & Mueller; Hickson & Khemka; Petitpierre & Tabin*) address issues of vulnerability to victimization in youth and adults with IDD, from a decision-making perspective, and advance new ideas for assessment and training that will be significant for consideration in future research. The adolescent period is a particularly crucial period in which regulatory skills required for social decision making are highly malleable to change via intervention during this time (Blakemore & Robbins, 2012). Furthermore, mental illness commonly has its onset during adolescence (Merikangas et al., 2010) and is likely to affect the development of decision making.

A longitudinal view to the development of adaptive decision-making attitudes and behaviors will help elucidate the mechanisms by which individuals with IDD can be supported from an early age to make successful transitions through the main developmental periods. This ascribes a positive psychology perspective to their roles and responsibilities in making decisions on their own that is normalized from the start. Our review also highlights the need for future work to identify crucial periods where persons with IDD diverge from their typically developing peers in the ability to make appropriate, long-term decisions and to build in more effective supports and training at those critical developmental points.

New Research Directions

Lifespan View of Decision Making

One of the key deterrents to people with IDD making decisions is the extent to which they have been previously denied the opportunity to make important decisions. As a result, we assert that making choices and decisions and becoming self-determining constitutes a sociocultural learning process with which an individual with IDD should be naturally involved from childhood. This requires a shift in cultural attitudes toward greater involvement and participation of individuals with IDD in independent decision making from an early age. We encourage readers to be more aware of childhood protective and risk factors that define early experiences with decision making for individuals with IDD that impact development of cognitive and executive functions and internal perceptions of control and self-determined independent decision making across their lifespan. To that effect, we propose attention to longitudinal studies that examine growth in decision-making processes and address the cognitive, psychosocial, and neurobiological features of development in IDD, across the lifespan of the individual, but especially from childhood to adolescence.

Impact of Comorbidities on Decision Making

Although delayed development and marked deficits in intellectual functioning are the most pervasive problems posed by intellectual disability (ID), research indicates that comorbid problems occur at high rates for individuals with ID as a subset within the larger IDD group. These difficulties could include challenging behaviors, mental health problems, health conditions, and other developmental disabilities (Matson & Cervantes, 2013). Although estimates vary, as many as 40% of individuals with ID could also have ASD, indicating that ASD is a common comorbidity with ID, discussed in more detail in the next section (La Malfa et al., 2004). Because of the considerable overlap of comorbid problems and with the presence of neuropsychiatric conditions, many individuals with ID are severely restricted in their language, communication, and social abilities. They may require extensive medical care, psychological services, or mental health supports to address their complex needs. As a result, opportunities for decision making in their daily lives might be restricted, and they may not have the necessary skills to engage in planful decision making. Given such unique needs, we posit that general programs designed for individuals with ID (or ASD) might not be sufficient to address the complexity of their needs (Wilkins & Matson, 2009) and that some overlap with mental health or psychiatric supports might be necessary in the design of future decision-making programs. Therefore, the study of comorbid conditions among persons with ID and their impact on decision making deserves more direct attention in future research.

High prevalence of psychiatric comorbidities (e.g., anxiety, obsessive compulsive disorder, self-harm) are also associated with ASD and ADHD (Ohnishi et al., 2019; Romero et al., 2016). In a recent volume by Bertelli et al. (2021) on the diagnosis and treatment of psychiatric conditions in people with ID/ASD, we can gain important insights into the specificities of psychiatric disorders (as well co-occurring behavioral syndromes, personality disorders, and disorders due to psychoactive substance use) in the context of ID/ASD and how best to meet the exact needs of these individuals. *Adams and Malone* and *Komeda* in this volume highlight the progress that they have made in their research to examine the role of anxiety in decision making by individuals with ASD. In summary, a conceptual understanding of how mental illness impacts decision making (see Cáceda et al., 2014, for a review for the general population) in individuals with IDD could provide useful information in understanding decision-making patterns and for future investigations of decision-making interventions that also focus on improvements in mental well-being and overall quality of life outcomes.

Etiological Understanding of Decision Making and ID/IDD

Recent conceptualizations of ID (Bertelli et al., 2018; Schalock et al., 2021) support a model of intelligence that goes beyond the diagnostic criteria of measuring ability by IQ tests, to assessing levels of individual impairment in terms of executive and specific cognitive functions that provide an individual profile of deficits in learning and adaptive functioning. Such an approach allows for better identification of specific cognitive functions and adaptive capacities that relate closely to decision-making functioning and to a more precise measurement of functions or processes that are malleable and can be targeted for improvement. This perspective also supports considering aspects of intelligence, such as risk unawareness and general commonsense abilities in handling everyday situations (*see Greenspan*, in this volume), that are more directly relevant to the study of decision making in individuals with ID, than a measure of cognitive ability in general. Decision making in individuals with ID, as illustrated in this volume, is inextricably associated with a range of distinct but related processes in the cognitive and noncognitive domains, as well as regulatory mechanisms. With the robust influence of etiology on the specific nature of difficulties in decision making of individuals with ID, focusing on individual profiles of decision functioning in light of etiological underpinnings is necessary. The chapter by *Goscicki* et al. in this volume pushes this thinking forward and explores the impact of specific etiologies.

Advances in genetics are rapidly transforming how we look at the classification and diagnostic categories of individuals with IDD, with a greater emphasis on the high rates of comorbidity and phenotypic overlap among the diagnostic categories due to shared genetic and environmental risk factors. Morris-Rosendahl and Crocq (2020) have proposed considering a genetic neurodevelopmental continuum and reason that "Thus, childhood neurodevelopmental disorders (ID, ASD, ADHD) and adult psychiatric disorders (including both bipolar disorder (BPD) and schizophrenia) could better be conceptualized as lying on an etiological and neurodevelopmental continuum, rather than being defined as discrete entities" (p. 90). This genotype-first approach for examining individual characteristics at the etiological level will necessitate looking more closely at similarities among the different groups, at the same time identifying needs that might be unique to specific groups. *Lyon* et al. and *Levin* et al. in this volume offer extensive reviews of their original research examining decision making in individuals with ADHD and ASD, respectively, applying cognitive science and individual difference perspectives and highlighting critical areas for future research.

As a result, the future study of decision making in individuals with IDD will need to keep pace with such new thinking, especially in the design of intervention methods for improving decision making. Training methods will perhaps need to be developed so that the strategies precisely address the shared underlying neurodevelopmental features and therefore may be applicable more easily across diagnostic groups on the neurodevelopmental continuum. As needed, additional inputs could be considered for personalizing to individual needs. These advances might be fruitful in identifying etiology-specific profiles of strengths and weaknesses underlying decision-making development that can be critical for forward planning of intervention strategies based on detailed knowledge of specific causes and mechanisms. We therefore underline the importance of cross-categorical, longitudinal research with individuals with IDD which can help elucidate the impact of all levels of development on individual decision-making functioning. Difficulties with decision making have been described as a key concept in understanding neurodevelopmental disorders, including ASD (Mussey et al., 2015). As we better understand the neural and cognitive mechanisms of decision making, there is also the potential to improve the diagnosis and treatment of these disorders (Lee, 2013).

Expanding Decision-Making Efforts to Include Individuals with Severe Disabilities

Largely excluded from research studies are individuals (including those with ASD) who have severe to profound intellectual disability (S/PID). This group deserves priority in future research with a focused need on addressing communication and behavioral skills as part of developing decision-making competency. In the continuing debate over notions of capacity and competence for individuals with severe and multiple disabilities with communication challenges, important insights for researchers on how to re-conceptualize access to communication and capacity building for greater decision-making possibilities are of value (e.g., Dee-Price et al., 2021; McNaughton et al., 2019). As Dee-Price (2020) claims "Fundamental to all of these features is a deeper re-conceptualizing of inclusive research to disrupt underlying concepts of people who 'can' and 'cannot' communicate. Research itself must explore notions of its own capacity and competence; to challenge its ability to shape ethics and standards to accommodate the 'voices' of people who communicate outside the status quo" (p. 141). The CRPD also emphasized "the freedom to seek, receive and impart information and ideas on an equal basis with others and through all forms of communication of their choice" (art. 21, para. 1, UN General Assembly, 2007). We therefore urge that new methods for promoting inclusive research, assessment, and valid and effective decision-making strategies for individuals with severe needs and communication differences be more responsive to and leverage new communication modes and assistive technology tools, including development of apps, smart phone-supported adaptive devices, and the evolving social media.

Aging and Decision Making

People with IDD are now living longer and are increasingly at risk of serious health conditions (Pimlott, 2019). The development of long-term medical and end-of-life plans, either at elder care facilities or within the family, has traditionally not included individuals with IDD, especially those having more complex needs and severe IDD. Decision making at end of life relates to many important decisions that define the overall quality of life experienced by individuals with IDD, ranging from medical issues to funeral planning, self-care, recreation, and level of engagement with friends and family (Bischoff et al., 2013). Such decisions are complex, and lack of a consistent approach to determining capacity makes the involvement of individuals with IDD in their own end-of-life decision-making process relatively sparse (Kirkendall et al., 2017). Research studies and anecdotal evidence have shown that elder individuals with IDD can successfully understand and consent to participate in healthcare decision making. By proactively supporting this process, community and healthcare settings may be able to directly facilitate more contribution from individuals with IDD, therefore better meeting the goal of person-centered support. *Sullivan* et al. (2019 and in this volume) offer some important research perspectives on the development of skills and supports to engage individuals with IDD in decisions relating to health, medical consent to procedures, and end-of-life decisions. We refer readers to significant research that has emerged in recent years on how to effectively assess and navigate end-of-life decision making with people with IDD (see Watson et al., 2017). Kirkendall et al. (2017) advocate for earlier discussions about end-of-life planning among individuals with ID before the onset of a life-limiting medical situation. In addition, new research on aspects of aging and its impact on decision-making competency open new frontiers for research and support building for elder populations with IDD. Bruine de Bruin et al. (2020) propose that age-related declines in fluid intelligence which impact decision-making ability could be counteracted by noncognitive factors such as emotion regulation and experience-dependent improvements in decision making. These theoretical advances carry promise for making progress in training and therapeutic approaches for supporting older adults with IDD. The 12th Edition of the AAIDD Manual provides an updated discussion on these components of intelligence for individuals with ID (Schalock et al., 2021).

Defining Capacity and Decision-Making Competency

We propose that future research around decision making should utilize a clearer, more consistent definition of decision making as a process or skill set that is distinct and not overlapping with other related functions such as choice making, problem-solving, social functioning, etc. Although decision making is a key component of self-determination, we often find the two terms used interchangeably. Research studies have also included aspects of decision making within a broad range of social skills (e.g., Boujarwah et al., 2010), and a few have used the term decision making imprecisely, making it difficult to draw clear conclusions from this body of work. Therefore, well-defined measures of decision-making functioning are needed. The Decision-Making Competency Scale (see Bruine de Bruin et al., 2007) in the general domain could provide a model to consider for future assessments for individuals with IDD. *Hickson and Khemka* in this volume provide some insight on developing assessment tools with examples from their own research work. Assessing decision making in individuals with IDD will require greater conceptual clarity and communication of the terms that assess different dimensions of decision making – *capacity*, *capability*, and *performance* in decision making. Capacity is a measure of a person's overall ability and available resources (or those that can be accessed) for decision making, while capability, or competency, generally refers to the skills and knowledge required for a particular decision task. Decision performance on the other hand is a measure of actual functioning and indicates how well a person decides in her/his environment, based on given criteria for effective decision making. Assessing and defining these interrelated concepts in decision-making measurement is a specific area worthy of more research and collaboration across different domains/sectors (e.g., healthcare, legal). By examining the influences of culture and diversity on aspects of decision making for individuals with IDD, *Vanegas* et al. in this volume have moved the research forward in acknowledging culture to be a vital part of an individual's social-ecological system.

One of the many complex aspects of establishing *capacity* for decision making stems from the fact that competencies that define capacity to make decisions may not be uniform – variations may exist even for a given individual, depending on the setting or context of the decision to be made. Further, capacity for decision making is not a static indicator of functioning and can change over time and under conditions of decision making (e.g., time, certainty, level of support). The fact that an individual is unable to make some decisions does not mean that she/he is unable to make any decisions. Therefore, presumptive ideas of limited capacity based on type and extent of a person's disability should be challenged, and whether training for independent decision making is necessary and appropriate should be determined by an assessment of the type of decision to be made.

The principle of *legal capacity* in the CRPD provides a direct path to all people with disabilities to make their own decisions in all domains of life while simultaneously obligating society to provide for the highest level of accommodations and appropriate supports to facilitate a person's legal right to autonomous decision making. However, controversy exists in how this legal right gets exercised when an individual is determined to have limited *mental capacity* due to an impairment or disturbance that severely restricts their ability to participate in decision making. In such specific situations, facilitators or substitutes most able to safeguard the rights and preferences of the person will need to get involved in decision making with the individual with IDD and in keeping with protecting the individual's right as per the principles of the CRPD. More clarity is needed in how these determinations are to be made and in defining and measuring mental capacity from the viewpoint of *skills* development. Strengths-based conceptualizations and research are necessary to investigate how these determinations are to be proportional and tailored to an individual's circumstances and level of decision-making competencies.

Conclusion

This volume connotes much enthusiasm and support for the future of decision-making research and practice in the field of IDD, backed by many new legislative and social changes globally in the direction of increased opportunities and supports for favorable long-term decision-making outcomes. However, since opportunities to promote decision making have been somewhat limited by the lack of a practical understanding of how to include individuals with IDD, the components that are necessary to increase availability of decision opportunities need to be more clearly articulated. We propose that implementation of best practices for promoting decision making, as supported in current research and self-advocacy and deliberated in this volume, will require substantial commitment and growth in funding, training, and policy at the grassroots and national level, along with continued involvement, training, and support by different stakeholders in the lives of individuals with IDD. *Burke* et al. in this volume emphasize the role of supporters within the family, primarily siblings and parents, and outline the structures of natural supports that could help facilitate beneficial arrangements for both the individual with IDD and his/her family while navigating minor and major life decisions. Long-term progress will require continuing supportive policy and law reform toward greater self-determination and decision-making independence that necessitates more economic freedom and financial protection, including more options for independent housing, employment, etc. In all, the environmental barriers and gaps in services and supports, if not monitored, can undermine the progress and benefits of increased decision making and personal agency acquired by individuals with IDD. Further, we note that most research and guidance for improving decision making among individuals with IDD has involved the work of professionals, service providers, or guardians/family members; real progress will require individuals with IDD themselves taking more charge of what the future holds for them while integrating research and policy work. This is consistent with the widespread acceptance of the CRPD rights afforded to individuals with IDD to take greater leadership and control in all aspects of their lives, including inclusive research (see Jones et al., 2020; Nind, 2011). According to Johnson (2009), "Underlying each of these articles is the common theme of increased power and active participation by people with disabilities in their societies" (p. 251).

The concept of capacity varies globally, presenting the need for having more uniformity and clarity in definitions across cultures, systems, and countries. We have tried to provide a comprehensive overview of the complexity and diversity of decision making and related issues for individuals with IDD as a strong foundation to future research, with contributions from some of the most outstanding scholars in the field globally. With diverse perspectives on how to approach the study of decision making presented by a worldwide (Australia, Canada, Japan, Netherlands, Switzerland, United Kingdom, United States) group of contributors, we intend for this volume to offer an up-to-date survey of research in the field of decision making and IDD by applying the best theories and research methods available. We recognize and appreciate the similarities (and differences) in the efforts and changes happening worldwide, especially in keeping with Article 12, CRPD. We urge that there be more discussions and critical examination of international progress made in research, policy, and law reform toward the success of Article 12. It is essential to include representation from countries (e.g., in Asia, Africa, South and Central America) where the overall systems of supports and services for individuals with disabilities might be generally under-resourced and restricted in opportunities for decision making.

In closing, we would like to extend our sincere thanks to series editors Michael Wehmeyer and Karrie Shogren for the opportunity to reflect upon our research in decision making and IDD conducted over the past two decades and to bring together this important work on decision making by renowned researchers from across the globe in this edited volume. Their vision on positive psychology has emerged as a unifying theme in this volume and underlies the various perspectives shared in the volume on how best to advance and support decision-making participation of individuals with IDD in all areas of life. The volume demonstrates that there already has been a noticeable shift in the field of decision-making research toward valuing a culture of positive psychology and strengths-based considerations, by the extensive application of the social-ecological model seen throughout the volume in understanding and assessing the decision-making experiences of individuals with IDD. Further, the chapters together are a measure of the important progress that has been achieved in the understanding of evidence-based practices and supports for improving decision-making outcomes and psychological well-being of individuals with IDD. It moves the field considerably forward in being able to interpret and articulate the need for effective actions and systems for the full realization of the principles set forth in the CRPD, a recurring theme in this volume. We hope the volume will propel a deeper understanding of decision making in individuals with IDD and will serve as a comprehensive guide for both researchers and practitioners to build on the progress made and tackle many of the gaps and challenges that remain.

We thank all chapter authors for their outstanding contributions and hard work in preparing their chapters to best align with our overall vision for the volume. The research, the ideas, and the suggestions for ongoing inquiry around effective decision making presented in this volume are a direct result of their expertise and dedication to decision-making research for individuals with IDD. We express deep gratitude to them and look forward to continuing this research in close collaboration. Importantly, we recognize that the thinking, writing, and decision making has all happened during this unprecedented challenging time of the global pandemic, and for that we are ever so grateful to our contributors for their time, effort, and determination to work with us and to see us through to the completion of this volume. We also acknowledge the loss of many lives of individuals with IDD in our societies during this time and, personally, the passing away of our two good friends Aikaterini Chatzistyli and Gay Culverhouse who cared deeply for people with disabilities and played significant roles in contributing to our research work. To them, we dedicate this volume.

References

Bertelli, M. O., Cooper, S. A., & Salvador-Carulla, L. (2018). Intelligence and specific cognitive functions in intellectual disability: Implications for assessment and classification. *Current Opinion in Psychiatry, 31*(2), 88–95. https://doi.org/10.1097/YCO.0000000000000387

Bertelli, M. O., Deb, S., Munir, K., Hassiotis, A., & Salvador-Carulla, L. (Eds.). (2021). *Text volume of psychiatry for intellectual disability and autism spectrum disorder*. Springer Nature.

Bigby, C., & Douglas, J. (2020). Supported decision making. In R. Stancliffe, M. Wehmeyer, K. Shogren, & B. Abery (Eds.), *Choice, preference, and disability* (Positive psychology and disability series) (pp. 45–66). Springer. https://doi.org/10.1007/978-3-030-35683-5_3

Bischoff, K. E., Sudore, R., Miao, Y., Boscardin, W. J., & Smith, A. K. (2013). Advance care planning and the quality of end-of-life care in older adults. *Journal of the American Geriatrics Society, 61*(2), 209–214. https://doi.org/10.1111/jgs.12105

Blakemore, S., & Robbins, T. W. (2012). Decision-making in the adolescent brain. *Nature Neuroscience, 15*(9), 1184–1191.

Boujarwah, F. A., Hong, H., Arriaga, R., Abowd, G. D., & Isbell, J.S. (2010). Training social problem-solving skills in adolescents with high functioning autism spectrum disorders (HFASD). In *4th International Conference on Pervasive Computing Technologies for Healthcare*, Munich, pp. 1–9. https://doi.org/10.4108/ICST.PERVASIVEHEALTH2010.8807

Browning, M., Bigby, C., & Douglas, J. (2020). A process of decision making support: Exploring supported decision making practice in Canada. *Journal of Intellectual & Developmental Disability*. https://doi.org/10.3109/13668250.2020.1789269

Bruine de Bruin, W., Parker, A. M., & Fischhoff, B. (2007). Individual differences in adult decision making competence. *Journal of Personality and Social Psychology, 92*(5), 938–956. https://doi.org/10.1037/0022-3514.92.5.938

Bruine de Bruin, W., Parker, A. M., & Fischhoff, B. (2020). Decision-making competence: More than intelligence? *Current Directions in Psychological Science, 29*, 186–192.

Cáceda, R., Nemeroff, C. B., & Harvey, P. D. (2014). Toward an understanding of decision making in severe mental illness. https://doi.org/10.1176/appi.neuropsych.12110268

Dee-Price, B. J. M. (2020). Social researchers and participants with intellectual disabilities and complex communication (access) needs. Whose capacity? Whose competence? *Research and Practice in Intellectual and Developmental Disabilities, 7*(2), 132–143. https://doi.org/10.1080/23297018.2020.1788418

Dee-Price, B. J. M., Hallahan, L., Bryen, D. N., & Watson, J. M. (2021). Every voice counts: Exploring communication accessible research methods. *Disability & Society, 36*(2), 240–264. https://doi.org/10.1080/09687599.2020.1715924

Johnson, K. (2009). No longer researching about us without us: A researcher's reflection on rights and inclusive research in Ireland. *British Journal of Learning Disabilities, 37*(4), 250–256. https://doi.org/10.1111/j.1468-3156.2009.00579.x

Jones, K. E., Ben-David, S., & Hole, R. (2020). Are individuals with intellectual and developmental disabilities included in research? A review of the literature. *Research and Practice in Intellectual and Developmental Disabilities, 7*(2), 99–119. https://doi.org/10.1080/23297018.2019.1627571

Kirkendall, A., Linton, K., & Farris, S. (2017). Intellectual disabilities and decision making at end of life: A literature review. *Journal of Applied Research in Intellectual Disabilities, 30*(6), 982–994. https://doi.org/10.1111/jar.12270

La Malfa, G., Lassi, S., Bertelli, M., Salvini, R., & Placidi, G. F. (2004). Autism and intellectual disability: A study of prevalence on a sample of the Italian population. *Journal of Intellectual Disability Research, 48*(Pt 3), 262–267. https://doi.org/10.1111/j.1365-2788.2003.00567.x

Lee, D. (2013). Decision making: From neuroscience to psychiatry. *Neuron, 78*, 233–248. https://doi.org/10.1016/j.neuron.2013.04.008

Luckasson, R., & Schalock, R. L. (2020). A balanced approach to decision making in supporting people with IDD in extraordinarily challenging times. *Research in Developmental Disabilities, 105*, 103719. https://doi.org/10.1016/j.ridd.2020.103719

Matson, J. L., & Cervantes, P. E. (2013). Comorbidity among persons with intellectual disabilities. *Research in Autism Spectrum Disorders, 7*, 1318–1322.

McNaughton, D., Light, J., Beukelman, D., Klein, C., Nieder, D., & Nazareth, G. (2019). Building capacity in AAC: A person-centered approach to supporting participation by people with complex communication needs. *Augmentative and Alternative Communication*, Advanced online publication. https://doi.org/10.1080/07434618.2018.1556731

Merikangas, K. R., He, J. P., Burstein, M., Swanson, S. A., Avenevoli, S., Cui, L., Benjet, C., Georgiades, K., & Swendsen, J. (2010). Lifetime prevalence of mental disorders in U.S. adolescents: Results from the National Comorbidity Survey Replication--Adolescent Supplement

(NCS-A). *Journal of the American Academy of Child and Adolescent Psychiatry, 49*(10), 980–989. https://doi.org/10.1016/j.jaac.2010.05.017

Morris-Rosendahl, D. J., & Crocq, M. A. (2020). Neurodevelopmental disorders-the history and future of a diagnostic concept. *Dialogues in Clinical Neuroscience, 22*(1), 65–72. https://doi.org/10.31887/DCNS.2020.22.1/macrocq

Mussey, J. L., Travers, B. G., Klinger, L. G., & Klinger, M. R. (2015). Decision-making skills in ASD: Performance on the Iowa Gambling Task. *Autism Research: Official Journal of the International Society for Autism Research, 8*(1), 105–114. https://doi.org/10.1002/aur.1429

Nind, M. (2011). People with intellectual disabilities: Towards a good life ? *British Journal of Learning Disabilities, 39*(1). https://doi.org/10.1111/j.1468-3156.2010.00671.x

Nygren, M. A., & Lulinski, A. (2020, December). State of the science on COVID-19 and people with IDD [Research brief]. *American Association on Intellectual and Developmental Disabilities*.

Ohnishi, T., Kobayashi, H., Yajima, T., Koyama, T., & Noguchi, K. (2019). Psychiatric comorbidities in adult attention-deficit/hyperactivity disorder: Prevalence and patterns in the routine clinical setting. *Innovations in Clinical Neuroscience, 16*(9-10), 11–16.

Parker, A. M., & Fischhoff, B. (2005). Decision-making competence: External validation through an individual-differences approach. *Journal of Behavioral Deciion Making, 18*(1), 1–27. https://doi.org/10.1002/bdm.481

Perske, R. (1972). The dignity of risk and the mentally retarded. *Mental Retardation, 10*(1), 24–27.

Pimlott, N. (2019). Aging with intellectual and developmental disabilities: Family physicians can make a difference. *Canadian Family Physician, 65*(Suppl 1), S3.

Romero, M., Aguilar, J. M., Del-Rey-Mejías, Á., Mayoral, F., Rapado, M., Peciña, M., Barbancho, M. Á., Ruiz-Veguilla, M., & Lara, J. P. (2016). Psychiatric comorbidities in autism spectrum disorder: A comparative study between DSM-IV-TR and DSM-5 diagnosis. *International Journal of Clinical and Health Psychology: IJCHP, 16*(3), 266–275. https://doi.org/10.1016/j.ijchp.2016.03.001

Schalock, R. L., Luckasson, R., & Tasse, M. J. (2021). *Intellectual disability: Definition, diagnosis, classification, and systems of supports* (12th ed.). American Association on Intellectual and Developmental Disabilities.

Shogren, K. A., Wehmeyer, M. L., Martinis, J., & Blanck, P. (2019). *Supported decision making: Theory, research and practice to enhance self-determination and quality of life*. Cambridge University Press.

Sullivan, W. F., Heng, J., Perry, A., Bach, M., & Casson, I. (2019). Aging that includes an intellectual and developmental disability: A time to flourish? *Canadian Family Physician, 65*(Suppl 1), S5–S7. (Eng).

UN General Assembly, Convention on the Rights of Persons with Disabilities : resolution / adopted by the General Assembly, 24 January 2007, A/RES/61/106, available at: https://www.refworld.org/docid/45f973632.html [accessed 19 July 2021]

Watson, J., Wilson, E., & Hagiliassis, N. (2017). Supporting end of life decision making: Case studies of relational closeness in supported decision making for people with severe or profound intellectual disability. *Journal of Applied Research in Intellectual Disabilities: JARID, 30*(6), 1022–1034. https://doi.org/10.1111/jar.12393

Weller, J. A., Moholy, M., Bossard, E., & Levin, I. P. (2015). Preadolescent decision making competence predicts interpersonal strengths and difficulties: A 2-year prospective study. *Journal of Behavioral Decision Making, 28*, 76–88.

Wilkins, J., & Matson, J. L. (2009). History of treatment in children with developmental disabilities and psychopathology. In J. L. Matson, F. Andrasik, & M. L. Matson (Eds.), *Treating childhood psychopathology and developmental disabilities* (pp. 3–28). Springer Science + Business Media. https://doi.org/10.1007/978-0-387-09530-1_1

Index

I. Khemka, L. Hickson (eds.), *Decision Making by Individuals with Intellectual and Developmental Disabilities*, Positive Psychology and Disability Series,
https://doi.org/10.1007/978-3-030-74675-9

D

Q

R

T

The manufacturer's authorised representative in the EU is Springer
Nature Customer Service Centre GmbH, Europaplatz 3, 69115 Heidelberg,
Germany. If you have any concerns regarding our products, please
contact ProductSafety@springernature.com

Printed and bound by CPI Group (UK) Ltd, Croydon, CR0 4YY
28/09/2026
02224065-0016